SOIL SCIENCE
Principles & Practices

SOIL SCIENCE
Principles & Practices

Third Edition

R. L. Hausenbuiller
Washington State University

wcb

Wm. C. Brown Publishers
Dubuque, Iowa

Cover photo by F. Roiter/The Image Bank

Copyright © 1972, 1978, and 1985 by Wm. C. Brown Company Publishers. All rights reserved

Library of Congress Catalog Card Number: 84–71942

ISBN: 0–697–05856–5

Printed in the United States of America
10 9 8 7

Contents

Preface

This text is intended for use in introductory soils courses at the college level. It deals with fundamental chemical, physical, and biological properties of soils and the relation of these properties to soil classification and use. Principal uses considered are plant culture, engineering works, and the abatement of environmental pollution, as in the disposal of organic wastes. An important theme throughout much of the text is the need for careful husbandry of the soil as a vital natural resource essential to all life on earth.

In many respects, this edition contrasts sharply with the first two. Major changes in both organization and content have been made. A number of discussions have been simplified, particularly those on soil mineralogy, ion exchange properties, soil water, soil temperature, and plant nutrition and fertility. Through reorganization and the removal of some of the more complex discussion on fertilizers, the number of chapters on nutrition and fertility has been reduced from six to four, and the material covered decreased by about one-third. It is believed that this has been done without eliminating important fundamentals of the nature and control of plant nutrition.

Considerable new material has been added. Soil horizons, discussed in Chapter 3, are defined according to the system of horizon classification and symbolization officially adopted by the U.S. Department of Agriculture in 1981. Old horizon symbols are retained to show their relationship to the new system, however. This edition also contains new material on land use planning as applies to both urban and agricultural environments. Example calculations demonstrating the application of various physical and chemical mathematical formulas have been substantially increased, as has the number of illustrations. The text now contains approximately twice as many illustrations as before.

Two other additions are worthy of special note. One is an introduction to the Standard International Units (SI units) of measure, which have been adopted for use in publications of the Soil Science Society of America. The second is a description of the system of soil classification used in Canada. These discussions are, respectively, in Appendixes C and D.

In this edition, as in the first two, every effort has been made to keep the material concise, readable, and logically ordered. To a greater degree than in earlier editions, however, I have attempted to continually bear in mind the need for many students to study soils without an adequate background in chemistry and physics. Accordingly, discussions requiring such a background have been reduced in number and scope. At the same time, the amount of supportive information on fundamental physical and chemical concepts has been increased.

R. L. Hausenbuiller
Washington State University
Pullman, Washington

Acknowledgments

I am indebted to the publisher, the Wm. C. Brown Company, and to editor Robert Stern for constant guidance and particularly for arranging extensive reviews of the text material prior to and following the initial phases of the current revision. Involved in the reviews were K. A. Barbarick, Colorado State University, Richard Zartman, Texas Tech University, Tom Ruehr, California Polytechnic University, Daniel DeVitis, Shippensburg (Pa) State College, and J. A. Robertson, University of Alberta, Canada. The reviews, all thoughtfully prepared, have been most helpful. Largely because of a time constraint and other considerations, not all recommendations for change could be followed. The majority were, nonetheless, and they account for the marked differences between this edition of the text and its predecessors.

I also express my appreciation to Dr. L. M. Lavkulich, University of British Columbia, Vancouver, for kindly reviewing the Appendix section on the Canadian System of Soil Classification, and to Carol Rogers and Joyce Mikelson for their care and helpfulness in typing the final manuscript. Further, I owe much to the many individuals who supplied new photographs or other illustrative material for the current edition. In this regard, I am especially grateful to Jody Kurty, Photo Librarian, U.S.D.A. Soil Conservation Service, Washington, D.C., who spent many hours seeking out a large number of photographs for me.

Finally, my greatest indebtedness is to my wife, whose seemingly endless support, patience, and understanding made this project much more bearable.

R. L. H.

Conversion factors for English and metric units

To convert column 1 into column 2, multiply by	Column 1	Column 2	To convert column 2 into column 1, multiply by
		Length	
0.621	kilometer, km	mile, mi	1.609
1.094	meter, m	yard, yd	0.914
0.394	centimeter, cm	inch, in	2.54
		Area	
0.386	kilometer², km²	mile², mi²	2.590
247.1	kilometer², km²	acre, acre	0.00405
2.471	hectare, ha	acre, acre	0.405
		Volume	
0.00973	meter³, m³	acre-inch	102.8
3.532	hectoliter, hl	cubic foot, ft³	0.2832
2.838	hectoliter, hl	bushel, bu	0.352
0.0284	liter	bushel, bu	35.24
1.057	liter	quart (liquid), qt	0.946
		Mass	
1.102	ton (metric)	ton (U.S.)	0.9072
2.205	quintal, q	hundredweight, cwt (short)	0.454
2.205	kilogram, kg	pound, lb	0.454
0.035	gram, g	ounce (avdp), oz	28.35
		Pressure	
14.50	bar	lb/inch², psi	0.06895
0.9869	bar	atmosphere, atm	1.013
0.9678	kg(weight)/cm²	atmosphere, atm	1.033
14.22	kg(weight)/cm²	lb/inch², psi	0.07031
14.70	atmosphere, atm	lb/inch², psi	0.06805
		Yield or Rate	
0.446	ton (metric)/hectare	ton (U.S.)/acre	2.240
0.892	kg/ha	lb/acre	1.12
0.892	quintal/hectare	hundredweight/acre	1.12
		Temperature	
$\left(\dfrac{9}{5}\,°C\right) + 32$	Celsius, C	Fahrenheit, F	$\dfrac{5}{9}\,(°F - 32)$
	−17.8°	0°	
	0°	32°	
	20°	68°	
	100°	212°	
		Water Measurement	
8.108	hectare-meters, ha-m	acre-feet	0.1233
97.29	hectare-meters, ha-m	acre-inches	0.01028
0.08108	hectare-centimeters, ha-cm	acre-feet	12.33
0.973	hectare-centimeters, ha-cm	acre-inches	1.028
0.00973	meters³, m³	acre-inches	102.8
0.981	hectare-centimeters/hour, ha-cm/hour	feet³/sec	1.0194
440.3	hectare-centimeters/hour, ha-cm/hour	U.S. gallons/min	0.00227
0.00981	meters³/hour, m³/hour	feet³/sec	101.94
4.403	meters³/hour, m³/hour	U.S. gallons/min	0.227

Source: Soil Science Society of America Journal. By permission of the Soil Science Society of America.

SOIL Principles & Practices SCIENCE

1

Introduction

Soil is a substance of essentially universal occurrence. Perhaps because it is so commonplace, we often fail to fully appreciate the impact it makes on our daily lives. No one, of course, overlooks the dependence we place on soil as a medium for plant growth, for in this role it serves as our primary source of food and fiber. Soil is important for many other reasons, however. It is a basic material with which the engineer must deal in the building of roads, dams, and similar works. Without the soil there would be little protection from the devastating effects of floods, since as a porous mantle over the earth's surface, it acts as a tremendous reservoir for the collection and storage of water from rain and melting snow. The slow release of the water stored in the soil helps maintain a steadier, more dependable flow of streams and rivers, which is of great importance to electrical power generation in dams, to irrigation, and to the use of these waterways in commerce. For a number of reasons, then, the soil must be viewed as a vital natural resource worthy of our personal consideration and study. Understanding the soil to the fullest extent possible is essential to its efficient use and preservation.

The various types of soil seldom function with equal effectiveness when subjected to different uses. For instance, some are excellent media for plant growth yet are totally unsuitable for use in engineering works of one type or another. Whether or not a soil yields desirable results under a certain application depends upon its properties or the extent to which they can be readily modified. Satisfactory utilization of soils requires an understanding of these properties and their relationship to soil behavior. Equally important is a knowledge of techniques that can be employed to change the soil when it fails to measure up to an accepted standard.

Soil as a three-phase system

The soil is a system consisting of three phases: *solid, liquid* and *gas*. The solid phase is a mixture of mineral and organic particles lying in intimate contact with each other and enclosing a system of pores that is continuous with the outer atmosphere. The pore space in soils is shared jointly by water and air. Whereas the solid phase of soils is relatively stable, the other two phases exist in a state of essentially constant flux, the result of the continual interchange of air and water between soil pores and the outer atmosphere. Typical is the loss of water by evaporation and plant use and its periodic replacement by rain or irrigation.

The solid, liquid, and gaseous phases of soils act both independently and interdependently in determining the properties and behavior of the whole soil. Certain of the properties used to characterize soils are, in reality, properties of but one of the phases without reference to the other two. Color of the soil typifies this kind of property. Other properties, particularly those identified with soil behavior, reflect the interaction of two or all three of the phases. A prime example is the capacity of the soil to support plant growth, for this depends on the ability of the soil to supply plants with essential elements (*nutrients*) from the solid phase as well as with adequate water and air from the liquid and gaseous phases.

The solid phase of soils

The principal component making up the solid phase of soils is fragmented mineral matter produced by the weathering of hard rock at the earth's surface. Mineral particles produced by weathering vary widely in size, but in soils they occur mainly in the three size groups of *sand*, *silt*, and *clay*. Sand particles are large, ranging from 0.05 to 2.0 mm in diameter; they can be seen by the unaided eye and can be perceived by touch when rubbed between the fingers. Silt particles (0.002 to 0.05 mm) cannot be perceived individually when rubbed between the fingers, and their visibility falls generally within the range of the ordinary light microscope. Clay consists of submicroscopic particles with a diameter of less than 0.002 mm. Although sand, silt, and clay are the dominant size fractions of soil mineral particles, larger fragments, such as *gravels*, *cobbles*, and *stones*, also occur in many soils. The presence of these large fragments is usually undesirable, particularly in soils used for agriculture. They make cultivation of the soil difficult, while reducing its ability to store and supply water and nutrients for plant use.

The nature of soils is strongly influenced by the proportionate quantities of sand, silt, and clay they contain. The relative contents of these three size fractions determine the *texture*, or the degree of coarseness or fineness, of soils. Soils of different texture differ in pore size and configuration and, therefore, in their ability to transmit air and water. Whether a soil is sticky and plastic (moldable) when wet, and whether it forms hard clods or remains in a loose, friable condition on drying also relates to texture. These latter properties are primarily a function of the clay content of soils.

The behavior of the soil is also influenced by *soil structure*, or the way in which soil particles are grouped together in aggregate form. Massive soil is transformed into a structured, or aggregated, state by the wedging action of ice or plant roots and from the cracking of the soil at frequent intervals as it shrinks on drying. The formation of aggregates tends to loosen hard soils, thus making them easier to cultivate. Further, the pores between aggregates are comparatively large, which aids water and air flow in the soil. Structure is not likely to occur in soils unless they have enough clay to bind the soil together in aggregate form. Gumlike substances from soil organic matter can also contribute to the binding process.

Organic matter usually makes up only a small part of the solid phase of soils. It accumulates within the mineral framework of the soil body as a result of the decomposition, or decay, of plant root, leaf, and stem tissue. The greater part of the organic fraction occurs as a blackish, finely divided material termed *humus*. Humus consists of highly altered organic compounds derived from plants along with new compounds synthesized by organisms active in the decay process. Compared to fresh plant tissue, humus is fairly stable in the soil environment. It succumbs to continued attack by decay organisms, however, and will gradually disappear if not replenished by the periodic addition of fresh plant residues to the soil. Under natural conditions, humus is maintained at a relatively constant level by annual gains from plant residues that offset its slow disappearance through decay. In much agricultural land, on the other hand, serious erosion has reduced the humus content of the soil because of the loss of topsoil in which the bulk of this highly valuable soil component has accumulated.

A number of soil properties are due to adsorptive or attractive forces exhibited at the surfaces of clay and organic particles. These forces are responsible for the retention of water and ions, and for the ability of the soil to cling to itself or to other objects, as evident in the stickiness of many soils when wet and in their ability to form hard clods on drying. Since these forces are associated with the surfaces of solids, they are most pronounced in soils high in clay and humus, for these particles exhibit far more surface per unit of weight than do comparable weights of coarser silt and sand particles.

Most clay and humus particles have a negative charge, which gives them the ability to adsorb positively charged ions (*cations*) from the surrounding solution. The adsorbed cations are held loosely at the particle surfaces, however, and for this reason can exchange with other cations from solution. The ability to adsorb and retain cations in exchangeable form is of great importance to plants, for it allows the retention of nutrient cations in a form plants can use but limits the ease with which they can be lost in water percolating through soil pores.

Liquid phase of soils

Water as the principal component of the liquid phase is important primarily because of its effect on plant growth. It is the major constituent of cell sap and is used directly in the synthesis of numerous organic constituents required for normal growth and maturation of plants. In the soil, water is involved in the release of essential plant nutrients and other elements from minerals and decaying organic matter. With these dissolved substances, water comprises the *soil solution* and provides the principal pathway for the transfer to plant roots of nutrients necessary for plant growth.

Plants have a comparatively high requirement for water. However, since soil water is lost continuously, and often rapidly, by the combined processes of transpiration[1] and evaporation, it must be replenished frequently to sus-

---

[1]Transpiration is the loss of water vapor from living plants. For the most part, the loss takes place through special openings, the stomata, which occur in greatest concentration on the underside of leaves.

tain plant growth. The abundance of vegetation over the surface of the earth is probably limited by the lack of water more than by any other factor.

Living plants can be adversely affected by a surplus as well as by a shortage of soil water. Excesses result in the blockage of soil pores, so that the movement of air within the root zone is restricted. If the supply of oxygen in the soil air is depleted because of a limited replenishment factor, root growth and other essential plant functions are greatly curtailed. Thus, while efforts in dry regions are usually directed toward the conservation of all available water for use by crop plants, water disposal is often a problem in humid areas. Minimal success in the utilization of many agricultural soils relates to the inability or unwillingness of man to control moisture relationships in these soils.

More often than not, excessive wetness in soils is associated with *waterlogging,* a condition caused by the accumulation of water over an impermeable subsurface layer to saturate all or part of the soil above. Water within the saturated zone is called *ground water.* Ground water moves freely, although slowly in most cases, in a lateral direction through the soil. In wet regions, flow is ordinarily toward natural drainageways, and it provides a means of carrying away excess water that passes downward through soils in so-called *internal drainage.* In dry regions, ground water may be sustained by seepage from drainageways into the neighboring soil.

Water is important to the physical behavior of soils. For example, sandy soils are loose and fall apart easily when dry, but cohere, or cling together, better when moist. As opposed to sandy soils, those high in clay may be very hard when dry. Water softens such soils and, if not present in excess, makes them easier to cultivate. If too wet, however, fine soils become sticky and plastic and are again difficult to work. Nor do they have much strength, as is evident in how little weight a muddy soil will support.

Gaseous phase of soils

Like the outer atmosphere, soil air consists of nitrogen (N_2) and oxygen (O_2) in combination with lesser amounts of carbon dioxide (CO_2) and other gases. Nitrogen is the most abundant gas in soil air. It is also an essential plant nutrient but cannot be directly used in gaseous form by higher plants. However, some soil microbes can utilize N_2 gas, and in so doing, incorporate it into proteinaceous substances that ultimately appear in the organic fraction of soils.

Oxygen, the second most abundant element in the soil atmosphere, is essential to all life processes in the soil. Plant roots and other soil organisms need a continuous supply of oxygen for *respiration,* the process that supplies them with energy (see page 6). The rate of respiration is a function of the concentration of O_2 in soil air, which is, in turn, controlled by the rate of gaseous interchange between the soil and the outer atmosphere. Since gas flow in soils depends on the average size of pores and the extent to which they are filled with water, the state of aeration, as defined by the concentration of O_2, is controlled by the liquid and solid phases of the soil.

Soil-plant relationships

One of the more important functions of the soil is as a medium for plant growth. For the most part, the association between plants and soils is mutually beneficial. The soil provides mechanical support for the plants, and it supplies nutrients and water necessary for their growth. The plants, in turn, protect the soil from erosion, and they return dead tissue in residues to sustain the all-important organic fraction of the soil. Plants are also a prime source of food and nutrients for other organisms living within the general soil environment.

Unique to plants is their ability to carry out *photosynthesis*. In photosynthesis, carbon dioxide from air and water from the soil meet in the plant, where they combine to form sugar. The process occurs in the presence of *chlorophyll*, the green pigment of plants, and is driven by energy from the sun. Free oxygen is given off by the reaction:

$$6CO_2 + 6H_2O + Energy \rightarrow \underset{\text{Sugar}}{C_6H_{12}O_6} + 6O_2 \uparrow \qquad (1.1)$$

Plants use sugar to synthesize other, more complex organic compounds found in structural and protoplasmic tissue. Structural tissue occurs in the cell wall, and it gives rigidity to plant stems and leaves. Protoplasmic tissue is concentrated within cells, and especially in the cell nucleus, which is the center for plant life processes. One of these processes is respiration. Respiration supplies energy to living organisms through the oxidation of organic carbon to carbon dioxide. In plants, sugar supplies the carbon used in respiration. The reaction is:

$$\underset{\text{Sugar}}{C_6H_{12}O_6} + 6O_2 \rightarrow 6CO_2 \uparrow + 6H_2O + Energy \qquad (1.2)$$

Note that this reaction is the reverse of the one shown for photosynthesis in equation (1.1).

To maintain satisfactory growth, plants must be supplied with at least 16 nutrients. The elements known to be essential are listed in table 1.1 according to source and relative amounts required by plants. *Macronutrients* are needed in comparatively large amounts; *micronutrients*, in smaller amounts. Among the 16 essential elements, carbon, hydrogen, and oxygen are required in greatest quantity. These three elements are the principal components of structural and protoplasmic tissue, which may make up more than 90 percent of the dry weight of plant tissue. Plants obtain carbon from carbon dioxide absorbed through their leaves and hydrogen and oxygen from soil water. These nutrients are assimilated by the plant during photosynthesis.

Plant nutrients other than carbon, hydrogen, and oxygen are supplied by either the mineral or organic fraction of soils. Organic matter is an important source of nitrogen, phosphorus, and sulfur. The reason for this is that once these nutrients are absorbed by plants they are incorporated into or-

Table 1.1. The essential plant nutrients listed according to the relative amounts required by plants and their principal source of supply. Chemical symbols of the elements are shown in parentheses.

Macronutrients	Micronutrients[a]
From water and air	*All from minerals*
Carbon (C)	Chlorine (Cl)
Hydrogen (H)	Copper (Cu)
Oxygen (O)	Boron (B)
	Iron (Fe)
From minerals	Manganese (Mn)
Calcium (Ca)	Molybdenum (Mo)
Magnesium (Mg)	Zinc (Zn)
Potassium (K)	
From minerals and organic matter	
Phosphorus (P)	
Sulfur (S)	
From organic matter	
Nitrogen (N)	

[a]/Cobalt is required by animals and certain microorganism, including microscopic plants known as algae. The necessity of cobalt for higher plants has not been conclusively demonstrated, however.

ganic compounds that are ultimately returned to the organic fraction of the soil on death of the plants. Other elements taken in by plants either do not form organic combinations, or if they do, it is in relatively small amounts.

Whereas phosphorus and sulfur occur in both organic and mineral form in the soil, there are no minerals from which nitrogen is derived. The nitrogen in soils originally came from the atmosphere. Atmospheric nitrogen makes its way into the soil by means of chemical or biological transformations that convert it into a form plants can use. Once in the plant, the nitrogen is assimilated into organic compounds that are later added to the soil. The accumulation of nitrogen is slow and closely parallels the build-up of organic matter in the soil.

The ability of soils to meet the nutrient requirement of plants depends in large part on the rate at which the nutrients are released from mineral and organic combination. Under natural conditions, native vegetation rarely suffers from a lack of nutrients. Shortages occur frequently on agricultural land, however, and the maintenance of satisfactory plant growth commonly requires the addition of fertilizers and manures to supplement the nutrient supply of the soil. Indeed, farmland that cannot benefit from the addition of supplemental nutrients is rare. The one added most often and in greatest amounts is nitrogen. This reflects in part the high demand of plants for nitrogen, but it is also due to the gradual depletion of soil organic matter, and nitrogen, by cropping and erosion of topsoil.

Soil as a biological system

Although the soil consists primarily of seemingly stable inorganic and organic substances, it is in fact a dynamic, life-sustaining system. Living organisms are a very important part of the soil and contribute greatly to its general properties and behavior. Often, the contributions made by soil-inhabiting organisms provide a more suitable environment for continued biological activity. This appears not too different from certain aspects of the relationship between man and his environment.

Plants contribute more to soil development and properties than does any other biological factor. Through the absorption of water and nutrients and the release of CO_2, roots of plants alter the composition of the soil solution and affect other components that tend to equilibrate with it. In addition, vegetation is the major contributor of organic residues to the soil. Thanks to such effects, plants exert a strong influence on mineral weathering, organic matter accumulation, and other processes important to soil formation. Indeed, few soils exist that do not exhibit some obvious indication of the role plants have played in their development.

In addition to plants, a multitude of other organisms, both macroscopic and microscopic, inhabit soils. Because of the joint activity of these organisms, principally in the decay of plant residues, the soil is maintained in a highly dynamic state. Changes induced by biological processes go on in an essentially unending fashion, and though centered in the organic fraction for the most part, they can ultimately have a profound effect on all soil components.

Biological change is generally desirable, for it normally has a favorable effect on the nutritional and physical state of the soil. From the point of view of most soil scientists, a high level of biological activity is the sign of a healthy soil; it reflects the right combination of characteristics that have a bearing on life processes associated with the soil environment. A principal measure of the suitability of these conditions is the capacity of soils to sustain an abundant growth of plants.

Soil as a product of the environment

Rocks at the surface of the earth, exposed to the elements, undergo considerable change with time. By a series of processes, they are converted to the more finely divided material we call soil. Initially, the conversion involves the breakdown, or weathering, of hard rock to a deposit of loose mineral matter referred to as *soil parent material*. Change in the parent material produces soil. Processes involved include continued mineral weathering and the synthesis of clay, the accumulation of organic matter, and the redistribution of clay by water percolating down through the parent material. These changes introduce new properties characteristic of the forming soil. The degree of change in the parent material depends on the nature and intensity of imposed physical, chemical, and biological reactions involved in soil formation, and on the length of time they have been in operation.

When soil-forming processes cause significant change over time, they tend to produce soil with characteristics that reflect the nature of the local environment. These characteristics are recognized by the way in which the appearance of the soil changes with depth. For example, the surface soil may be darkened by organic matter, or it may show principally the effect of weathering or of clay loss. Another feature may be a dense subsurface

Figure 1.1.
Diagrams of two soil profiles showing different degrees of development. These differences may be due to differences in the rate of soil formation, as controlled by the soil-forming environment, or to differences in time for soil development.

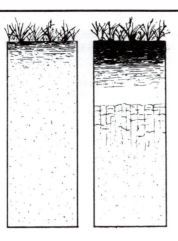

layer created by the accumulation of clay that has been washed downward from the surface soil by percolating water. Known as *horizons*, layers of this type are the mark of soil formation; they provide individuality to soils and serve as the basis for their identification and classification. When considered together, horizons make up the *profile* of a soil.

Most changes attending soil formation are contingent on the availability of water and heat in the soil-forming environment. Thus, the climate, as described by precipitation and temperature data, is an important factor in determining the types of changes involved in the conversion of parent material to soil. The contribution of climate relates to its impact on mineral weathering, on plant growth and organic matter accumulation, and on various other reactions important to soil development. Climate alone does not control soil formation, however, for the climatic influence is often modified significantly by the kind of mineral matter present and by local topographic and vegetational features. These several factors, then, climate, vegetation, topography, and parent material, collectively describe the environment in which soil formation takes place. As environments over the surface of the earth differ, so do their associated soils, provided time has been ample to allow for significant change to have taken place in the original mineral matter (see fig. 1.1).

Soil classification and survey

Information about soils important to their use is provided by soil classification and survey. The classification of soils is ordinarily based on profile properties, with soils forming under comparable conditions and having similar profile characteristics being placed together in a common class. In survey, soils belonging to various classes are located and their distribution on the landscape shown on a map. Soil survey and mapping therefore supply an inventory of soil resources and serve as an invaluable aid in land use planning, both for agricultural and nonagricultural purposes.

The distinction among soils in survey and classification is based on relatively few profile properties. It consists principally of characterizing soil profiles on the basis of the kinds and sequence of horizons present, with each horizon being described in terms of color, texture, structure, and general physical behavior when worked by hand in both a dry and moist condition. From these few properties, much can be judged about the behavior of the soil under use, as, for example, its ease of cultivation, strength, and its ability to transmit air and water to roots or to store water for plant use. However, making such judgments is not an intuitive process but comes from a broad knowledge of soils gained through long study and practical experience. The text that follows is a starting point for gaining knowledge basic to this process.

Summary

The soil occurs as a thin mantle over most land surfaces. It consists of mineral particles mixed with small, though varying, quantities of organic matter. The mineral fraction is derived from the weathering of hard rock; the organic, or humus, fraction is produced by the decay of plant and animal residues. Together, mineral and humus particles provide a relatively stable soil skeleton that encloses a system of interconnected pores, which is shared jointly by air and water. The ability of soils to supply air (oxygen) and water, along with plant nutrients, determines in part their suitability as a medium for plant growth.

Plant growth also depends on the physical state of the soil. The principal determinant of the physical state of soils is texture, or the proportions of sand, silt, and clay in the mineral fraction. The amount of clay is especially important to the physical state. In addition, clay, with humus, holds ions in adsorbed, exchangeable form, which is important to plant nutrition as well as to various physical and chemical properties of the soil.

Plants sustain life in the soil environment. Through photosynthesis, they produce food for animals and microorganisms, and residues for humus formation and accumulation. However, continued decay causes humus to disappear, and it can be maintained only if fresh plant residues are periodically added to the soil.

Soils are natural bodies with profile properties that depend on environmental conditions and the time involved in their development. Individual soil bodies are recognized and classified on the basis of their profile properties. The classification of soils organizes information about them important to their efficient use and management.

Review questions

1. Why are soil descriptions limited primarily to a description of the solid phase?
2. How are clay and humus similar, and how do they differ?

3. Why is the maintenance of soil humus dependent on the continual addition of fresh plant residues to the soil?
4. Why are pore space relations in soils a function of both texture and structure?
5. Why does waterlogging limit plant growth?
6. What physical characteristics indicate the presence of clay in soils?
7. Distinguish between transpiration and respiration, and explain why respiration is the more important of the two.
8. Show that respiration is chemically the reverse of photosynthesis.
9. Distinguish between macronutrients and micronutrients.
10. Why is soil organic matter important to the nitrogen nutrition of plants?
11. What is the advantage of classifying soils on the basis of their profile characteristics?

2

Rocks, rock weathering, and the formation of soil parent materials

Hard rocks are the original source of the mineral matter in soils. The minerals in soils are produced by weathering at or near the surface of the earth. Weathering converts massive rock to particles of comparatively small size, and it may be caused by either physical or chemical processes. If chemical weathering prevails, the original minerals may decompose and transform to new compounds within the weathering environment. Thus, similarities in the mineralogical composition of soil and the original rock depend on the extent to which chemical processes have dominated weathering reactions. Although the course of weathering is strongly influenced by climatic factors that determine the availability of water and heat in the weathering environment, it also depends on the nature of the rock being weathered.

General classes of rocks and minerals

The bulk of mineral matter in the earth's crust occurs in three kinds of rocks: *igneous*, *sedimentary*, and *metamorphic*, which are distinguished on the basis of origin, or mode of formation. Igneous rocks form from high-temperature melts (*magma*) that solidify into various minerals on cooling. Sedimentary rocks are formed from the products of mineral weathering that are created in one location and then transported and deposited in another, usually on the ocean floor. Metamorphic rocks result from the transformation of igneous or sedimentary rocks under high heat and pressure, but most frequently, it is in the lower strata of deep, sedimentary deposits.

Initially, all the earth's rocks were igneous, but weathering and the redistribution of weathering products over the ages have caused most land surfaces to be covered by sedimentary materials. These may be hard sedimentary rock formed by cementation on the ocean floor, or they may be loose sediments scattered over the landscape by wind, water, and flowing ice (*glaciers*). The principal exposures of igneous rocks are in high mountains where sedimentary materials have been removed by erosion (see fig. 2.1). Igneous rocks continue to form through cooling and solidification of magma, either deep within the earth's crust, or in *lava*, which is magma that flows out over the earth's surface from vents or fissures in the earth's crust.

Figure 2.1.
The Teton Mountains in northwestern Wyoming produced by the upthrust of igneous rock (granite) through older sedimentary formations. The latter have been removed by erosion. (U.S.D.A. Soil Conservation Service photo.)

Table 2.1. Contents of the principal elements as a percent of the total weight and total volume of all mineral matter in the earth's crust.

Element	Ionic form	Percent by weight	Percent by volume[a]
Oxygen	O^{2-}	46.6	93.9
Silicon	Si^{4-}	27.7	0.8
Aluminum	Al^{3-}	8.1	0.5
Iron	Fe^{2-}, Fe^{3-}	5.0	0.4
Magnesium	Mg^{2-}	2.1	0.3
Calcium	Ca^{2-}	3.6	1.0
Sodium	Na^{-}	2.8	1.3
Potassium	K^{-}	2.6	1.8
	Total	98.5	100.0

[a]Rounding causes these values to be slightly high in total.

Important minerals of rocks and soils

Minerals of the earth may be divided into two general groups based on origin: *primary* and *secondary*. Primary minerals form from molten magma and are therefore the kinds of minerals contained in igneous rocks. Secondary minerals are new compounds produced as a by-product of mineral weathering. Compounds that cement hard sedimentary rocks are secondary minerals, as are most of the clays in soils.

One means of classifying minerals into specific groups is by chemical composition; that is, according to the kinds and proportions of elements they contain. For a general classification, only the 8 elements in table 2.1 need be considered, for in various combinations they comprise virtually all of the earth's minerals. Of these elements, silicon and oxygen are by far the most abundant (see table 2.1); together, they make up about 75 percent of the weight and 95 percent of the volume of all mineral matter. It is little wonder, then, that most minerals, both primary and secondary, are compounds of silicon and oxygen. These minerals are called *silicates* to identify their major elemental components.

Based on composition silicate minerals occur in three major groups: *quartz*, the *aluminosilicates*, and the *ferromagnesium* minerals. As shown in table 2.2, quartz contains only silicon and oxygen. The aluminosilicates contain aluminum, and the ferromagnesium minerals contain iron and magnesium as major components in addition to silicon and oxygen. Calcium, sodium, and potassium are important constituents of minerals within the aluminosilicate and ferromagnesian groups.

An important distinguishing characteristic of primary minerals is color. Both quartz and the aluminosilicates are light in color, whereas the ferromagnesian minerals are dark, if not black. The black color is due to divalent iron (Fe^{2+}), which is the form iron assumes in the ferromagnesian minerals.

Over 95 percent of the earth's crust, a layer about 15 km thick, is made up of primary minerals, with more than 90 percent of these consisting only of the five types listed in table 2.3. Secondary minerals, with those in table 2.4 predominating, tend to be concentrated more in relatively thin surficial deposits at the outer boundary of the crust. Secondary aluminosilicates and the hydrous oxides occur mainly as clay-sized particles in sedimentary rocks and soils. The aluminosilicate clays and gibbsite are products of primary aluminosilicate mineral weathering; hydrous iron oxides are derived from ferromagnesian minerals.

Most lime minerals are formed as precipitates from sea water; they are the principal cementing agent in sedimentary rocks. Lime is produced when calcium and magnesium, derived as soluble products from mineral weathering on land, combine with carbonate (CO_3^{2-}) ions present in sea water. Lime is an important component of many soils, but principally those of drier regions. It is a fairly soluble mineral, so tends to be dissolved and leached from soils of the wetter regions. Gypsum, which is usually formed in dry regions as water containing calcium and sulfate ions is lost by evaporation, is even more soluble than lime. Its occurrence in soils is also limited to the drier regions, but it is not nearly so common a soil mineral as is lime.

Major rock classes

Rocks within the igneous, sedimentary, and metamorphic groups are classified on the basis of two properties: *mineralogy* and *texture*. Mineralogy refers to the kinds and proportions of minerals present; texture identifies the size of grains in which the minerals occur. Two textural ranges, coarse and fine, are commonly recognized. Mineral grains in coarse-textured rocks are clearly visible to the naked eye, but in fine-textured rocks, the grains can be seen only with the aid of magnification.

A classification of the more important igneous rocks is given in figure 2.2. Three groupings according to mineralogy are shown, with approximate proportions of different kinds of minerals designated graphically across the

Rocks, rock weathering, and the formation of soil parent materials

Table 2.2. Important primary minerals.

Mineral type	Chemical formula
Quartz	SiO_2
Aluminosilicates	
Calcium feldspar	$CaAl_2Si_2O_8$
Sodium feldspar	$NaAlSi_3O_8$
Potassium feldspar	$KAlSi_3O_8$
Muscovite mica	$KAlSi_3O_{10}(OH)_2$[a]
Ferromagnesian minerals	
Olivine	$(Fe,Mg)_2SiO_4$[b]
Augite	$Ca_2(Al,Fe)_4(Mg,Fe)_4Si_6O_{24}$
Hornblende	$Ca_2Al_2Mg_2Fe_3Si_6O_{22}(OH)_2$
Biotite mica	$KAl(Mg,Fe)_3Si_3O_{10}(OH)_2$

[a] In some minerals, hydrogen (H^-) ions are associated with certain structural oxygens to yield hydroxyl (OH^-) ions.
[b] Elements in parenthesis and separated by a comma vary relative to each other within the mineral type.

Table 2.3. Approximate content of major primary minerals in the earth's crust.

Mineral	Content
	%
Feldspars	59
Hornblende and Augite	17
Quartz	12
Biotite	4
Others	8
	100

Table 2.4. Important secondary minerals.

Mineral type	Composition
Clays	
Aluminosilicates	Variable
Hydrous oxides[a]	
Hematite	Fe_2O_3
Goethite	$Fe_2O_3 \cdot H_2O$
Gibbsite	$Al(OH)_3$
Lime	
Calcite	$CaCO_3$
Dolomite	$CaMg(CO_3)_2$
Gypsum	$CaSO_4 \cdot 2H_2O$

[a] Also called sesquioxides.

top of the figure. For example, those in the left-hand column have potassium feldspar and quartz as dominant components; for those in the right-hand column, calcium feldspar and ferromagnesian minerals predominate.

Among igneous rocks, granite and basalt are by far the most abundant. These two rocks are illustrated in figure 2.3. Granite is a coarse-grained rock with feldspar and quartz as its principal minerals. Granite has a coarse texture because it forms from magma that cools and solidifies slowly while still deep within the earth's crust; slow cooling of the magma permits crystals of

Figure 2.2.
A classification of igneous rocks based on mineralogy and texture.

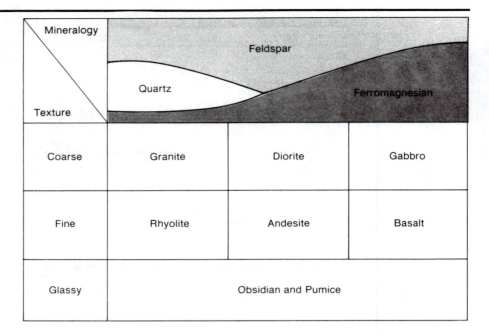

Mineralogy / Texture	Quartz / Feldspar	Feldspar / Ferromagnesian	Ferromagnesian
Coarse	Granite	Diorite	Gabbro
Fine	Rhyolite	Andesite	Basalt
Glassy	Obsidian and Pumice		

Figure 2.3.
Basalt (left), a dark-colored, fine-grained igneous rock high in ferromagnesian minerals, and granite (right), a light-colored, coarse-grained igneous rock containing feldspars and quartz predominantly. The dark crystals in granite are hornblende and biotite.

Rocks, rock weathering, and the formation of soil parent materials

large size to form. Basalt contains no quartz but has calcium feldspar and ferromagnesian minerals as dominant components. Basalt is the usual rock that forms from lava, and it has a fine texture. The reason for this is that lava cools and solidifies rapidly after flowing out over the earth's surface, which gives ions in the lava little time to organize into crystals of large size.

Granite has been the prime source of materials from which sedimentary rocks have formed. On weathering, the feldspar and ferromagnesian minerals in granite are converted primarily to clay. The quartz, however, being very resistant to chemical breakdown, is released as sand-sized grains at the weathering site. Through erosion, massive quantities of sand and clay produced by weathering, especially of granite, have been transported and laid down in sedimentary deposits. Where deposition has been in the ocean, cementation of the sediments, principally by lime, has converted them to hard sedimentary rocks. The more important of the rocks produced in this way are *shale*, formed from clays, *sandstone*, containing quartz sand in most cases, and *limestone*, which is a rock made up mostly of lime. Where calcite is dominant, *calcitic limestone* results. An accumulation of dolomite produces *dolomitic limestone*. Because of their mode of formation, sedimentary rocks have a layered appearance as shown in fig. 2.4.

Where sedimentary rocks have accumulated in very deep beds, heat and pressure may have caused transformation of the lower strata to metamorphic rock. The metamorphosis of sandstone yields of rock of almost solid quartz called *quartzite*. Shales are transformed to *slate*, whereas the metamorphosis of limestone yields *marble*. Metamorphosis of igneous rocks also occurs. Here, little change takes place in the mineralogy of the rock, although under high heat and pressure mineral crystals become stretched, or elongated, which gives the resulting metamorphic rock a banded appearance. *Gneiss* is the most important metamorphic rock produced in this way.

Sedimentary and associated metamorphic rocks are made a part of continental land masses by shifts in the earth's crust that lifts them above sea level, sometimes to elevations of several thousand kilometers (see fig. 2.5). Both sedimentary and metamorphic rocks are common to the higher mountains in the western United States and Canada, although erosion has removed them to expose underlying igneous rocks in many places. Sedimentary rocks are also widely distributed throughout the central part of the United States and Canada, but they occur in most locations as bedrock underlying loose mineral deposits carried in by wind, water, or glacial ice. Uplifted and folded sedimentary and metamorphic rocks are the dominant rock form in the Alleghany Region of the eastern United States.

Mineral weathering

Weathering is essential to the formation of soil parent material and to the transformation of parent material to soil. In the vast majority of cases, weathering involves both the physical and chemical breakdown of rock and mineral matter, with the relative influence of either depending on the prevailing conditions. In regions that are either dry or cold, physical forces tend

Figure 2.4.
Face of a thick sedimentary rock formation exposed by erosion. The layered nature of the rock provides evidence of its origin. (U.S.D.A. Forest Service photo.)

to dominate, but in warmer, wetter regions, the reverse is true. Under the latter conditions, physical weathering occurs, but its effects are often overshadowed by chemical change in the minerals.

Physical weathering

This type of weathering occurs either in place or as fragmented mineral matter is moved from one location to another by water, wind, or glacial ice. An important process of physical weathering in place is *unloading,* which is the removal of overlying material from deeply buried rock formations by erosion. Loss of the overburden reduces pressure in the rock, which allows it to expand and crack (see fig. 2.6). The cracks can then be enlarged by the expansion of ice or plant roots.

Figure 2.5.
Remnants of an eroded sedimentary rock formation in the Lewis and Clark National Forest, Montana. The elevation of the mountains is about 2.5 km above sea level. (U.S.D.A. Forest Service photo.)

Figure 2.6.
Rock, exposed at the earth's surface by the removal of overlying material through erosion, showing cracks formed by unloading. (U.S. Geological Survey photo by G. K. Gilbert.)

Rocks, rock weathering, and the formation of soil parent materials

Figure 2.7.
Rock showing the
effects of physical
weathering by
exfoliation. (U.S.
Geological Survey
photo by W. D.
Johnson.)

Exposed rock is also weathered physically by a process termed *exfoliation*, which is a peeling of the outer layers of the rock as demonstrated in figure 2.7. Exfoliation results from a softening of the outer skin of a rock by water penetrating microcracks or pores to a shallow depth. Strong *hydration* forces, due to the attraction of mineral surfaces for water, cause the outer layer to swell and soften. Freezing of the water aids in separating the surface layer from the firmer, unweathered rock beneath.

Although substantial reduction in particle size can result from physical weathering in place, it occurs more rapidly where previously fragmented materials are moved from place to place. Glacial action is especially effective in abrading and pulverizing rock material, and its influence on surface geology is widespread in the Northern Hemisphere. In North America, most of Canada and Alaska and a belt across United States just south of the Canadian border show signs of intense glaciation in times past. Although glaciation is currently very restricted, wind and flowing water continue as active weathering agents, with water being the more important of the two. Interestingly, a substantial part of the materials moved by wind and water in cooler regions is pulverized debris left from earlier glacial action.

Chemical weathering

The basic cause of chemical weathering is the attraction of polar water molecules for charged ions in mineral structures. Because of this attraction, ions are pulled free of the minerals to become a part of the weathering solution.

Rocks, rock weathering, and the formation of soil parent materials

Some of the ions recombine to precipitate as secondary clays, but others remain soluble and can be lost by *leaching*, which is the downward transport of soluble materials in percolating water. The removal of ions either by precipitation or leaching keeps their concentration in the weathering solution low and thereby encourages the continued release of other ions from the weathering minerals.

Other reactions that are commonly associated with chemical weathering are *hydrolysis, carbonation*, and *oxidation*. Hydrolysis is the reaction with water and is typified by the hydrolysis of feldspar, an important component of granite. Products of the reaction are kaolinite, a clay common to many soils, and the base, potassium hydroxide:

$$KAlSi_3O_8 + HOH \rightarrow HAlSi_3O_8 + KOH \qquad (2.1)$$
$$\text{Kaolinite}$$

The kaolinite precipitates at the weathering site, but the potassium hydroxide remains soluble and can be lost by leaching.

Carbonation is the result of a reaction between a compound and carbonic acid, H_2CO_3. Carbonic acid is present in all aqueous environments, and it is produced when carbon dioxide from air dissolves in and combines with water:

$$CO_2 + H_2O \rightarrow H_2CO_3 \qquad (2.2)$$

Though carbonic acid is classed as weak because it does not dissociate extensively, it is essentially inexhaustible and, over long periods, has an important impact on mineral weathering because of its ability to acidify weathering solutions and thereby increase the solubility of many minerals.

Carbonation is particularly effective in promoting the weathering of lime minerals such as calcite, $CaCO_3$. On reaction with carbonic acid, calcium carbonate, a compound of fairly limited solubility, is transformed to more readily soluble calcium bicarbonate, $Ca(HCO_3)_2$, thus enhancing leaching loss:

$$CaCO_3 + H_2CO_3 \rightleftharpoons Ca(HCO_3)_2 \qquad (2.3)$$

A similar reaction yielding a mixture of calcium and magnesium bicarbonates results from the carbonation of dolomite in dolomitic limestone.

Most of the calcium and magnesium released through weathering is leached and transported to the ocean in bicarbonate form. In the ocean, the reaction responsible for transforming calcium and magnesium carbonates in lime to bicarbonates is reversed; that is, the carbonates are reformed with the release of carbon dioxide:

$$Ca(HCO_3)_2 \rightarrow CaCO_3\downarrow + H_2O + CO_2\uparrow \qquad (2.4)$$

This type of reaction is involved in the accumulation of lime minerals in oceanic sediments.

Oxidation is the loss of electrons by ions and is very important to the weathering of iron-bearing ferromagnesian minerals. Its effect is illustrated by the oxidation of iron following its release in relatively soluble form following the hydrolysis of iron olivine. On hydrolysis, iron olivine yields iron (ferrous) oxide and silicic acid:

$$Fe_2SiO_4 + 2HOH \rightarrow 2FeO + H_4SiO_4 \qquad (2.5)$$

Both of the products of the reaction are somewhat soluble and can be lost by leaching. However, in the presence of free oxygen, the iron oxide, FeO, is oxidized to very slightly soluble iron oxides such as hematite, Fe_2O_3, or its hydrated counterpart, goethite, $Fe_2O_3 \cdot H_2O$:

$$4FeO + O_2 \rightarrow 2Fe_2O_3 \qquad (2.6)$$

Because of the extremely low solubility of these iron oxides, very little of the iron released from weathering minerals is lost from the weathering environment.

The chemical weathering of rocks follows patterns dictated in part by their texture and mineralogy. Most rapid weathering is of lime-cemented sedimentary rocks, which is due to the relatively high solubility of lime minerals. Once lime is removed, however, residues that remain resist further change, because they consist predominantly of highly stable quartz sand or secondary clay minerals.

Primary minerals are much more slowly weathered by chemical means than are lime minerals. Differences exist among the primary minerals, however, for quartz is much more resistant to chemical breakdown than are the aluminosilicate and ferromagnesian minerals. In the weathering of granite, for example, the feldspars and ferromagnesian minerals may be totally broken down and replaced by clay without much happening to the quartz that is present. This accounts for sand and clay occurring together at the site of granite weathering. Basaltlike rocks can weather totally to clay, since they contain no quartz. The residue of basalt weathering is typically red. This color is due to iron oxides formed from the ferromagnesian minerals that occur copiously in basalt or other materials of volcanic origin.

Weathering is involved in the formation of both soil parent material and soil. In the normal view, the formation of parent material precedes that of the soil, and this is indeed the case where products of weathering are moved to a new location before soil formation is initiated. Where the products of weathering remain in place, however, the soil and parent material form at the same time and a clear distinction cannot be made between them.

Classification of soil parent materials

In its normal concept, soil parent material is viewed as a residue of unconsolidated mineral debris resulting from a previous cycle of weathering. The formation of soils results from changes in the parent material. The kinds and extent of changes that take place during soil formation depend on factors that determine the soil-forming environment. Parent material is, in itself, one of these factors. Parent material influences soil formation to the degree that its two principal characteristics, texture and mineralogy, control the kinds of properties that are created in the forming soil.

Figure 2.8.
A soil forming in residual parent material derived from weathering limestone. The shallow, stony nature of this soil greatly limits its usefulness for most purposes. (U.S.D.A. Soil Conservation Service photo by W. A. Wade.)

Two types of parent materials derived through mineral weathering are recognized. These are *residual* and *transported*. *Organic deposits* comprise a third general type of parent material. Of the three, transported materials are the most widespread. Although very important on a local basis, organic materials are much more limited in extent than are either of the other two.

Residual parent materials

The weathering of consolidated rock in place produces residual parent material (see fig. 2.8). The nature of residual parent material depends on the nature of the original rock (parent rock) and the manner in which it has been changed by weathering. In general, residual parent materials are of greatest extent within stable landscapes where weathering has been both long and of moderate to high intensity. This combination of conditions tends to be most common in tropical and subtropical regions. In the United States, residual parent materials are of greatest importance in the East and Southeast, including Puerto Rico, and in the wetter parts of the Pacific Northwest and Hawaii. They are also important in mountainous regions, where they and the soils formed from them, though often very thin, are highly critical to the maintenance of a vegetative cover.

The rate of residual parent material formation is usually slow, often requiring several tens of thousands of years to affect hard rock to appreciable depths. Yet, under intense weatherng, soft limestone and other lime-cemented sedimentary rocks may yield significant deposits of residual parent material over much shorter periods. Indeed, the relatively rapid rate at which some lime-bearing rocks weather accounts for the presence of residual parent materials where little if any weathering of more resistant rocks would be expected. If the intensity of weathering is quite low, however, as in arid to semiarid regions, the formation of residual parent material from any type of rock is unlikely. Under these conditions, transported parent materials may provide essentially the only sites for soil formation.

Transported parent materials
Parent materials of this type differ widely in both origin and form. They are usually named to show the principal agency responsible for their transportation and deposition. Although classes established on this basis vary widely in character, each can be identified with at least a few properties that have an important bearing on soil-forming processes. A common feature of transported parent materials is that they occur in an unconsolidated state, which distinguishes them from hard igneous, sedimentary, and metamorphic rocks as sources of soil-forming material.

Deposits from running water   Sediments deposited from running water are called *alluvium*. Deposition occurs when water loses velocity and its ability to carry material in suspension. It may be in *alluvial fans*, as where steep mountain streams empty on to flatter land (see fig. 2.9), in *deltas* where rivers empty into still water, or as *flood-plain alluvium* laid down parallel to rivers when they overflow their banks. In the latter instance, coarser materials are dropped immediately to form somewhat elevated *levees* next to the river, as shown in figure 2.10. Finer materials settle out on flatter areas beyond the levees.

Parent materials of alluvial origin may consist of freshly weathered mineral matter, or they may be made up of older material, such as topsoil eroded from the neighboring watershed. Alluvial materials show no consistent textural pattern, although within a continuous drainage system, they may grade in average particle size from coarse to fine as they move further from the source. Waterlogging, a condition where ground water fills the pores in the lower part, if not all, of the profile, is typical of floodplain deposits that lie at about the same level as the water in permanent drainageways. In comparison to floodplains, natural levees are usually better drained, primarily because they have a coarser texture and are at a somewhat higher elevation.

Changes in the earth's crust sometimes occur that increase the gradient of a river and cause down-cutting and the formation of a new, deeper channel in an established floodplain. This, in effect, elevates the original floodplain relative to the river and converts it into a *terrace* (see fig. 2.11). In the

Rocks, rock weathering, and the formation of soil parent materials

Figure 2.9.
An alluvial fan formed where a sediment-laden stream loses velocity on reaching the more nearly level valley floor in the foreground. (Geological Survey of Canada photo.)

Figure 2.10.
Diagram illustrating a natural levee along the Mississippi River. The Caruthersville soil on the levee has a coarser texture and is better drained than the Commerce soil lying at a slightly lower elevation beyond the levee and consisting of finer textured material. (From the Soil Survey of Cape Girardeau, Mississippi, and Scott Counties, Missouri. Washington, D.C.: Superintendent of Documents.)

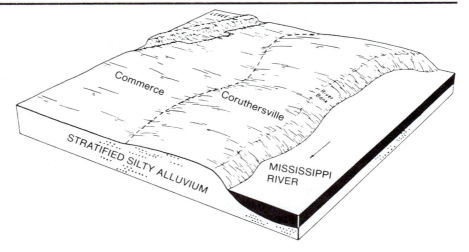

Figure 2.11.
The flood plain of the Lava River, west of Aniakchak Crater on the Alaska Peninsula, bordered by higher-lying, broad, flat terraces of glacial outwash material. (U.S. National Park Service photo by M. Woodbridge Williams.)

new position terrace alluvium is more adequately drained and is no longer subject to flooding. Soils formed on terrace alluvium are frequently productive and well-adapted to agricultural uses.

Marine sediments   These sediments accumulate in an ocean environment and are later exposed by natural or artificial means while still in an unconsolidated state. Marine sediments vary widely in mineralogical and textural composition. Many of the soils occurring along the Atlantic Seaboard and in the states bordering the Gulf of Mexico are derived from sediments of marine origin.

Lacustrine deposits   Lacustrine deposits are materials that have settled from lake water. They vary texturally and, more often than not, are stratified. The most extensive occurrence of lacustrine materials in the United States is in the northern glaciated areas.

Rocks, rock weathering, and the formation of soil parent materials

Figure 2.12.
A dramatic view of a glacial ice sheet flowing seaward from Ellesmere Island, off northern Canada. The fan shaped formation in the foreground is an ice shelf, or mass of floating ice permanently attached to the land. (From Post, Austin, and LaChapelle, Edward R., *Glacier Ice.* Seattle: University of Washington Press, 1971. U.S. Geological Survey photo by Austin Post.)

Glacial deposits Glaciation has made an important impact on the surface geology in each continent of the Northern Hemisphere. North America has been affected by four glacial periods, the latest occurring from 7,000 to 10,000 years ago. Physical weathering is the principal means of parent material formation by glaciation. It results from tremendous pressure produced by thick sheets of flowing ice, which cause severe gouging and abrasion of surface rock structures. Broad ice sheets are no longer as prevalent as during the Ice Age, but they still persist in Antarctica, over much of Greenland, and in Arctic islands north of Canada (see fig. 2.12).

Parent material laid down by glaciers or glacial melt water is termed *glacial drift;* that deposited directly from the ice, is known as *till.* Material carried beyond the glacial ice front by water is *glacial outwash.* Till is usually a heterogeneous mixture of particles ranging in size from clay to gravels and sometimes boulders (see fig. 2.13). It is deposited in various landforms beneath, to the side of, and at the front of glaciers. The deposition of glacial outwash may be in broad *outwash plains* laid down in the wake of the receding glacier, or it may be in lacustrine or alluvial deposits.

Figure 2.13.
Morainal deposits of glacial till between Mounts Deception and Fricaba in the Olympic Mountains of Washington State. The fragmented materials range from clay-sized particles to boulders weighing several tons. Finer particles removed by the meltwater in the foreground may be deposited elsewhere as glacial outwash material. (Photo courtesy of William A. Long, Crescent City, CA.)

In this country, all of the states just south of the Canadian border have been directly affected by ice flows moving southward out of Canada (see fig. 2.14). The most extensive incursions have been in the northeastern and northcentral states. All of New England and New York have been covered by glacial ice. In the center of the country, the southernmost boundary of the glacial advance is marked approximately by the Ohio and Missouri Rivers. The five Great Lakes, along with thousands of other smaller lakes concentrated mainly in Minnesota, Wisconsin, and Michigan, are the result of glacial action. Much of the outwash materials derived from the huge glacial sheets has been carried farther south and deposited in alluvial formations by the Missouri, Ohio, Mississippi, and associated rivers. Extensive deposits of glacial origin have also been laid down, largely as lacustrine materials, in the Columbia Basin of central Washington, in southern Idaho, and as far south as Utah and northern Nevada.

The mineralogy of glacial deposits is determined by the kinds of rock formations from which they have been derived and follows no consistent pattern. Many of the North American deposits are calcareous (lime-bearing), however, which indicates that limestone has been one of the important source rocks.

Rocks, rock weathering, and the formation of soil parent materials

Figure 2.14.
Centers of origin and extent of glaciation on the North American continent during the Pleistocene epoch, which lasted about 1 million years and ended some 10,000 years ago.

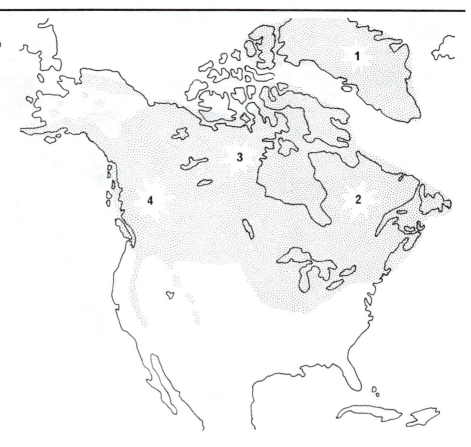

Wind-laid sediments   Extensive deposits of wind-transported materials are found in many parts of the world. These are *eolian* materials and occur in one of three principal forms: (1) slowly shifting, coarse-textured *sand dunes*, (2) *loess*, which consists largely of silt and very fine sand particles, and (3) *tephra*, which is air-borne volcanic material of any size.

Sand dunes, which are illustrated in figure 2.15, form because of the limited ability of wind to move large mineral grains. Movement may occur only during short, gusty periods, with the sand accumulating behind obstructions where the wind velocity is comparatively low. Unless stabilized by vegetation or other means, sand dunes tend to migrate slowly over land surfaces. This movement is due to the repeated transfer of sand from the face of the dune to the opposite, downwind side. The sand occurring in dunes may come from sites of crystalline rock weathering, from sandy alluvium or similar deposits, or from beach sand concentrated by wave action along the shores of large bodies of water.

Unlike dune sand, mineral particles making up loess are of a size that can be picked up and transported by wind for deposition at some distant point. These particles appear to be primarily of glacial origin, and their ac-

Figure 2.15.
Continually shifting
sand dunes in Death
Valley National
Monument, California.
(U.S. National Park
Service photo.)

cumulation as loess is thought to occur in two stages: (1) the removal, sorting, and redeposition of glacially produced silts and very fine sands by melt water, and (2) their ultimate removal and redeposition by wind. These deposits may be a hundred meters or more thick in places (see fig. 2.16).

In the United States, the largest body of loess occurs in a broad belt extending from Colorado to Ohio in an east-west direction and in a narrower belt paralleling much of the Mississippi River (see fig. 2.17). The loess in this body apparently comes from alluvial deposits laid down by the Missouri, Ohio, and Mississippi Rivers in interglacial and postglacial periods. The loess is distributed to either side of these great rivers, but because of prevailing winds, the deposits are both deeper and more extensive in an easterly direction from the rivers. The depositional pattern is most clearcut along the Mississippi, where a belt of loess up to 150 km wide parallels the river on the east. The depth of the loess, which may be 50 m or more near the river, decreases with increasing distance from the river, as does average particle size.

In some places deposits of glacial outwash materials may contain particles of both sand and silt size. Reworking of these deposits by wind sorts out the finer particles, which may be carried aloft and deposited elsewhere

Figure 2.16.
A deep roadcut
through Peorian loess
in Washington County,
Nebraska. (U.S.D.A.
Soil Conservation
Service photo by R. K.
Jackson and W.
McKinzie.)

Figure 2.17.
Shaded areas show
the major loess
deposits in the United
States; the broken line
shows the
approximate limits of
glacial ice flow out of
Canada. (From a
U.S.D.A. map.)

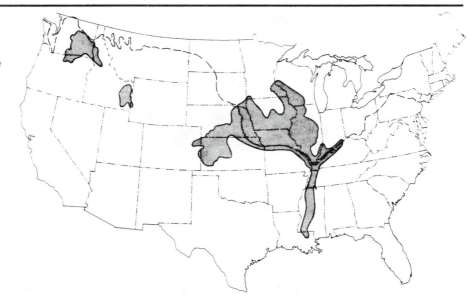

Rocks, rock weathering, and the formation of soil parent materials

Figure 2.18.
Mt. St. Helens,
Washington, in
eruption on May 18,
1980. (U.S. Geological
Survey photo by J. C.
Rosenbaum.)

as loess. The coarser particles remain behind to accumulate as dune sand. Under these circumstances, the depositional pattern results in a continuum characterized by a gradation in texture from coarse to fine with increasing distance from the source.

Tephra is an important soil-forming material in many places in the western United States and Hawaii. For example, soils downwind of Mt. St. Helens, in Washington (see fig. 2.18), were coated with 15 cm or more of volcanic ash following its eruption in May, 1980. Because of the way in which it is laid down, tephra is often interbedded with other transported materials, or where it is of recent origin, it may occur as the topmost layer of parent material. Most tephra contains a high proportion of poorly structured glass particles. This type of material tends to weather more rapidly than do crystalline mineral grains present in igneous rock such as granite.

Colluvium   This type of parent material is produced by the movement of materials down a slope under the direct influence of gravity. The process, sometimes referred to as *mass wasting,* is typified by landslides, mudflows, or the slow shifting of a soil mass (*solifluction,* or *soil creep*) downslope.

Rocks, rock weathering, and the formation of soil parent materials

Figure 2.19.
Extensive talus deposits skirting Medicine Bow Peak in southern Wyoming.

Where water erosion contributes to the accumulation process, the resulting deposit may be called *alluvial-colluvial* material. Regardless, this type of parent material may range from finely divided mineral particles to coarse, fragmented rock (*talus*), as shown in figure 2.19. The most common occurrence of colluvial materials is along the borders of river valleys, particularly where they are bounded by steep, if not precipitous, slopes.

Organic deposits The greater part of organic parent materials accumulates in lake or swamp water where decomposition of plant residues is retarded by the limited supply of oxygen. The degree of decomposition of these materials is highly variable. Sometimes, organic deposits contain substantial proportions of mineral matter, either intimately mixed or interbedded with the organic materials. Soils derived from organic parent materials are widely distributed throughout the wetter regions of the world, although their total area is not great. One of the more extensive areas is in western Ireland, where the soil is dug, dried, and used as fuel (see fig. 2.20). Due to their highly leached nature, organic soils are often lacking in some essential plant nutrients and may not support abundant plant growth unless fertilized. Plant growth may also be limited because of excessive wetness.

Figure 2.20.
A deep peat soil typical of the eternally wet, low-lying areas of Connemara, western Ireland. The peat has been dug and laid out to dry for use as fuel.

Significance of parent material classification

A number of things can be judged about a parent material from knowledge of its class. One such inference is the approximate age of the material. As an example, major glacial deposits in North America are known to be some 7,000 to 10,000 years old, as are the outwash, lacustrine, and loessial materials derived from them. Similarly, many deposits of volcanic dusts and alluvial materials are known to be of very recent origin, so recent in fact, that only minimal change has been made in them by soil formation.

Sometimes the geographic or topographic position of a parent material can be judged from its class name. For example, on the North American continent, soils developing in glacial drift are limited to the northern latitudes; they are widespread in Canada and Alaska, but they extend southward into the contiguous 48 United States to only a limited extent, as has been shown in figure 2.14.

Topographically, glacial and loessial materials are located principally on sloping terrain, such as rolling or hilly landscapes. Typical examples are shown in figure 2.21. Glacial outwash materials, like others laid down by

Rocks, rock weathering, and the formation of soil parent materials

Figure 2.21.
(A) an irregular,
glaciated landscape in
Alaska produced by
both the removal and
deposition of materials
by flowing ice. Lacking
a normal surface
drainage pattern,
much of the area is
covered by lakes that
form as lower lying, or
depressional,
positions fill with
water. (B) Hills of the
Palouse Region, in
eastern Washington,
produced by natural
erosion of deep loess
deposits. (U.S.D.A.
Soil Conservation
Service photos by
W. J. Watts and E. R.
Baker.)

A

B

running water, as well as lacustrine and marine sediments, appear mostly on nearly level landscapes. As will be seen shortly, the position of a parent material on the landscape can be extremely important to the pattern of change in the parent material as it is gradually converted to soil over time.

On occasion it is possible to judge the texture of a parent material from its class name. This is most often the case with eolian loess and sand dune materials, either of which tends to fall within a fairly narrow textural range. The heterogeneous character of glacial till is also a reasonably consistent property. Little can be said about fixed textural relationships for other types of parent materials, however.

Summary

The first step in soil formation is the weathering of hard igneous, sedimentary, and metamorphic rock to more finely divided soil parent material. Weathering of these rocks is by both physical and chemical means. Physical weathering results only in the reduction of mineral particle size, but chemical weathering brings about the partial or complete destruction of the minerals. The dissolution of minerals is the principal process of chemical weathering, but it is aided by hydration, hydrolysis, carbonation, and oxidation reactions.

Most chemical weathering is accompanied by the synthesis of new (secondary) clay minerals from weathering products. The clay normally forms at the site of weathering, where it accumulates as other weathering minerals disappear. Quartz in weathering rocks tends not to disappear rapidly, however. Rocks containing quartz may therefore weather to a mixture of clay and quartz sand. Rocks that lack quartz, on the other hand, are likely to weather totally to clay. Parent materials high in silt will probably have been produced largely by physical weathering.

Weathering yields two classes of soil parent material: residual and transported. Residual parent materials are formed in place; transported parent materials are moved to a new location following initial weathering. The principal agents of transport are water, wind, and glacial ice. Most soils have formed in transported parent materials, the principal types being glacial, alluvial, and eolian. Lacustrine and marine deposits are also important in some areas. A small proportion of soils have formed in deposits of organic materials that have accumulated under conditions of poor aeration in bogs or swamps.

Review questions

1. Why have most soil parent materials probably come from the weathering of sedimentary rather than of igneous or metamorphic rocks?
2. What accounts for the vast majority of the earth's minerals being classed as either quartz, aluminosilicate, or ferromagnesian minerals?
3. Distinguish between primary and secondary minerals.

4. Why are both the texture and mineralogy of rocks factors of chemical weathering?
5. Why does table 2.3 suggest that, on weathering, primary minerals in the earth's crust are likely to yield a greater proportion of sedimentary shale than sandstone?
6. Compare secondary clay and lime minerals with respect to their usual site of formation.
7. On the basis of origin, distinguish among alluvium, glacial till, loess, lacustrine, and marine types of parent material.
8. What environmental conditions promote the accumulation of organic parent materials, and why?
9. Why might fresh deposits of loess or glacial parent materials be expected to contain a higher proportion of minerals unaffected by chemical weathering than would be expected for average alluvial parent materials?

Selected references

Evans, R. C. *An Introduction to Crystal Chemistry.* Cambridge: At the University Press, 1964.

Garrels, R., and MacKenzie, F. T. *Evolution of Sedimentary Rocks.* New York: W. W. Norton & Company, Inc., 1971.

Geological Institute, American. *Dictionary of Geological Terms.* Garden City, N.Y.: Doubleday & Company, Inc., 1962.

Gilluly, J.; Water, A. C.; and Woodford, A. O. *Principles of Geology.* San Francisco: W. H. Freeman and Company, 1975.

Grim, R. E. *Clay Mineralogy.* New York: McGraw-Hill Book Company, Inc., 1968.

Marshall, C. E. *The Physical Chemistry and Mineralogy of Soils. Vol. 1: Soil Materials.* New York: John Wiley & Sons, Inc., 1965.

Mason, B., and Berry, L. G. *Elements of Mineralogy.* San Francisco: W. H. Freeman and Company, 1968.

3

Soil formation

Soil occurs as a comparatively thin mantle over much of the earth's land surfaces. Although it forms a continuous system and is used everywhere for many of the same purposes, the soil varies considerably from place to place. Because of this variation, the soil continuum is viewed as a collection of soil bodies, or individuals, each occupying a small segment of the landscape and differing from other soil individuals by virtue of a unique set of profile properties. Not only do these properties allow a distinction to be made among different kinds of soil, they also serve as a basis for judging soil behavior under one or more of a wide range of uses.

An important concept of soils is that they are natural bodies produced over time by a set of processes acting on mineral matter at the surface of the earth. Since these processes function under a wide range of environmental conditions, they produce soils of divergent character. An explanation of differences among soils therefore depends on the relationships between soil properties and the conditions under which they form. The study of these relationships is known as *pedology* (Gr. *pedon*, ground, and *logos*, discourse).

General nature of soil formation

The keynote of soil formation is change. It involves the conversion of previously weathered mineral matter, the parent material, to a new product, the soil. Some phases of soil formation are destructive, for they cause continued weathering and wastage of minerals in the parent material. Yet, soil formation can also be constructive where it causes the synthesis of new mineral compounds and humus. Through changes such as these, new properties are introduced that replace or mask the properties of the parent material. The degree of change depends on both the intensity and time of soil formation.

The effect of soil development is judged from the relative expressions of two kinds of properties in a soil: *inherited* and *acquired*. Inherited properties are those contributed directly from the parent material; that is, they appear in the soil much as they did in the original parent material. Acquired properties are those produced by change in the parent material, as for example, a texture, structure, or color that is different in the soil than in the parent material. Where soil formation has produced little change in parent material, soil properties will be inherited for the most part. Drastic change yields a preponderance of acquired properties, on the other hand. In most soils, acquired properties are expressed to at least a small degree; in some, they are overriding. The kinds of acquired properties in a soil depend largely on which of the several soil-forming processes have been most active in its formation.

Processes of soil formation

The soil-forming processes consist of a combination of physical, chemical, and biological reactions that transform parent material into soil. The more important of these are: (1) continued mineral weathering and clay synthesis, (2) organic matter accumulation, (3) exchange of ions adsorbed to clay and organic particles, (4) translocation of solid and soluble components within the profile, (5) structure formation, and (6) mechanical or biological mixing. The effectiveness of these processes depends on the conditions under which soil formation takes place. A prime factor is the availability of water, which is essential to virtually all reactions involved in soil formation.

The processes of soil formation tend to affect one part of the profile more than another, which accounts for the development of different kinds of horizons in the profile. The general nature of the processes and their effects in horizon formation are discussed in the sections that follow.

Mineral weathering and clay synthesis

Weathering and clay synthesis cause changes in the texture, mineralogy, and color of parent materials through the breakdown of coarse mineral grains and their replacement by clay. These processes are most active in warm, moist climates. In tropical regions, for example, they may cause the complete destruction of rock and its conversion to clay to depths of many meters. In arid climates, however, weathering is limited and may consist of nothing more than the dissolution of lime and its downward translocation to a shallow depth below the surface.

In moist climates, weathering is usually accompanied by leaching, or the removal of soluble materials in water percolating downward through the soil. Leaching removes ions that are released in soluble form from weathering minerals, many of which are plant nutrients. This causes a loss in fertility and explains the general inability of strongly weathered and leached soils of tropical regions to sustain abundant plant growth for long.

Organic matter accumulation

Organic matter in soils is a product of plant-residue decay. The pattern of organic matter accumulation depends largely on the source of the residues; that is, whether they come from grasses or trees. These plants yield different patterns of accumulation because of dissimilarities in their root systems (see fig. 3.1). Grass roots are relatively shortlived, and, through death and regeneration, contribute substantial residue directly to the mineral soil each

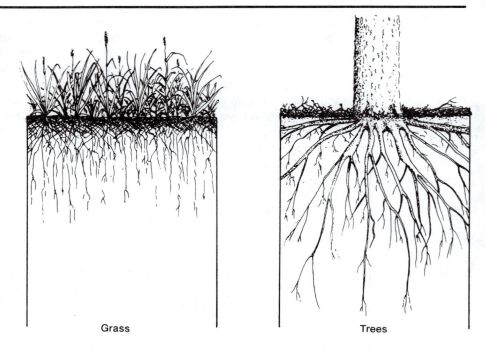

Figure 3.1.
Comparison of the nature and density of roots under grassland and forest vegetation.

Grass

Trees

year. The build-up of humus from decaying grass roots produces a dark-colored horizon at the surface of the mineral soil. The color and thickness of this horizon depends on the depth of the root system and the amount of residue it adds to the soil each year.

Unlike grasses, trees have a perennial root system that adds little residue directly to the mineral soil on an annual basis. Instead, trees supply residues almost totally as leaf, needle, and twig fall. In time, these residues accumulate to form an organic horizon upon the mineral-soil surface. A similar pattern of organic matter accumulation is produced by mosses and other aquatic plants growing in swamps or bogs (muskeg, in northern latitudes), and if thick enough, results in the formation of unique organic (peat or muck) soils.

Ion Exchange

Ion exchange is the displacement of ions adsorbed to clay and humus particles by other ions from the soil solution. In young, unweathered soils, the predominant adsorbed ions are calcium and magnesium. This is preferred, for calcium and magnesium are plant nutrients. Further, they tend to maintain a near neutral to slightly basic soil reaction that is ideal for many plants.

In humid regions, a normal trend in soil development is the displacement of adsorbed calcium and magnesium by hydrogen ions added to the soil in rainwater or supplied by biologically produced acids, as from organic matter decay. Once displaced, the calcium and magnesium are usually lost through leaching, the net result being a build-up in soil acidity. Soil acidification is slowed if the parent material is high in lime (lime neutralizes acidity) or if slow permeability of the soil limits the leaching loss of calcium and magnesium.

With decreasing precipitation, ion exchange becomes progressively less important to soil formation. An exception is in waterlogged soils of dry regions where sodium derived from salt-bearing ground water accumulates as an adsorbed ion by displacing calcium and magnesium from soil particle surfaces. The build-up of adsorbed sodium is undesirable and is usually suspect in soils with a strongly basic reaction.

Translocation

Translocation is the movement of a soil component from one level to another in the profile by water. One example is the upward movement of soluble salts from ground water, which allows them to accumulate at the surface of the soil. In most soils, however, translocation involves the movement, or *eluviation*, of a component out of a horizon at or near the surface and its redeposition, or *illuviation*, in the horizon immediately below. Lime, gypsum, silicate clays, humus, and iron and aluminum are most often affected. The translocation of lime and gypsum occurs primarily in soils of the drier regions and results in subsurface layers enriched in one or both of these two materials. The eluviation-illuviation of humus, iron, and aluminum is restricted principally to forest soils of the cooler, wetter regions and yields a typifying, highly pigmented subsurface horizon in these soils. The translocation of silicate clays is a widespread phenomenon and can occur in essentially any readily permeable soil material.

Structure formation

The formation of structure in soils is in most cases due to physical forces that separate an otherwise cohesive mass of soil into aggregates of various sizes and shapes. Forces involved may be the wedging action of plant roots or expanding ice crystals, or they may be contractive forces that cause the soil to shrink and crack on drying. In some instances, flat platelike structural forms in a soil may result from textural layering of the parent material, as is typical of lacustrine deposits.

Structure is not a universal characteristic of soils. Structure may be lacking if there is too little clay to bind soil particles into aggregate form, or if a particular part of a profile has never been subjected to a force that produces structure. Where structure does exist, the kind is related in a very general way to depth in the profile. For example, smaller structural forms tend to be concentrated in the surface horizon, whereas those of large size are limited principally to horizons in the subsoil. In many soils, subsurface horizons high in illuvial clay are easily recognized because of a distinctive structural form.

Soil mixing

There are several ways in which soils are mixed by natural forces. Highly localized mixing of surface soils is often caused by worms or burrowing insects, and it may result in the complete turnover of soil to the depth of mixing in a matter of a few years. A similar effect is produced in certain high-clay

Figure 3.2.
Photos illustrating
processes of soil
mixing. (A) Soil
slippage downslope,
or solifluction, (B) tree
throw, and
(C) cryoturbation. In
the latter, the
formation of large ice
wedges has caused
severe distortion of
the organic layers in a
peat soil. (Credits: (A)
and (B), U.S.D.A. Soil
Conservation Service
photos; (C), U.S.
Geological Survey
photo by T. L. Pewe.)

A

B

C

Soil formation

soils that shrink and crack on drying. Mixing results as particles or aggregates lying loose on the surface are washed into the cracks, where they become entrapped as the soil subsequently expands on wetting. In some forest soils, biological mixing of surface litter with the mineral soil below is responsible for the development of a darkened horizon at the surface of the mineral profile.

Other forms of soil mixing include *tree throw*, or the scattering of soil by roots of trees as they are felled by wind; *solifluction*, or soil slippage on slopes; and *cryoturbation*, which is soil churning caused by swelling and contraction of wet, cold-region soils as they undergo cyclic freezing and thawing (see fig. 3.2). Each of these types of soil mixing is more likely to destroy rather than aid in the formation of soil horizons.

Effects of the soil-forming processes vary from place to place, for differing conditions in the soil-forming environment result in different combinations of processes that are most active in soil development. The processes responsible for a given soil can usually be judged from the kinds of horizons present in its profile.

Classification of soil horizons

Soil horizons are identified, or named, by symbols consisting of combinations of upper- and lower-case letters. Each symbol recognizes a property or combination of properties introduced into the soil by soil formation, or *genesis*. Because of their origin, these properties are often called *genetic properties*, and the horizons they produce are called *genetic horizons*.

Based on the effects of soil genesis, well developed soil profiles are divided into two parts. The upper part is termed the *solum* (L. *solum*, soil), or sometimes, *true soil*, and is the part of the profile showing effect of organic matter accumulation, weathering, and the eluviation and illuviation of clay, humus and other relatively immobile components. Soil below the solum shows minimal effects of soil formation, and in some cases, its properties are like those of the parent material from which the overlying solum has formed.

The classification of horizons listed below is of recent origin; it was adopted for use in the United States in October, 1981. In this system, horizon symbols consisting of upper- and lower-case letters replace older symbols made up of letter-number combinations. Three letters in the older system, A, B, and C, denoted the sequence of horizons down through the mineral profile, with the A and B comprising the solum. Subhorizon symbols within the A and B were established by adding Arabic numerals to the capital letters. Six such combinations were used in recognition of six specific kinds of horizons: A1, A2, A3, B1, B2, and B3. Since these symbols appear in most of the current literature dealing with soils in the United States, they are shown in parentheses in the classification scheme that follows:

O (O)   Horizon dominated by organic material. It may be a litter layer on the surface of a mineral soil, or it may be organic material that has accumulated in water under conditions of poor aeration, as in *peat* or *muck* soils.

A (A1)	Either (1) a horizon at the surface of a mineral soil characterized by an accumulation of humified (well decomposed) organic matter, as produced by root decomposition in place or by incorporation of O-horizon material in the mineral soil by burrowing or other mixing activities, or (2) a surface horizon, regardless of origin, that has been disturbed by cultivation, pasturing, or other means, and designated as an Ap horizon.
E (A2)	A mineral horizon in which the main feature is loss of silicate clay, iron, aluminum, or some combination of these, leaving a concentration of sand or silt particles, usually of quartz or other resistant minerals. Conceptually, an *eluvial* horizon.
B (B2)	A horizon formed beneath an A, O, or E horizon and distinguished on the basis of illuviated humus, silicate clay, or sesquioxides (hydrous oxides), alone or in combination, or by a typifying structure acquired as a result of soil formation. Accumulations of humus or sesquioxides need not be large, but they should be sufficient to give a color that is stronger, darker, or redder than in horizons immediately above or below.
C (C)	Mineral horizon, excluding bedrock, that may or may not be like the parent material from which overlying A, E, or B horizons were formed. The C horizon may show negligible change as a result of profile development, or it may show effects of weathering, cementation, or the accumulation of salts, but not of humus or clays as occur in A or B horizons.
R (R)	Consolidated bedrock, which may or may not have been the source of parent material for the overlying soil.

In many profiles, boundaries between neighboring A, E, and B horizons are sharp and easily distinguished. Just as often, however, changes down through the profile are gradual, with the result that the master horizons described above are separated by transitional zones, or horizons. The following transitional horizons are recognized:

AB (A3)	Transition from an A to an underlying B horizon in which properties of the A predominate.
BA (B1)	Same as AB, but properties of the B predominate.
AC (A3)	Transition from an A to an underlying C horizon in which properties of the A predominate. May occur in soils lacking a B horizon.
EB (A3)	Transition from an E to an underlying B horizon in which properties of the E predominate.
BE (B1)	Transition from an E to an underlying B horizon in which properties of the B predominate.
BC (B3)	Transition from a B to an underlying C horizon in which properties of the B predominate.

On occasion, soils have so-called mixed horizons in which material characteristic of one horizon is scattered as islands within material of a neighboring horizon. Five possibilities exist, and they are symbolized A/B, B/A, E/B, B/E, and B/C.[1] The letters designate the horizons in a mixture, with the first one listed providing the matrix in which the other is scattered.

Modified horizon symbols are used to show more information about a horizon than is indicated by capital-letter symbols alone. For example, the symbol B recognizes an horizon containing illuviated material (*e.g.*, silicate clay, lime, humus, sesquioxides), but it does not indicate which type of material is present. This information is supplied by adding one or more of the following lower-case letters to the master horizon (capital letter) symbols. A number of these symbols have been only recently adopted, and some of them replace older symbols, which are shown in parenthesis:

a	Highly decomposed organic material (humuslike).
b	Buried genetic horizon
c	Concretions or hard nonconcretionary nodules cemented by materials harder than lime.
e	Organic material at an intermediate stage of decomposition.
f	Frozen soil (permafrost).
g	Strong gleying.[2]
h	Illuvial accumulation of organic matter, as in a B horizon.
i	Slightly decomposed organic matter, as fresh litter or fibrous peat.
k (ca)	Accumulation of carbonates, commonly calcium carbonate.
m	Cementation (induration), typically in a continuous, hard layer.
n	Accumulation of sodium as an adsorbed (exchangeable) ion.
o	Residual accumulation of sesquioxides.
p	Plowing or other disturbance. Used only with A or O, as in Ap or Op, to indicate a surface horizon disturbed by plowing or other means.
q	Accumulation of silica (SiO_2).
r	Weathered, or soft, bedrock. Used with C, as Cr.
s (ir)	Illuvial accumulation predominantly of sesquioxides, but with some organic matter.
t	Accumulation of silicate clay.
v	Plinthite (see page 71).
w	Development of color or structure. Used with a minimally developed B to indicate color or structure where illuvial material is lacking.
x	Fragipan character (see page 70).
y	Accumulation of gypsum.
z (sa)	Accumulation of salts more soluble than gypsum.

[1]Formerly symbolized A&B, B&A, E&B, B&E, and B&C.

[2]Gleyed soil has a dull gray or blue-gray color, sometimes with rust-colored mottles, produced by the chemical reduction of iron and other metals under poorly aerated, waterlogged conditions. More often than not, gleying affects only subsurface horizons.

Figure 3.3.
Diagrams of typical grass- and forestland soil profiles. An A horizon darkened by humus from grass roots and tops dominates the grassland profile. Deposition of surface litter, and the weathering and translocation of particles in the profile are responsible for strongly developed O, E, and B horizons in the forest soil.

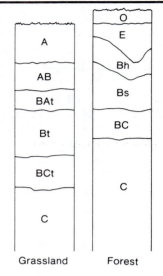

Arabic numerals continue to be used as a part of horizon symbols. One use is where a horizon is subdivided for reasons other than can be identified by capital or lower-case symbols. For example, an A horizon may differ in its upper and lower parts because of a difference in structure. The A would therefore be subdivided into A1 and A2 horizons. The reasons for the subdivision would be given in a detailed description of the profile.

Arabic numerals are also used to note changes in parent material down through the profile. Such changes are called *lithologic discontinuities*, and they are indicated by placing Arabic numbers, starting with 2, before the horizon symbol. For example, if a soil has an A, E, and B horizon formed in one parent material and a C horizon in another, the sequence of horizon symbols would be A-E-B-2C, with the 2 indicating a second type of parent material. The numeral 1 is omitted, since it would impart no additional information.

Where conditions for development are similar, soils tend generally to have the same kinds and sequence of horizons. Where conditions for soil formation differ sharply, on the other hand, soils with substantially different kinds of horizons are likely to result. Such differences are exemplified in the diagram of figure 3.3, which illustrates two diverse profiles, one for a grassland and one for a forest soil. In general, the grassland soil shows the effect of grass roots on the accumulation of organic matter in the surface soil and the effect of eluviation and illuviation on the redistribution of silicate clay in the profile. Processes important to development of the forest soil have been litter accumulation on the surface, strong acid weathering and leaching, which has resulted in a high-quartz E horizon, and eluviation and illuviation of humus and sesquioxides.

The horizon symbols discussed above are used primarily with mineral soils, but two of them, O and C, are also used for organic soils. The symbol O, in combination with the lower-case letters a, e, and i, designate horizons of well decomposed (Oa), moderately decomposed (Oe), and slightly decomposed (Oi) organic materials. The symbol C is used to identify mineral or mineral-like layers that may occur interbedded with organic horizons. These layers consist of such things as eroded soil, volcanic tephra, or *limnic* materials, the latter consisting of deposits laid down by water-inhabiting organisms. Principal limnic materials are *marl*, a calcareous, or lime-enriched, deposit, *coprogenous* earth, a fecal-enriched deposit, and *diatomaceous earth*, an accumulation of silicious skeletal material from diatoms.

Factors of soil formation

Factors of soil formation are the agents or conditions that determine the rate, direction, and extent of soil development. Five general soil-forming factors are recognized: (1) parent material, (2) climate, (3) living matter, (4) topography, and (5) time, and they function both independently and interdependently in soil formation. Independent influences are manifest in soil properties that correlate directly with a given factor, as for example, the way in which the texture and mineralogy of a soil relates to the texture and mineralogy of the parent material from which it forms. Interdependency among the factors is seen in the way one factor modifies the influence of others in soil development. Thus, even though a given factor contributes in its own way to soil formation, the ultimate effect it has on soil properties will vary depending on its relationship to other factors in the soil-forming environment.

Parent material and soil formation

The formation of soils starts with and results from changes in parent material. The nature of these changes is strongly influenced by the parent material, since, through its texture and mineralogy, it is a determinant of various physical and chemical attributes of the soil-forming environment. Texture, for example, by virtue of its control over pore space relationships, is important to water flow and storage. These factors, in turn, influence plant growth and organic matter accumulation, weathering, leaching, and eluviation-illuviation, all of which are processes capable of initiating horizon formation in soils.

The mineralogy of parent materials is a prime determinant of both the rate and products of weathering. For example, highly calcareous (lime-bearing) parent materials often weather rapidly and, due to the loss of lime by leaching, may leave little residue at the weathering site. Siliceous parent materials weather more slowly, with products of weathering tending to accumulate as soil material at the weathering site; quartz persists as sand, whereas other silicates tend to convert to clay. Thus, whether an extensively weathered soil has a coarse or fine texture depends largely on the ratio of quartz to other silicates in the original parent material.

Soils may be very much like their parent material where a lack of time has limited soil development. Time limits soil formation most often where the parent material is of recent origin. Examples are glacial deposits and associated outwash, lacustrine, and loessial materials, many of which are not more than 10,000 years old. Most alluvial deposits, save those on terraces, are much younger than this, and they usually support soils showing only minimal profile development. The same may be said for soils forming in dune sand, in many volcanic deposits, or in materials on steep, highly eroded slopes.

Climate and soil formation

All of the various components of climate play some role in the formation of soil. Precipitation, by affecting the amount of water that enters the soil, influences the chemical and physical processes involved in mineral weathering, eluviation, and ion movement. Biological processes that determine the abundance of residues returned to the soil and the rate at which the residues undergo decomposition are also dependent upon the moisture relationships within the soil-forming environment. Whereas the growth of plants is regulated to a large degree by the availability of water, the rate of decay of residues reflects more strongly the indirect effect of soil water on aeration.

Temperature, a second important climatic component, is a measure of the heat available for all physical, chemical, and biological reactions involved in soil development. Where water is not limiting, the rate of mineral weathering increases with the temperature and attains a maximum in tropical regions. Heat also increases evapotranspiration and thus decreases the average amount of water available for plant growth. At the same time, higher temperatures encourage rapid decay and disappearance of organic residues from the soil. As a consequence, under conditions of equal precipitation, an increase in the mean temperature is usually associated with a decrease in the average organic matter content of soils.

Two aspects of climate are important to soil formation. One is the *gross*, or *external*, climate as measured by normal meteorological data. The other is the *effective*, or *soil*, climate as expressed by the relative availability of moisture and heat in the soil-forming environment. Major differences in soils over the earth's surface are attributable to variation in the external climate, soils of deserts showing less genetic change than do soils of humid regions, for example. However, soils on local landscapes can also differ markedly in climatically related properties, this in spite of their having formed under the same external climate. Such variability relates to difference in effective climate. As will be shown in a later discussion, variation in effective climate within a local landscape is, more often than not, due to interactions between the external climate and topography.

Living matter and soil formation

The principal biotic factor in soil formation is vegetation. Growing plants provide a protective cover that limits runoff and erosion while increasing the infiltration of water into the profile. Plants are also the primary source of soil organic matter, evident in the development of dark A horizons from mineralized root tissue of grasses and other forbs, or in O horizons produced from litter accumulating under forest vegetation. The degree to which these horizons develop is a function of the quantity of residue added each year as well as the rate at which it undergoes decay. Decay of plant residues is least rapid under cool, wet soil regimes.

Root activity is also important to soil formation. Roots aid in forming soil structure and, by creating acidic environments, in the weathering of minerals. Further, the absorption of nutrients and their return in residues is a major factor in fertility maintenance, especially in regions of intense mineral weathering and leaching. Where calcium and magnesium are involved, their absorption by plants and return to the soil surface in residues slows the rate of soil acidification. Because of their ability to counteract soil acidity, calcium and magnesium are called *bases*, and their return to the soil in plant residues, *base cycling*. The capacity of vegetation to cycle bases varies depending on type: it is greatest for grasses, intermediate for deciduous trees, and lowest for coniferous trees. This is consistent with the fact that the most strongly acid soils are usually associated with coniferous forests. This type of plant seems to be the most well-adapted to leached, acid soils, which are relatively incapable of supplying abundant nutrients, including bases.

Organisms other than higher plants also contribute significantly to soil properties. Rodents, earthworms, and various other animals and insects assist in the decomposition of organic substances and, at death, become a part of the residues that undergo decay. Through burrowing activities, these organisms also mix the soil, particularly at or near the soil surface. Mixing by these animals may be responsible for the formation of A horizons in some forest soils.

Microscopic organisms contribute to soil formation mainly through organic matter decay and humus formation. In addition, however, certain specialized organisms, through their ability to alter the chemical form of nutrients such as nitrogen, sulfur, iron, and manganese, affect plant growth and thereby have an indirect influence on the role plants play in soil development. By various means, then, organisms have a profound influence on soil development. Indeed, without the impact of biological change, there would be no true soil as normally conceived. Instead, mineral transformation at the surface of the earth would be limited largely to weathering and allied processes usually considered to be geologic in nature.

General relationships between climate and vegetation

Climate is the chief factor controlling the distribution of plants over the earth, its influence being traceable in most instances to the effects of either temperature or precipitation on plant growth. Where temperatures are not particularly limiting, the type of vegetation and total growth depend strongly on the availability of water. In equatorial regions, for example, temperatures are high and show little variation, either on a daily or annual basis. Under these conditions, vegetational patterns are strongly correlated with precipitation; it varies from dense jungles, as in northern South America and Southeast Asia, to the barren landscapes of the Saharan and Saudi Arabian deserts.

In northern latitudes and at high elevations, plant growth is so often restricted by low temperature that wide variation in precipitation can occur without causing a material change in the vegetational pattern. Except where it is extremely cold, trees are dominant under these conditions. They also provide the principal vegetation in warmer regions where water is available the year around. If water is not available throughout the year, grassland vegetation, or desert vegetation where it is very dry, is more likely to be dominant.

Table 3.1. General types of vegetation related to different climatic conditions.

Vegetation type or class	Climatic region
Vegetation type determined largely by moisture conditions[a]	
Rain forest, deciduous or coniferous trees, complete canopy and dense undergrowth	Superhumid, warm to hot
Forest, deciduous and coniferous trees, partial to complete canopy, moderate to limited undergrowth	Humid, cool to warm
Prairie, tall grasses	Subhumid, cool to hot
Steppe, short grasses	Semiarid, cool to hot
Desert, shrubs and short bunch grasses	Arid, cool to hot
Vegetation type determined largely by temperature	
Boreal forest, largely spruce and fir	Cool to cold
Tundra, principally mosses, sedges, and grasses	Cold
None	Perpetual frost

[a]*Savannahs* are also included in this group. Savannahs consist of areas largely in grass but with scattered trees or shrubs. They occur in warm climates characterized by cyclic dry periods, in drier climates where brush and grass are intermingled, and in transition zones between prairies or steppes and forested areas.

In table 3.1 several major vegetational groups are shown in relation to the climatic range with which each is most often associated. Five of these groups are identifiable with temperate and tropical regions where the type of vegetation depends on average soil moisture conditions for the most part. The two remaining vegetational groups are common to zones where temperature exerts the dominant influence on plant growth. These occur chiefly in cool to cold areas. The distribution of these vegetation types on the North American Continent is shown in figure 3.4. Examples of some of the different vegetational types appear in figures 3.5 through 3.8.

Topography and soil formation

The configuration of the land surface, or topography, affects soil formation primarily by modifying climatic influences. By controlling runoff, topography influences the effectiveness of precipitation and the extent to which erosion removes the forming soil. Similarly, the effectiveness of solar radiation varies with the topography, for the direction and degree of slope determines how effectively the sun's rays warm the soil. By affecting soil temperature and evaporation, topography alters the effectiveness of precipitation even after it has entered the soil, evident in the occurrence of drouth-sensitive plants on only the more protected parts of some landscapes.

Differences in soil moisture regimes induced by topographic influences are often large. Low-lying areas that receive runoff, or broad, flat reaches on the landscape that retain most of the water they receive from precipitation usually are significantly wetter than neighboring slopes. Variation in the water relationships induced by slope differences is often the main reason for variation in soil properties over a landscape receiving the same amount of precipitation each year.

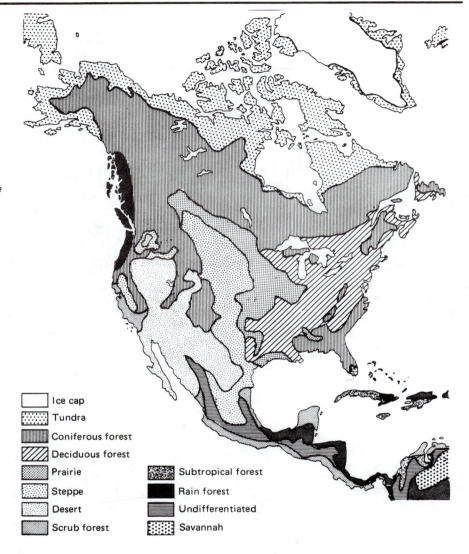

Figure 3.4.
Vegetation zones of the North American continent. The northern coniferous forests are commonly referred to as boreal forests. Tropical and subtropical rain forests are predominantly deciduous, whereas conifers dominate the rain forest of the Pacific Coast. (Adapted from *Climate and Man, 1941 Yearbook of Agriculture,* Washington, D.C.: Government Printing Office.)

Ice cap
Tundra
Coniferous forest
Deciduous forest
Prairie
Steppe
Desert
Scrub forest

Subtropical forest
Rain forest
Undifferentiated
Savannah

The terms *upland* and *lowland* are sometimes used to distinguish between soils that differ because they have formed on different topographic positions. Here, soils of the lowlands are considered to be those developing primarily on floodplain or similar alluvium. Parent materials of this type are ordinarily of rather recent origin and tend not to have been modified much by soil formation. As a rule, materials in upland positions have been subjected to change for longer periods and may support strongly developed soils for this reason. Terrace soils, by virtue of their elevated position above the valley floor, are frequently old and sufficiently well developed to be classed with soils of the uplands (see fig. 3.9).

A   B

Figure 3.5.
Either very cold or very dry climates limit vegetative growth.
(A) Tundra and scattered spruce trees stunted by cold on an
Alaskan landscape, and (B) shrub vegetation on the desert in
southeastern California. (U.S. National Park Service photos.)

A   B

Figure 3.6.
Semiarid to subhumid climates usually result in grassland
vegetation. (A) Bison grazing on native short grass, or steppe,
vegetation in Wind Cave National Park, SD, and (B) a tall grass
prairie in Oklahoma. (U.S.D.A. Soil Conservation Service photos by
R. S. Cole and Chester Fry.)

A

B

Figure 3.7.
Forest vegetation of the mild to cool, moist climates. (A) Stand of mixed hardwood trees in the Nicolet National Forest, Wisconsin, and (B), highland white spruce and balsam fir in Keweenaw County, Michigan, representative of boreal forests. Note the extensive litter under the conifers. (U.S.D.A. Forest Service photos by L. J. Prater and R. C. Starling.)

Soils with minimal profile development may be observed in upland positions if they occur on steeply sloping topography where there is a substantial loss of precipitation as runoff. The loss of water as runoff not only reduces the supply available for various soil developmental processes, it may also result in the erosion of surface soil material at a rate equal to or exceeding the rate of soil development. Even where water induces little erosion, gravity can cause slow mass movement or creep downslope and thereby erase the effects of previous soil formation. Either of these effects retards the development of clearcut profile features, and if the soil is forming from residual parent material, its depth over bedrock may be limited.

The classification of soils into upland and lowland categories often permits a meaningful judgment of internal soil drainage. Lowland soils are the more poorly drained, especially if they are subject to seepage from adjacent waterways. Drainage problems in upland areas are generally temporary. They are commonly associated with soils having slowly permeable substrata, or with those occurring in low-lying positions that receive runoff water from surrounding slopes.

A

B

C

Figure 3.8.
Forests of the mild to hot, wet regions. (A) A lush Pacific rain forest
on the Olympic Peninsula, Washington, and tropical rain forests in
(B) Puerto Rico and (C) Guyana. (Photos A and B, U.S.D.A. Soil
Conservation Service; C, Field Museum of Natural History, Chicago.)

Figure 3.9.
A series of glacial outwash terraces in Teton National Park, Wyoming, which are well elevated above the Snake River. Soil on the flood plain, at the lower left, is subject to erosional-depositional influences at flood time. (U.S.D.A. Soil Conservation Service photo by Jack Young.)

Time and soil formation

The four soil-forming factors described in previous sections determine the nature of the environment for soil development; in combination they dictate both the kinds and rate of change in parent material. Time, the fifth factor, determines the total potential for change under an imposed set of environmental conditions. The effect of time is judged by the extent to which parent material has been altered during the course of soil formation. Since the change depends on both the kinds of reactions involved and the rate at which they take place, the time factor must be evaluated on a relative rather than on an absolute basis.

Some processes modify the character of parent material in shorter times than others. Those that can result in rapid change include ion exchange and the loss or accumulation of readily soluble materials. These processes sometimes produce measurable effects in a matter of a few years.

Compared to the above, the accumulation of humus or of illuvial clay, or changes due to mineral weathering take place much more slowly. Humus may noticeably darken the soil after a few decades, but its full effect may require up to several hundred years under some circumstances. The effect of eluviation and illuviation, as manifest in B-horizon formation, tends to be even more gradual. However, since eluviation and illuviation can vary markedly depending on the conditions of soil development, it is very difficult to express the rate at which they occur in absolute terms.

Of all the soil-forming processes, weathering takes place at the slowest rate. The extensive degradation of highly siliceous parent materials requires several tens of thousands of years under moderate to intense weathering conditions. Even so, evidence of weathering may appear early in soil formation, as for example, where precipitates of iron oxides, occurring as thin coatings over other mineral grains, mask the original color of the parent material. However, the degree of weathering is not judged from color changes of this type; it is determined by the magnitude of change in the mineralogy of the parent material.

Comparisons of the several processes of soil formation within a specific time frame is possible only where conditions allow each of the processes to function. For example, the relation of time to the weathering of silicate minerals is of little significance in very dry climates. Under these conditions the effect of time must be judged from such things as the solution and translocation of lime or other more readily soluble salts, the accumulation of humus, and in some cases, the extent eluviation and illuviation. With an increase in effective precipitation, however, mineral weathering becomes a much more important process, and under hot, wet conditions, as in tropical regions, may provide the principal basis for judging the effect of time on soil formation.

The general relationship among the soil-forming factors

Variation in any soil-forming factor can produce a change in soil properties. Sometimes the factors vary in such a way that their combined influences are additive; at other times they interact so that the effect of one compensates for or nullifies the effect of another. For example, two areas of different total annual precipitation may show almost the same degree of soil development if, in the area of higher rainfall, increased average slope has resulted in greater water loss through runoff so that the effective precipitation in the two areas is essentially the same.

When considering the broad spectrum of soils over the surface of the earth, the most apparent changes in properties can be identified with differences in climate and its closely associated factor, vegetation. Whereas parent material and topography do differ markedly from place to place, locally or regionally, when considered in terms of properties of major importance to soil development, they tend to appear more uniform than climate and vegetation. For example, most landscapes are characterized by sloping rather than flat topography, and the majority of parent materials are transported and have been initially derived from siliceous rocks. Similarly, time for development has been sufficient in most instances to bring about significant change in the nature of the parent material. No such generalizations can be made with regard to climate and vegetation.

One may conclude from the foregoing that the greatest differences among the majority of soils have to do with the properties that show the influence of climate and vegetation on soil development. Broadly based comparisons of soils are therefore made largely with reference to climatically-related properties. In classification, key separations of soils recognize these properties. Other separations then depend on recognition of the way in which unusual parent material or topography, or a limited time for development, modify the effect of climate and vegetation on acquired properties.

Summary

Soil formation is the result of change in parent material induced by a set of soil-forming processes, namely: (1) continued mineral weathering and clay synthesis, (2) organic matter accumulation, (3) exchange of ions, (4) translocation of soluble and solid soil components within the profile, (5) structure formation, and (6) mixing of soil materials. The specific processes dominating soil formation determine the kinds of horizons that develop at a given location.

The rate and extent of change that takes place during soil formation is controlled primarily by five interrelated soil-forming factors: (1) parent material, (2) climate, (3) living matter, (4) topography, and (5) time. The first four of these factors determine the nature of the soil-forming environment. The fifth factor, time, determines the extent of change that takes place under an imposed set of environmental conditions.

Soil properties may be either inherited from the parent material, or they may be acquired as the result of soil formation. Acquired properties are apparent in the kinds of horizons that develop in the profile. The nature of the horizons and, to a degree, their relative position in the profile is indicated by letter symbols O, A, E, B, and C, either alone or in combination with one or more lower-case alphabetical symbols. The symbols identify specific properties or conditions introduced into parent material by the several processes of soil formation.

Review questions

1. Describe the general nature of each of the processes of soil formation, and indicate the type of change they are likely to produce in soil parent material.
2. What do A and O horizons have in common, and in what way do they differ?
3. If a soil profile has an E horizon, why will it also likely have a B horizon?
4. Identify the principal illuvial material in each of the following horizons: Bt, Bh, and Bs.
5. How are lithologic discontinuities noted in soil profile descriptions?
6. Which process of soil formation is most affected by the cycling of bases by plants, and why?
7. Explain why each soil-forming factor, other than time, is a determinant of the environment of soil formation.
8. Distinguish between upland and lowland soils, and explain why soils in these two positions will normally differ in relative age and in the potential for waterlogging.

Selected references

Birkeland, P. W. *Soils and Geomorphology.* New York: Oxford University Press, 1984.

Buol, S. W., Hole, F. D., and McCracken, R. J. *Soil Genesis and Classification.* Ames, IA: The Iowa State University Press, 1980.

Jenny, H. *Factors of Soil Formation.* New York: McGraw-Hill Book Company, Inc., 1941.

Simonson, R. W. What Soils Are. In *Soil, 1957 Yearbook of Agriculture,* pp. 17–31. Washington, D.C.: Government Printing Office.

Soil Survey Staff. *Soil Survey Manual.* U.S.D.A. Handbook 18. Washington, D.C.: Government Printing Office, 1951.

4

Physical properties of soils

Having considered the origin and general nature of soils, we may now turn our attention to specific properties important to soil use and behavior. Physical properties, so named because they are measured by physical means, are the first to be considered. Although there are many different physical properties important to soil use and behavior, three stand out because they can be used to predict a number of other things about the soil. These are texture, structure, and color, which are, in themselves, rather easily judged from the way the soil looks or how it feels or behaves when worked by hand. Once texture, structure, and color are known, meaningful estimates of pore-space relations, soil strength, organic matter content, and the extent of mineral weathering are often possible. The latter characteristics, in turn, are important in the appraisal of water- and air-transmission characteristics, the ease with which the soil can be cultivated, its suitability for various engineering works, and the ability of the soil to supply nutrients stored in the organic and mineral fractions of the soil.

Among the properties used to predict other soil characteristics and behavior, texture is of greatest overall value. This is because such a large number of other soil properties vary with texture. These properties, as with texture, relate to the proportions of different size groups of mineral particles within the soil.

The size classification of mineral particles

The classification of mineral particles into different size fractions is in recognition of different sets of size-related properties. The classification is wholly arbitrary, with the choice of size limits assigned to each fraction being conditioned by the judgment of the classifier. As a result, geologists, engineers, and soil scientists, all of whom are interested in size-related properties of mineral matter, have devised separate systems of particle-size classification. Although these systems categorize particles into roughly the same size fractions, there is some variation in the size limits assigned to each fraction.[1]

[1]The system of particle-size classification used by engineers is discussed in Chapter 20.

Table 4.1. Size fractions of soil mineral particles according to the U.S. Department of Agriculture system of particle-size classification.

Size fraction	Diameter
	mm
Coarse fragments[a]	
Boulders	Above 600
Stones	250–600
Cobbles	75–250
Gravels	2–75
Soil separates	
Very coarse sand	2.00–1.00
Coarse sand	1.00–0.50
Medium sand	0.50–0.25
Fine sand	0.25–0.10
Very fine sand	0.10–0.05
Silt	0.05–0.002
Clay	Below 0.002

[a]A partial listing only and limited to coarse fragments that are rounded. A more complete listing of coarse fragments is found in table A1, page 544.

Our immediate concern is with the system used by soil scientist in this and certain other countries. It has been developed by the U.S. Department of Agriculture and is based primarily on properties important to soil use in agriculture.

The U.S. Department of Agriculture system of particle-size classification recognizes two broad groups of mineral particles: *coarse fragments* and *soil separates*. The first of these includes such subclasses as *stones*, *cobbles*, and *gravels*. Soil separates are the sand, silt and clay fractions. The range in size for each of these fractions is shown in table 4.1. Note the five subdivisions of sand; these are included because of the wide variation in the properties of sand throughout its full range in size. One such property is erodibility by wind, with the finer sand grains being much more erodible. Thus, for sandy soils in regions of high wind it is often desirable to know if the sand fraction is made up predominantly of coarse or fine sand particles.

According to U.S. Department of Agriculture guidelines, a deposit of mineral matter should contain 10 percent or more of sand, silt, and clay to be called soil, otherwise it will likely be unsuitable for most plant growth. For agricultural use, soils made up totally of the separates are preferred. Small amounts of the finer coarse fragments cause little trouble, but large amounts are undesirable because they limit the ability of the soil to supply water and nutrients to plants, and they interfere with cultivation. For cultivation especially, stones create the worst problems, cobbles less, and gravel the least (see fig. 4.1).

Size-related properties of the soil separates

Properties used to characterize the soil separates are exhibited by mineral particles both on an individual basis and in bulk. Properties of individual particles that vary with size include mass, volume, surface area, and shape. The surface area of particles influences the ability of the soil to retain adsorbed water and nutrients for plant use, and in general, the smaller the

Figure 4.1.
Gloucester very stony
fine sandy loam in
Worchester County,
Massachusetts.
Stones make normal
cultivation of this soil
impossible and even
interfere with its use
as pasture. (U.S.D.A.
Soil Conservation
Service photo.)

average particle size the greater the total surface area per unit quantity of soil (see fig. 4.2). Particle shape is important because, with size, it is a determinant of the *geometry* (size and shape) of pore spaces in the soil. Pore geometry is a major factor controlling the *permeability* of the soil; that is, its ability to transmit water and air.

Particle shape is also of great importance to the *cohesive* and *adhesive* properties of soils. These properties express the force with which the bulk soil clings to itself (cohesion) or to other objects (adhesion). Particle shape affects these properties by determining the extent of contact between adhering and cohering surfaces. Greatest contact occurs between flat surfaces that lie parallel to each other.

Cohesive and adhesive properties in soils are evaluated through the measurement of such traits as the firmness or hardness of aggregates or clods when the soil is dry, or by the degree of *plasticity* or *stickiness* displayed by the soil when wet. Plasticity denotes the ability to be molded, and it is measured by the force required to deform or remold the soil mass when in a wet, pliable condition. Stickiness, which becomes manifest when the soil is wetted beyond the plastic state, is measured by the force required to pull the wet soil apart.

Figure 4.2.
Relation between diameter of spherical mineral particles of average density and the number of particles or the square centimeters of surface area per gram. A diameter of 0.001 mm falls in the midrange of clay, and 1.0 mm, the upper limit of coarse sand. Note that a decrease in size over the range 0.001 to 0.1 mm increases the total surface area by 100 million times, approximately.

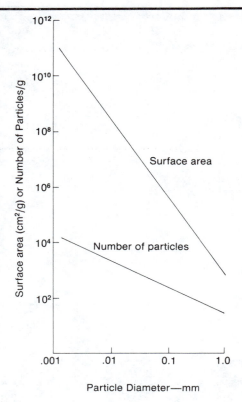

Most of the above properties are expressed by each of the three soil separates. The degree of expression varies, however, as shown by the brief descriptions of the separates, which follow:

Sand

Sand grains are large enough to be seen by the naked eye, and they can be felt individually when rubbed between the fingers. Sand imparts a gritty feel to soil, and its presence is frequently detected on this basis. Sand grains can be perceived most readily by touch if they have sharp edges and corners.

Because of their large size and uneven surfaces, sand grains make only limited contact with other surfaces. For this reason, sand in bulk is neither sticky nor plastic. Sand that is molded when wet and then dried falls apart under very gentle pressure. Because of the failure of sand grains to cling to each other, they do not form stable aggregates in soils. Pores formed between neighboring sand grains are large, and permeability to water and air is high.

Silt

Individual silt particles can neither be seen with the unaided eye nor felt when rubbed between the fingers. The size of silt particles corresponds roughly to the range of the ordinary light microscope. In bulk, silt is more cohesive and adhesive than sand. Even so, it displays only limited stickiness

Physical properties of soils

Figure 4.3.
The extensive shrinking and cracking on drying of soil in this field near Las Cruces, New Mexico, is due to a high content of expanding clay. (U.S.D.A. Soil Conservation Service photo by R. E. Neher.)

and plasticity. On drying, molded samples of silt are easily crushed to a powder, which, when rubbed between the fingers, has a floury feel. Compared to sand, silt produces relatively small pores, but like sand, it contributes little to stable aggregate formation in soils.

Clay

Particles of clay are so small that they are visible only under the electron microscope. In general, bulk samples of clay are both sticky and plastic when wet, and they tend to form hard clods on drying. However, the degree to which each of these characteristics is expressed varies among different kinds of clay. For example, some clays, referred to as *expanding* types, swell and shrink markedly on wetting and drying, whereas others do not. Clays that expand and contract on wetting and drying tend to be very sticky and plastic when wet, and they tend to form very hard clods on drying. Soils high in expanding clay are often recognized by their ability to form frequent, large cracks as they shrink on drying (see fig. 4.3).

Soil texture

Texture is defined as the percentage by weight of sand, silt, and clay in the mineral fraction of soils. Texture thus predicts the relative contributions of the three separates to overall soil properties. The predictions are approximate, however, for the separates in themselves are variable in properties. For example, two soils with the same clay content will not behave the same if one has expanding clay and the other doesn't. In spite of such limitations, judgments based on texture are highly valuable in assessing the general character and behavior of soils, especially when coupled with information on the structural state and organic matter content of the soil. Structure and organic matter need to be taken into account in the characterization, for they can modify the relationship between texture and other soil properties, often significantly. Texture is the more permanent feature of the soil and will not ordinarily change over a man's lifetime. No such statement can be made for the structural state or for the organic matter content of the soil.

Textural analysis of soils

The texture of soils is determined by a laboratory procedure known as *particle-size* or *mechanical analysis*. Two steps are normally involved: in the first step, sieving is used to separate sand and coarser fragments from silt and clay, and to subdivide coarser fractions into their respective subfractions (see fig. 17.2). In the second step, silt and clay are separated by sedimentation in water, the separation depending on the faster rate of settling of the larger silt particles. Prior to sedimentation, the sample is dispersed (aggregates broken down) so that the particles can settle independently. For precise work it is necessary to remove organic matter from the sample, for organic materials stabilize aggregates and may prevent their complete dispersion prior to sedimentation.

The contents of sand, silt, and clay are expressed on a weight-percent (dry-soil) basis. For example, if a soil sample contained 20 g each of sand and silt and 10 g of clay, the total weight of the three fractions would be 50 g, and the percent content of each, or the texture of the soil, would be 40 percent sand, 40 percent silt, and 20 percent clay. Coarse fragments, if present, would be measured and recorded separately.

Soil textural classes

Soil textural classes are groupings of soils based on specific ranges in texture. Twelve such classes are recognized in the United States, and they are shown by name in the textural triangle of figure 4.4. A soil is placed in its appropriate textural class by plotting a point in the triangle corresponding to its respective percentages of sand, silt, and clay. The box in which the point falls identifies the textural class of the soil. The method of plotting textures in the textural triangle is described in the caption of figure 4.4.

Except for loam, textural class names identify the separate or separates that dominate the properties of a soil. In loams, properties appear to be intermediate in character and, therefore, cannot be attributed to any one particular separate. The concept of loam is a soil having properties that are as nearly ideal as possible; these include mellowness and easy workability, a suitably high permeability to air and water, and a reasonably high water-retention capacity. Texturally, loams contain somewhat more silt or sand than clay. The clay content of loams never exceeds 27 percent, and it may be as low as seven percent, as may be judged from figure 4.4.

Textural class names other than loam recognize the dominating influence of only one or two of the separates on soil properties. For example, in clay loams, silt loams, and sandy loams, one separate has an overriding influence on properties, but only to a degree. In the textural classes of clays,

Figure 4.4.
Triangle showing the twelve textural classes in the U.S.D.A. system of textural classification. To determine the textural class of a soil, locate its percent-clay line, which extends horizontally from the left axis. Locate either the percent-silt line and follow it down toward the bottom axis, or the percent-sand line and follow it up toward the left axis. The box in which either of these lines intercepts the clay line gives the textural class name of the soil. (U.S.D.A. diagram.)

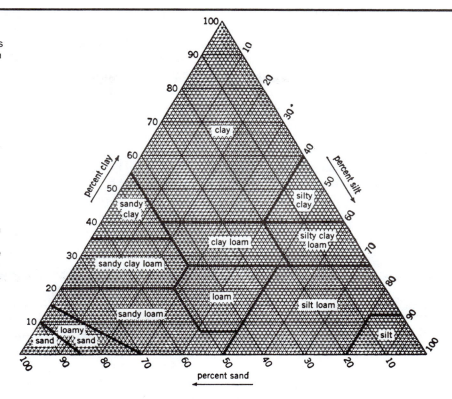

sands, and silts, on the other hand, one separate totally dominates properties. In comparison, names such as sandy clay loam, silty clay loam, or sandy or silty clay recognize that the properties of the soil are derived mainly from two of the three separates.

On occasion, it is convenient to group soils into only three broadly defined categories of coarse, medium, and fine texture. Specific textural classes normally included under each of these three headings are:

Coarse	**Medium**	**Fine**
Sands	Loams	Textural classes with
Loamy sands	Silt loams	clay in their name
Sandy loams except very fine sandy loams	Silts	
	Very fine sandy loams	

Silts, silt loams, and very fine sandy loams are naturally medium-textured because silt and very fine sand, the dominant separates in these classes, are intermediate in size and therefore contribute an intermediate range of texturally-related properties to the soil.

Modified textural class names

Textural class names are modified where it is deemed necessary to recognize the influence of coarse fragments or organic matter on soil properties and behavior. If the modifying factor is organic matter, its presence is indicated by adding the prefix *mucky* to the textural class name. Use of this prefix is limited to soils containing from 15 to 20 percent organic matter. Above this range, organic matter usually so dominates properties that textural class names are not used. Instead, the soil is called a *muck* if the organic matter is well decomposed (humuslike), or *peat* if it is not.

Textural class names are sometimes modified to show the predominant influence of a subfraction of sand on soil properties, as exemplified by the names very fine sandy loam, loamy fine sand, or coarse sandy clay. The presence of coarse fragments is indicated by adjectives used as prefixes to the textural names. Some examples are gravelly sandy loam, cobbly clay, or stony loam. The modification of textural class names to indicate the presence of coarse fragments in soils is somewhat complex. Because of its rather limited application, the procedure is not discussed here but rather in Appendix A on page 543.

Soil structure

Soil structure is defined as the manner in which soil particles are assembled in aggregate form. Aggregation in soils depends primarily on the cohesive nature of the finer particles and on natural forces that organize and retain them in specific structural units, or *peds*, of definable shape and size. Soil peds vary appreciably in size, ranging from a fraction of a centimeter to several centimeters along their greatest dimension.

In concept, structural units result from the development of a network of permanent cleavage planes (planes where bonding forces between soil particles are weak) within an otherwise coherent soil mass. The pattern and frequency of distribution of these surfaces determine the size and shape of the structural units.

Types of structure

The classification of soil structure is based largely on shape.[2] Four general types of structure, as illustrated in figure 4.5, are recognized: (1) *platy*, evident in the separation of soil into flat, horizontal units of varying thickness; (2) *prismlike*, consisting of peds having longer vertical than horizontal axes and surfaces that are comparatively flat; (3) *blocklike*, in which axes are of approximately equal length in both vertical and horizontal directions, and the bounding ped surfaces are either relatively flat or somewhat rounded; and (4) *spheroidal*, which applies mainly to the smallest aggregates. These tend to be rounded but may have irregular surfaces.

Several of the major structural forms are subdivided into subtypes. Prismlike forms that have either flat or indistinct upper surfaces are *prismatic*; if the upper surfaces are rounded, they are *columnar*. The subtypes of blocklike structures are *angular* blocky, wherein bounding surfaces are fairly flat and form sharp corners where they intersect, and *subangular*

[2]The classification of structure is discussed in greater detail on pp. 225-30 of the *Soil Survey Manual*, U.S.D.A. Handbook 18, 1951.

Physical properties of soils

A

B

C

D

E

Figure 4.5.
Types of soil structure. (A) Columnar, underlying
massive (unstructured) soil; (B) platy, caused by parent-
material layering; (C) prismatic; (D) angular blocky; and
(E) granular. The bar markers are 2.5 cm long. (Photo C
courtesy of U.S.D.A. Soil Conservation Service.)

Physical properties of soils   67

blocky, in which both faces and intersects tend to be rounded. Finally, spheroidal is subdivided into *granular*, which is more dense and therefore less porous than the second subtype of *crumb*.

Though shape is the main distinguishing characteristic of soil peds, they also differ in size. For example, granules and crumbs are small in comparison to most other structural forms, normally having a diameter ranging from 1 to 10 mm, with a common size falling about midway in this range. Blocklike and prismlike structures, on the other hand, often range between 20 and 50 mm in their smallest dimension, with variation beyond these limits also being possible. Platelike structural units are the most diverse in size. They vary from thin, flakelike units no more than perhaps a millimeter thick to continuous layers that may be very thin at times, but also ranging up to several centimeters in thickness on occasion. Platelike structures of the latter type are typical of soils forming in layered parent materials, such as lake-laid sediments.

Structure formation in soils

There are two basic requirements for the development of structure in soils: one is a degree of cohesiveness that will allow the soil to hold together in aggregate form; the second is a means or mechanism that separates the soil into aggregate form. In some instances, both of these requirements may be fulfilled simultaneously, as for example, where cohesion causes the soil to shrink and crack into aggregate form on drying. Soils that shrink and crack on drying normally contain either an expanding clay or gumlike organic materials. Organic gums, which are natural products of organic matter decay, are most important to structure development in the surface horizon of soils.

Roots and ice appear to be especially important to soil structure formation. As roots or ice crystals grow in soil pores, they not only wedge or pry the soil apart, they also compress particles making up aggregates and force them into closer, firmer contact. Further, the removal of water through absorption by roots or by ice formation contributes to structural development because these processes promote soil drying and shrinkage.

The development of structure is benefited, at least to a degree, by the burrowing and mixing activities of worms, insects, and other soil-inhabiting animals. The main effect of these organisms is in the formation of channels through the soil.

Structural form in relation to soil depth

Certain types of structure tend to occur more in one part of the soil profile than in another. The development of the finer granular or crumb structures in the surface soil is typical. This is due mainly to the high density of roots as well as to the comparatively high frequency with which freezing and thawing or wetting and drying cause multiple separations of the surface soil into small aggregates. Aiding these processes in small aggregate formation and stabilization are the burrowing activity of soil animals and the production of organic gums by decay organisms, either of which is maximum in the surface soil.

Processes of structure formation are both less intense and of lower frequency in the subsoil than in the surface soil. As a consequence, structure formation in the subsoil, if it occurs at all, results generally in peds of larger size, as exemplified by the prism- or blocklike forms. Ordinarily, these structural types are associated with B horizons where accumulated clay washed down from an A or E horizon allows the soil to shrink and crack on drying. For the most part, shrinkage in the B horizon produces vertical cracks necessary for the separation of the soil into prismlike units. Under some conditions, formation of the prismlike structures is followed by the erosion and rounding of their upper surfaces, which converts them to the columnar form. This type of transformation is relatively rare, however, and is limited to soil forming under rather select conditions in regions of low rainfall.

Vertical cracking in a soil containing or developing horizontal planes of weakness explains the formation of some blocklike structures. The cause of horizontal cleavage is not always clear, but it may be due to variation in texture with depth, or it may result from wetting and drying to different depths in the profile.

Variation in texture with depth is an important cause of platy structure in soil. Platiness is typical of soils forming from water-laid deposits consisting of interbedded coarse- and fine-textured materals. Often, these plates are comparatively thick. Fine platy structure is sometimes seen in the surface layer of soils that have not been disturbed for a long time. In this case, the plates are separated by closely spaced, horizontal cleavage planes that are probably formed by repeated wetting and drying or freezing and thawing to different, though shallow, depths in the soil.

Nonstructural states in soils

Soil materials of low inherent cohesiveness tend to remain structureless. Sands, for instance, occur normally in a *single-grained* nonstructural state. When handled, sands separate readily into individual grains, for they do not cohere sufficiently to form identifiable aggregates. Silts and nonswelling clays also often fail to establish well defined structures, largely because they do not shrink on drying. Instead, they form a continuous system wherein the particles are bound together into a weakly to moderately coherent *massive* nonstructural state. Rather than yielding identifiable structures, the crushing of massive materials produces fragments of unpredictable size and shape, for splitting of the soil does not take place along established planes of weakness.

Working a soil while it has too high a moisture content tends to destroy its structure and to return it to a massive state. This effect is termed soil *puddling*. A puddled condition can result from cultivation, from compaction by animals or machinery, or from rain falling with full force on an exposed soil surface.

Pan structures in soils

Many soils contain dense layers, or *pans*, which may have a profound, usually undesirable, influence on soil profile properties (see fig. 4.6). The principal problem with pan layers is interference with root and water penetration; in effect, they produce shallow soils. The high density of pans results from compaction or from the filling of larger pores with clay or chemical precipitates, and if they are very firm, they are called *hardpans*. Brief descriptions of several kinds of pans follow.

Figure 4.6.
Pan layers in soils.
(A) A spade slice
showing a plowpan
about 2.5 cm thick,
and (B) a caliche layer
(between arrows).
Either layer restricts
root penetration into
the soil below. (Photo
(A) by Merle Ramey,
U.S.D.A. Soil
Conservation Service.)

A

B

Claypan	Claypans are dense soil layers normally produced by the downward migration of clay and its accumulation in the subsoil as B-horizon material. Claypans occur where clay accumulation is pronounced.
Duripan	These are layers of soil cemented by precipitated silica, either alone or in combination with precipitated iron oxides or calcium carbonate. Duripans occur at various depths in the soil.
Fragipan	The term fragipan comes from the Latin word, *fragilis*, which means brittle. Fragipans are dense subsoil layers in which the soil is bonded into a hard, brittle form, apparently by clay. They normally occur at least 50 to 60 cm beneath the soil surface.
Caliche	Caliche is a hard, lime-cemented layer. Most caliche is white and occurs primarily in arid-region soils.

Physical properties of soils

Plinthite	This type of layer, formerly called *laterite*, contains precipitated sesquioxides as a cementing agent. It occurs in highly weathered soils of many tropical regions. If continuously moist, plinthite tends to remain relatively soft, but if it dries, it converts irreversibly to a bricklike hardpan. Plinthite layers form at a depth of the B horizon, but they may occur at the surface if exposed by erosion.
Plowpan	Plowpans are artificially produced and are widespread in agricultural soils. They are normally attributed to soil compaction caused by the weight of plows or other cultivation implements. Plowpans can sometimes seriously interfere with root, air, and water movement below the depth of plowing.
Stability of structure	Structural stability refers to the ability of soil aggregates to resist physical breakdown. Primary concern over structural stability is in maintaining the surface layer of medium- to fine-textured soils in a well granulated state. In soils of this type granulation is important to aeration, water penetration, and seedling emergence, but it may be difficult to maintain where the soils are repeatedly compacted or abraded by machinery or animals. With the loss of structure, the surface soils can be easily puddled by falling rain. They may also be more easily eroded or more inclined to form surface crusts on drying.

An important factor influencing the stability of soil structure is the kind of ions countering the negative charge of clays. Maximum bonding and stability are afforded by ions of high valence, principal among them being Ca^{2+}, Mg^{2+}, or Al^{3+}. Monovalent ions, especially Na^-, provide relatively weak bonding forces between particles. Soils high in adsorbed sodium consistently occur in a puddled, easily eroded condition and are often poorly aerated because of the lack of large, structure-related pore spaces. The physical state of these soils can normally be improved if the sodium is replaced by calcium.

The presence of adsorbed sodium appears to be responsible for the development of columnar structure in soils. Weak bonding by sodium is thought to allow the erosion of the upper surfaces of soil prisms, leaving them with rounded tops characteristic of columnar structural forms. Where exposed at the surface by erosion or deep plowing, columnar structures are easily destroyed, often by nothing more than the puddling action of rain. This, in turn, encourages erosion and soil crusting. For reasons such as these, the presence of a columnar structure is usually taken to mean that the soil has, at least potentially, some very serious management problems.

Cementation may at times be an important factor in structural stability. The principal cementing agents are precipitated lime, silica, or sesquioxide compounds. Cementation by chemical precipitates may be harmful if it blocks soil pores against the entry of air and water into aggregates or if it results in the formation of soil pans. Cementation may be of value, however, if there is no other means of stabilizing structure against the destructive and erosive influences of rain.

Organic substances, where present, can make a vital contribution to structural stability. They not only aid in minimizing the breakdown of structure but also in its re-formation should breakdown occur. A highly important

Figure 4.7.
Diagram illustrating pore space relationships in granulated soil. The size of smaller pores within the granules depends on texture; they provide the principal sites for water retention. Large structurally-related pores between the granules drain readily and provide passageways for rapid air movement through the soil.

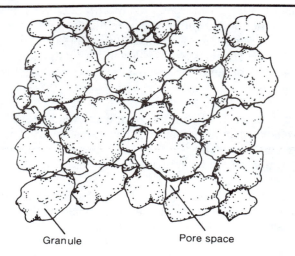

Granule   Pore space

aspect of soil management is the return of plant residues or other organic materials to sustain biological activity and the production of microbial gums in soils. It is a well known fact that where the return of fresh organic materials is stopped, deterioration of the structural state will likely follow. In contrast, one of the best means of maintaining surface soil structure is with permanent grass sod. Sod crops not only have dense root systems that aid in the formation and preservation of aggregates, they provide a continuing source of fresh root residues that sustain a high level of microbial activity in the soil.

The importance of structure to plant growth

An important function of structure is to modify properties that are otherwise controlled by the texture of the soil. Principally, benefits derived from structure relate to the effect it has on the ratio of large to small pores in the soil. Large pores are essential for air flow that supplies O_2 for root and microbial growth. Small pores serve mostly to retain water for plant use. A proper balance in pore size therefore reflects in the best possible relationship between the supply of both water and air.

Greatest concern over the structural state is with soils of fine texture, for if these soils are not well structured, they tend to contain too few large pores for aeration. Granules or crumbs are preferred, since their formation introduces an extensive network of large pores without causing a marked change in the water-retention characteristics of the soil (see fig. 4.7). Measures that help sustain soil granules and crumbs include the periodic use of sod crops, the return of crop residues to the land, and careful cultivation of the soil.

Cultural practices designed to improve structure relations in soils affect the plow layer for the most part. Subsoil structure is not influenced much except over rather long periods of time. Nonetheless, structural conditions of the subsoil are important to most plants and often explain poor growth where no other cause is evident.

Soil consistence

Soil consistence is a measure of soil workability, and it is expressed in terms of the resistance the soil offers to deformation or rupture when subjected to a compressing, shearing, or pulling force. Consistence is the manifestation of cohesional and adhesional properties, and it varies with any factor that affects these two properties. Variation in consistence among different soils is, more often than not, due to differences in texture and reflects the influence of different levels of clay on cohesive and adhesive properties. Consistence is also affected by the structural state and by the organic matter content of the soil. For example, well granulated soils, or those high in organic matter, tend to crumble more readily than do those in a massive, unstructured state or that contain relatively little organic matter. Granulation reduces the apparent cohesive and adhesive character of the soil because of the weak force with which these structural units are bound to each other. Organic matter has a similar effect, in part because of its contribution to structural stability, and in part because much of the humic material in soil is in itself only weakly cohesive and adhesive. Organic matter is well noted for its ability to loosen dense, tight soils that are otherwise often difficult to cultivate.

Consistence is also a function of the water content of soils, evident in the softening effect water has when used to moisten a hard, dry clod. Water softens soil as it moves in and separates soil particles, thereby reducing the force of cohesion between them. Unless a hard soil is cemented, a slight moistening tends to convert it to a more crumbly, friable state. With the continued addition of water, the soil becomes plastic and then sticky. When in a moldable, plastic state, cohesion is dominant over adhesion; that is, the soil coheres to itself more strongly than to other objects. When in the sticky state, however, the reverse is true.

Evaluation of consistence properties

Soil consistence is judged for various reasons: it may be used to rate the stability of natural peds in a soil profile, the suitability of a soil for plowing, or its susceptibility to erosion. Hand-working soil at different moisture contents is commonly used to estimate texture, for an evaluation of consistence, along with grittiness,[3] allows an experienced person to approximate the percentage of sand, silt, and clay contained in a soil.

Procedures used for assessing consistence properties depend on the purpose of the determination. As an example, when texture is being estimated, the soil is usually worked almost entirely in the plastic or sticky range, since the effects of clay and the presence of sand are more readily detected under these conditions (see fig. 4.8).[4] However, when a judgment as to its suitability for plowing is being made, a soil is examined just as it occurs in the field.

[3]Grittiness is not a consistence property; it relates to the size of particles, not to their ability to cohere or adhere.

[4]It should be noted that the relationship between texture and consistence varies widely among soils of diverse origin. Variations in such things as the type of clay, the amount of organic matter, or the degree of cementation of soil aggregates can so alter consistence properties that they appear to be unrelated to the proportions of sand, silt, and clay present.

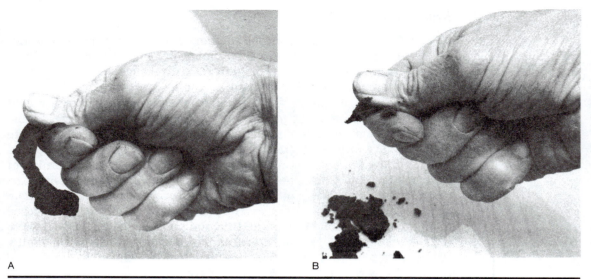

A   B

Figure 4.8.
Judging consistence (plasticity) by ribboning the soil. (A) A silty clay
loam with about 25 percent clay forms a relatively long ribbon when
squeezed between the thumb and forefinger. (B) A sandy loam with
less than 10 percent clay and very low cohesion does not form a
ribbon. It also has a gritty feel.

Consistence is commonly evaluated at three arbitrarily defined moisture states: *wet*, *moist*, and *dry*. The first of these is represented by a recently wetted soil in which little water loss has resulted from drainage, plant
use, or evaporation. For some fine-textured soils this condition may persist
for a number of days following a soaking rain. Most soils are soft and pliable
when wet, and the consistence properties for which they are examined are
plasticity and *stickiness*. Stickiness is judged at a moisture content above
that required to produce a plastic condition.

The moist state is defined as a moisture content approximately midway
between wet and air-dry. It is intended to represent a soil in suitable condition for plowing. Consistence terms used for soils at this moisture content
are *loose*, *friable*, and *firm*. A loose consistence is generally limited to non-
coherent, coarse-textured soils. Friable soils tend either to have medium texture or to be very well aggregated if their texture is fine. Soils described as
firm when moist are usually relatively dense, often the result of compaction.
A firm consistence as a natural property may be due to limited cementation.

Consistence when dry is determined on air-dry soil. Terms such as *loose*,
soft, and *hard* are used to describe consistence at this water content. A loose
consistence again applies to sandy materials. Soft suggests that the soil would
crush to a powder when worked in a dry condition. Cultivation of dry soil
having a hard consistence, if it could be carried out practically, would leave
the surface rough and cloddy.

Soil density

Density is a standard physical term defined as the weight (mass) per unit volume of a substance, or:

$$\text{Density} = \text{Weight}/\text{Volume} \qquad (4.1)$$

The value of density of a material depends on the units used to express weight and volume. In soils work, the units commonly used are grams and cubic centimeters, with density thus being expressed in grams per cubic centimeter (g/cc).[5]

Two density values are important in the physical characterization of soils. These are *particle density* and *bulk density*. As the name implies, particle density is the weight per unit volume of individual particles making up the soil. Bulk density is the weight per unit volume of whole soil, pore space included. Since the weight of a volume of soil equals the weight of the particles present, the density of the particles is a factor controlling bulk density values. Except for highly organic soils, the particle density is a function of the kinds of minerals a soil contains.

Particle density

The density of soil particles can be rather easily determined. The procedure consists, basically, of pouring a known weight of soil into water and then measuring the increase in water volume that results. The increase in volume is equal to the volume of the particles. Dividing this volume by the sample weight gives the density of the particles.

The density of minerals that make up soil particles can vary considerably, but for those most common to soils, particularly quartz and the feldspars, the range is usually between 2.6 and 2.7 g/cc. Silicate clays may fall slightly below this range, but even so, an average that can be safely applied to soils made up principally of silicate minerals is 2.65 g/cc.[6] Exceptions occur where iron minerals are present in relative abundance. Ferromagnesian minerals, for example, have densities ranging from 2.9 to 3.5 g/cc, and the density of iron oxides may exceed 4 g/cc. Where precise values are needed, particle densities should be determined by separate analyses.

Bulk density

The bulk density, or weight per unit volume of soil, is a function of the particle density and the proportionate volumes of solid particles and pore space in the soil. For soils of similar particle density, variation in the total pore space volume, or *porosity*, is the principal cause of variation in bulk densities. For individual soils, variation in porosity normally reflects differences in the structural state at different levels in the profile, the porosity increasing with an increase in structurally related pore space. Since structural de-

---

[5]A current trend is to adopt the International System of Units, referred to as SI units, which expresses densities in kilograms per cubic meter. In this system, a gram per cubic centimeter would become 1000 kilograms per cubic meter. Since some of the current literature of soils uses these units, they are discussed briefly in Appendix C.

[6]The density of water is approximately 1.0 g per cc. This means, then, that the mineral matter in soils is about 2.65 more dense than water.

Table 4.2. General relationship among texture, bulk density, and porosity of soils.

Textural class	Bulk density	Porosity
	g/cc	%
Sand	1.55	42
Sandy loam	1.40	48
Fine sandy loam	1.30	51
Loam	1.20	55
Silt Loam	1.15	56
Clay loam	1.10	59
Clay	1.05	60
Aggregated clay	1.00	62

Figure 4.9.
Core sampler for determining soil bulk density. The sampler yields a core of fixed volume. The core is dried and weighed. The weight divided into the volume gives the bulk density of the soil.

velopment tends to be greatest in the surface soil, it is at this level in the profile that porosity is usually the greatest and bulk density the lowest. Lower porosities and higher bulk densities in the subsoil are usually associated with a relatively low degree of structural development and a higher state of compaction due to the weight of the overlying soil.

Differences in texture are a major cause of variation in bulk density from one soil to another. As shown in table 4.2, soils of coarse texture have the lowest porosity and, therefore, the highest bulk density. Coarse soils have large pores, but the pores are few in number. The net result is a comparatively low total pore space volume for soils of coarse texture. In comparison, fine-textured soils have small pores, but they occur in such great numbers that their total volume exceeds that of coarse-textured soils. Although the highest bulk density value shown in table 4.2 is 1.55 g/cc for sands, compact subsoils layers, including soil pans, may have bulk densities as high as 1.7 to 1.8 g/cc.

The bulk density of soils is usually determined with sampling equipment of the type illustrated in figure 4.9. In the procedure, a core of soil is obtained by driving the cylindrical sampler into the soil, being careful to preserve its structure and porosity. The soil core, whose volume is equal to the internal volume of the sampler, is then oven-dried and weighed. The bulk density is obtained by dividing the oven-dry weight of the core by its volume:

Bulk density = Weight of soil/Volume of soil   (4.2)

For example, if the oven-dry weight of a soil core having a volume of 480 cc were 576 g, the bulk density would be 1.2 g/cc (576/480 = 1.2).

Bulk density measurements are made for several reasons. One is to assess the effect of cultural treatments applied to increase pore size, and porosity, where soil compaction, such as that due to machinery, interferes with plant growth. Engineers use bulk density measurements to determine the degree to which soils have been intentionally compacted, as in road fills or earthen dams. For soil uses of this type, high density and low porosity are desired. Highly compacted soil has greater strength, which is wanted in road fill under highways, and it also has low porosity, which is preferred in earthen dams or other structures for water retention.

A bulk density value often used in practical considerations is the weight of a hectare furrow slice, which is a soil volume one hectare (10,000 sq m) in area to the depth of plowing. For a bulk density of 1.0 g per cc, the weight of a soil volume with an area of 1 ha and a depth of 1 cm is 100 million g, or 100,000 kg. Accordingly, the weight of a hectare furrow slice (hfs) for any bulk density or any depth of plowing can be computed from the equation:

Weight/hfs (kg) = 100,000 $\times$ Bulk density $\times$ Plowing depth   (4.3)

where the bulk density is in grams per cubic centimeter and the plowing depth in centimeters. For example, a loam with a bulk density of 1.2 g per cc and a plowing depth of 17 cm, a commonly accepted average, would have a weight per hectare furrow slice of:

Weight/hfs = 100,000 $\times$ 1.2 $\times$ 17 = 2,040,000 kg   (4.4)

For the range of bulk densities in table 4.2, weights per hectare furrow slice, 17 cm deep, vary between 1,700,000 and 2,635,000 kg. In practical work, an arbitrarily selected average of 2,000,000 kg is commonly used.

Pore space relations in soils

The highly diverse nature of soil pores makes their precise description difficult. However, one of their more significant features is their occurrence in an interconnected system that is continuous with the outer atmosphere. Only because of this relationship is the ready interchange of water and gases between the soil and the outer atmosphere possible. In general, the interchange of gases and water between the soil and the outer atmosphere is most rapid in soils of coarse texture, which have large pores. However, large pores may be introduced into fine soils by the development of structure or by the burrowing activity of worms, insects, and other animals.

Although the size of pores is highly important to the behavior of soils, this property is rarely measured in absolute terms. Instead, it is assessed indirectly by measuring other closely related soil properties or behavioral characteristics. A principal measure is the ability of the soil to transmit air or water. This capacity in a soil is extremely important to plant growth, and it relates more to the size of pores than any other soil property.

Generally speaking, pore size is more important than porosity so far as soil behavior is concerned. Yet, porosity values do have certain fundamental applications at times, and they can be rather easily determined where the bulk and particle densities of a soil are known. The determination is made through use of the equation:

$$\text{Porosity } (\%) = \left(1 - \frac{\text{Bulk density}}{\text{Particle density}}\right) \times 100 \qquad (4.3)$$

In this equation, the ratio of the two density values expresses the fraction of the total soil volume occupied by discrete soil particles. If the volume of the whole soil is set equal to unity and the fraction taken up by particles subtracted from it, the difference represents the fraction of the soil taken up by pore space. Multiplying the pore space fraction by 100 converts it to a percentage of the total soil volume. Either fractional or percentage values may be used to express soil porosities.

The pore space of soils is filled jointly by air and water, and the higher the water content the lower the amount of space left for the circulation of air through the soil. The freedom of air movement, as it affects the supply of oxygen to roots and beneficial soil organisms, is critical to plant growth. For this reason, it is of value to know the fraction or percentage of the total soil volume that is air-filled at a particular moment. This value is known as the *aeration porosity*. In general, if the aeration porosity falls much below about 10 percent of the total soil volume, the growth of many plants can be hindered because of the slow rate of delivery of oxygen to roots.

Soil tilth

Tilth describes the physical state of the soil as it relates to plant growth. Soil tilth is a variable characteristic; it is subject to natural change as well as to modification by artificial means such as plowing and cultivation. In concept, good tilth infers a soil state that provides for an adequate supply of both air and water. The penetration of roots and the emergence of shoots from the soil are more easily accommodated when tilth is at its best.

Most often, poor tilth is associated with an extreme in the average size of pores. The pores may be too large, either because the soil has a very coarse texture or, if it is fine-textured, because it occurs in a cloddy condition. Large pore size affects plant growth adversely by allowing rapid air circulation that hastens drying of the surface soil. Seeds planted in an open, porous bed may fail to germinate because water is not supplied throughout the germination period.

Plant growth and the germination of seeds in soils of excessively small pore size is limited because of a lack of O_2 or because root and shoot penetration is restricted. The transfer of O_2 from the outer atmosphere into the soil through small pores is slow at best. The problem is greatly amplified when the soil is moist, for water tends to fill the pores and further interferes with air passage. It is usual for pores of small size to remain water-filled longer after soil wetting than do pores of large size.

Puddling of fine-textured soils induces poor tilth. One reason for this is considered above, since puddling results in a reduction in pore size and aeration. Puddled soils also may form hard, impenetrable crusts on drying. Seedlings sometimes never emerge from a soil where planting has been followed by a rain that puddles the surface layer and allows it to crust over on drying.

Tillage and tilth   Although soils are tilled for a number of reasons, the main ones are (1) to prepare seedbeds for planting and (2) to control weeds. Tillage for these or for any other reason should be under conditions that offer the best opportunity to preserve good tilth.

Seedbed preparation normally involves plowing and one or more subsequent operations that crumble the soil and leave it in a smooth, well granulated condition. A properly prepared seedbed favors the mechanical planting of seeds and provides a suitable environment for germination. Sometimes these conditions already pertain in a soil, but it may have to be cultivated anyway, either to bury plant residues that would interfere with seeding or to destroy existing vegetation that would compete with the crop being planted.

Plowing lifts, twists, and shears the soil so as to cause cleavage along the surfaces of naturally occurring structural forms (see fig. 4.10). The operation is most successful when soil moisture conditions are optimum; enough water so that aggregates separate from each other easily but not so wet that puddling occurs (see fig. 4.11). The relationship between moisture and plowing time is more critical for soils of fine than of coarse texture. The latter soils tend not to assume aggregate form, and the texture, which determines pore space relationships in them, is not altered by cultivation.

In most soils plowing results in a roughened condition that needs to be modified prior to planting; plowing is therefore customarily followed by discing or harrowing, or by both (see figs. 4.12 and 4.13). The blades of a disc or the teeth of a harrow are closely spaced so that they disturb most of the soil as they pass through it at a shallow depth. Cultivation with these implements causes further soil crumbling and leaves the surface in a smooth condition as well. One fault with this is that the finely divided soil may be relatively easily eroded by wind or water. This problem can be minimized by seeding into old crop residues without plowing or into soil that has only been plowed and is therefore in a rough, cloddy condition.

Figure 4.10.
The action of the moldboard plow on soil. The plow blade, which is moving to the right in the photograph, underrides the surface layer of soil, cutting away a slab that is then tilted upward and to the right. As the soil rides over the blade surface it tends to shear into thin blocks that break into smaller fragments as they tumble into the neighboring furrow. (National Tillage Machinery Laboratory, U.S.D.A. photograph.)

Figure 4.11.
Manifestations of consistence in a silt loam soil in three states of moisture: dry (left), wet (center), and moist (right). When dry the soil has a hard consistence. Crushing the dry soil requires considerable pressure and causes it to break into large, irregular fragments. The wet soil is plastic; moderate pressure is needed to deform it, with the deformation resulting in a tendency to smear. In the moist state the soil is friable; it crumbles readily into fine aggregates when subjected to a relatively weak pressure. The latter condition would be preferred for plowing or other cultivation, since it would allow the soil to be worked into a more nearly ideal physical state with a minimum of effort.

Figure 4.12.
Plowing leaves soil
with a rough surface
and may need
smoothing before
seeding to a crop.
(Photo courtesy of
Deere and Co.)

Cultivation, if practiced with discretion, has a very useful role in crop production, but it can be harmful if carried out under improper conditions or to excess. Fine-textured soils, if plowed when too wet, may puddle and dry into hard clods that are difficult to break down during subsequent seedbed preparation. Excess cultivation, even at the most suitable moisture level, can destroy natural granulation by compaction or abrasion. Such an effect not only lowers the immediate value of the soil for plant growth but also increases its susceptibility to erosion.

The matter of soil tilth cannot be closed without considering the physical state of the subsoil. In this respect textural and structural conditions as well as compaction or cementation of the deeper soil layers can have a marked influence on plant growth regardless of surface soil tilth. This fact is illustrated in figure 4.14, which shows differences in the root growth of corn and soybeans caused by differences in subsoil conditions. Good root growth is evident in deep open, well aerated soil, but it is very poor in soils with dense, poorly aerated subsoils.

Figure 4.13.
Smoothing plowed
land by discing. The
degree of
pulverization depends
on the number and
size of disc blades,
and the depth and
angle at which they
are drawn through the
soil. (Photo courtesy
of Deere and Co.)

Soil color

Color is a physical property that has little effect on soil behavior beyond influencing the gain and loss of radiant energy. Soil color may be judged qualitatively from memory, or it may be evaluated quantitatively by comparison with standard color charts based on the Munsell system of color notation. In this system color is expressed by letter-number symbols that denote three variables, namely: *hue, value,* and *chroma.* The hue relates to a spectral color such as red, yellow, or blue, and varies with the dominant wavelength of light reflected from a colored surface. Value denotes the degree of lightness or darkness of the color, whereas chroma refers to the strength or purity of the dominant color. High chroma means a strong color, but if the chroma is low, the color will tend toward black if the value is low or toward white if the value is high.

The color of a soil can be determined by comparison with cards containing a series of color chips, each identified with a specific hue, value, and chroma (see fig. 4.15). Once a match has been made, the color for the soil

Figure 4.14.
The effect of subsoil conditions on root growth of corn and soybeans in Illinois soils. The Muscatine and Flanagan soils have deep, comparatively uniform profiles. Dense subsurface layers in the Huey and Cisne soils restrict the depth of root penetration and the utilization of water stored in the deep subsoil. (Photo courtesy of J. B. Fehrenbacher, University of Ilinois, Urbana.)

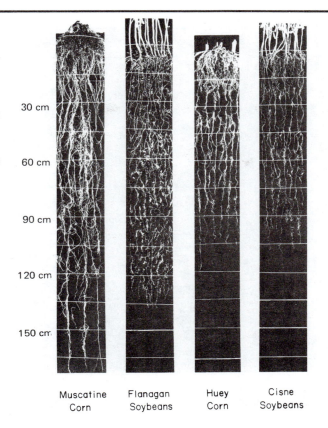

30 cm

60 cm

90 cm

120 cm

150 cm

| Muscatine | Flanagan | Huey | Cisne |
| Corn | Soybeans | Corn | Soybeans |

can be read directly off the card. An example notation of color is 10YR 6/4, which may be described as a light yellowish brown with a hue of 10YR, meaning yellow-red, a value of 6, and a chroma of 4.

If a dominating spectral color is lacking, the chroma is zero and the color will range from black through gray to white depending on the value. The value can range from zero for absolute black to 10 for absolute white. Since they lack both chroma and hue, these colors are called neutral and are denoted by a symbol in which N replaces the designation for hue and the space for chroma is left blank. Some example symbols for neutral colors are N2/, N5/, and N8/, which are respectively for black, gray, and white colors.

Color is particularly useful for making a number of meaningful predictions about the soil. It is a guide to such things as the extent of mineral weathering, the amount and distribution of organic matter in the soil, and the state of aeration.

For the most part, changes in color that result from weathering are due to the formation of secondary iron oxides and, in some instances, manganese oxides. The distinct red, yellow, and brown colors in soils are normally attributable to the iron compounds. Thin coatings of iron oxides over other mineral grains may impart a light brown or buff color to the soil body. Pro-

Figure 4.15.
Munsell color chart for hue 10YR. Colors become darker from top to bottom and brighter from left to right. (Adapted from a U.S.D.A. diagram.)

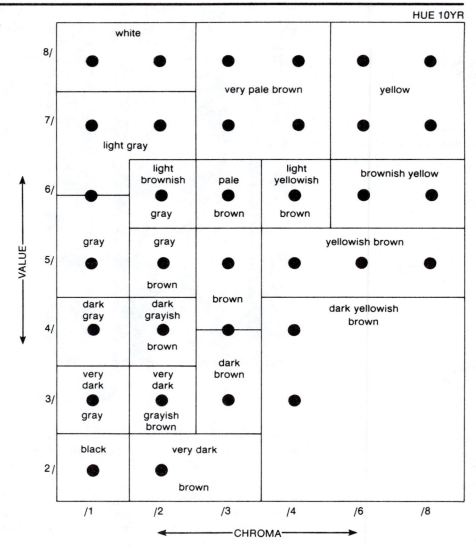

nounced red and yellow colors usually identify mineral substances that have been extensively weathered. Such colors are typical of many tropical and subtropical soils.

The extent to which organic matter darkens soils depends upon the chemical nature of the organic material, the quantity present, and its form, that is, whether it occurs as large fragments or as finely divided coatings over mineral grains. Generally, organic matter accumulating under grassland vegetation has the darkest color. It may cause the soil to be almost black, even when it constitutes no more than 4 to 5 percent of the soil by weight. The same quantity of organic matter has less of a darkening effect in forest soils. Apparently, the lack of dark color is due to the specific nature of the decomposition products retained as humic substances in these soils.

Physical properties of soils

The interrelationship between color and aeration usually reflects the obstructing influence of water on air flow through the soil. Poor aeration results when pores remain water-filled for prolonged periods. When this occurs, iron and manganese assume chemically reduced forms that impart grayish or bluish colorations typical of gleying. If poor aeration due to excess water is intermittent, iron and manganese that are solubilized during the wetter part of the year tend to oxidize during periods of improved aeration and precipitate as more brightly colored oxide coatings over other mineral grains. Frequently, these coatings appear as red, yellow, or brown mottles (splotches) or streaks, or they may form a horizontal band where ground water is present. The presence of these stains is a clear indication that the soil is probably waterlogged during at least part of the year and should therefore be avoided as a site for dwellings or other structures with basements, or for septic tanks used for on-site sewage disposal.[7]

Light-colored horizons may form in soil if continuous weathering and leaching deplete the mineral matrix of Fe compounds. The residual mineral matter may be quartz or whitish clay. Some white tropical soils contain an abundance of gibbsite, which appears to be more resistant to weathering than Fe oxides under the prevailing conditions. A whitish cast in arid-region soils is often indicative of accumulated lime.

In this country the U.S. Department of Agriculture has the responsibility of collecting and publishing information about soils, and usually it is in standardized soil survey reports. In the reports, information is given on the conditions under which each soil occurs, as well as on its potential use with agricultural crops and on the types of conservation practices needed to preserve it against serious deterioration, as by erosion. In addition, a description of the profile for each soil is supplied (see fig. 4.16). One such description follows; it is for the Cisne soil, which occurs in the State of Illinois.[8]

A— 0 to 11 inches, very dark grayish-brown (10YR 3/2) silt loam; moderate, fine, granular structure; friable when moist, slightly sticky and slightly plastic when wet; medium acid; gradual, smooth boundary.

E1— 11 to 18 inches, grayish-brown (10YR 5/2) silt loam; weak, fine, granular structure; friable when moist, slightly sticky and slightly plastic when wet; strongly acid to medium acid; clear, smooth boundary.

E2— 18 to 21 inches, light-gray (10YR 6/1) silt loam; few, fine, prominent, dark yellowish-brown (10YR 4/4) mottles; massive; friable when moist, not sticky and not plastic when wet; strongly acid to medium acid; abrupt, smooth boundary.

Bt1— 21 to 28 inches, dark grayish-brown (10YR 4/2) silty clay loam to silty clay; many, fine distinct, dark yellowish-brown (10YR 4/4) mottles; weak to moderate, medium and fine, angular blocky structure; dark-gray (10YR 4/1) clay films; firm when moist, very plastic and very sticky when wet; medium acid; gradual, smooth boundary.

[7]Soil drainage requirements for septic tank installation are discussed on page 531.
[8]See the Soil Survey of Montgomery County, Ill. (1969)

Figure 4.16.
A soil scientist examining a soil profile from a pit. He determines texture, structure, color, and other characteristics required for soil descriptions. A book containing Munsell color charts is at the right. (U.S.D.A. Soil Conservation photo by R. L. Meeker.)

Bt2— 28 to 40 inches, gray (10YR 5/1) silty clay loam; few, fine, prominent, yellowish-brown (10YR 5/8) mottles; weak to moderate, coarse, angular blocky structure; very plastic when wet; neutral in reaction; gradual, smooth boundary.

BCt— 40 to 60 inches, same as the B2t horizon, except that the texture is light silty clay loam.

As may be noted above, descriptions of soil profiles are based largely on physical properties including horizon thickness, texture, color, structure, and consistence. These descriptions are used primarily to aid identification of different soils in the field, but they also allow judgments to be made about certain behavioral characteristics of the soil. For example, the material in the B horizons of the Cisne soil described above is very sticky and plastic when wet. From this, one would suspect that these horizons contain expanding clay and are potentially slowly permeable to water. This is indeed the fact, for a problem with this soil is poor internal drainage. The B horizon of the Cisne soil is commonly considered to be a claypan and, as shown earlier in figure 4.14, interferes with root growth into the deeper subsoil.

Summary

Physical properties of soils are measured by physical, as opposed to chemical or biological, means. The more important of these properties are (1) texture, (2) structure, (3) consistence, (4) the covariable properties of density and porosity, and (5) color. Among these properties, texture is viewed as being the most fundamental. The reason for this is that many other properties tend to vary with and can therefore be judged from texture.

Physical properties are important to both agricultural and nonagricultural use of the soil. For agricultural use, namely plant production, physical properties of greatest importance are pore space relationships that determine the supply of water and oxygen to plant roots, and consistence properties that influence the ease of cultivation and the penetration of the soil by roots and shoots. The physical state of soil as affects plant growth is referred to as soil tilth.

One aspect of soil tilth is soil density, which varies inversely with changes in porosity. Density, which is easily measured, can therefore be used in estimating pore space relationships in the soil. Variation in density and porosity is often the result of compaction on the one hand and the loosening of the soil by cultivation on the other.

Except as it affects the absorption or reflection of solar radiation, color has no direct effect on normal soil use. Color can be used as an index of other soil properties or conditions, however. Among these are organic matter content, mineralogy, and drainage state.

Review questions

1. Why can the physical behavior of soils be predicted more accurately from their texture (percent sand, silt, and clay) than from their textural class?
2. Explain the purpose of modified textural class names.
3. Why are peat and muck not textural class names?
4. Explain the importance of granular or crumb structures in the plow layer of medium- to fine-textured soils.
5. Why are soil pans usually undesirable in agricultural soils?
6. Why is grittiness due to sand not a consistence property?
7. What is the general relationship between soil texture and consistence properties such as hardness of clods in dry soil and the degree of plasticity and stickiness in wet soils?
8. Why do coarse-textured soils normally have a higher bulk density than do fine-textured soils?
9. Why will the conversion of a puddled soil to a well granulated state likely reduce the bulk density of the soil?

10. Explain why very coarse and very fine textures are sometimes the cause of poor tilth in soils.
11. Explain the statement: Soil color can be used as an index of the organic matter content, the mineralogy, and the state of aeration in soils.

Selected references

Baver, L. D.; Gardner, W. H.; and Gardner, W. R. *Soil Physics.* New York: John Wiley & Sons, Inc., 1972.

Hillel, Daniel *Fundamentals of Soils Physics.* New York: Academic Press, Inc., 1980.

Rose, C. W. *Agricultural Physics.* New York: Pergamon Press, 1966.

Russell, M. B. Physical Properties. In *Soil, 1957 Yearbook of Agriculture,* pp. 31–38. Washington, D.C.: Government Printing Office, 1957.

Soil Survey Staff. *Soil Survey Manual.* U.S.D.A. Handbook 18, pp. 189–234. Washington, D.C.: Government Printing Office, 1951.

5

Organisms and organic matter in soils

.

An outstanding feature of soils is their ability to support a wide variety of plant and animal life. The kinds of organisms involved range from higher land plants and animals to a myriad of microscopic organisms that live primarily within the soil. Among these organisms, plants have a particularly important role because of their ability to combine carbon and other nutrients into organic compounds that other organisms can use for food. Most of these compounds have their origin, directly or indirectly, in photosynthesis, but there are some produced *chemosynthetically* by certain soil-inhabiting microorganisms. Like green plants, these microbes obtain carbon from carbon dioxide, but they contain no chlorophyll and can function in the absence of sunlight.

The great majority of soil organisms that do not synthesize carbon compounds through photo- or chemosynthesis contribute beneficially to soils through the decomposition of plant and animal residues. In the process, they convert carbon from the residues into carbon dioxide which, along with other nutrients released during decay, may then be reused in a subsequent cycle of plant growth. In addition, through the transformation of residues and the synthesis of new tissue, these organisms are responsible for the accumulation of humus, which can be of long-term benefit to both the nutritional and physical state of the soil. The physical state of the soil is also enhanced by the burrowing and mixing activities of rodents, insects, worms and other macroanimals.

Not all organisms associated with soils are beneficial. Those that aren't are mainly parasites that attack, and often devastate, living plants and animals. Such organisms include chewing and sucking insects, and a host of microorganisms that induce disease in plants and animals. Viruses, though classed neither as plant nor animal, are also harmful to the soil environment because of their ability to induce disease in living things.

Common soil organisms

Countless organisms live in the soil environment, but only those having a major impact on the soil need to be considered. Examples of some of the more important of these, identified within the separate categories of macro- and microanimals and plants, are as follows:

Macroanimals
Mammals: Mice, moles, gophers, and other burrowing species
Insects: Ants, termites, mites, flies, and grubs and other larvae
Gastropods: Slugs and snails
Earthworms
Microanimals
Nematodes
Protozoa
Macroplants
Higher land plants
Microplants
Fungi
Bacteria
Actinomycetes
Algae
Viruses

Generalized descriptions of the above groups follow.

Macroanimals

This group of organisms affects soils in various ways, but one of the most important is by means of burrowing and mixing activities. Through such activity, these animals loosen the soil and provide numerous channels that enhance air, water, and root movement, often to appreciable depths in the profile. At times, they are also responsible for the mixing of substantial quantities of organic surface litter, such as that under forest vegetation, with the mineral soil below. Because they use organic materials for food, macroanimals are also involved in organic matter decay.

Of the various macroanimals in the soil environment (see fig. 5.1), earthworms probably have the greatest overall effect on soil properties, the result of both their burrowing and feeding habits. Where they occur in large numbers, earthworms may cause complete mixing of the surface soil in a matter of a few decades. By ingesting and passing soil through their bodies, they promote the break-down of organic components and the solution of minerals. Earthworms prefer moist, well drained soils that are rich in organic matter. Where such requirements are met, earthworms may occur in numbers exceeding 2.5 million per hectare.

The soil ingested by earthworms is excreted in part as molded *casts* around the openings they leave at the surface of the soil. The casts tend to be relatively high in available nutrients and are structurally stable. However, where the casts build up in quantity, they may create a roughened sur-

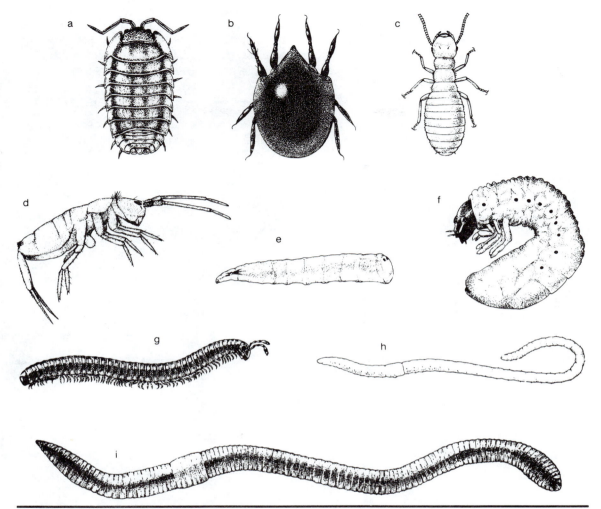

Figure 5.1.
Typical small, soil-inhabiting macroanimals. (A) Wood louse,
(B) orbated mite, (C) termite, (D) springtail, (E) fly larvae, (F) beetle
larvae, (G) millipede, (H) enchytraeid worm, and (I) earthworm. (From
Edwards, C. A. *Scientific American* 220(4):88–99. Copyright © 1969
by Scientific American, Inc. All rights reserved.)

face that is undesirable in lawns and golf greens. A similar effect is also often
produced by the burrowing and mounding activities of gophers, moles, and
other rodents.

Certain macroanimals gain particular attention because of their attack
on living plants. Great damage can be done to plants by insects, both in the
adult and larval stages, along with slugs, snails and other higher animals that
use plants for food. Shrugged off as essential to the balance of nature in the
natural environment, such attacks can have serious consequences when they
are on economic agriculture and forest crops.

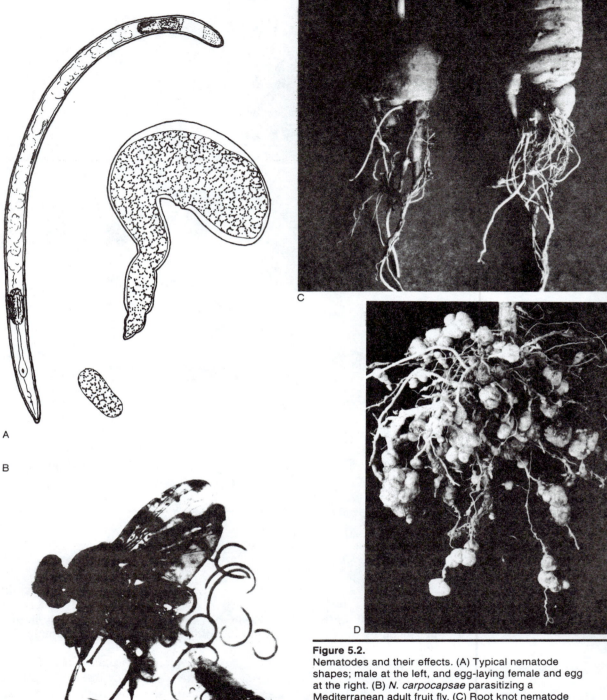

A

B

C

D

Figure 5.2.
Nematodes and their effects. (A) Typical nematode shapes; male at the left, and egg-laying female and egg at the right. (B) *N. carpocapsae* parasitizing a Mediterranean adult fruit fly. (C) Root knot nematode damage in carrots, and (D) in peas. (Credits: (B), G. O. Poinar, Jr., University of California, Berkeley; (C and D), G. S. Santo, Washington State University, Prosser.)

Microanimals   There are two groups of soil microanimals of importance: *protozoa* and *nematodes*. The protozoa are simple, single-celled animals that appear able to ingest submicroscopic particles, including bacteria, algae, and other microorganisms. Though not nearly so abundant as other microscopic life in soils, the protozoa contribute some to humus formation and accumulation. One protozoa, an *amoeba*, is responsible for a form of dysentary that afflicts humans, mostly in hotter regions, and is relatively difficult to cure.

Nematodes, sometimes called eelworms or threadworms (see fig. 5.2A), comprise a class of worms that parasitize plants and animals,[1] or they may occur as free-living organisms in soil or water. Those inhabiting soil normally range from about 0.5 to 4mm in length. Heavily infested soil may contain as many as 10 nematodes per cubic centimeter.

Nematodes use either living or dead tissue as a source of food; often it is other microorganisms, including other nematodes. In some agricultural soils, nematode damage to plant roots seriously reduces crop yields (see figs. 5.2C and D). Yet, some nematodes are beneficial, for, as shown in figure 5.2B, they attack the larvae or adults of certain insects and thereby provide one means of biological pest control. One species of nematode is even a participant in the fermentation of fruit juices to vinegar.

Macroplants   Macroplants comprise the vegetation normal to most land surfaces, and they contribute more to soil properties and behavior than do all other living organisms combined. The effect of plants is both direct and indirect. Indirect effects relate primarily to the return of root and top tissue for incorporation into the humus fraction of the soil. The direct influence of plants on soils comes mainly from roots, and it is both physical and chemical in nature.

Roots affect physical properties by forming a network of channels through the soil and by aiding structure formation. They influence the chemistry of the soil through the absorption of nutrients and the release of carbon dioxide and organic substances, some of them acidic, into the soil solution. Acidity, which is produced both by the organic acids and the interaction of carbon dioxide with water, increases the solubility of some minerals and thereby encourages the release of nutrients for absorption by roots.

Chemical and biological activity in the soil is greatest in the zone immediately surrounding actively growing roots. This zone, called the *rhizosphere*, is enriched by organic exudates and sloughed root tissue, with the result that it supports a much higher level of microbial activity than characterizes the surrounding soil. The rhizosphere also tends to be more acidic than the main body of soil. The ability of roots to grow profusely and make extensive contact with soil particle surfaces is basic to the continuous, ample supply of some nutrients to plants.

Microplants   There are four main groups of microplants: *fungi*, *actinomycetes*, *bacteria*, and *algae*. Members of the first three groups are of particular importance because they make up the principal soil organisms involved in the decom-

---

[1]For example, nematode infestation of man is responsible for two serious diseases, elephantiasis and trichinosis, the latter often caused by eating undercooked meat, especially pork.

position of organic residues. They assume this role because of their ability to utilize a wide range of organic substances as a source of energy and structural carbon. Organisms with this capability are termed *heterotrophs* (literally, various feeding), which distinguishes them from *autotrophs* (self-feeding) that obtain carbon from carbon dioxide. Certain bacteria, which form a highly diversified group, are autotrophs. Sometimes referred to as special purpose organisms, they obtain energy by oxidizing inorganic ions and use the carbon from carbon dioxide for the chemosynthesis of structural and protoplasmic tissue. Algae are also autotrophic. Further, they contain chlorophyll. Thus, like higher plants, they are able to utilize carbon from carbon dioxide both for energy and tissue synthesis. As will be seen in the following discussions, distinctions among the microplants are based primarily on morphological features.

The fungi

There are many different types of fungi. They range in form from single-celled yeasts, through filamentous or threadlike *molds*, to the more familiar *mushrooms*. The common bread molds and the smuts and rusts that cause serious damage as they parasitize grain crops are examples of mold-type fungi. Among the fungi the molds are by far the most active in organic matter decay.

The most outstanding morphological feature of the mold fungi is their threadlike form (see fig. 5.3A). The threads, termed *hyphae*, are multicelled and are the vegetative parts of these organisms. Groups of single-celled spores, which develop on small projections from the hyphae, are one means of reproduction. The spores, being pigmented, are the cause of the blue, green, or black colors seen on moldy bread, or they may impart a black, orange, or yellow color to the stems and seed heads of smutted or rusted grain crops. Spores are not the only means of reproduction of fungi, for the extension of hyphae into unexplored parts of the soil or the dispersal of fragments of these threadlike structures by other means also account for their rather widespread distribution.

The fungi attack virtually any organic residue in the soil and are amenable to a wide variety of environmental conditions. Some species thrive well under strongly acid conditions, a fact that makes fungi especially important to the decay processes in leached, acid soils. Fungi do not function well under conditions of poor aeration (low O_2 supply), as may occur in waterlogged soils or in the deep subsoil.

Some mold fungi serve an exceedingly important role in soils because of their ability to form *mycorrhiza*, which is a mutually beneficial (symbiotic) association with plant roots (see fig. 5.3B). In the association, fungal hyphae penetrate roots to obtain carbonaceous compounds for energy and growth. The fungi benefit the plants by absorbing and transporting nutrients to the roots, often in greater quantity than the roots could absorb on their own. Many plant species, both herbaceous and woody, and including numerous economically significant field and forest crops, appear to need well developed micorrhiza for satisfactory growth under some circumstances.

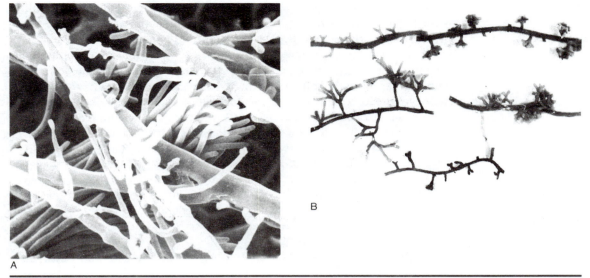

Figure 5.3.
(A) A scanning microphotograph of mold fungi showing the multicellular structure and branching of the threadlike hyphae. The short branches are reproductive structures (conidiophores), some of which have egg-shaped spores (conidia) at the end. (B) Five types of ectomycorrhizal fungi associated with pine roots. (Credits: (A), Y. Hiratsuka, Canadian Forestry Service, Northern Forest Research Center, Edmonton; (B) D. H. Marx, U.S. Forest Service, Athens, Georgia.)

Mycorrhiza are classified into two groups: *ectomycorrhiza* and *endomycorrhiza*. The distinction between the two depends on the way in which the fungal hyphae penetrate the root tissue. Ectomycorrhizal hyphae do not enter root cells but only the spaces between. Endomycorrhiza penetrate the root cells as well as the space between. Many mushroom-type fungi form ectomycorrhizal associations with trees. Endomycorrhizal associations occur both with trees and with many common agricultural crops.

Mycorrhiza seem to be essential for the proper growth of some plant species, especially of woody plants that grow on soils typically low in certain nutrients. Such plants native to an area usually have a well established relationship with mycorrhizal fungi. Where a new species of tree is introduced for the first time, however, it may grow poorly unless the soil has been artificially inoculated with the proper mycorrhizal organism. Poor growth in forest nurseries used for seedling production has occurred many times because of this limitation.

Bacteria

Bacteria include some of the smallest of living things, certain species measuring only a fraction of a micron in diameter. As shown in figure 5.4A, they assume several shapes and occur as tiny spheres and as rods and spirals of varying length. Frequently, bacterial cells possess hairlike appendages that give them greater mobility and permit the ready invasion of new environments. The freedom of movement accorded many bacteria accounts in part for their effectiveness in promoting biological change.

Not only do bacteria assume a number of different forms, they also adapt to a wide variety of environmental conditions and perform numerous specific functions in soils. Some of the specialized bacteria are autotrophs capable of oxidizing inorganic ions or elements for energy. Ammonium, nitrite, iron, and manganese ions are all subject to microbial oxidation in the soil. Sulfur, either elemental or in combined form, as in sulfide minerals, can also be oxidized by a separate group of sulfur-oxidizing bacteria. In the process, the sulfur is transformed to sulfate (SO_4^{2-}). Hydrogen ions are also released in the process and, with the sulfate, form sulfuric acid. The biological production of sulfuric acid accounts for the extreme acidity that sometimes develops in sulfur-bearing coal mine spoil or in reclaimed tidal-flat soils. It also explains why elemental sulfur is sometimes added to soils to increase the level of acidity.

Certain bacteria are of great importance because of their ability to assimilate gaseous nitrogen from the atmosphere. This process, known as *nitrogen fixation*, has been the principal source of the nitrogen stored in soil organic matter. Some nitrogen-fixing bacteria live as independent agents in the soil, and they supply fixed nitrogen to the soil when they die. Other nitrogen-fixing bacteria live in a symbiotic relationship with certain plants. These organisms colonize specialized structures, called *nodules*, on roots. Like mycorrhizal fungi, symbiotic nitrogen fixers obtain carbon for energy and tissue synthesis from the plant. They benefit the plant by supplying it with fixed nitrogen. Nitrogen fixation in legumes such as alfalfa, clover, beans, and peas is particularly outstanding, but it is carried out in other plants as well.

A number of bacteria can exist in the absence of free oxygen. These are known as *anaerobes*. Still others, the *facultative* bacteria, survive either in the presence or absence of free oxygen. Bacteria with these characteristics are important to organic matter decomposition in poorly aerated soil or in swamps. However, most organic matter decay is carried out by *aerobic* heterotrophs. Aerobes require free oxygen, and they include all fungi and actinomycetes and most bacteria. Aerobic decay converts organic carbon to carbon dioxide; anaerobic decay converts it to methane.

Soil anaerobes obtain oxygen by reducing inorganic oxides. Typical is the reduction of sulfate to sulfide, notably to hydrogen sulfide, and nitrate (NO_3^-) to free nitrogen (N_2) gas. The latter reaction is important to soils, for it may greatly limit the usefulness of nitrate fertilizers added to poorly aerated soils. Yet, where excess nitrate has accumulated from fertilizers or decomposing animal manures, its conversion to inert nitrogen gas under anaerobic conditions reduces the potential for the nitrate to leach and pollute ground water.

Many bacteria survive periods of adversity, as during severe drouth or when temperatures are abnormal, by assuming a protective *spore* form. The spores convert to an active, vegetative form when conditions are more fa-

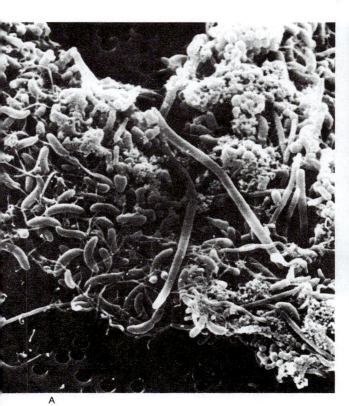
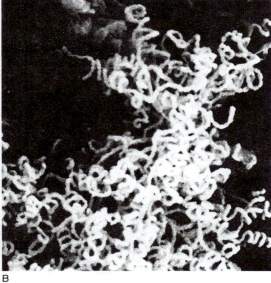

B

A

Figure 5.4.
(A) Three types of bacteria: round cocci, rod-shaped bacilli, and spiral-shaped spirochetes. (B) Actinomycetes in a fruiting stage. (Credits: (A) D. Birdsell and W. Fischlschweiger, University of Washington, Seattle; (B), P. H. Tsao, University of California, Riverside; from *Soils for Management of Wastes and Waste Waters,* pp. 115–69, 1977, by permission of the American Society of Agronomy.)

vorable. Probably more important to the maintenance of a continuous microbial population, however, is the ease with which organisms are reintroduced into a soil and the rapid rate at which they reproduce and grow in suitable environments. Reproduction of bacteria is by cell division.

Actinomycetes

Members of the third principal group of soil microbes important in organic matter decay, the actinomycetes, resemble both the fungi and bacteria (see fig. 5.4B). Like the bacteria, they are single-celled, but as with the fungi, they are threadlike and branched. They reproduce either through the formation of tiny spores or through cell division or fragmentation. Actinomycetes appear to form and grow more slowly than do bacteria.

Actinomycetes do not tolerate a particularly wide range of environmental conditions. By and large they function poorly in strongly acid environments and do not adapt well to conditions of limited aeration. They are relatively resistant to drouth, on the other hand. Because of this and because of their preference for a neutral soil reaction, actinomycetes may be important to decay in soils of dry regions. Conversely, they tend to be less active in acid soils or in those affected seriously by waterlogging.

Algae

This group of organisms consists of pigmented, single-celled plants that photosynthesize their own energy materials. The green and blue-green groups, and diatoms, which have a highly silicified cell wall, are the prevailing types in soils. Since algae require light for photosynthesis, they occur in greatest abundance at or near the soil surface, often as a thin film over the surface. Algae also infest ponded or slowly moving water, where they may accumulate as thick scums. Algae in water are an important link in the food chain, and through the release of oxygen during photosynthesis, are of major importance in maintaining the oxygen balance in the earth's atmosphere.

Algae perform a number of significant functions in soils. Through the photosynthetic production of organic tissue they contribute to the supply of organic matter, although in very small amounts on an annual basis. In addition, some blue-green algae fix atmospheric N_2 gas. Since algae grow only at the surface of the soil, and then in only limited numbers if shaded by higher plants, they fix relatively little nitrogen on an annual basis. However, they have a significant effect on nitrogen accumulation in soils over long periods of time. This source of N is important to the nutrition of rice grown in paddies, where algae often occur in relative abundance (see page 353).

Plant disease organisms and viruses

There is a wide variety of diseases that infect plants through the soil. Example diseases include crown gall, root rot, potato scab, smut, wilt, and damping off, the latter a disease that destroys plants during or soon after germination. In most cases, organisms responsible for these diseases are dormant when in the soil, but they become active in the presence of a suitable host plant. Eliminating the host for a period of time, as through crop rotation, is one means of controlling certain of the soil-borne diseases. Other control measures include adjustment of soil pH, changing soil drainage and aeration conditions, and soil sterilization. Since sterilization is expensive, it is usually limited to small tracts, such as nursery land, or to greenhouse soils. For many plants, successful disease control has been through the breeding of resistant varieties.

Agents of disease in plants include the three widely distributed microbial groups of bacteria, fungi, and actinomycetes, as well as viruses. Bacteria are responsible for crown gall, a disease most prevalent in woody plants, and they cause certain types of vascular wilt by plugging or destroying the conductive tissue of plants. Some wilt diseases are also produced by fungi, and they are largely responsible for seed decay, damping off, and root rot. Smut, which destroys the developing seed of grain crops, is a fungal disease.

Probably the most widely known disease associated with actinomyces is potato scab, which disfigures the skin of potato tubers and reduces their marketability. The responsible genus is *Streptomyces*, the source of the antibiotic, streptomycin. This organism is pH-sensitive and can be controlled by maintaining the soil pH below 5.

Viruses harbored in the soil are another cause of plant disease. Further, they sometimes attack and destroy beneficial soil microbes. Plant diseases produced by viral infection may be limited to the discoloration or disfiguration of plant tissue and a reduction in growth, or they may cause death of the plant. Since viruses often show a high specificity for host plants, breeding of resistant varieties is one of the more successful approaches to viral disease control.

The soil organic fraction

The organic fraction of soils consists of plant and animal residues in various stages of decomposition. The principal organic component is the product of long decay, humus, although at times, living organisms, including roots and their partially decomposed remains make up a significant part of this fraction. *Peat* soil is unique in this regard; it consists largely of organic materials that have accumulated under conditions of poor aeration and have therefore been altered to only a limited degree by decay. Most peat accumulates in bogs or swamps and is often derived from *Sphagnum* moss.

The accumulation of organic matter in soils is a function of two opposing processes: the addition of residues from plants and animals, and the decay of these residues by microbes and other soil-inhabiting organisms. In the earliest stages of soil development, the loss of organic substances through decay tends not to keep pace with residue addition, so there is a gradual build-up of organic matter with time. Over long periods of accumulation, however, a point is ultimately reached where the gains and losses of organic materials come into balance. At this point, the organic matter level appears to stabilize and undergoes little further change so long as conditions in the soil environment stay about the same.

Role of plants in organic matter accumulation

All plants contribute to the organic regime of soil, but to varying degrees depending on the type of plant and its potential for growth. Major differences in the plant effect occur between herbaceous and woody species, these differences relating primarily to the diverse nature of their respective root systems. Woody plants have relatively permanent root systems, so they add little residue directly to mineral soils on an annual basis. In contrast, herbaceous plants, with grasses being the most outstanding example, tend to die back each year. Thus, essentially all the tissue they produce during a single season is returned to the soil as fresh residue. A substantial portion of this residue comes from roots; certain grasses, for instance, yield more root tissue in the surface 25 cm of soil than is produced by the entire above-ground portion of the plant.

Plants differ not only in form and growth habit but also in conditions required for normal growth and maturation. Thus, differences in the environment are ordinarily accompanied by differences in the kind of vegetative cover. The principal environmental variable is climate, and especially the precipitation pattern, or the regularity of the moisture supply throughout the year. Areas that suffer periodic drouth, for example, are not so likely to support forest vegetation. Drouth interferes with the establishment of forest vegetation, usually by causing the death of shallow-rooted seedlings. For this reason, forest species tend to concentrate in the more humid regions where a continuous supply of available moisture is more nearly assured. Continuity of moisture supply is not a requirement for the survival of adapted grassland and desert plants, however. These plants are able to survive drouth either by completing their lifecycle during the time span when moisture is available, or by converting to a dormant or semidormant state when the supply of available water has been exhausted.

The principal limitation to plant growth in arid and subhumid desert and grassland regions is the low moisture supply. Commonly, the yield of residues in these regions ranges from only a few hundred kilograms to no more than about five metric tons per hectare per year. In comparison, forest vegetation of the temperate regions may produce up to seven metric tons of leaf and needle fall per hectare annually, and in tropical rainforests, yields of residue during the same time interval may be as high as 10 to 20 metric tons per hectare.

The kind of vegetation is not only a key to the abundance and pattern of residue addition to soils, but it also determines the chemical composition of these residues. Composition, in turn, particularly as it relates to the principal organic compounds in plant tissue, has an important bearing on decay and requires some consideration for this reason.

Organic composition of plant materials

The principal component of plant tissue is C, which occurs in combination with H, O, N, and a number of other elements to form a wide variety of compounds. The numerous inorganic ions contained in plant tissue make up the *ash* that sometimes accounts for as much as 10 percent of the dry weight of the tissue. Although much of the ash consists of accumulated essential plant nutrients, nonessential elements are also present.

Although a large number of specific organic compounds comprise plant tissue, only a few are present in quantity. These consist primarily of three general organic groupings: (1) *carbohydrates*, (2) *proteins* or proteinlike substances, and (3) *lignin*. Also occurring in lesser amounts are fats, oils, waxes, and resins. The nature of these substances and the quantity in which they are present in plant residues are highly important to decay reactions and to the subsequent accumulation of residual organic substances in the soil.

The carbohydrates

There are four general groups of carbohydrates synthesized by plants: *sugars*, *starches*, *hemicelluloses*, and *cellulose*. All are combinations of C, H, and O. The simplest of these compounds are sugars consisting of short carbon chains as illustrated for the six-carbon sugar *glucose:*

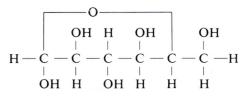

Glucose, which is the most common simple sugar in nature, has an empirical formula of $C_6H_{12}O_6$.

The other principal groups of carbohydrates, namely, starch, cellulose, and hemicellulose, have relatively complex structures. They are combinations (*polymers*) of many simple sugar molecules linked together in long, chainlike molecules. The formation of these large molecules by *polymerization* is accompanied by the release of water. The water forms from hydrogens and hydroxyls displaced at the point where two sugar molecules become linked together.

The carbohydrates occur throughout the plant kingdom. Sugars and starches, for instance, are components of all plants, although they are found in greater amounts in some species than in others. These substances tend to concentrate in specific plant parts. Typical is the accumulation of starch in seeds and in potato tubers. Sugar is stored in large quantities less often than starch, but it is an important component of sugarcane stems and the roots of sugar beets. The sugar in these plants is sucrose.

Cellulose is the most abundant organic component of plant tissue, with the hemicelluloses being next in abundance. Both of these materials occur together as a constituent of the cell wall, and they increase in quantity as plants age. The gradual accumulation of cellulose and hemicelluloses accounts in part for the stiffness acquired by maturing plant tissue. The straw of small grains and the stems of woody plants are particularly high in these two carbohydrate compounds. Cotton fibers are essentially pure cellulose, as are the fibers extracted from wood for paper manufacture. Hemicelluloses make up a more diverse group, but among the better known examples are gelatinlike pectin and other, somewhat similar gumlike substances.

The properties of the carbohydrates change significantly with variation in molecular complexity. The simple sugars, for instance, are readily soluble in water. Starch and certain of the hemicelluloses imbibe water or become dispersed in it but are not soluble in the strict sense of the word. Other hemicelluloses and cellulose are insoluble.

Proteins

Proteins are complex combinations of *amino acids*, which are small C-chain compounds made up primarily of C, H, and N with lesser amounts of O. Some also contain S. The N in amino acids occurs as an *amino* ($-NH_2$) group attached to the C chain, as illustrated by the structure of the amino acid glycine:

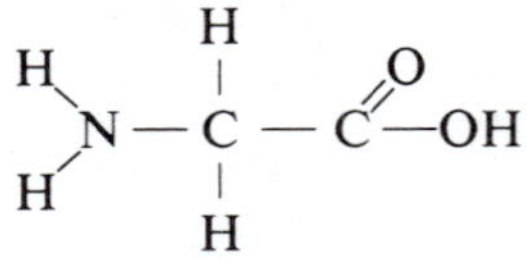

$$
\begin{array}{c}
\quad\quad\quad H \\
\quad\quad\quad | \\
H \\
\;\;\diagdown \quad\quad\quad\quad\quad O \\
\quad\quad N - C - C{\diagup}\!\!\diagup \\
\;\;\diagup \quad\quad\quad\quad\quad\; OH \\
H \quad\quad\quad | \\
\quad\quad\quad H
\end{array}
$$

The acid part of the molecule is the terminal C linked to an oxygen and an OH group, this combination shown symbolically as $-COOH$ and called a *carboxyl* group. The H of the carboxyl group is capable of entering into neutralization reactions with bases, and thus gives the compound a property of acids.

Like complex carbohydrates, proteins are formed by polymerization. They consist of many amino acid molecules, some of them sulfur-bearing, linked together through the amino and carboxyl groups. The union of amino acids in protein formation also results in the release of water.

Many examples of proteins can be given. The protoplasm of cellular tissue, which is the site of life functions in all organisms, both plants and animals, consists of this type of substance. All plant cells contain protein, but as plants mature, protein formation becomes centered primarily in seeds.

Although most of the N in plants is contained in proteins, it is also an important constituent of chlorophyll and of nucleic acids.[2] Most proteins also contain S, derived from a small complement of S-bearing amino acids.

Lignin

The third most abundant constituent of plant tissue is lignin, which occurs in the cell wall in association with cellulose and hemicellulose. Like cellulose and hemicellulose, the quantity of lignin increases with plant age. It is a very important constituent of woody tissue and is present in relative abundance in most mature plants. Lignin is insoluble in water.

Lignin has a large and very complex molecule containing C in six-atom rings as well as in short chains. Opinions differ as to the specific organization of the lignin structure. One concept is of a molecule containing a large number of basic C units, each made up of C in both chain and ring form, linked into a continuous network through oxygen atoms, as follows:

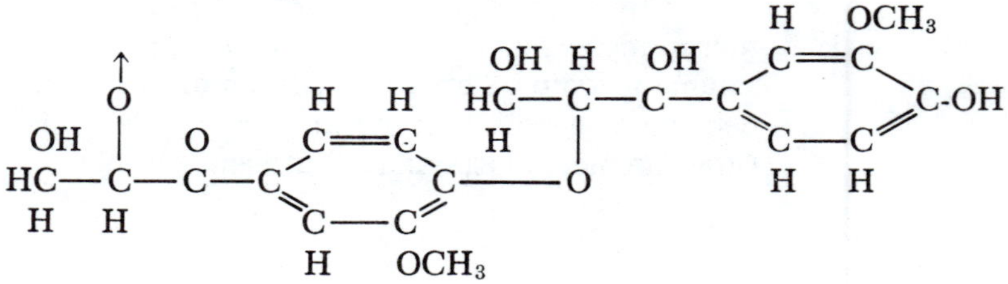

[2]Ribose nucleic acid (RNA) and deoxyribose nucleic acid (DNA), which are fundamental genetic components of cell nuclei, are examples. They contain P as another important inorganic constituent.

The arrow in the above diagram indicates the point of linkage to the remainder of the lignin molecule.

An important feature of the lignin molecule is the presence of $-OH$ (phenolic) groups, which are linked to the C rings as shown to the right in the above diagram. The H in the phenolic groups behaves much like that in the carboxyl groups of organic acids; that is, it can enter in neutralization reactions with bases. In reactions of this type, the H is removed and its place taken in the organic radical by Ca^{2+}, Mg^{2+}, K^+, or other cations supplied by the base.

Fats, oils, waxes, and resins

With but few exceptions these substances are not produced in large quantity by herbaceous plants, although they are important constituents of coniferous trees. These compounds have diverse and relatively complex molecular structures. They occur principally as combinations of C and H in large molecules, often in long chains. They are insoluble in water.

The simpler oils and fats are stored as energy materials in plant tissue. They are found in small quantities in vegetative cells and in higher concentrations in seeds. Waxes are commonly distributed as thin coatings over leaves or other plant parts, where they retard the loss of water to the atmosphere. Resins are formed in largest amounts in the wood and needle tissue of conifers. *Rosin* is an example of this type of compound. It is obtained commercially as a residue from the distillation of turpentine extracted from the wood of a number of pine species.

Decay of organic residues

In its overall effect, the decay of organic substances in soils is a destructive process caused by a variety of organisms, mainly microscopic, as they seek carbon and other nutrients for energy and growth. Some of the nutrients are utilized by the organisms in the synthesis of new tissue and are thereby retained in the soil, at least temporarily. To the extent that the nutrients are converted to new types of compounds, processes of decay are constructive as well as destructive. Eventually, however, all nutrients present in soil organic matter at a particular time are eventually converted to simple end products that can be lost from the soil. These products include such things as carbon dioxide and water, which can be lost by volatilization, and various soluble ions or compounds, which can be removed from the soil by leaching. A natural consequence of decay is therefore a reduction in the level of soil organic matter, an effect that can be countered only by the frequent addition of fresh residues to the soil.

Important features of the decay process

Organisms differ in their mode of attack on organic materials. Larger animals, along with microscopic nematodes and protozoa, ingest organic tissue, whereas heterotrophic fungi, actinomycetes, and bacteria require soluble materials that can be absorbed through the cell wall. Among the major organic groups in plant tissue, only sugars and amino acids are readily soluble and, thus, immediately available to decay microbes. Since the more complex carbohydrates, proteins, lignin, resins, and the fats, oils, and waxes are not soluble in water, they must be simplified to soluble form prior to their utilization by microbes. The ease of simplification is a major factor determining the rate at which these compounds undergo decomposition in the soil.

One method of simplification of large, complex organic molecules is *hydrolysis*, or the reaction of the molecules with H^+ and OH^- ions from water. Among the major compounds synthesized by plants, only proteins and the complex carbohydrates, starch, cellulose, and the hemicelluloses, are subject to hydrolysis. In the process, proteins are transformed to amino acids and the carbohydrates to simple sugars. These products can then be absorbed by the decay microorganisms. Unlike the carbohydrates and proteins, lignin, fats, oils, waxes, and resins cannot be simplified by hydrolysis, and as a result, they decay much more slowly than do proteins and complex carbohydrates.

Of the various reactions involved in organic matter decay, respiration, which supplies energy to heterotrophic decay organisms, is probably of greatest overall significance. In respiration, organic carbon is oxidized (combined with oxygen) to form carbon dioxide (CO_2), as shown for the reaction where sugar is the carbon source:

$$C_6H_{12}O_6 + 6O_2 \rightarrow 6CO_2 \uparrow + 6H_2O + \text{Energy} \qquad (4.1)$$

Since the carbon dioxide produced by this reaction escapes and is lost, the net effect of respiration is to reduce the carbon content and, thus, the content of organic matter in the soil. Any carbon compound, whether in fresh residues or in humus, can be decomposed and its carbon used in respiration. Since respiration goes on relentlessly, the organic matter content of soils is bound to decline unless sustained by the addition of fresh residues, at least on occasion.

Whereas carbon used in respiration is lost as carbon dioxide, that assimilated into microbial tissue is retained as a component of the organic fraction of the soil. As much as 50 percent of the carbon in compounds attacked by decay organisms may be retained as reconstituted structural and protoplasmic tissue. Appreciable quantities of nitrogen (N), phosphorus (P), and sulfur (S) are also retained in microbial tissue. However, if the amounts of these elements in decaying tissue exceed the needs of the organisms, they are released in simple ionic form. The N appears as the NH_4^+ ion, whereas S and P form SO_4^{2-} and HPO_4^{2-} or $H_2PO_4^-$ ions. Inorganic components such as Ca^{2+}, Mg^{2+}, and K^+ are liberated as soluble cations. The ionic forms of these elements, along with CO_2 and H_2O, represent the simple end products of organic matter decomposition.

The conversion of elements held in organic combination to an inorganic form is referred to as *mineralization*. The reverse of this, or the conversion of inorganic ions or compounds to organic form, is called *immobilization*, with the general relationship between the two processes being expressed as:

$$\text{Organic form} \underset{\text{Immobilization}}{\overset{\text{Mineralization}}{\rightleftharpoons}} \text{Inorganic form}$$

Through mineralization, nutrients stored in organic form are released for use by plants or decay organisms. Immobilization occurs if these nutrients are reabsorbed and assimilated by the living organisms, but if they are not, they are subject to loss from the soil environment. Mineralization can affect elements from any source, whether it be soil organic matter, fresh plant residues, or manure. Similarly, immobilization can affect inorganic ions or compounds from any source. They may be products of previous mineralization, of mineral weathering, or they may have been supplied by fertilizer added to the soil.

Sometimes, the supply of a nutrient in inorganic form, which is the form plants utilize, is too small to satisfy the simultaneous demand of both plants and decay organisms. Where this occurs, any immobilization by the decay organisms is undesirable, for it may create a deficiency of the nutrient for the plant. Where it prevents the leaching loss of nutrients, immobilization serves a very useful function, on the other hand. This effect is considered to be useful because the nutrients immobilized and conserved in organic form will eventually be released in a form plants can use.

Factors affecting the decay process

The ease with which a fresh residue undergoes decomposition is dependent upon a number of interrelated factors. These include: (1) aeration, (2) moisture, (3) temperature, (4) availability of nutrients essential for microbial growth, and (5) composition of the residue, the last item determining in part the availability of C as well as other nutrients to the organisms. These factors are discussed in order.

Aeration

As shown in equation (5.1), oxygen is required for respiration by heterotrophic decay organisms, and decay within the soil can be greatly restricted if oxygen from the outer atmosphere cannot flow in to replace that consumed by the decay organisms. Open, porous soils therefore favor the rapid breakdown of organic matter, other factors permitting. The degree of aeration reflects in the organic matter content of soils in their undisturbed state, for those that are naturally well aerated display a lower organic matter content than do those that are inherently more poorly aerated.

Moisture

Although the amount of water consumed by microbes is small, the soil must be adequately moist at all times if decay is to proceed without interruption. Optimum conditions for decay are represented by a relative humidity in excess of 98 percent. Lower humidity causes the desiccation and death of some microbes, although many assume a resistant spore form that permits survival during drouthy periods.

Excess moisture hampers decay by blocking soil pores and preventing the resupply of O_2 necessary for aerobic biological processes. Although water in the soil may initially contain dissolved O_2, the quantity is too small to support a high level of microbial activity for more than a few hours. Even in open bodies of water, where convection currents aid in maintaining the O_2 supply, the decomposition of submerged residues is greatly restricted because of a lack of this essential gas.

| Temperature | Microbial activity is temperature-dependent. Although there are some exceptions, decay organisms function most effectively witin the temperature range of 25 to 40° C. The upper limit of this range is generally exceeded only at the very surface of exposed, dry soil. Below 25°C the rate of decay gradually decreases until the freezing point is reached. Below freezing, decomposition proceeds very slowly and is carried on by only a limited number of organisms. |

Because of the dependence of decay organisms on heat, a decrease in the mean annual temperature from one climatic zone to another slows the average rate at which plant residues disappear. Plants adapt to different temperature conditions rather readily, however; thus plant growth and the addition of residues are not as seriously affected by lowered temperatures as is the decomposition rate. Further, plant growth in cool climates is less often limited by the lack of water than is plant growth in warm climates. Principally because of the unique interaction between temperature on the one hand and plant growth and decay on the other, the organic matter content of soils naturally tends to increase with decreasing average temperature, other factors remaining equal.

| Nutrients | All decay organisms need a number of nutrients in addition to the C used for energy and for structural and protoplasmic tissue. These nutrients may be obtained from the soil solution, or they may be derived directly from the organic materials undergoing decay. By and large the element N is most often lacking to the point of limiting decay. At times, and particularly in acid soils, an inadequate supply of available Ca or other nutrients may also interfere with decay. |

| Organic composition of the residues | The rate at which residues decay often depends on the type of carbon compounds they contain. In general, simple, readily soluble organic components succumb at the fastest rate. Protein, starch, cellulose, and the hemicelluloses are not absorbed until they have been hydrolyzed, this process taking place in the presence of enzymes excreted by the decay microbes. The need for preliminary hydrolysis does not appear to seriously delay microbial utilization of starch and protein, but it may be a factor in the slower decay of cellulose and hemicellulose. Lignin, resins, and fatty substances are not simplified by hydrolysis; instead, their destruction appears to be almost atom by atom. The transformation of lignin may initially involve modification of its original structure, with new forms evolving that are also resistant to microbial change. |

Within environments appropriate for rapid decay, the utilization of sugars, starches, and amino compounds may be complete within a matter of a very few days. Hemicelluloses and cellulose decay more slowly, but not nearly so slowly as lignin, resinous materials, and the fats, oils, and waxes. Indeed, as much as half of the lignin added in a given residue may remain after one year of decomposition under optimum conditions. During this same period most of the original hemicelluloses and cellulose will have disappeared. However, if environmental conditions do not favor rapid decay, even the simple compounds may remain relatively unaltered for prolonged periods.

Table 5.1. The approximate composition of various types of plant materials. Values shown are percentages of the total dry material. (S. A. Waksman and F. G. Tenney, *Soil Sci.* 24:275–83, 1927, and 26:155–71, 1928).

Components	Young rye plants	Mature rye straw	Corn stalks	Alfalfa tops	Dead oak leaves	Dead pine needles
Water-soluble[a]	29.5	7.3	14.1	17.2	13.0	7.3
Hemicelluloses	12.7	24.8	21.9	13.1	18.0	17.1
Cellulose	17.8	33.9	28.7	23.6	12.8	14.8
Lignin	10.6	12.4	9.5	8.9	24.8	21.9
Protein	12.3	0.9	2.4	12.8	4.2	2.1
Fats, waxes, resins	2.3	5.3	5.9	10.4	9.2	23.9
Ash	12.5	4.2	7.5	10.3	5.1	2.5
C/N[b]	20/1	100/1	40/1	20/1	100/1	200/1

[a]Includes sugars, starches, and small amounts of amino acids.
[b]The ratio of total C to total N. Values are typical for the types of tissue shown.

The contents of the principal organic groups and ash for a number of common plant materials are shown in table 5.1. Two sets of data for rye plants are included to show the effects of aging on organic composition. The most notable effects of aging are the reduction in water-soluble compounds and protein, and this is accompanied by an increase in the more decay-resistant hemicellulose, cellulose, lignin, fats, oils, waxes, and resins. Other values of note in the table include the high content of protein in alfalfa and young rye, and the high content of lignin in oak leaves and pine needles. Pine needles also have a much higher content of fats, waxes, and resins than do the other materials.

Representative carbon-nitrogen (C/N) ratios are also listed for the different plant materials in table 5.1. These values are based on the total C and total N present in plant tissue. The C/N ratio is used to judge the rate at which a residue will decay and whether it will release or tie up (immobilize) inorganic nitrogen after incorporation into the soil.

The C/N ratio and organic matter decay

There is a reasonably consistent relationship between the C/N ratio of an organic tissue and the rate at which it will decay. In general, decay proceeds more rapidly where the ratio is narrow than where it is wide. Further, tissue of narrow C/N ratio is more likely to release mineralized N soon after decay starts. Wide-ratio materials tend to cause immobilization of inorganic N, on the other hand, and in soils, this may seriously reduce the level of N that is available to plants. Customarily, narrow C/N ratios fall in a range of from 15/1 to 30/1, with wide ratios tending toward 100/1 or more.

Variation in the C/N ratio of plant tissue is due more to variation in N than in C contents. Carbon usually makes up from 45 to 55 percent of the dry weight of plant tissue. In comparison, nitrogen contents may vary from below 0.5 percent to more than 3.0 percent, which is far greater, relatively, than the variation for carbon. Thus, for a tissue with a 50–percent C content, the C/N ratio would be 100/1 if the N content were 0.5 percent, but it would be just a bit over 16/1 if the N content were 3.0 percent.

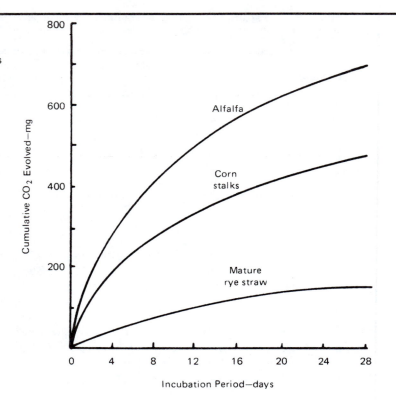

Figure 5.5.
The rate of decay of plant materials of different C/N ratios as indicated by the cumulative release of CO_2. (From S. A. Waksman, and F. G. Tenney, *Soil Sci.* 26: 155–71, 1928.)

Organic residues with narrow C/N ratios decay rapidly for two reasons. First, they have a comparatively high N (protein) content, so the microbes will not likely suffer a shortage of this element as they attack the residue. Second, tissues with narrow C/N ratios normally have a higher proportion of the total C in the more easily decomposable compounds. The reverse is true for tissues of wide C/N ratio. These materials not only supply little N in relation to the total C they contain, a higher proportion of the C occurs in more resistant compounds, such as cellulose and lignin. Thus, even if wide-ratio materials are supplied with extra N, as from fertilizer, they tend not to decay as rapidly as narrow-ratio materials because of the kinds of carbon compounds they contain.

The C/N ratio of plant tissue depends on the type of plant and on its state of maturity, as suggested by table 5.1. As shown in the table, plants with a narrow C/N ratio (young rye, alfalfa) are highest in protein and have relatively low contents of cellulose and lignin. In general, the reverse is true for those materials of higher C/N ratio. Based on the information in the table, one would judge that young rye should decompose more rapidly than mature rye, which is consistent with the change in the C/N ratio for this plant as it matures.

The general relationship between C/N ratios and the rate of organic matter decay is demonstrated by the curves in figure 5.5. Here, the rate of decay is judged from the rate of CO_2 evolution measured for three of the

Organisms and organic matter in soils

materials listed in table 5.1 when subjected to a period of moist incubation in the laboratory. A comparison of these data with the C/N ratios in table 5.1 shows that the wider the ratio the slower the release of CO_2 and, therefore, the slower the rate of decay. The evolution of carbon dioxide is a fairly good index of decay activity because the ratio of C used for energy and given off as carbon dioxide to the total C utilized by decay organisms does not vary greatly under normal circumstances.

Effects of decay on the composition of residues

The organic components of plant residues undergo great change as they are subjected to decomposition. All are degraded, some being affected more rapidly than others. Simple soluble components disappear first, but their loss results in little change in the outward appearance of the tissue being decayed. As the remaining, more resistant structural compounds decay, the form of the tissue is gradually lost. There appears in its place, though in appreciably smaller quantity, a darkened residue of decay products, some consisting of altered plant substance and some consisting of new matter synthesized by the decay organisms. These products, which are infused with a mass of living microorganisms, undergo further change at a comparatively slow rate. In this state they are considered a part of the semistable humus fraction of soil organic matter.

There is no wholly satisfactory definition of humus. As a product of decay it must not retain structural characteristics identifiable with the plant and animal tissue from which it is derived. Some similarities in chemical makeup persist, for both the source materials and humus contain many of the same basic organic compounds.

The composition of humus

Chemically, humus is a mixture of highly altered or resynthesized products of decay. Its principal components appear to be proteinaceous substances, primarily of microbial origin, along with complex degradation products derived from lignin. Admixed with these are smaller amounts of microbially produced hemicellulosic gums, as well as a wide range of compounds of variable and sometimes poorly defined composition. A number of the products of decay have a dark brown to black color, which is imparted to the humus.

Because of their rapid disappearance, sugar, starch, amino acids, and similar readily digestible components of fresh plant residues are essentially totally lacking in humus. In contrast, the continual resynthesis of proteinaceous materials and the extremely slow rate at which lignin and its alteration products disappear accounts for their accumulation as the major components of humus. Up to a third of the humus fraction may display proteinlike characteristics, whereas a slightly larger proportion may have properties normally associated with lignin. Where cellulose is added to the soil in relatively large amounts, it may also appear as a part of humus, but in concentrations of no more than a few percent. Small amounts of fats, waxes, and resins, which are slow to decay, may also be present.

Some important properties of humus

Essentially the only resemblance between humus and the plant materials from which it forms is that both are organic. Physically, humus is a finely divided, or colloidal, material. It is generally dark brown to black in color, and when present in sufficient quantity, may alone determine the color of the soil. Because of the relationship to color, the amount of organic matter can sometimes be gauged approximately by the degree to which humus darkens the soil body. Inconsistencies occur in the relationship, however. For example, the organic matter content of many forested soils may be comparatively high without a corresponding darkening of the mineral matrix, presumably because the predominant products of organic decomposition are light in color.

Since humus is a complex mixture of organic substances, it has a variable influence on the soil, affecting both its chemical and its physical properties. The effect of humus on physical properties relates largely to its ability to bind soil particles together. By virtue of its binding action, humus helps stabilize loose soils against erosion. Further, humus increases the friability of stiff, tight soils; that is, it makes them crumbly and easier to work. It also tends to provide pore space relationships that favor the ready movement of air and water through these soils.

Like mineral clays, humus particles are negatively charged and are therefore able to hold cations in adsorbed form. This is of benefit to plant growth where the adsorbed cations are essential plant nutrients. A major site of negative charge appears to be the phenolic ($-OH$) and carboxyl ($-COOH$) groups of altered lignin and protein. These groups provide sites for the retention of cations such as Ca^{2+}, Mg^{2+}, K^+, and others by displacement of hydrogen bound to the negative charge of the oxygens.[3] Humus also contributes to plant nutrition by releasing stored nitrogen, phosphorus, and sulfur through mineralization. Because of these effects, it is little wonder that humus is such a highly valued part of the soil!

Humus does not comprise all of the organic matter in soils, although it is the principal component. Second in importance is extensively decayed tissue, largely microbial, that has yet to be altered to a semistable form characteristic of humus. It has been estimated that this kind of substance makes up as much as one third of the total organic matter in some soils. Fresh or slightly altered plant residues generally provide a very small part of soil organic matter at any one time. Even so, because of their influence on microbial activity, these residues can play a predominant role in controlling the biology of the soil immediately after they have been added in fresh, unaltered form.

The organic matter cycle in soils

The decay of a residue is but one phase of an unending cycle in which the components involved return to the same state again and again. In the soil organic matter cycle, C is transformed by decay from plant to microbial tissue and humus, but its ultimate disposition is as CO_2. Eventually, the CO_2 is reabsorbed by growing plants and, through photosynthesis, is once again converted into plant tissue. By this means the cycle is reinitiated. In addition to C, other elements in soil organic matter pass a similar cyclic existence.

---

[3]The strength of bonding between hydrogen and oxygen in phenolic and carbolylic groups in organic matter is not nearly as great as in inorganic hydroxyl (OH^-) ions. Conditions that allow the displacement of hydrogen from the organic groups are discussed in Chapter 6.

Organisms and organic matter in soils

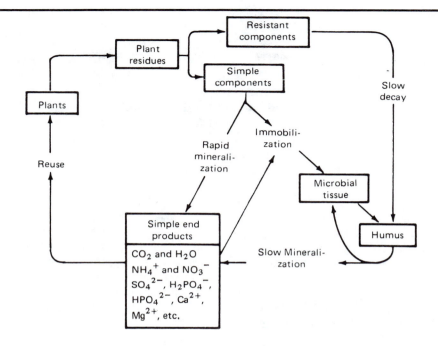

Figure 5.6.
The organic matter cycle in soils.

The organic matter cycle is shown in generalized form in figure 5.6. The diagram indicates that the simpler nonresistant compounds in plant residues are broken down rapidly, with the constituent elements appearing as mineralized end products or as new microbial tissue. Alteration products of the more resistant components, along with resynthesized substances of microbial origin, persist for longer periods and make up the humus fraction of the soil. Eventually, however, these intermediate products are transformed into simple molecular or ionic forms, for no organic substance remains indefinitely in the soil environment.

The organic matter content of soils

Knowledge of the organic matter content of soils often allows one to make meaningful judgments about their nutritional or physical state. Because of this, it is a common practice to determine the level of organic matter in soil samples sent to laboratories for fertility or other analyses. The amount of organic matter reflects primarily the humus content of the soil.

Expressing the organic matter content of soils

The amount of organic matter in soils is customarily indicated on a weight-percentage basis. In theory the content can be determined as the loss in weight of a sample of soil when its organic constituents are quantitatively removed. In practice, however, the separation of inorganic and organic soil components is difficult and often far from precise. As a consequence, estimates of the organic matter content are usually obtained by simpler, less direct means.

Indirect methods of determining organic matter contents are based on the analysis of a single elemental component common to all organic substances in the soil. The element most widely used is C. Several reasons account for this: (1) as an essential part of all organic compounds, C is a major component of soil organic matter, (2) it comes closer than other elements to comprising a constant fraction of the organic materials contained in a wide range of soils, and (3) its determination is comparatively easy. Analyses for N are sometimes used for estimating organic matter contents, but they are less dependable because this element makes up a rather small part of soil organic matter and is a more variable component than C. The N content of soil organic matter often ranges near five percent but may run as low as half this concentration. In comparison, C usually makes up from 55 to 60 percent of the weight of the soil organic fraction, with 58 percent being the accepted average.

Climate and soil organic matter content

Climate is a major factor determining how much organic matter accumulates over long periods of soil development. Climate exerts this control because it influences both the addition and decay of plant residues. Other factors may modify, or even override, the climatic effect at times. The state of aeration is one such factor, for it can also strongly influence the rates of residue return and decay in soils. For this reason, comparisons to show the relationship between climate and organic matter accumulation are limited to soils that are well aerated.

The general relationship between climate and soil organic matter accumulation is illustrated by data in figure 5.7, which are for three soils formed under quite different climatic conditions. The two diagrams on the left in the figure are for soils that have developed in association with grassland vegetation. One is for a soil that has formed under semi-arid conditions in Utah; the second for a soil found under the more moist, cooler climate of Minnesota. The higher organic matter content of the Minnesota soil, as compared to that of the Utah soil, reflects the return of greater quantities of residue as well as a slower rate of decay caused by moist, cool soil conditions. The depth of organic matter accumulation also differs between these two soils, this difference resulting in an A horizon that extends almost to a 50–cm depth in the Minnesota soil but to less than a 25–cm depth in the Utah soil.

The third diagram in figure 5.7 is for a forest soil of Indiana. Although, at maximum, the organic matter content of this soil approaches that of the Minnesota grassland soil, little organic matter has accumulated below the surface 5 to 10 centimeters. The organic matter in the surface few centimeters is assumed to come from mixing surface litter with the mineral soil by burrowing organisms.

Cultivation and the soil organic matter content

The accumulation of organic matter is important to the development of most mineral soils. The rate of accumulation is comparatively rapid at first, but as shown in figure 5.8, it decreases with time until an essentially constant level is attained. This level then tends to persist up to the time the soil is first

Figure 5.7.
The distribution of organic matter in three soils formed under differing climatic-vegetational regimes. The general type of vegetation and the approximate average annual precipitation at each site are: Utah, short grasses, 50 cm; Minnesota, tall grasses, 75 cm; Indiana, deciduous forest, 100 cm. (Data from Soil Survey Staff, *Soil Classification: A Comprehensive System*, Washington, D.C.: Government Printing Office, 1960).

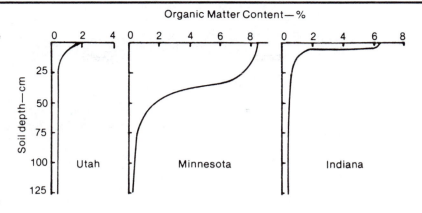

Figure 5.8.
The relative change in the content of soil organic matter with time. Introduction of cultivation at time **t** alters environmental conditions, usually with the result that it lowers the level of organic matter established during an earlier period of soil development.

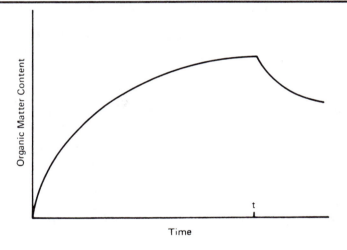

cultivated, unless other major changes in the environment occur. When cultivation is introduced, the organic matter content normally declines (see fig. 5.8), the rate and extent of reduction depending on how well the soil is managed.

One reason why cultivation lowers the organic matter content of soils is because of the loss of a permanent plant cover normally associated with the virgin state. Two things happen in the absence of a continuous plant cover that affect the level of organic matter in the soil. First, the average temperature is raised. With an increase in temperature, both added residues and the residual humus in the soil tend to disappear more rapidly as a result of decay. Second, erosion is increased, with a surface soil of the highest organic matter content being the first to go. We may add to these the effect of harvesting, for the removal of plants or plant parts at this time often results in the return of very little residue to the soil, save perhaps that left by roots.

From the foregoing, one may conclude that management practices capable of minimizing erosion while providing a maximum return of residues are fundamental to the maintenance of a high level of organic matter in cultivated soils. Important to these practices is fertility control. Sustaining abundant plant growth through the use of fertilizers not only provides more cover for erosion control but also increases the production of roots and other unharvested plant parts for retention by the soil.

Summary

Soils support a wide variety of life, including both macro- and microplants and animals. The existence of these organisms depends basically on the availability of carbon and other nutrients supplied by residues from a plant cover. Utilization of these residues results in their decay and conversion to humus.

Numerous associations exist between plants and other organisms that live in the soil environment. Some are harmful, as in the case of parasitic or infectious associations, or where needed plant nutrients are immobilized by decay organisms. Others are beneficial, however. Examples are mineralization of organically bound nutrients by decay organisms, and the increased supply of nitrogen and other nutrients by symbiotic associations established between plants and certain microorganisms.

Microorganisms are largely responsible for the decay of plant residues and their conversion to humus. Decay is a spontaneous process that depends on the type of residue, its nutrient content, and the supply of water and oxygen in the soil. The rate of decay, which is most often limited by the availability of nitrogen, can be approximated reasonably well from the carbon-nitrogen ratio of the decaying residue.

Decay, along with erosion, reduces the organic matter content of soils, unless the organic matter is replenished by periodic additions of fresh residues. In most agricultural soils, the return of organic matter in residues does not keep pace with its loss through decay and erosion, so that over time the tendency is for the organic matter content of these soils to decline.

Review questions

1. Outline the principal effects of burrowing insects and animals on the physical and chemical properties of soils.
2. Compare algae with the heterotrophic microorganisms (fungi, actinomycetes, and bacteria) as to source of energy carbon and role in the accumulation of humus in soils.
3. Define these terms: mycorrhiza, symbiotic, heterotrophic, autotrophic, and anaerobic.

4. Discuss the relationship between climate and the accumulation of organic matter in soils, indicating why maximum organic matter accumulation occurs in soils of the wetter, cooler regions.
5. Compare the ease of decomposition of sugars, starch, cellulose, protein, and lignin by common decay organisms, and explain any differences in their relative decay rates.
6. Why does decay of soil organic compounds tend to reduce the carbon and organic matter contents of soils.
7. Define mineralization and immobilization, and explain the importance of these two processes to plant growth.
8. Explain why maturing (aging) of plants slows the rate at which their residues decay once added to the soil.
9. Why do carbon/nitrogen ratios provide an index to the rate of plant-tissue decay?
10. Describe humus, and identify its major effect on the chemical and physical properties of soils.
11. Why is organic carbon used as the basis for determining the organic matter content of soils?

Selected references

Alexander, M. *Introduction to Soil Microbiology.* New York: John Wiley & Sons, Inc., 1977.

Broadbent, F. E. The Soil Organic Fraction. *Advances in Agron.* 5:153–83. New York: Academic Press, Inc., 1953.

————. Organic Matter. In *Soil, 1957 Yearbook of Agriculture*, pp. 151–57. Washington, D.C.: Government Printing Office, 1957.

Clark, F. E. Living Organisms in the Soil. In *Soil, 1957 Yearbook of Agriculture*, pp. 157–65. Washington, D.C.: Government Printing Office, 1957.

Russell, E. W. *Soil Conditions and Plant Growth*, pp. 135–264. London: Longmans, Green and Co., Ltd., 1974.

6

The mineral fraction of soils

The mineralogical make-up of soils is highly variable, ranging from relatively coarse, physically weathered materials to fine soils containing essentially only secondary clays produced by chemical weathering. The kinds of minerals present are prime determinants of both the physical and chemical characteristics of the soil. An understanding of the basic properties of minerals is therefore important to understanding the contribution they make to the properties and behavior of soils.

Fundamental properties of mineral matter

Differences among minerals depend primarily upon two fundamental characteristics: *chemical composition* and *structure*. The composition of a mineral is determined by the kinds and proportions of ions it contains, whereas the structure relates to the spatial arrangement of these ions with respect to each other. Changes in composition and structure from one mineral to another are accompanied by variation in such properties as color, hardness, crystal shape, and cleavage, the last of these being the tendency to split or break in a given direction. Since these properties are easily detected and measured, they provide the principal means for the identification of minerals in the field.

Mineral matter may be either *crystalline* or *amorphous* (Gr., *amorphous*, without form). Crystalline materials have a structure in which ions are arranged in an orderly and repeated spatial pattern. These materials form slowly, which allows the well ordered ionic pattern, and they develop as crystals with a shape that reflects the arrangement of ions in the structure.

Amorphous materials lack a well organized structure; they form too quickly to allow an orderly distribution of ions to develop. Glassy rocks that form in rapidly cooling lava are an example. Amorphous materials of this type are ordinarily less stable (more soluble) against chemical weathering than are crystalline minerals. Amorphous silica and crystalline quartz illustrate this point; both have the same composition, expressed as SiO_2, but amorphous silica is some 10 times more soluble than quartz.

Table 6.1. The radius and coordination number of the important cations in silicate minerals. The radius of oxygen is shown for comparison.

Ion	Radius[a]	Coordination number
	1×10^{-10}m	
O^{2-}	1.45	—
Si^{4+}	.41	4
Al^{3+}	.45	4 or 6
Fe^{2+}	.75	6
Mg^{2+}	.65	6
Ca^{2+}	.94	8
Na^+	.98	8
K^+	1.33	8 or 12

[a]Expressed in meters as a recommended SI unit. A length of 1×10^{-10} meters is equal to one Angstrom unit.

Oxygen in mineral structures

As noted in Chapter 2, the principal minerals in the earth's crust are made up mostly of oxygen ions in combination with the seven cations listed in table 6.1. Because of their abundance and large size, oxygens determine the general structure of these minerals; they occur in closely packed arrays with the cations fitted in the spaces between. In the majority of cases, the oxygens completely enclose the individual cations. However, since the cations differ in size, the number of oxygens needed to enclose them varies. This number (*coordination number*) is shown for the principal cations in minerals in table 6.1, and it ranges from four for Si^{4+}, and sometimes for Al^{3+}, to twelve for K^+ ions.

Oxygens occur in fixed types of spatial arrangements if they are in four-fold or six-fold coordination with cations. In four-fold coordination, the oxygens form a four-sided *tetrahedron*, as pictured in diagrams A and B of figure 6.1. The enclosed, or *central*, cation in tetrahedra is either Si^{4+} or Al^{3+}. When in six-fold coordination, oxygens form an eight-sided *octahedron*, as shown in diagrams C and D of figure 6.1, and they enclose either Al^{3+}, Fe^{2+}, or Mg^{2+} as the central cation. If larger Ca^{2+}, Na^+, or K^+ ions are present, they occur at the center of clusters of tetrahedra, with each tetrahedron supplying a part of all the oxygens needed for eight-fold or twelve-fold coordination. In this arrangement, the larger cations provide a center of positive charge that attracts and holds the clusters of tetrahedra together. The cations occurring in this position, that is, outside or between neighboring tetrahedra, are called *accessory cations* to distinguish them from the central cations within tetrahedra or octahedra. Figure 6.2 illustrates one of the ways in which accessory cations can be positioned with respect to neighboring tetrahedra.

Charge relationships in mineral structures

Stability in minerals requires their structure to be electrically neutral; that is, the negative charge of the oxygens in the structure must be exactly balanced by the positive charge of the cations. The source of positive charge may be either the central cations in tetrahedra or octahedra or accessory cations between neighboring tetrahedra. In some minerals, hydrogen (H^+) ions neutralize part of the oxygen charge. These hydrogens occur in com-

Figure 6.1.
Diagrams of closely packed oxygens in (A) a tetrahedron and (C) an octahedron, with (B,D) expanded views to show the location of the central cations within each structural unit.

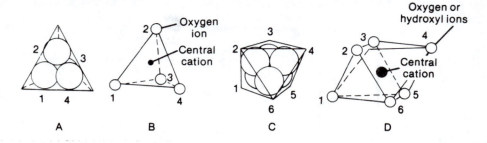

Figure 6.2.
An accessory cation positioned between neighboring tetrahedra. This cation is in six-fold coordination with oxygens, and by exerting a pulling force on them, binds the two tetrahedra together.

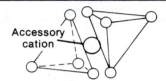

Figure 6.3.
Diagrams showing a single oxygen shared between two tetrahedra (left) and two oxygens shared between two octahedra (right).

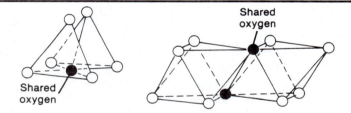

bination with oxygens as monovalent hydroxyl (OH^-) ions. Hydroxyl ions have the same size as oxygens and may take the place of oxygens in the octahedra of some minerals.

The total need for cations as a source of positive charge depends in part on the degree to which oxygens are *shared* in mineral structures. The sharing of oxygens is illustrated in figure 6.3, and it is the result of the oxygen occurring as a part of more than one tetrahedron or octahedron. The charge of the oxygen is therefore neutralized by (shared between) the central cations of the neighboring units. Sharing reduces the total number of oxygens required for a given number of tetrahedra or octahedra and, as a consequence, the net negative charge needing neutralization. The sharing of oxygens also strengthens mineral structures, because it increases the degree to which oxygens are bonded directly to the central cations of tetrahedra and octahedra. The central cations are principally Si^{4+} and Al^{3+}, either of which is small and of high charge (valence). In general, the smaller the cation and the higher its valence the stronger the bond between it and the oxygens.

Ion substitution is the replacement of the *normal* central cation of tetrahedra and octahedra by some other kind of cation at the time of mineral formation. Ion substitution is attributed to a high concentration of substituting ions in a mineral-forming medium so as to increase their chance of entering the mineral structure in place of the normal cation. The pattern of substitution is relatively consistent; Al^{3+} is the only important ion that substitutes for Si^{4+} in tetrahedra, and Fe^{2+} and Mg^{2+} are the usual cations to substitute for Al^{3+} in the octahedra of aluminosilicates.

Ion substitution upsets the electrical balance in mineral structures, for if it did not occur, the presence of the normal Si^{4+} or Al^{3+} ions would result in electrically neutral structures. Electrical imbalance occurs because the valence of the substituting ions is lower than that of the ions they replace. Each substitution by Al^{3+} for Si^{4+} or by Fe^{2+} or Mg^{2+} for Al^{3+} results in the loss of a positive charge and leaves an unsatisfied negative charge that must be neutralized to maintain a neutral system. Neutralization of the excess negative charge is accomplished by the inclusion of accessory cations in the structure. In the primary minerals, calcium, sodium, and potassium are the principal accessory cations that neutralize the negative charge resulting from ion substitution. In secondary clays, it is by cations adsorbed to the clay particle surfaces.

The structure of primary minerals

The majority of primary minerals consist of oxygens in tetrahedra, and they range from compounds in which no sharing of oxygens occurs to those in which all tetrahedral oxygens are shared. Primary minerals with the simplest structure are the olivines, which are ferromagnesian minerals. In the olivines, oxygens occur in independent tetrahedra linked together by Fe^{2+} and Mg^{2+}. These two elements, as accessory cations, are located between the tetrahedra as illustrated in figure 6.2. The tetrahedra function as SiO_4^{4-} anions. As indicated by the formula for magnesium olivine, Mg_2SiO_4, there must be two divalent cations for each tetrahedron present in the structure.

Primary minerals other than the olivines contain tetrahedra linked together through shared oxygens. The tetrahedra in these minerals occur in chain- or sheetlike arrangements, as shown in figure 6.4, or in a three-dimensional *framework structure*, which is described below. The two most abundant ferromagnesian minerals, augite and hornblende, contain tetrahedra in chains. In these minerals, the chains lie parallel to each other and are tied together in bundles by Fe^{2+} and Mg^{2+} ions. Augite contains single tetrahedral chains (model A, fig. 6.4), and hornblende, double chains (model B, fig. 6.4). Tetrahedra linked through oxygen sharing in a sheetlike arrangement (model C, fig. 6.4) occur in mica minerals, which are discussed shortly.

Two primary mineral groups, quartz and the feldspars, have a framework structure. The framework structure consists of an interlinked, three-dimensional network of tetrahedra tied together through the complete sharing the tetrahedral oxygens. In quartz (see fig. 6.5), every tetrahedron contains a Si^{4+} ion, which gives a ratio of one Si^{4+} for every two O^{2-} ions and a

Figure 6.4.
Tetrahedral linkage
through shared
oxygens to form
(A) single chains,
(B) double chains, and
(C) tetrahedral sheets.

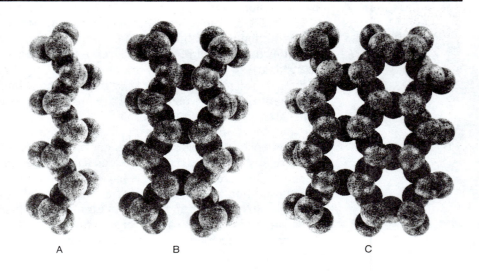

A   B   C

Figure 6.5.
A large quartz
fragment. Note its
uneven edges and
surfaces, the result of
irregular breakage, or
fracture. Quartz sand
grains commonly
show a similar
irregular shape.

formula of SiO_2 for quartz. In the feldspars, either one-fourth or one-half of
the tetrahedra in the framework contain Al^{3+} rather than Si^{4+}. To compensate for the loss of positive charge due to the substitution of Al^{3+} for Si^{4+},
calcium (Ca^{2+}), sodium (Na^+), or potassium (K^+) are included as accessory
cations in the feldspar structures. In calcium feldspar, $CaAl_2Si_2O_8$, half of the
tetrahedra contain Al^{3+} instead of Si^{4+}; in sodium or potassium feldspar,
$NaAlSi_3O_8$ or $KAlSi_3O_8$, one out of every four tetrahedra contain Al^{3+} rather
than Si^{4+}.

The last primary minerals to be considered are the micas. These minerals are distinctive because they contain oxygens in octahedra as well as
in tetrahedra, with both occurring in a sheetlike arrangement. As illustrated
in figure 6.6, the mica structure contains thin, flat units, called *platelets*, in

Figure 6.6.
The structure of muscovite mica in which two tetrahedral sheets enclose a single octahedral sheet to form a platelet. The upper four tetrahedra are a part of a second platelet. Potassium ions occur between and bind adjacent platelets together. Ions in the structure are shown at the right, with those in parentheses being substituting ions. Biotite mica has the same structure, but it contains iron and magnesium rather than aluminum at the center of the octahedral sheet.

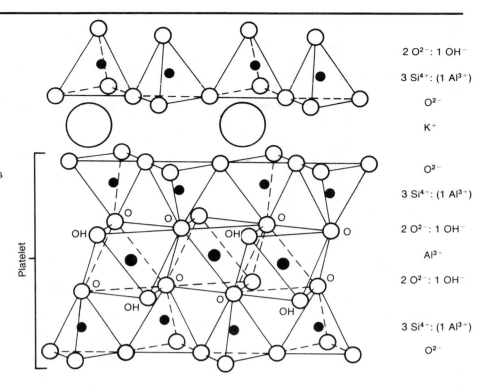

$2 O^{2-} : 1 OH^-$

$3 Si^{4+} : (1 Al^{3+})$

O^{2-}

K^+

O^{2-}

$3 Si^{4+} : (1 Al^{3+})$

$2 O^{2-} : 1 OH^-$

Al^{3+}

$2 O^{2-} : 1 OH^-$

$3 Si^{4+} : (1 Al^{3+})$

O^{2-}

which two tetrahedral sheets enclose a single octahedral sheet. The tetrahedral and octahedral sheets are linked together through shared oxygens. Unshared oxygens are combined with H^+ ions and occur as octahedral OH^- ions. Because of the ratio of two tetrahedral sheets to each octahedral sheet in the platelets, the micas are called *2:1 layer* minerals. Several important silicate clays have this type of structure.

The diagram in figure 6.6 is for muscovite (white mica). It is a colorless aluminosilicate mineral with only Al^{3+} ions in the central octahedral sheet of the platelets. Biotite (black mica) has Fe^{2+} and Mg^{2+} in the octahedral sheet, which makes it a ferromagnesian mineral. The divalent iron (Fe^{2+}) ions cause the black color in biotite.

About one-fourth of all tetrahedra in micas contain Al^{3+} as a substituting ion in place of Si^{4+}. Although the central octahedral sheet in the platelets is neutral, substitution causes the tetrahedral sheets to have a residual negative charge. In the micas, this charge is neutralized by potassium (K^+) ions positioned in the *interlayer space* between neighboring platelets (see fig. 6.6). The K^+ ions occur in twelve-fold coordination, for they are enclosed by two six-oxygen rings, one from each platelet.[1] As centers of positive charge, the K^+ ions bond the platelets together, but with less force than that due to oxygen sharing within the platelets. As a consequence, micas can be split rather

[1]These rings are formed by the bottom layer of oxygens in model C of figure 6.4.

Figure 6.7.
A fragment of muscovite mica. The upper surface is flat, the result of breakage (cleavage) parallel to the platelets making up the mica crystal. The lower edges of the fragment give evidence of the layered, or platelike, nature of the mineral.

easily into thin, platelike cleavage fragments, as shown in figure 6.7. The splitting takes place along the planes of relatively weak potassium bonding.

Structure, composition, and weathering of primary minerals

The rate at which primary minerals succumb to chemical weathering reactions relates to a degree to their composition and structure. The main features that affect weathering include the silicon content, the extent of oxygen sharing, and the amount of iron in the structure. The content of iron is a factor because, as a divalent ion, it is subject to oxidation and a change in valence to the trivalent form. Oxidation of iron upsets the electrical balance in a mineral and reduces the stability of its structure.

Oxygen sharing increases mineral stability because it reduces the need for accessory cations, which provide relatively weak bonds in minerals. There is no oxygen sharing in the olivines, the bonding between tetrahedra being totally through accessory cations. Because of this, and because of the presence of divalent Fe^{2+}, olivines are the most easily weathered of all primary minerals. Oxygen sharing is greater in the chain minerals, augite and hornblende, and they are more resistant than are the olivines. Still more sharing occurs in the micas, and they show a fairly high degree of weathering resistance. However, potassium can be rather easily removed from the interlayer space in micas, but this does not affect platelet stability. Of the two micas, biotite is least resistant, which can be explained by the presence of Fe^{2+} as a principal ion in the platelet structure.

All oxygens are shared in the framework minerals, which includes the feldspars and quartz. Differences in the weatherability of these minerals reflect primarily the extent to which Al^{3+} substitutes for Si^{4+}. Substitution is greatest in calcium feldspar, less in sodium and potassium feldspars, and does not occur in quartz. Among these minerals, calcium feldspar is the least resistant to weathering and quartz the most. Indeed, quartz is the most stable of all the more abundant primary minerals.

Differences as described above explain variability among igneous and similar rocks in their ability to resist chemical breakdown. Granite, for example, is more resistant than basalt. This relates to the high content of resistant potassium feldspar and quartz in granite (see fig. 2.2, page 16), and the high content of calcium feldspar and ferromagnesian minerals, such as olivine, augite, and hornblende, in basalt. The fact that granite has a coarser texture than basalt also contributes to its higher stability in a weathering environment.

Clay minerals

Clay is a general term used to identify mineral particles of submicroscopic size. Both primary and secondary compounds occur as clay-sized particles, but it is usual for those of secondary origin to be dominant. Serious error is not likely if one assumes the clay in a soil to consist principally of secondary mineral compounds.

When considered on the basis of chemical composition, clay minerals consist principally of (1) silicates, (2) hydrous oxides (sesquioxides), and (3) a variable group of poorly structured, amorphous materials. The more abundant silicate clays are layer-type minerals, and they tend to dominate the clay fraction of soils in cooler or drier regions. Intensely weathered soils of hot, wet regions are more inclined to have hydrous oxides as the dominant clay. The hydrous oxides, which are largely compounds of iron and aluminum, have many of the same characteristics and are normally discussed as a single group. The silicates, on the other hand, comprise a highly diverse group, so they are normally discussed individually. There are five major types of silicate clay: (1) the smectites, with montmorillonite as the principal type, (2) kaolinite, (3) vermiculite, (4) illite, and (5) chlorite.

Montmorillonite

This clay mineral forms by crystallization from solutions high in soluble silica (SiO_2) and magnesium. It is a common product of mineral weathering under conditions where ions released in soluble form are not easily lost by leaching. Montmorillonite has a 2:1 layer structure, as illustrated in figure 6.8. All tetrahedra in the platelets contain Si^{4+} ions. Aluminum is the normal ion in the central sheet, but about one-eighth of the octahedra contain Mg^{2+} as a substituting ion for Al^{3+}. The negative charge caused by substitution is neutralized by various hydrated cations adsorbed to the surfaces of the platelets.

Much like potassium in the micas, cations adsorbed by montmorillonite attract neighboring, negatively charged platelets and hold them together. The force of bonding is not nearly as strong as in the micas and depends on the amount of water present. In dry montmorillonite, the bonding force is relatively strong, apparent in the ability of this clay to form very hard clods on drying. When water is added to dry montmorillonite, however, it is readily drawn into the interlayer space between platelets and causes the clay to swell dramatically. Because of this behavior, montmorillonite is called an *expanding clay*.

Figure 6.8.
The structure of montmorillonite, a 2:1 layer clay. Magnesium, shown in parentheses, occurs as a substituting ion in the octahedral sheet, which results in a negatively charged platelet. The upper four tetrahedra are a part of a second platelet. Cations in the interlayer space neutralize the negative charge of the platelets while bonding them together, although weakly.

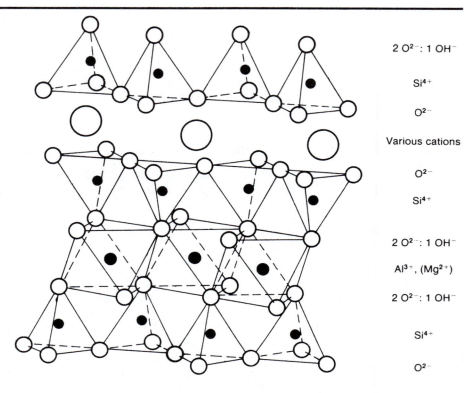

2 O^{2-} : 1 OH$^-$

Si^{4+}

O^{2-}

Various cations

O^{2-}

Si^{4+}

2 O^{2-} : 1 OH$^-$

Al^{3+}, (Mg^{2+})

2 O^{2-} : 1 OH$^-$

Si^{4+}

O^{2-}

Figure 6.9.
A photomicrograph of montmorillonite particles generally no more than a single platelet thick. Several are folded back over themselves, showing their substantial flexibility. The angular shape indicates that the particles are crystalline. The line at the right is 1 micron long. (From Nixon, H. L. and Weir, R. H., *Mineralogy Magazine* 31:413–16, 1957.)

When wet, montmorillonite can be easily separated, or dispersed, into very thin, platelike particles (see fig. 6.9). Because of this, montmorillonite exhibits very extensive surface for the adsorption of water and ions. The quantity of ions adsorbed is, of course, a function of the magnitude of the negative charge in the platelets, but compared to other clays, the total negative charge is fairly high. The amount of water that can be drawn in between montmorillonite platelets is considerably higher than for other clays, however, evident in the ability of soils high in this mineral to swell on wetting and to shrink and crack extensively on drying (see fig. 4.3).

Kaolinite

This mineral is the most important member of a group of layer silicates made up of platelets in which only one tetrahedral sheet is linked to the octahedral sheet. Because of this, kaolinite is classified as a 1:1 type layer silicate. The structure of kaolinite is shown in figure 6.10. As may be determined from this figure, the two surfaces of a 1:1 mineral platelet are formed by different ions; one consists of tetrahedral oxygens and the other of OH^- ions belonging to the octahedral sheet. When the 1:1 platelets occur in stacks, the OH^- ions of one platelet lie next to and in close contact with the O^{2-} layer of its neighbor. Because of this arrangement, the positive charge of the H^+ ions in the OH^- layer exerts a strong attraction for the negative oxygens of the neighboring platelet. In this way the platelets of kaolinite are tightly bound together.

The strong bond afforded by H^+ encourages the growth of comparatively large kaolinite particles consisting of many individual 1:1 platelets (see fig. 6.11). It also results in a nonexpanding mineral, for kaolinite particles are unable to absorb water into the interlayer position between adjacent platelets.

Kaolinite has a small negative charge, and although the exact cause is not known, it is assumed to be due to a low level of ionic substitution within the platelets. The negative charge is countered by cations adsorbed on the outermost surfaces of the particles. Internal adsorption of cations is not possible in kaolinite because it is nonexpanding.

The nonexpanding character of kaolinite explains the failure of soils high in this clay to swell or shrink much on wetting and drying. It also accounts for the widespread use of kaolinite as a pottery clay, since the shape into which it is molded when moist and pliable is retained on drying (see fig. 6.12).

As with montmorillonite, kaolinite is formed by crystallization from solution. Kaolinite has a lower silicon content, however, so it can form where the concentration of soluble silica is less than that required for montmorillonite formation.

Figure 6.10.
The structure of a kaolinite platelet consisting of one octahedral and one tetrahedral sheet. The upper four tetrahedra are a part of a second platelet. The upper surface of the platelet is a layer of hydroxyl (OH⁻) ions that lie immediately adjacent to the outer oxygen (0²⁻) layer of the platelet just above. The H⁺ ions in the hydroxyls exert a pull on the negatively charged oxygens above and thereby bind the platelets to each other.

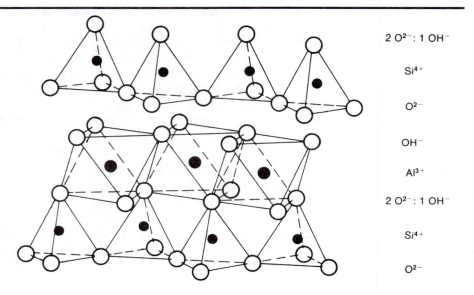

$2\ O^{2-} : 1\ OH^{-}$

Si^{4+}

O^{2-}

OH^{-}

Al^{3+}

$2\ O^{2-} : 1\ OH^{-}$

Si^{4+}

O^{2-}

Figure 6.11.
Photomicrograph of kaolinite clay (3000X magnification) showing blocklike particles formed from thick stacks of tightly bound 1:1 platelets. The hexagonal shape of the particles attests to the crystalline nature of kaolinite. (Photo courtesy of W. D. Keller, Department of Geology, University of Missouri, Columbia.)

The mineral fraction of soils

Figure 6.12.
A potter deftly transforms a shapeless mass of kaolinite clay into a visually pleasing and useful vase. On drying, the vase will have the same size and shape as when wet.

Vermiculite

This clay has the basic 2:1 layer structure of primary mica minerals; that is, it contains either Al^{3+} or Mg^{2+} and Fe^{2+} as normal octahedral ions, and tetrahedral sheets in which Al^{3+} occurs as a substituted ion in place of some of the Si^{4+}. Vermiculite differs from the micas in that it contains hydrated cations rather than unhydrated K^+ in the interlayer space. The weak bonding afforded by these ions allows vermiculite to expand on wetting, a property not displayed by micas. Expansion is less than in montmorillonite, however. This can be explained by the magnitude and site of negative charge, for it is larger and nearer the platelet surfaces and the interlayer cations in vermiculite than in montmorillonite. The larger negative charge also accounts for the adsorption of more cations per unit of platelet surface area in vermiculite than in montmorillonite.

Unlike montmorillonite and kaolinite, vermiculite does not form by crystallization from solution but, instead, is formed by *alteration*, or the selective replacement of ions in a structure without destroying the structure. Vermiculite is produced by the replacement of interlayer potassium in micas and similar minerals by hydrated cations from a weathering solution. Ions must be taken in to replace potassium, otherwise the system would become electrically unbalanced.

Illite

This term identifies a group of clay minerals with a 2:1 layer structure and sufficient interlayer K^+ to limit expansion on wetting. The K^+ content of illite is less than that of micas, however, which means that it is inadequate to neutralize all the negative charge of the platelets. Charges not neutralized by K^+ are countered by hydrated cations.

It is believed that illite formation takes place in K-rich sediments subjected to low-temperature metamorphism. In addition to K, these sediments also contain expanding clays such as vermiculite or montmorillonite. The process of illite formation is initiated as K^+ replaces some of the interlayer cations in these clays, and is completed when heat and pressure causes the dehydration and collapse of the clays into a nonexpanded form. Illitic clays are widespread in soils.

Chlorite

As with illite, the term chlorite embraces a range of minerals that have certain outstanding characteristics in common. All have a basic 2:1 layer structure, and they are usually nonexpanding. Chlorites differ from other 2:1 layer minerals in one unique respect, however; they contain a stable, positively charged octahedral sheet rather than adsorbed cations in the interlayer space (see fig. 6.13). The octahedral sheet consists of two layers of OH^- ions that enclose either Mg^{2+} or Al^{3+} as the central octahedral cations. The number of Mg^{2+} or Al^{3+} ions exceeds that required to neutralize the OH^- ions and is the cause of the positive charge of the sheet. By virtue of its positive charge, the interlayer sheet neutralizes the negative charge of the 2:1 platelets and binds them to each other in a comparatively rigid arrangement (see fig. 6.14). Because chlorite contains two octahedral sheets, it is called a 2:2 layer mineral.

Sometimes, octahedral materials in chlorite neither totally fill the interlayer space between platelets nor completely neutralize the negative charge of the platelets. This unsatisfied charge is neutralized by various cations adsorbed to the particle surfaces from the solution phase.

Figure 6.13.
Structure of chlorite
showing the location
of an octahedral sheet
in the interlayer space
between platelets. An
excess positive
charge in the
interlayer octahedral
sheet neutralizes the
negative charge of the
platelets while binding
them together. The
negative charge of the
platelets results from
Al^{3+}, shown in
parenthesis,
substituting for Si^{4+} in
a portion of the
tetrahedra.

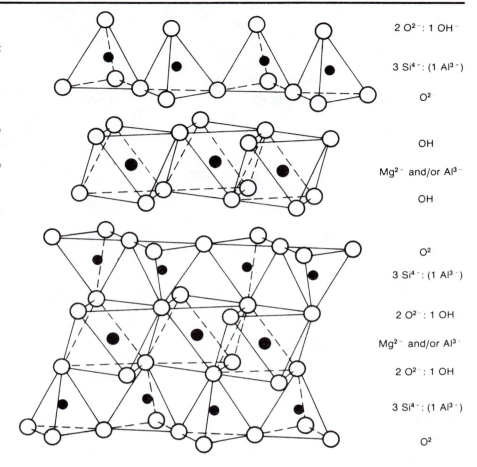

$2\ O^{2-} : 1\ OH^-$

$3\ Si^{4-} : (1\ Al^{3-})$

O^{2-}

OH

Mg^{2-} and/or Al^{3-}

OH

O^2

$3\ Si^{4-} : (1\ Al^{3-})$

$2\ O^{2-} : 1\ OH$

Mg^{2-} and/or Al^{3-}

$2\ O^{2-} : 1\ OH$

$3\ Si^{4-} : (1\ Al^{3-})$

O^2

Chlorites in soils come from two sources. One is by the accumulation of octahedral material, through chemical precipitation, in the interlayer space of pre-existing expanding clays, such as vermiculite or montmorillonite. Here the octahedral material ordinarily consists of Al^{3+} held in six-fold coordination between two layers of OH^- ions. A second method of chlorite formation is the metamorphic transformation of biotite mica or other ferromagnesian minerals in rocks such as basalt. Chlorite formed in this way will have interlayer octahedral material that is rich in magnesium, and it becomes a part of the soil mineral complex through rock weatherng.

Figure 6.14.
A transmission electronmicrograph of a thin chlorite particle (side view) showing a rigid, parallel arrangement of platelets. Spacing between the parallel lines equals the thickness of a single 2:2 unit. (Photo courtesy of J. L. Brown, Georgia Institute of Technology, Atlanta, and M. L. Jackson, University of Wisconsin, Madison.)

Hydrous oxides

Hydrous oxides (sesquioxides) are widely distributed compounds of iron and aluminum. They occur in greatest abundance where weathering has been extensive, mainly as hematite, Fe_2O_3, goethite, $Fe_2O_3 \cdot H_2O$, and gibbsite, $Al(OH)_3$. Iron in the oxides is trivalent (Fe^{3+}), as opposed to its occurrence in divalent (Fe^{2+}) form in the primary ferromagnesian minerals.

Hydrous oxides may be either crystalline or amorphous, the latter probably formed by very rapid precipitation reactions. Crystalline forms consist of layers of O^{2-} or OH^- ions organized in an octahedral pattern. The Fe^{3+} or Al^{3+} ions are positioned within the octahedra, the number depending on the number of positive charges required for electrical neutrality. For example, the number of Al^{3+} ions in $Al(OH)_3$ is only half the number of Fe^{3+} ions in Fe_2O_3, for half the oxygen charge in $Al(OH)_3$ is neutralized by H^+.

The hydrous oxides are invariably products of weathering. They are very stable minerals and, once formed, seem to persist almost indefinitely in most environments. Because of the highly resistant nature of hydrous oxides, they tend to accumulate where weathering has been long and intense. Their presence in soils is indicated by yellow, brown, or red colors attributable to iron oxides. The color of gibbsite, which is white, is usually masked by the iron oxides.

Hydrous oxides lack a negative charge and, therefore, do not retain cations in adsorbed form. This is unfortunate, because retention of cations

Table 6.2. Some important properties of clay minerals.

Clay type	Properties					
	Structural form	Degree and site of substitution	Magnitude of platelet charge	Means of platelet bonding	Strength of bonding and implication	Usual source of clay
Kaolinite	1:1 layer	Small; site unknown	Very low	H^+ from structural OH^- ions	Strong bond; nonexpanding; forms large particles	Precipitation at weathering site
Illite	2:1 layer	Same as montmorillonite or vermiculite	High to moderately high	Mostly by interlayer K^+ ions	Strong; nonexpanding mineral	Alteration of vermiculite or montmorillonite
Vermiculite	2:1 layer	Extensive; Al^{3+} for Si^{4+} in tetrahedra	High	Various interlayer cations	Weak; expands on wetting	Alteration of mica or illite
Montmorillonite	2:1 layer	Moderate; Mg^{2+} for Al^{3+} in octahedra	Moderately high	Same as vermiculite	Weak; expands more than vermiculite	Precipitation at weathering site
Chlorite	2:2 layer	Moderate to extensive; site varies	Moderate to high	Positively charged octahedral layer	Strong; nonexpanding	Metamorphic rock; alteration of vermiculite or montmorillonite
Hydrous oxides (Sesquioxides)	Oxygen and hydroxyl ions in octahedra	None	Does not apply	Does not apply	Does not apply	Precipitation at weathering site

such as calcium, magnesium, and potassium by clay is important to plant nutrition, and minerals supplying these nutrients are generally absent in soils where hydrous oxides are the dominant clays.

Amorphous clays   In addition to the crystalline minerals discussed above, the clays of many soils contain substantial proportions of poorly structured, amorphous compounds. Principally, these materials consist of silica (SiO_2), hydrous oxides, and a less well defined secondary substance known as allophane, the latter having an approximate empirical formula of $Al_2O_3 \cdot SiO_2 \cdot nH_2O$. The principal occurrence of allophane is as a product of rapid weathering of volcanic materials in tropical or semitropical regions. Allophane has a high internal negative charge of unknown origin, which allows it to adsorb large numbers of cations from solution.

Clays form a varied group and, as a consequence, can affect soils differently depending on the kind and amount present. Some of the more important properties of these minerals are listed in summary form in table 6.2.

Factors of weathering and clay formation

Weathering is the result of exposure of mineral matter to an environment in which it is unstable. Through reactions or processes involved in weathering, minerals are transformed to products of greater stability. These may be ions that remain soluble and are lost from the system by leaching, or they may be new compounds, mainly clays, that form and persist in the weathering environment because they are stable under the prevailing conditions. The stability of these compounds is determined primarily by the composition of the weathering solution.

The composition of weathering solutions is strongly dependent on the kinds of minerals that are undergoing weathering. As original minerals dissolve and disappear, however, there is a change in the composition of the solution and in the kinds of secondary minerals that can form from it. Over long periods, then, the clays that form first eventually become unstable and are, in their own turn, decomposed and replaced by other secondary compounds of seemingly greater stability. An increase in stability is normally accompanied by a decrease in the structural and compositional complexity of the minerals.

Clay mineral sequences in weathering

Major changes that occur in the composition of the weathering solution over time are due principally to the relatively rapid leaching of elements such as calcium, magnesium, sodium, and potassium, and the more gradual, but inevitable, decrease in soluble silica. The loss of the first of these elements prevents the accumulation of readily soluble salts in weathering environments where leaching is a significant process. The gradual loss of soluble silica results in the formation and disappearance of clays in an ordered sequence, starting with those of highest silica content and ending with those containing no silica, namely, hydrous oxides.

In a long cycle of silicate mineral weathering, the first clays to form are 2:1 layer types. So long as magnesium is in ample supply, montmorillonite will be a major initial product of primary aluminosilicate mineral weathering. Vermiculite may form as mica or illite are altered by the displacement of interlayer K^+ ions. Iron oxides may also appear early, for they form directly from ferromagnesian minerals. Once formed, iron oxides seem to persist almost indefinitely in the weathering environment, which attests to their great stability under most conditions

As weathering proceeds in its normal course, 2:1 clays give way to simpler 1:1 layer types, mainly kaolinite. Eventually, this clay decomposes, the silica released from it is leached, and the aluminum transforms to a hydrous oxide, usually gibbsite. Gibbsite has roughly the same stability against chemical breakdown as do iron oxides. Together, these minerals tend to persist as the final products of long and intense silicate mineral weathering.

Soil geography and mineralogy

Weathering and clay synthesis are time-rate functions; that is, they depend on the intensity, or rate, of weathering reactions and the length of time these reactions have been in progress. The rate of weathering, while conditioned by the mineralogy and texture of the weathering material, is controlled principally by the climatic factors of temperature and precipitation. Silicate mineral weathering and clay synthesis are limited under either dry or cold conditions, but they proceed rapidly under hot, wet conditions, as in tropical regions. Normally, it is only within tropical environments that a full cycle of silicate mineral weathering is ever completed. The time span required for such a cycle is several tens of thousands of years even under intense weathering.

The kinds of clays contained in soils can often be related to the weathering factor of soil formation. For example, 2:1 clays tend to dominate soils of the drier and cooler regions of the North American continent, where weathering has been limited. Montmorillonite is widespread in these soils. Illite and vermiculite are also common, especially in young soils forming in parent materials derived from mica- and illite-bearing sedimentary or metamorphic rocks. Rocks of this type have been an important source of minerals in glacial deposits and, therefore, in soils forming in glacial parent materials. Many of these soils have been undergoing formation, and weathering, for as little as 10 thousand years, a short span so far as silicate mineral weathering is concerned.

Not all soils of the colder, drier regions are dominated by 2:1 type clays, for many of them have formed in parent material produced by a previous cycle of chemical weathering. For example, parent materials derived from sedimentary shales may contain large amounts of clay, and it may be of any type, including 1:1 clays or hydrous oxides.

Soils with clays high in kaolinite and hydrous oxides tend to be restricted to older landscapes in regions that are both warm and wet. For example, soils in the warmer, wetter southeastern part of the United States are frequently high in kaolinite. Hydrous oxides and aluminum-chlorite are also common. Iron oxides cause intense red or yellow colors in many of these soils. Clays of the 2:1 type are also present, particularly in soils on younger landscapes.

Soils containing hydrous oxides as the dominant clays are limited to tropical regions. Such soils occur in Puerto Rico and the Hawaiian Islands. Soils forming in volcanic materials, as in Hawaii or along the west coast of the continental United States may contain significant amounts of the amorphous clay, allophane.

The mineralogy of coarse soil fractions also relates to the extent of weathering during soil formation. For example, where 2:1 minerals remain as principal components of the clay fraction, weathering will not likely have had much of an impact on even the more easily weathered primary minerals in the coarse fraction. Further, if leaching has not been too extensive, lime may remain in the profile if the parent material was calcareous.

Soils in which weathering intensity and time have been sufficient to produce kaolinite as the principal clay usually lack lime, but they may have substantial quantities of the more easily weathered primary minerals in the coarser soil fractions. In soils dominated by hydrous oxides, however, coarse particles, if present, will likely consist mainly of quartz.

Summary

Mineralogically, soils are highly diverse, the nature of their mineral fraction depending on the nature of the parent material and the extent to which it has been altered by weathering. The alteration of minerals during weathering is a function of their chemical composition and structure, as well as the conditions for weathering and the length of time it has been in progress.

Most soil minerals are crystalline compounds consisting of closely packed oxygen ions with various cations located in the spaces between. The oxygens usually occur in either tetrahedra or octahedra linked together through accessory cations or shared oxygens. The majority of primary minerals contain oxygens in tetrahedra, which occur as independent units, in long chains, or in a three-dimensional framework pattern. In layer minerals, tetrahedra arranged in sheets occur in combination with sheetlike arrangements of octahedra.

Soils usually contain a mixture of primary and secondary minerals, with the latter being predominantly clays. The clays are either layer silicates or hydrous iron and aluminum oxides and hydroxides (sesquioxides). Some clays are amorphous and therefore lack the well ordered ionic structure typical of crystalline minerals.

Clays are products of chemical weathering, and the kind that prevails depends on the original, or source, minerals and their stage of weathering. Where weathering has not been extensive, clays of the 2:1 layer type tend to dominate. With advanced weathering, these give way to 1:1 layer types, and ultimately to the hydrous oxides and hydroxides. Soils that have undergone long and intense weathering may consist totally of the latter types of clay materials.

Review questions

1. Distinguish between the two basic mineral properties of chemical composition and structure.
2. What is the coordination number of a cation, and what is the general relationship between this number and the strength of bonding between cations and the oxygens in mineral structures?
3. Why is the bonding between tetrahedra through accessory cations weaker than bonding through shared oxygens?
4. Describe the principal types of ionic substitution in mineral structures, and explain their general effect on charge relationships in minerals.
5. Why can we say that quartz and the feldspars are similar structurally but dissimilar chemically?
6. Explain the differences in the ease with which the various primary minerals succumb to chemical weathering.
7. What is the basic structural difference between 2:1 and 1:1 types of layer silicate minerals?
8. Why do kaolinite, illite, and chlorite not expand on wetting?
9. Describe the alteration process that is responsible for the conversion of illite and mica minerals to vermiculite.
10. Why are hydrous oxides not classed as layer clays?
11. Why are kaolinite and hydrous oxide clays more important in soils of the tropics than in soils of cooler or drier regions?

Selected references

General

Evans, R. C. *An Introduction to Crystal Chemistry*. Cambridge: At the University Press, 1964.

Garrels, R., and MacKenzie, F. T. *Evolution of Sedimentary Rocks*. New York: W. W. Norton & Company, Inc., 1971.

Geological Institute, American. *Dictionary of Geological Terms*. Garden City, N.Y.: Doubleday & Company, Inc., 1962.

Gilluly, J.; Water, A. C.; and Woodford, A. O. *Principles of Geology*. San Francisco: W. H. Freeman and Company, 1975.

Grim, R. E. *Clay Mineralogy*. New York: McGraw-Hill Book Company, Inc., 1968.

Marshall, C. E. *The Physical Chemistry and Mineralogy of Soils. Vol. 1: Soil Materials*. New York: John Wiley & Sons, Inc., 1965.

Marshall, C. E. *The Physical Chemistry and Mineralogy of Soils. Vol. II: Soils in Place*. New York: John Wiley & Sons, Inc., 1977.

Mason, B., and Berry, L. G. *Elements of Mineralogy*. San Francisco: W. H. Freeman and Company, 1968.

Construction of atomic models

Armstrong, E. E., and Drew, J. V. Simplified Clay Mineral Models for Classroom Use. *Proc. Soil Sci. Soc. Amer.* 25:251–52, 1961.

Foth, H. D. A Modification in the Construction of Clay Mineral Models. *Proc. Soil Sci. Soc. Amer.* 28:297–98, 1964.

Wear, J. I., et al. Clay Mineral Models: Construction and Implications. *Soil Sci.* 66:111–17, 1948.

7

Ion exchange in soils

Ion exchange in soils is a phenomenon based on the surface charge of clay and organic particles. By virtue of this charge, ions released from weathering minerals or decaying organic matter, or those added to the soil in rain, irrigation water, or fertilizers can be adsorbed to the particle surfaces and their movement in or with soil water reduced. Yet, retention of these ions is not with such force that they cannot exchange with other ions from the soil solution nor be taken up by plants. This is of great importance to plant nutrition, for it allows the retention of nutrient ions in a form that is available to plants but not subject to much loss by leaching.

As a process, ion exchange is the replacement of adsorbed ions by others from the surrounding solution. However, the ease or speed with which adsorbed ions can be replaced is highly variable, and this has led to the recognition of two types of adsorbed ions: *exchangeable* and *nonexchangeable*. Ions that are weakly held and in direct contact with the soil solution so that they can be quickly replaced by others from solution are classed as exchangeable. Ions that are adsorbed with such tenacity or are located in positions of such low accessibility that they are displaced with great difficulty, or only very slowly, are termed nonexchangeable. The potassium held tightly between platelets of nonexpanding illite clay and mica minerals is classed as nonexchangeable. Although this potassium can be gradually displaced by other cations, as during weathering, the process is much too slow to be viewed as an exchange reaction.

Cation exchange in soils

Soil clays and organic matter tend generally to be negatively charged, although under some circumstances certain clays develop sites of positive charge capable of adsorbing anions in exchangeable form. By and large, however, the negative charge is the principal cause of ion adsorption and exchange in soils. For this reason, cation exchange is viewed as having greater overall significance and is the first to be considered.

Source of negative charge in soils

The negative charge of soil clays and organic matter is developed in two ways: (1) by ionic substitution within the platelets of layer silicate clays, and (2) by the removal of H^+ ions from $-OH$ groups at the platelet edges or from phenolic ($-OH$) or carboxylic ($-COOH$) radicals in organic matter. The charge due to ionic substitution is internal to the clay platelets and is neutralized primarily by cations adsorbed to the broad, planar surfaces of the platelets. Adsorption of cations at platelet edges or to organic particles is directly to the oxygens remaining after the removal of H^+ from the $-OH$ groups or radicals. However, since the bond between oxygen and hydrogen in $-OH$ radicals is strong, removal of the H^+ is difficult, occurring only under certain specified conditions. Once the H^+ is removed, other cations adsorbed in its place behave very much like those adsorbed by the internal charge of clay platelets.

Basis for cation-exchange reactions

Cation exchange occurs between exchangeable and soluble cations, the distinction between these two ionic forms depending on the source of negative charge they neutralize. By definition, exchangeable cations neutralize the negative charge at soil particle surfaces, whereas soluble cations neutralize the negative charge of anions in the soil solution. Exchangeable cations tend to concentrate at or near the particle surfaces. Soluble cations remain somewhat separated from particle surfaces, on the other hand, for they associate with anions that are repelled by the negative charge of the particles. Since soluble cations and anions are not attracted by the particles, they move freely in or with soil water; they can be totally removed from the soil by leaching, for example. In contrast, exchangeable cations cannot be removed by leaching, for they must always remain in sufficient quantity near particle surfaces to exactly balance the negative charge of the particles.

In spite of their attraction to soil particles, exchangeable cations are in a state of constant, heat-induced motion. Because of this, exchangeable cations are able to break away, or *dissociate*, momentarily from the adsorbing surface and move outward for short distances into the neighboring soil solution. During this moment of separation, a cation from solution can move in and take the place of, or exchange for, the dissociated cation.

The dissociation and displacement of exchangeable cations goes on continuously. As a consequence, exchangeable and soluble cations tend to maintain an equilibrium wherein the kinds and proportions of cations present in exchangeable form depend on the kinds and amounts of cations in the solution phase. Because of this, any change in the composition of the solution causes a spontaneous and essentially instantaneous shift in the kinds and proportions of exchangeable ions as a new equilibrium condition is established. The principal causes for change in the composition of the soil solution are the removal of cations by leaching or plant absorption, or the addition of cations from weathering minerals or decaying organic matter, or in rain, irrigation water, or fertilizers or chemical amendments added to the soil.

Due to the continuous exchange and intermingling of soluble and exchangeable cations near soil particle surfaces, no clearcut boundary exists between them in the soil solution. Instead, as depicted in figure 7.1, they

Figure 7.1.
The distribution of cations (open circles) and anions (solid circles) in the solution near a negatively charged soil particle. The total charge of the cations is equal to the combined negative charge of the particle and the anions. Anions are repelled by the particle but may approach its surface because of thermally induced motion.

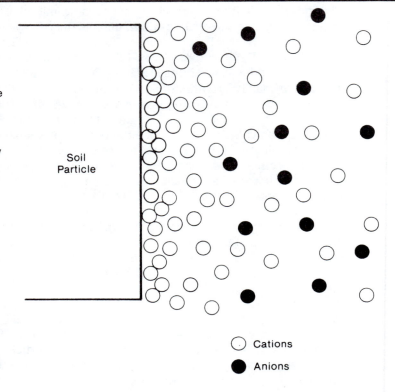

Soil Particle

○ Cations

● Anions

form a continuous system in which the combined charge of all the cations is equal to the total negative charge of the soil particles and the soluble anions. Yet, physical separation of soluble from exchangeable cations does occur if the soil is either leached or dried. On leaching, the soluble cations move out of the soil with the soluble anions, while the exchangeable cations remain behind at or near the particle surfaces. When the soil is dried, exchangeable cations are drawn firmly to adsorption sites at the particle surfaces, and the soluble cations and anions combine to precipitate or crystallize as salts.

General nature of cation-exchange reactions

The kinds of exchange reactions most often considered are those occurring spontaneously when a new cation or combination of cations is introduced into the soil solution. These ions, as they contact soil particle surfaces, undergo rapid exchange with adsorbed ions and cause an instantaneous shift in the pre-existing exchange equilibrium. For example, If H^+ ions were added to a soil containing only adsorbed Ca^{2+}, the reaction between these two ions might be as follows:

$$\begin{array}{c} Ca\ Ca\ Ca \\ \boxed{\ Soil\ }Ca \\ Ca\ Ca\ Ca \end{array} + 4H^+ \rightleftharpoons \begin{array}{c} Ca\ Ca\ Ca \\ \boxed{\ Soil\ }{}^H_H \\ Ca\ Ca\ Ca \end{array} + 2H^+ + Ca^{2+} \qquad (7.1)$$

where the distribution of ions prior to the reaction is shown on the left, and the equilibrium distribution of ions after they have reacted is shown on the right.

Three things may be noted about the reaction in equation (7.1). First, all the H^+ did not become adsorbed. Second, the exchange of ions was on an equivalent (charge-for-charge) basis; two monovalent H^+ ions were adsorbed as a single divalent Ca^{2+} ion was displaced. Finally, at equilibrium each species of ion occurred in both exchangeable and soluble form. However, the final distribution of ions depends on the number of H^+ ions added to the system; the greater this number the greater the proportion of H^+ in exchangeable form once a new equilibrium has been attained.

The principal exchangeable cations in soils

Only a few species make up the bulk of exchangeable cations in soils. Several of these are derived from mineral weathering, including aluminum (Al^{3+}), calcium (Ca^{2+}), magnesium (Mg^{2+}), potassium (K^+), and sodium (Na^+). Ammonium (NH_4^+), which comes largely from decaying organic matter, is also present but comprises only a small part of the naturally-occurring exchangeable cations in soils. This ion assumes greater prominence in fertilized soils. Hydrogen (H^+), another relatively important exchangeable cation, comes primarily from carbonic acid, H_2CO_3, but it is also supplied in small amounts by other acids, both organic and inorganic, which are produced by biological reactions in the soil.

Cations other than those listed above do not occur in large amounts in exchangeable form, because they either are present in extremely small total amounts in soils or they tend to form compounds of very low solubility under most soil conditions. Iron and manganese are prime examples of the latter type of ions.

In soils that have undergone little weathering and leaching, calcium and magnesium are by far the most abundant exchangeable cations.[1] As soils undergo long periods of leaching and weathering, however, exchangeable ions such as calcium and magnesium decline in quantity because they are replaced by other ions, notably by aluminum and hydrogen. This trend is primarily the result of three changes that accompany weathering, namely: (1) an increase in acidity, which supplies hydrogen ions as a factor of weathering and also for adsorption to exchange sites, (2) the gradual disappearance of primary minerals and lime, which lowers the potential for maintaining a high level of soluble and, therefore, exchangeable calcium and magnesium, and (3) an increase in the solubility of aluminum-bearing clay and other minerals that supply aluminum for adsorption to cation-exchange sites. Because of these paralleling changes, aluminum and hydrogen tend to replace calcium and magnesium as weathering proceeds, and for reasons to be discussed shortly, the build-up of exchangeable aluminum is normally greater than that of exchangeable hydrogen. This can result in one

[1]This is not to overlook other significant cations such as sodium and potassium, which are usually present in small amounts in all soils and may be present in relatively large amounts in some. In the great majority of soils, however, exchangeable calcium and magnesium are far more important as determinants of soil behavior than are sodium and potassium.

of the more undesirable consequences of weathering and soil acidification, for excess soluble and exchangeable aluminum can have serious toxic effects on plants.

Factors important to cation-exchange equilibria

The nature of specific cation-exchange equilibria depends in part on the source or site of negative charge in the soil and in part on differences among cations in their ability to compete for sites of negative charge (cation-exchange sites) at soil particle surfaces. The ability of cations to compete for exchange sites depends on two things: (1) the force of attraction between the cation and exchange sites, as is evident in the degree to which these ions dissociate from particle surfaces, and (2) the *effective* concentration of the cations in the soil solution.[2] These two factors are closely interrelated since the greater the concentration of an ion in the soil solution the lower the degree to which it will dissociate from soil particle surfaces.

Force of cation adsorption to exchange sites

The force with which cations are adsorbed to sites of negative charge is a function of their valence and the degree to which they *hydrate* in solution. In general, the higher the valence and the lower the degree of hydration the stronger the force with which a cation is attracted to a site of negative charge. Hydration of ions is due to the physical attraction between the charged ions and polar water molecules.[3] By means of hydration, ions in solutions become enclosed in a shell of water molecules that limits the closeness with which the ions can approach an adsorbing surface. This effect of hydration reduces the force of ionic bonding to exchange sites, the greater the degree of hydration the weaker the force of bonding and, therefore, the greater the ease, or degree, of dissociation of the ion when in adsorbed form.

The principal determinant of the force of attraction between cations and sites of negative charge at particle surfaces is the charge, or valence, of the cation. In general, the force of attraction decreases with valence, and for the more common exchangeable cations in soils is highest for trivalent Al^{3+}, intermediate for divalent Ca^{2+} and Mg^{2+}, and lowest for monovalent K^+, NH_4^+, and Na^+ ions, or:

$$Al^{3+} > Ca^{2+} = Mg^{2+} > K^+ = NH_4^+ > Na^+$$

[2]The effective concentration, technically referred to as *activity*, indicates the quantity of free (active) ions in a solution. This is important to numerous inorganic compounds which, though soluble, remain partly in combined, or molecular, form when dissolved in water. Only those compounds that split apart, or dissociate, contribute to the effective concentration of ions in the solution. Many compounds common to soils tend to dissociate completely, so that the total and effective concentrations of the constituent ions are the same. Since this is not always the case, one should normally interpret references to concentrations of ions in the soil solution as meaning effective concentrations.

[3]Polarity here refers to the unequal distribution of positive and negative charges in an otherwise electrically neutral molecule. Charge polarization in water (H_2O) molecules is due to the tendency for the positively charged H^+ ions in the molecule to concentrate on one side of the O^{2-} ion. The side of the molecule bearing the hydrogens is therefore residually positive and the other residually negative. Polarity explains the ability of water molecules to cling to each other, as in droplets; the molecules are held together because of the attraction of the positive side of one molecule for the negative side of its neighbor.

Among the three monovalent ions, Na^+ is adsorbed less tightly than either K^+ or NH_4^+. The reason for this is that Na^+ ions are more highly hydrated than is either of the other two.

Hydrogen is not shown in the above series because it is attracted to soil particles with a different force depending on the origin of the negative charge. Whereas H^+ is adsorbed to the planar surfaces of silicate clay platelets with about the same force as are monovalent K^+, NH_4^+, or Na^+ ions, it is held more tightly than any of the other cations at clay platelet edges or by soil organic matter, where bonding is directly to structural oxygens.

The force with which cations are adsorbed to cation-exchange sites relates to the degree to which they dissociate from the particle surfaces. The degree of dissociation, in turn, determines the relative ease of displacement by cations from the solution phase. Except for H^+ held by organic matter or at clay edges, Al^{3+} dissociates the least and is therefore more difficult to replace than are the other common exchangeable cations. Conversely, Na^+ dissociates to the highest degree and is the easiest to displace from cation-exchange sites. The extremely low degree to which H^+ dissociates from $-OH$ radicals explains the difficulty in displacing it from exchange sites in organic matter or at the edges of clay platelets.

Concentration effects

The ability of cations to compete for exchange sites depends not only on the force of attraction to sites of negative charge but also on their concentration in the soil solution. For example, if Ca^{2+} and Mg^{2+} were at the same concentration in the soil solution, they would also occur in about the same concentrations in exchangeable form, for the two are about equally attracted to the sites of negative charge. In contrast, if Ca^+ and Na^+ were in the same concentrations in the solution phase, the proportion of exchange sites taken up by Ca^{2+} would exceed that taken up by Na^+. However, if the concentration of Na^{2+} in the solution were made sufficiently large, its concentration in exchangeable form could become greater than that for Ca^{2+}. For this to occur would ordinarily require a concentration of soluble Na^+ many times greater than that of Ca^{2+}. This requirement reflects a much higher degree of dissociation of Na^+ than of Ca^{2+} from soil particle surfaces.

Under practical situations, either leaching or precipitation reactions can aid in the displacement of exchangeable cations from exchange sites. For example, cations added to the soil in water undergo immediate exchange with adsorbed ions the moment they enter the soil. As the displaced cations move out into the water flowing through the pores, they are carried downward leaving the added ions in sole possession of the exchange sites at or very near the point where they enter the soil. This explains why fertilizer cations, when added in salts that are spread over the surface and then washed into the soil in water, become adsorbed within a very thin layer at the surface of the soil. This limits the loss of the added nutrients by leaching, but it also limits their availability to plants where subsequent drying at the surface stops root activity in that part of the profile.

Precipitation reactions that remove competing ions from the soil solution also facilitate the adsorption of cations added to the soil. For example, Na^+ accumulating in salts in arid-regions soils can easily displace more tightly adsorbed Ca^{2+} and Mg^{2+} if carbonate (CO_3^{2-}) ions are accumulating with the Na^+. The effect of the carbonate is to cause the precipitation of Ca^{2+} and Mg^{2+} as lime. The nature of the reaction is shown by the equation:

$$\boxed{\text{Soil}}\;\;\Big|Ca + 2Na^+ + CO_3^{2-} \longrightarrow \boxed{\text{Soil}}\;\;\Big|^{Na}_{Na} + CaCO_3 \downarrow \quad (7.2)$$

If sufficient Na^+ and CO_3^{2-} accumulate, essentially all exchangeable Ca^{2+} and Mg^{2+} in a soil can be replaced by Na^+. The reason for this is that Ca^{2+} and Mg^{2+} are attracted more strongly to the carbonate than to cation-exchange sites, with the reverse being the case for Na^+.

Expressing concentrations of exchangeable and soluble ions

The amount, or concentration, of an ion in soluble or exchangeable form can be expressed in a number of ways. One is on a gravimetric (weight) basis, as for example, grams of an ion per liter of solution for soluble ions, or grams of ion per unit of dry soil weight for those in exchangeable form. Whereas gravimetric values are important and are the basis for determining the absolute amounts of ions present, it is often more useful to express concentrations in terms of *equivalent weights*, or simply *equivalents*, of ions. An equivalent weight of an ion is the amount of the ion, in grams, that will supply the same total charge as is supplied by 1 g of H^+.

To compute equivalent weights requires knowledge of the ionic weight and the valence of an ion. Ionic weights express the weight of ions relative to the weight of H^+, with the weight of H^+ being arbitrarily set at 1.0. Equivalent weights are then computed by dividing the ionic weight by the valence of the ion. For example, Na^+ ions weigh 23 times as much as H^+ ions. Like H^+, Na^+ is monovalent, so 23 g of Na^+ supplies a total charge equal to that in 1.0 g of H^+; its equivalent weight is therefore 23 g. Similarly, Ca^{2+} ions weigh 40 times as much as H^+, and this weight, when divided by two, the valence of Ca^{2+}, gives an equivalent weight of 20 g. Thus, it takes 20 g of Ca^{2+} ions to supply the same total charge as is supplied by 1.0 g of H^+. It follows, then, that solutions containing 1 g of H^+, 23 g of Na^+, or 20 g of Ca^{2+} would contain 1 equivalent of each ion. These solutions would also contain 1 equivalent of anion to balance the charge of the cations.

Because the concentrations of adsorbed or soluble cations in soils are ordinarily quite low, it is normal to express them in milliequivalent rather than equivalent quantities. A milliequivalent weight, abbreviated me, is equal to 1/1000th of an equivalent weight. The equivalent and milliequivalent weights for several cations and anions important to exchange and other chemical reactions in soils are listed in table 7.1.

Table 7.1 Ionic, equivalent, and milliequivalent weights of common soluble and exchangeable ions in soils.

Ion	Symbol	Weight[a]	Valence	Equivalent weight	Milliequivalent weight
		g		g	g
Hydrogen	H^+	1	1		
Sodium	Na^+	23	1	23.0	0.023
Potassium	K^+	39	1	39.0	0.039
Ammonium	NH_4^+	18	1	18.0	0.018
Calcium	Ca^{2+}	40	2	20.0	0.020
Magnesium	Mg^{2+}	24	2	12.0	0.012
Aluminum	Al^{3+}	27	3	9.0	0.009
Oxygen	O^{2-}	16	2	8.0	0.008
Hydroxyl	OH^-	17	1	17.0	0.017
Chloride	Cl^-	35	1	35.0	0.035
Nitrate	NO_3^-	62	1	62.0	0.062
Sulfate	SO_4^{2-}	96	2	48.0	0.048
Carbonate	CO_3^{2-}	60	2	30.0	0.030
Bicarbonate	HCO_3^-	61	1	61.0	0.061

[a]The weight of the ion relative to the weight of H^+, which is arbitrarily set equal to 1.0. These values have been rounded for convenience. For ions containing single elements, the ion weight is the same as the atomic weight.

The reaction, or pH, of soils

One of the more important properties of soils is their level of acidity or basicity, as judged from the relative concentrations of H^+ and OH^- ions in the soil solution. This property, sometimes referred to as *reaction*, is normally expressed in pH units. Specifically, pH is a measure of the effective concentration of H^+ ions in equivalents per liter of solution. Indirectly, pH is also a measure of the OH^--ion concentration, for there is a consistent, though inverse, relationship between the concentrations of H^+ and OH^- ions in aqueous (water) solutions, that is, the higher the concentration of one of these ions the lower the concentration of the other. There is no aqueous solution that totally lacks either H^+ or OH^-.

The relative concentrations of H^+ and OH^- ions determines whether a solution is *acidic, basic,* or *neutral.* Where the concentration of H^+ exceeds that of OH^-, the solution is acidic, but if the concentration of OH^- exceeds that of H^+, the solution is basic. An equal concentration of the two ions yields a neutral solution. Pure water has a neutral reaction, as do solutions of numerous salts (neutral salts) that have no effect on the concentrations of H^+ or OH^- when dissolved in water. Some compounds are able to increase the concentration of OH^- over that of H^+ when dissolved in water; these are called *basic* compounds, or simply *bases.* Others have the reverse effect and increase the concentration of H^+ over that of OH^-. These are called *acidic* compounds, or *acids.*

Pure water has a neutral reaction because, on dissociation, water molecules split into equal quantities of H^+ and OH^- ions:

$$H_2O \rightleftharpoons H^+ + OH^- \tag{7.3}$$

The degree to which water dissociates is extremely low, however; only about two out of every billion water molecules are dissociated at any given moment. The low degree of dissociation of water molecules attests to the extremely high affinity between H^+ and OH^- ions, a feature of great importance to the behavior of these ions in soils.

Because of the limited dissociation of water, it contains a very low concentration of H^+ and OH^- ions, only 0.0000001 equivalent of H^+ and OH^- per liter. Since it is inconvenient to deal with decimals of this type, concentrations of H^+ ions in aqueous solutions are usually transformed into more convenient units of pH. In the transformation, the decimals are converted to positive integers through the use of logarithms.[4]

Specifically, pH is the negative logarithm of the H^+-ion concentration expressed in equivalents per liter, or

$$pH = -\log[H^+] \qquad (7.4)$$

where the symbol $[H^+]$ is the concentration of H^+ in equivalents per liter. According to this equation, the pH of pure water, which has a H^+-ion concentration of 0.0000001, or 10^{-7}, equivalents per liter, is 7.0:

$$pH = -\log[H^+] = -\log 10^{-7} = -1(-7) = 7 \qquad (7.5)$$

In this computation, the log of 0.0000001, or -7, is converted to a positive integer (though a negative log) through multiplication by -1. Since the concentrations of H^+ and OH^- are equal in pure water, its pH, or 7, is taken as neutral. An increase in the H^+-ion concentration above that in water causes the pH to drop below 7 and results in an acid reaction. A decrease in the H^+-ion concentration causes the pH to rise above 7, thus producing a basic reaction.

The relationship between pH and the H^+-ion concentration is such that a 10-fold change in the concentration results in a 1-unit change in pH. Also a 10-fold change in the H^+-ion concentration is accompanied by a 10-fold change in the OH^--ion concentration, with the two varying inversely. The relationship between the concentration of these ions and pH is shown in table 7.2.

---

[4]Any number can be expressed as a logarithm (abbreviated log) or its equivalent power of 10. A power of 10 is frequently noted exponentially, as where 10 to the second power (10 squared) is written 10^2. The whole number expressed exponentially by 10^2 is 100. The power of 10, or the exponential numeral 2 in this instance, is also the logarithm of 100. Thus, log 100 = 2, or log 10^2 = 2.

For the number 10 or any of its multiples (100, 1000, etc.), either the power of 10 or the log is a positive whole number. For decimals consisting of the numeral 1, either alone or in combination with one or more zeros, either the power of 10 or the log is a negative whole number. For example:

Number	Power of 10	Log
1000	10^3	3
100	10^2	2
10	10^1	1
1.0	10^0	0
0.1	10^{-1}	-1
0.01	10^{-2}	-2
0.001	10^{-3}	-3

Accordingly, a one-unit change in either the power of 10 or the log corresponds to a 10–fold change in the number they represent.

Table 7.2. Relationship between concentrations of H^+ and OH^- ions and pH.

pH	Ion concentration (eq/l)		Reaction range
	H^+	OH^-	
0	1.0	0.00000000000001	↑
1	0.1	0.0000000000001	
2	0.01	0.000000000001	
3	0.001	0.00000000001	
4	0.0001	0.0000000001	Acid
5	0.00001	0.000000001	
6	0.000001	0.00000001	
7	0.0000001	0.0000001	Neutral
8	0.00000001	0.000001	
9	0.000000001	0.00001	
10	0.0000000001	0.0001	
11	0.00000000001	0.001	Basic
12	0.000000000001	0.01	
13	0.0000000000001	0.1	
14	0.00000000000001	1.0	↓

In most discussions of pH-related soil properties it is usually necessary to refer to pH in general terms only. Often, such broad connotations as acid, neutral, or basic are sufficient to indicate the state of a pH-related property. Where these generalizations are too broad, more narrowly defined ranges of acidity or basicity may be used. An example is as follows:

Neutral 6.6–7.4

Slightly acid	6.0–6.6	Slightly basic	7.4–8.0
Moderately acid	5.0–6.0	Moderately basic	8.0–9.0
Strongly acid	below 5.0	Strongly basic	above 9.0

Hydrolysis as a factor of pH

Certain compounds, because of their ability to selectively inactivate $H+$ or OH^- ions derived from the dissociation of water, can affect the pH of aqueous solutions. The reaction between these compounds and the H^+ or OH^- from water is termed *hydrolysis*. If hydrolysis of a compound added to water causes a reduction in the concentration of OH^- ions relative to that of H^+, an acid solution results, but if the reverse effect occurs, the solution will be basic. In soils, hydrolysis reactions may tend to increase either the acidity or the basicity of the solution phase, the direction of change depending on the kinds of hydrolyzable materials the soil contains. Examples of such reactions occur in the following discussions.

Acidity in soils

The fundamental cause of soil acidity is the accumulation of soluble acids, either organic or inorganic, at a faster rate than they can be neutralized or removed from the soil system. Dissociation of these acids provides most of the free H^+ occurring in the soil solution at any one time. Exchangeable H^+ and Al^{3+} can also contribute to the supply of free H^+ in the soil solution. Exchangeable H^+ makes its contribution through direct dissociation into the solution phase; for Al^{3+}, it is due to both dissociation and hydrolysis. Upon hydrolysis, Al^{3+} reacts with and ties up OH^- from water, which results in the release of an equivalent amount of H^+ in free ionic form. The reaction in simplified form is as follows:

$$Al^{3+} + H_2O \rightleftarrows AlOH^{2+} + H^+ \tag{7.6}$$

Because of their ability to increase the H^+-ion concentration of the soil solution, exchangeable H^+ and Al^{3+} are called *exchangeable acids*. However, neither of these ions occurs in significant quantity in exchangeable form unless the soil is already rather strongly acid. For example, a pH of 5.5 or less is required before much Al^{3+} appears in exchangeable form. Above this pH, the increasing concentration of OH^- causes Al^{3+} and OH^- ions to combine into hydroxy-Al ions or even to form essentially insoluble $Al(OH)_3$, the effect of such combinations being to reduce the concentration of free Al^{3+} ions in solution. In the slightly acid range, Al^{3+} and OH^- combine to form positively charged $AlOH^{2+}$ and $Al(OH)_2^+$ ions. These combinations do not remain in solution, however. Instead, they are deposited as positively charged coatings over clay particle surfaces. These coatings, in which the Al^{3+} and OH^- ions are in an octahedral sheet arrangement, neutralize part of the negative charge of the clays and thereby block sites that might otherwise hold cations in exchangeable form.

Hydrogen seems not to accumulate in exchangeable form unless the pH is about 4.0 or below. This relates to the low affinity of H^+ for the negative charge of clays. Unless the pH is low and the concentration of soluble H^+ comparatively high, this ion is unable to successfully compete for exchange sites with more tightly held Ca^{2+}, Mg^{2+}, and Al^{3+} ions. It follows, therefore, that the latter three ions tend to dominate exchange sites in acid soils, the majority of which lie above a pH of 4.0. In the acid range above pH 5.5, Ca^{2+} and Mg^{2-} are generally the dominant exchangeable cations, and as the pH drops below 5.5, Al^{3+} becomes progressively more important.

Basicity in soils
The principal cause of basic reactions in soils is hydrolysis, either of certain exchangeable cations or of basic salts, such as the $CaCO_3$ and $MgCO_3$ in lime, or Na_2CO_3. Exchangeable cations that appear most capable of producing a basic reaction on hydrolysis are Ca^{2+}, Mg^{2+}, K^+, NH_4^+, and Na^+. These ions, when present in exchangeable form, are termed *exchangeable bases*. Their hydrolysis is typified by the reaction between exchangeable Na^+ and free H^- derived from the dissociation of water:

$$\begin{array}{|c|} \hline Na~~Na \\ Soil \\ Na~~Na \\ \hline \end{array} Na + H_2O \rightleftarrows \begin{array}{|c|} \hline Na~~Na \\ Soil \\ Na~~Na \\ \hline \end{array} H + Na^+ + OH^- \tag{7.7}$$

In this reaction, H^+ is inactivated by exchange adsorption in the place of Na^+. The displaced Na^+ does not combine with and inactivate OH^-, however. Thus, the concentration of OH^- ions in solution is raised above that of H^+, and a basic reaction results.

The extent to which exchangeable bases hydrolyze depends on their ability to compete with H^+ for exchange sites. Weakly held ions such as K^+ and Na^+ are less able to compete than are more tightly held ions such as Ca^{2+} and Mg^{2+}. For this reason, exchangeable Na^+ and K^+ are more extensively hydrolyzed and produce a higher pH than do exchangeable Ca^{2+} and Mg^{2+}. In fact, hydrolysis of exchangeable Ca^{2+} or Mg^{2+} is so limited that it never causes more than a very weakly basic reaction in soils. In contrast, hydrolysis of exchangeable Na^+ or K^+ appears capable of producing a moderate to strongly basic reaction under certain soil conditions. Even here, the effect is limited if neutral salts suppress hydrolysis and thereby limit the rise in pH.[5] Leaching of neutral salts from soils usually causes a rise in pH.

Basic compounds such as $CaCO_3$, $MgCO_3$, or Na_2CO_3 hydrolyze as follows:

$$CaCO_3 + H_2O \rightleftharpoons Ca^{2+} + OH^- + HCO_3^- \qquad (7.8)$$

In this reaction, H^+ from water is inactivated through combination with carbonate to form bicarbonate (HCO_3^-) ions. Hydroxyl ions are not inactivated through combination with Ca^{2+}; thus, the resulting solution is basic.

The hydrolysis of either $CaCO_3$ or $MgCO_3$ is limited by their relatively low solubilities, so they tend to produce a pH in soils no higher than about 8.0 to 8.2. Soils containing Na_2CO_3 usually have a higher pH, however, frequently in the range of 10.0 to 10.5. This relates to the comparatively high solubility of Na_2CO_3, for the greater the solubility the greater the potential for hydrolysis. Soils containing soluble Na_2CO_3 are usually also relatively high in exchangeable Na^+.[6]

Measurement and significance of pH

The pH of soils is easily measured, either through the use of dye solutions or dye-impregnated test papers that change color with a change in pH,[7] or by means of electric pH meters. For each of these methods, the determination can be made on dilute soil-water suspensions, on an extract from such suspensions, or on a semi-fluid paste. Since pH values of soil pastes vary somewhat with soil water content, it is important to standardize the moisture level used for the measurement. In general, increasing the water content increases the pH by diluting the H^+ ions in the solution.

Knowledge of soil pH allows certain useful judgments about the chemical and nutritional status of the soil. Extremes in pH suggest problems that may need correction before good plant growth can be obtained. In the strongly

[5]Neutral salts suppress hydrolysis by supplying cations that compete with the H^+ from water for exchange sites. Recall that the adsorption of this H^+ is essential to the rise in pH when exchangeable bases hydrolyze. The more important neutral salts in soils are chlorides and sulfates of Ca^{2+}, Mg^{2+}, Na^+, and K^+. They are called neutral salts because they form neutral solutions with water.

[6]Infrequently soils may contain K_2CO_3 and a reasonably high level of exchangeable K^+. Such soils also tend to have high pH values.

[7]One type of test paper is impregnated with litmus, a dye that changes color from pink to blue at pH 7.0. It is commonly used to determine whether a solution is acid or basic. Some test papers or solutions contain a mixture of dyes chosen to give a gradual change in color throughout the relatively broad range of pH encountered in soils.

acid range, the problem may be excess soluble Al^{3+} or Mn^{2+}. The solubility of these ions increases with increasing acidity, and either may be toxic to plants. The availability of plant nutrients such as phosphorus and molybdenum may also be seriously reduced under strongly acid conditions.

Soil problems in strongly basic soils are normally associated with excess exchangeable Na^+. By and large, these problems are the result of poor tilth caused by the inability of exchangeable Na^+ to maintain the soil in a well structured state.

Where the soil pH falls outside the desired range, it can be adjusted by treating the soil with an *amendment* such as lime or gypsum.[8] Lime is used to overcome excess soil acidity; it not only supplies Ca^{2+}, either alone or in combination with Mg^{2+}, to displace exchangeable H^+ and Al^{3+}, but more importantly, it provides OH^- ions for the neutralization and inactivation of the H^+ and Al^{3+}. The OH^- for this reaction is produced by the hydrolysis of the $CaCO_3$ or $MgCO_3$ in the lime as shown in equation (7.8).

Excess exchangeable Na^+ can be removed and a high soil pH reduced by treating with an amendment that increases the supply of soluble Ca^{2+} in the soil solution. Through exchange, the added Ca^{2+} displaces the Na^+, which can then be removed from the soil by leaching. The amendment most often used for supplying soluble Ca^{2+} in sodium-soil reclamation is gypsum $(CaSO_4 \cdot 2H_2O)$. Lime is useless for this purpose because of its extremely low solubility under basic conditions.[9]

The cation-exchange capacity of soils

The cation-exchange capacity is the amount of negative charge per unit weight of soil that is neutralized by easily replaceable cations. The standard unit used to express the cation-exchange capacity is milliequivalents per 100 g of oven-dry soil, and it is measured by determining the quantity of cations the soil can hold in exchangeable form. Theoretically, the cation-exchange capacity is equal to the sum of the milliequivalent quantities of all exchangeable cations present. For example, assume that a 100-g sample of soil is found to contain only four kinds of exchangeable cations and in the following amounts:

Ion	Weight of ion	Milliequivalents of ion
	g	
H^+	0.001	1
Al^{3+}	0.018	2
Ca^{2+}	0.120	6
Mg^{2+}	0.024	2
		Total 11

[8]An amendment is a material other than fertilizer that is added to the soil to improve conditions for plant growth. The effect may be either physical or chemical.

[9]The correction of excessively acid- or basic-soil conditions is highly important to crop production in many situations. The procedures involved are discussed in detail in Chapters 17 and 18.

Since the total milliequivalent quantity of exchangeable cations is 11 and the quantity of soil is 100 g, the cation-exchange capacity of the soil would be 11 me per 100 g.

Determination of cation-exchange capacities by the above means is difficult, primarily because it means a separate, and often lengthy, analysis for each ion. Thus, a simpler technique is normally used. The usual procedure consists of displacing all original exchangeable cations from a soil sample with a single cation species. The latter cation is, in turn, displaced from the soil and determined in a single analysis. Cations used most often for this are ammonium (NH_4^+) and sodium (Na^+).

Soil pH and the cation-exchange capacity

The cation-exchange capacity of soils is not constant but varies with pH. Variation in the cation-exchange capacity is due to the displacement of nonexchangeable hydrogens from structural $-OH$ groups as the pH of the soil is raised. Such groups occur in organic matter, at clay platelet edges, and in the amorphous clay, allophane. If the pH of any of these materials is raised from some low value by adding a base such as calcium hydroxide, $Ca(OH)_2$, the reaction is:

$$R \overset{\diagup OH}{\underset{\diagdown OH}{}} + Ca^{2+} + 2OH^- \rightarrow R \overset{\diagup O}{\underset{\diagdown O}{}} Ca + 2H_2O \qquad (7.9)$$

The R in the equation symbolizes either an organic molecule or clay to which the $-OH$ groups are attached. In effect, raising the pH transforms nonexchangeable hydrogen to an exchangeable form and thereby increases the cation-exchange capacity of the material. The OH^- ions are essential to this reaction, for they attract the hydrogen in the $-OH$ groups and then inactivate it in undissociated water.

The negative charge produced by displacement of hydrogens from structural $-OH$ groups is called the *pH-dependent charge* of clays and organic matter. This type of charge is solely responsible for cation-exchange capacity in soil organic matter and allophane clay. Under strongly acid conditions, these materials have neither a negative charge nor a cation-exchange capacity. Both acquire a negative charge as the pH is raised, however, and it can be appreciable above the neutral point.

Layer clays also have a pH-dependent charge, although it is small in comparison to that in organic matter and allophane. In addition, they have an internal charge due to ionic substitution. Since this charge is fixed, it is called the *permanent charge*. In combination, the permanent and pH-dependent charges determine the cation-exchange capacity of layer clays. If the internal charge is low, as in the 1:1 clay kaolinite, the pH-dependent charge can account for a substantial part of the cation-exchange capacity. The effect of the pH-dependent charge is much less in clays with a high permanent charge, the prime examples being montmorillonite and vermiculite.

The general relationship between pH and the cation-exchange capacity is demonstrated by data for 60 Wisconsin soils in table 7.3. The relationship is shown for organic matter and clays as well as for the whole soil. Beyond

Table 7.3. Effect of pH on the cation-exchange capacity of 60 Wisconsin soils and their organic and clay fractions. (After Helling, C. S. *et al,* Soil Sci Soc Amer Proc 28:517–20, 1964).

pH	Organic matter	Clay	Whole soil
	Cation-exchange capacity—me/100 g		
2.5	36	38	5.8
3.5	73	46	7.5
5.0	127	54	9.7
6.0	131	56	10.8
7.0	163	60	12.3
8.0	213	64	14.8

the fact that all values increase with increasing pH, two points are of particular note. One is that with increasing acidity the cation-exchange capacity of the organic fraction appears to tend toward zero, but that of the clay fraction does not. This supports the view that the charge in organic matter is totally pH-dependent, whereas the charge in clays is at least partially permanent. The other point of note is the relatively high cation-exchange capacity attained by organic matter in the basic range; raising the pH from the minimum to the maximum shown in the table causes approximately a sixfold increase in the cation-exchange capacity of the organic materials, while the cation-exchange capacity of the clays does not even double.

Because the cation-exchange capacity varies with pH, its determination in the laboratory is ordinarily at a standard pH. For lime-free soils that are naturally neutral to acid in reaction, a pH of 7.0 is customarily used. For calcareous soils, or others that are naturally basic, the determination is more often made at pH 8.2. These choices of pH allow estimates of the highest cation-exchange capacities that are likely to occur under practical conditions in the field.

The cation-exchange capacity of clays and organic matter

Representative cation-exchange capacities, at pH 7.0, of the principal soil clays and organic matter are shown in table 7.4. According to the table, cation-exchange capacities are, in general, highest for organic matter, intermediate for the expanding clays and allophane, and lowest for the nonexpanding clays and the hydrous oxides. Hydrous oxides have no internal charge due to ionic substitution, and they do not develop a pH-dependent negative charge in the normal range of soil pH. Kaolinite has a small cation-exchange capacity, principally because it has a small permanent charge and a pH-dependent charge of roughly comparable magnitude.

Allophane, an amorphous clay, has a chemical composition near that of crystalline kaolinite, but with tetrahedra and octahedra in a poorly organized array. The octahedra apparently have a high content of hydroxyl ions that are capable of dissociating H^+ ions to produce a negative charge with increasing pH. This charge is totally pH-dependent and appreciable at pH 7.0.

Illite and chlorite have a high total charge internally but only a moderate cation-exchange capacity. The reduction in cation-exchange capacity below the total negative charge is the result blockage of the charge by in-

Table 7.4. Representative cation-exchange capacities of common exchange materials in soils as measured at pH 7.0.

Exchange material	Cation-exchange capacity
	me/100 g
Organic matter	100–300
Vermiculite	100–150
Allophane	100–150
Montmorillonite	60–100
Chlorite	20–40
Illite	20–40
Kaolinite	2–16
Hydrous oxides	0

terlayer K^+ ions in illite and the positively charged, interlayer octahedral material in chlorite. Similarly, any clay with positively charged hydroxy-Al coatings will have a cation-exchange capacity that is lower than the total negative charge, since these coatings neutralize, or block, a part of the internal negative charge of the clay. The effect of the coatings varies with pH, however. They do not form at higher pH levels, say above pH 8.0 or so. In this range, aluminum is more inclined to precipitate as insoluble $Al(OH)_3$. The coatings occur mainly in the acid range, but they decrease if the pH is raised, which allows the cation-exchange capacity to increase. With changing pH, therefore, the effect of the hydroxy-Al coatings on cation-exchange capacity is the same as that of the pH-dependent charge.

The cation-exchange capacity of soils

The cation-exchange capacity of whole soils is a function of the type and amount of clay and the amount of organic matter. Where these factors are known, a reasonable estimate of the cation-exchange capacity can be made using values in table 7.4 and the formula:

$$CEC_{soil} = [(\%Clay_a \times CEC_a) + (\% Clay_b \times CEC_b) + (\%OM \times CEC_{OM})] \times 0.01 \qquad (7.10)$$

where CEC is the cation-exchange capacity of the whole soil, the clays, and organic matter, with subscripts a and b representing different kinds of clay. The use of this equation can be shown by applying it to a hypothetical soil containing 10 percent each of montmorillonite and kaolinite (10 g clay in 100 g of whole soil) and 2 percent organic matter (2 g organic matter in 100 g of soil). From table 7.4 we assign average values of 80 and 7 me/100 g as the cation-exchange capacities for montmorillonite and kaolinite, and 200 me/100 g for the cation-exchange capacity of the organic matter. Through use of the equation (7.10), the estimated cation-exchange capacity for this soil would be 12.7 me/100 g of soil, or:

$$[(10 \times 80) + (10 \times 7) + (2 \times 200)] \times 0.01 = 12.7 \qquad (7.11)$$

Of the total 12.7 milliequivalents, 8.0 would come from the montmorillonite, 0.7 from the kaolinite, and 4.0 from the organic fraction. In this computation, the effect of silt and sand on the cation-exchange capacity of the whole soil is assumed to be zero. Such an assumption will usually be correct, except for

Table 7.5. The cation-exchange capacity of soils in relation to the quantity and type of clay present. (Data selected from Soil Survey Staff. *Soil Classification: A Comprehensive System.* Washington, D.C.: Government Printing Office, 1960).

Textural class	Clay	Cation-exchange capacity	Location
	%	(me/100 g)	
2:1 clays dominant			
Sand	1	2	Florida
Sandy loam	11	8	Arizona
Silt loam	28	26	Iowa
Silty clay loam	35	36	Arizona
Clay	58	49	Texas
1:1 and sesquioxide clays dominant			
Clay	56	14	Puerto Rico
Clay	72	19	Puerto Rico

Table 7.6. The influence of variation in organic matter (OM) or clay content on the cation-exchange capacity (CEC), in me/100 g, at different depths in soil profiles. (Data selected from Soil Survey Staff. *Soil Classification: A Comprehensive System.* Washington, D.C.: Government Printing Office, 1960).

Depth	Clay	OM	CEC	Depth	Clay	OM	CEC
cm	%	%		cm	%	%	
0–7.5	15.4	20.2	40.5	0–20	11.6	1.8	7.8
7.5–28	16.6	2.8	22.3	20–30	11.1	0.9	5.4
28–56	15.7	1.2	18.4	30–45	26.2	0.8	13.8
56–72	8.5	1.1	11.9	45–68	25.8	0.3	15.6
				68–90	28.6	0.1	19.1

some soils high in vermiculite. These soils may contain vermiculitic sands and silts, probably from mica weathering, that have significantly high cation-exchange capacities.

Some representative cation-exchange capacities (pH 7.0) of soils are listed in tables 7.5 and 7.6. Table 7.5 is for surface soils from different locations. These data have been selected to show two things: the general dependence of the cation-exchange capacity on the amount of clay present, and the effect that differences in the kind of clay can have on this relationship. Recall that 2:1 clays tend to induce a higher cation-exchange capacity in soils than does an equivalent amount of 1:1 clay, such as kaolinite, or the hydrous oxides. As shown in the table, soils high in the latter clays have comparatively low cation-exchange capacities.

The influence of organic matter on the cation-exchange capacity is indicated by the left-hand set of data in table 7.6. It may be judged from these data, which are for an Alaskan soil, that about half the cation-exchange capacity of the surface 0–7.5 cm layer is attributable to organic matter. The second set of data in table 7.6, for a soil from Illinois, illustrates a feature often observed of soil profiles; that is, a simultaneous increase in the clay content and in the cation-exchange capacity with increasing depth. Organic matter probably adds 2–3 me to the cation-exchange capacity associated with the surface 20-centimeters of the latter soil.

Buffering in soils

Buffering is a standard chemical term denoting the capacity of a system to resist changes in pH when treated with an acid or a base. Soils are viewed as buffered systems, although they vary markedly in this property. Where the pH is nearly ideal for the growth of plants, buffering is desirable, for it limits the potential for large changes in pH that might adversely affect growth. On the other hand, buffering is undesirable if there is a need to correct excess acidity or basicity through the addition of an amendment to the soil. Large amounts of amendment may be required on soils of high buffering capacity, which makes an adjustment in pH comparatively expensive.

Soils provide a buffered system to the extent that they inactivate added H^+ or OH^- ions and thus limit their effect on the pH of the soil solution. Buffering against added H^+ is afforded by adsorption to exchange sites or by neutralization reactions that convert the H^+ to water. One of the more notable neutralization reactions is between H^+ and lime in calcareous soils; so long as lime remains in these soils, it is virtually impossible to lower their pH by acid treatment. As shown by the following neutralization reactions, lime ($CaCO_3$ in this example) converts H^+ to carbonic acid, H_2CO_3, which then decomposes to carbon dioxide and water:

$$CaCO_3 + 2H^+ \rightarrow Ca^{2+} + H_2CO_3 \tag{7.12}$$

$$H_2CO_3 \rightarrow H_2O + CO_2 \uparrow \tag{7.13}$$

In lime-free soils, acid treatment can cause a significant drop in pH, the magnitude of which depends upon both the amount of acid supplied and the capacity of the soil to inactivate the added H^+ through exchange adsorption. However, if easily weatherable minerals other than lime are present, they too will slowly react with and neutralize added H^+, and thereby prevent a permanent change in pH. This explains why a soil treated with a heavy application of acid-forming fertilizer[10] may experience an increase in acidity, but then gradually return to a more normal pH over a period of time.

Resistance to change in pH when a base is added to acid soil is due principally to neutralization of the base by adsorbed H^+, Al^{3+}, or hydroxy-Al ions. These ions, when adsorbed to particle surfaces, comprise the *reserve acidity* of a soil, whereas free H^+, which determines the pH of the soil solution, is referred to as *active acidity*. Because active and reserve acidity are in equilibrium, both must be neutralized if the pH is to rise. However, most of the base added to acid soil is used in neutralizing reserve acidity, since its content is far greater than that of active acidity. To illustrate, consider the theoretical relationship between reserve and active acidity in a soil at pH 5 and containing 20 percent water. Under these conditions, the quantity of active (solution) H^+ in a hectare-furrow slice weighing 2,000,000 kg would amount to only 4 g. If the level of reserve acidity were 1 me per 100 g, the

---

[10]Amonium fertilizers are especially noted for their ability to produce acidity in soils. This effect is caused by the oxidation of ammonium ions to nitric acid (HNO_3) by a special group of soil organisms. The oxidation of ammonium is discussed on p. 412.

total quantity in the same soil would be equivalent to, roughly, 20 kg of H^+. Neutralization of the reserve acidity would therefore require some 5000 times more base than would be needed to neutralize free H^+ in the soil solution. In some strongly acid soils, total acidity may be much greater than the amount indicated in this example. The amount of base required to raise the pH of a soil therefore depends largely on the amount of reserve acidity that has accumulated, and, if the soil has a high cation-exchange capacity, the quantity of reserve acidity can be large.

Base saturation in soils

A method that has long been used to characterize the status of exchangeable cations in soils is *percent base saturation*, which is the total content of exchangeable bases (Ca^{2+}, Mg^{2+}, K^+, and Na^+) expressed as a percent of the cation-exchange capacity at pH 7.0 or 8.2, or:

$$\text{Base saturation } (\%) = \frac{\text{Sum of exchangeable bases}}{\text{CEC at pH 7.0 or 8.2}} \times 100 \qquad (7.14)$$

In the above equation, both the sum of the bases and the cation-exchange capacity are expressed in milliequivalents per 100 g of soil. The principal application of this concept has been in the evaluation of plant growth relationships in acid soils. The concept has little application for neutral or basic soils, because they are inclined to have a high and relatively constant level of bases regardless of conditions. Under nearly all circumstances, soils above a neutral pH may be assumed to be base saturated.

The percent base saturation in acid soils decreases with decreasing pH, and it is the result of (1) a reduction in the net negative charge holding exchangeable cations and (2) the increasing displacement of exchangeable bases by accumulating H^+ and Al^{3+}. Regardless of the specific cause, the net effect of decreasing pH is a reduction in the concentration of exchangeable bases, both in absolute terms and in relation to the quantity held at pH 7.0 or 8.2. In general, low concentrations of exchangeable bases reflect in reduced plant growth. However, the degree to which the percent base saturation must be reduced before plant growth is impaired varies substantially from soil to soil. As a consequence, there is no fixed relationship between these two variables that can be applied universally to soils.

There are two principal uses made of data on the level of base saturation in soils. One is in the assessment of lime needs on acid soils, and the other is in soil classification. For example, in one kind of classification, soils are placed into different categories depending on whether they are above or below 50 percent base saturation. Soils in the group with less than 50 percent base saturation have, on the average, a lower level of inherent fertility and are ordinarily more responsive to treatment with lime.

Judgments of lime needs for the correction of acid-soil conditions are normally based on curves of the type shown in figure 7.2. Through use of such curves, one can estimate the change in the level of base saturation required to produce a desired increase in soil pH. The quantity of liming material that must be supplied to cause the required change in base saturation

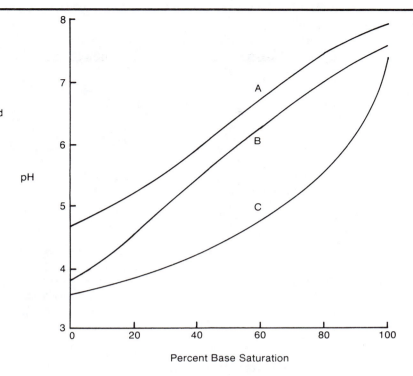

Figure 7.2.
The relationship between pH and percent base saturation for three types of cation-exchange materials in soils: (A) kaolinite, (B) organic matter, and (C) montmorillonite. (From Coleman, N. T. and Mehlich, A. In *Soils, 1957 Yearbook of Agriculture,* pp. 72–79. Washington, D.C.: Superintendent of Documents, 1957.)

Percent Base Saturation

can then be computed. The benefit derived from liming differs from place to place, but it may be from an increase in the level of plant-available calcium and magnesium directly supplied by the lime, from an increase in the availability of nutrients such as phosphorus and molybdenum, or from a decrease in the concentration of toxic elements such as aluminum or manganese. Changes of the latter two types relate to the influence of pH on the solubility of the elements involved.

Changing the status of exchangeable cations in soils

Chemical treatment of soils often changes the status of the exchangeable cations present. These changes may be intentional, as when adsorbed H and Al are removed by soil liming, or they may be unintentional, as where the formation of nitric acid from ammonium fertilizers increases the level of acidity in the soil. Regardless of the reason for these changes, it is often important to know the extent to which they will take place when a given quantity of a chemical compound is added to the soil. Approximate judgments of this can be made from data such as those shown in table 7.7.

Table 7.7 lists the important exchangeable cations in soils and the quantity in which they occur, in kilograms per 2,000,000 kg of soil, when present in the ratio of 1 me per 100 g of soil. Shown at the right in the table are quantities of various compounds that will supply the amounts of five of the elements given in the second column. For example, it takes 1000 kg of $CaCO_3$ to supply 400 kg of Ca, because the Ca content of $CaCO_3$ is 40 percent.

Table 7.7. Weights of ions per 2,000,000 kg of soil equivalent to 1 me of ion per 100 g of soil, along with amounts of selected compounds containing or otherwise capable of supplying the ions in the weights specified.

Ion	kg/2 million	Equivalent weight of compound
Al^{3+}	180	
H^+	20	320 kg S; 660 kg $(NH_4)_2SO_4$ fertilizer[a]
Ca^{2+}	400	1000 kg $CaCO_3$; 1720 kg Gypsum $(CaSO_4 \cdot 2H_2O)$[b]
Mg^{2+}	240	830 kg $MgCO_3$
Na^+	460	
K^+	780	1740 kg K_2SO_4; 1500 kg KCl fertilizer
NH_4^+	360[c]	1320 kg $(NH_4)_2SO_4$; 1600 kg NH_4NO_3; or 340 kg gaseous NH_3 fertilizer

[a]Sulfur oxidizes to H_2SO_4 yielding 2 H^+ ions for each S atom; oxidation of NH_4^+ to HNO_3 yields 2 H^+ ions for each NH_4^+.
[b]Gypsum is used as an amendment to supply soluble Ca^{2+} for the replacement of exchangeable Na^+ in strongly basic soils.
[c]Equivalent to 280 kg of elemental N.

Weights of fertilizers required to supply potassium or ammonium in a ratio of 1 me per 100 g of soil are listed primarily as a matter of interest. Normal applications of these fertilizers rarely exceed a few hundred kilograms per hectare. Such applications may therefore be expected to change the level of potassium or ammonium by no more than a fraction of a milliequivalent per 100 g of soil.

Anion exchange in soils

The ability of soils to hold anions in exchangeable form is attributed to a pH-dependent positive charge that can develop at clay surfaces under strongly acid conditions. The site of positive charge is presumably exposed hydroxyls at the clay-platelet edges and on surfaces of hydrous oxide minerals. Whereas these hydroxyls dissociate H^+ ions under basic conditions, they attract H^+ under conditions of high H^+-ion concentration. The binding of H^+ to these groups results in a residual positive charge capable of attracting negatively charged anions. The effect is most outstanding in clays of low total negative charge, which includes kaolinite and the hydrous oxides. The 2:1 clays of high permanent negative charge strongly repel anions, thus limiting their adsorption. Soil organic matter appears incapable of developing a positive charge under strongly acid conditions.

Interest in anion adsorption and exchange centers mainly on the chloride (Cl^-), nitrate (NO_3^-), sulfate (SO_4^{2-}), and phosphate (HPO_4^- and $H_2PO_4^{2-}$) ions, principally because of their importance to plant nutrition. Of these ions, chloride and nitrate appear to be the most weakly held and, therefore, the most easily displaced by other anions. Sulfate and phosphate ions are retained in much larger amounts. Furthermore, once retained after being added to soils, they are not fully recoverable by exchange with ions such as chloride and nitrate. The reason for this appears to be that, in strongly acid soils, sulfate and phosphate are not only adsorbed but combine with dissolved iron and aluminum and precipitate as very slightly soluble iron and aluminum sulfate and phosphate compounds. Because of this complication, measurement of anion exchange capacities is not clearcut, for it varies depending on the kind of anion used for its determination.

Anion adsorption and exchange is of little importance in soils where 2:1-type clays, such as montmorillonite, vermiculite, and illite, dominate. Normally, these clays are the principal kinds in slightly to moderately weathered soils. Where weathering has been more intense and of long duration, as is common in tropical and subtropical regions, clays of the 1:1 and hydrous oxides types become more prevalent. As a consequence, these latter soils naturally have a low cation-exchange capacity, and under some circumstances, may show a greater ability to retain anions than cations in exchangeable form.

Summary

Clays and organic matter in soils are able to attract and retain ions in a loosely held, or exchangeable, form. For the most part, exchangeable ions are the cations hydrogen, aluminum, calcium, and magnesium. Sodium may be an important exchangeable cation in salt-affected soils. In some strongly acid soils, development of a positive charge permits the retention of exchangeable anions.

Whether hydrogen, aluminum, calcium, or magnesium occur in large quantities as exchangeable ions depends on the extent to which soils have been weathered and leached. Young, unweathered soils tend to be high in calcium and magnesium, but these ions are replaced by hydrogen and aluminum under prolonged weathering and leaching. The build-up of hydrogen and aluminum causes an acid reaction in soils. Exchangeable calcium and magnesium cause a neutral to slightly basic soil reaction. Strongly basic reactions can result if sodium carbonate or large amounts of exchangeable sodium are present.

The ability of soils to hold cations in exchangeable form is expressed by the cation-exchange capacity, which varies with type and amount of clay and amount of organic matter. The cation-exchange capacity varies with pH, so it is measured at a fixed value, usually pH 7.0 or 8.2. At either of these values, the cation-exchange capacity is highest for organic matter, intermediate for expanding clays, and lowest for nonexpanding clays.

The preferred exchangeable cations are calcium and magnesium. Where these ions are lacking, as in strongly acid or strongly basic soils, they may be increased by the addition of amendments that supply them in readily soluble form.

Review questions

1. Distinguish between exchangeable and nonexchangeable cations in soils.
2. Identify the sources of negative charge in clays and soil organic matter, and explain why this charge varies with pH.

3. Why is the dissociation of ions important to cation-exchange reactions?
4. Explain the build-up of exchangeable hydrogen and aluminum in soils as they undergo continued weathering and leaching.
5. Why is a soil sample capable of holding 1 g of hydrogen ion in exchangeable form able to hold 20 g of calcium or 23 g of sodium in exchangeable form?
6. Why do exchangeable hydrogen and aluminum yield an acid soil reaction, whereas exchangeable calcium, magnesium, and sodium yield basic soil reactions?
7. Show why a 10-fold change in the concentration of hydrogen ions causes a 1-unit change in the pH of a solution.
8. Why is the negative charge that retains exchangeable cations in illite and chlorite less than the permanent charge of these clays?
9. Explain the differences in the cation-exchange capacities (pH 7.0) of the important soil clays.
10. Define buffering, and explain why this property is important to the amount of amendment needed to change the pH of strongly acid or strongly basic soils.
11. Define percent base saturation.
12. Under what conditions do some soils hold anions in exchangeable form, and why?

Selected references

Allaway, W. H. pH, Soil Acidity, and Plant Growth. In *Soil, 1957 Yearbook of Agriculture*, pp. 67–71. Washington, D.C.: Government Printing Office.

Bohn, H., McNeal, B., and O'Connor, G. Soil Chemistry. New York: John Wiley & Sons, 1979.

Coleman, N. T. and Mehlich, A. The Chemistry of Soil pH. In *Soil, 1957 Yearbook of Agriculture*, pp. 72–79. Washington, D.C.: Government Printing Office.

Pearson, R. W., and Adams, F., eds., *Soil Acidity and Liming*. Madison, Wis.: American Society of Agronomy, 1967.

Soil Survey Staff. *Soil Survey Manual*. U.S.D.A. Handbook 18, pp. 255–58. Washington, D.C.: Government Printing Office, 1951.

8

Water relationships in soils

Water performs a number of important functions in soils. It is essential for mineral weathering and organic matter decay, reactions that provide soluble nutrients necessary for plant growth. In addition water serves as the medium in which nutrients move to plant roots. When too much water is present, however, mobile nutrients can be lost from the root environment by leaching, or, where evaporation is high, dissolved salts may be carried to the surface layer of soil, often to accumulate in quantities that are damaging to plants. Excess water also limits the movement of air through the soil and may deny plant roots of much needed O_2. Water can therefore be either beneficial or detrimental to plant growth, depending on the amount that is present in the soil.

Water is also important to the physical behavior of soils. For example, it was shown earlier that the soil water content has a marked effect on consistence properties and on the suitability of the soil for cultivation. Similarly, variation in the water content can affect the compressibility of the soil, its load-bearing capacity, and its stability on sloping surfaces. These latter properties are often of great significance to the use of soils or soil materials for engineering purposes.

The retention of rain or irrigation water for plant use is one of the essential functions of soil. The soil supplies water for all phases of plant growth, including the synthesis of structural and protoplasmic tissue. Only a small fraction of all the water taken up is retained within or as a part of plant tissue, however. Indeed, more than 99 percent of the absorbed water may be lost to the atmosphere by transpiration that takes place through stomatal leaf openings. Thus, a plant's existence depends to a great extent on the ability of the soil to supply adequate water to take care of transpirational losses. Should the transpirational requirement not be met from soil sources, water will be sacrificed to the atmosphere from cellular plant tissue, which results in the loss of cell turgidity and in wilting. Prolonged wilting can ultimately lead to the death of plants.

Three interrelated functions are involved in supplying water to plants: (1) acquisition of water by the soil, (2) its temporary storage within soil pores, and (3) the delivery of the stored water to plant roots. The amount of water acquired depends in part on the ability of the soil to rapidly absorb and transmit surface-applied water downward. However, it is also controlled by external factors of water supply, such as the total annual precipitation and its distribution throughout the year. For the most part, water storage and its related function, delivery to roots, reflect the pore space configuration of the soil.

Expressing the water content of soils

The quantity of water held by a soil can be indicated in a number of ways. Relative terms such as wet and dry are often used, both of which represent indefinite ranges in water content and are thus subject to rather broad interpretation. Somewhat similar are the terms of *saturation* and *unsaturation*. Saturation defines a soil condition wherein all pores are filled with water, and unsaturation corresponds to any water content less than saturation (see fig. 8.1).

The absolute amount of water held by a soil is expressed on either a weight or volume basis. The preliminary determination is the loss in weight caused when a moist sample is dried at about 105°C for a 24- to 48-hour period. The loss in weight represents the water originally present in the sample. The water content of the soil, however, is customarily expressed as a percent of the oven-dry weight of a sample and is computed according to the equation:

$$\text{Water content (Wt.-\%)} = \frac{\text{Moist wt.} - \text{Oven-dry wt.}}{\text{Oven-dry wt.}} \times 100 \quad (8.1)$$

where the weights are of a soil sample before and after oven-drying. The oven-dry weight is used as the basis for the computation because it expresses the absolute quantity of soil present and also establishes a direct, linear relationship between absolute and percentage water contents. Thus, if the absolute amount of water in the soil is doubled the percentage water content will also double.

Whereas the determination of soil water content is usually by gravimetric means, it is more convenient to express the content on a volumetric basis, either as the *water ratio*, sometimes called the *volume fraction*, or as a *volume percent*. The water ratio is the ratio of the volume of water to the volume of the soil in which the water is retained, or:

$$\text{Water ratio} = \text{Volume of water/Volume of soil} \quad (8.2)$$

Multiplying the water ratio by 100 then converts it to a volume-percent:

$$\text{Water content (volume-\%)} = \text{Water ratio} \times 100 \quad (8.3)$$

In practical considerations of soil water contents, water ratios are used far more often than are volume-percent values.

The water ratio is a volumetric unit, but it also expresses the water content as an equivalent depth of water per unit of soil depth. Consider, for example, a one-centimeter depth of soil that has 50 percent pore space and

Figure 8.1.
Illustrating the effect of two different water contents on the degree of filling of soil pores. On the left, a saturated condition wherein all pores are filled. At the right, an unsaturated condition wherein water is retained principally in the narrower necks of pores and as very thin films over soil particle surfaces.

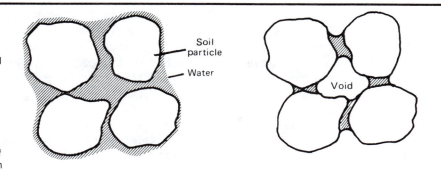

is saturated with water. Since all the pore space is filled with water, the volume of water is equal to half the soil volume. However, since the soil is a layer 1 cm deep, the water it contains will be equivalent to a layer 0.5 cm deep. In this instance, then, the water ratio of 0.5 identifies a depth of water per unit of soil depth. Similarly, if the soil with a water ratio of 0.5 were 1 meter deep, then the equivalent water depth would be one-half meter. The use of depths is convenient, since additions to or losses of water from the soil are usually expressed in depth units, as, for example, centimeters of rainfall or evaporation.

For soils of known bulk density, it is possible to compute the water ratio from gravimetric water-content values. The equation is:

$$\text{Water ratio} = \text{Water content (wt.-\%)} \times \frac{\text{Bulk density}}{100} \qquad (8.4)$$

where the bulk density is in grams per cubic centimeter.

Retention of water by soils

Water is held in soils by two means. One is over an impermeable subsurface layer that allows the water to accumulate when it cannot flow rapidly from the soil by normal drainage. Such an accumulation results in saturation of the soil, a condition often referred to as *waterlogging*. Waterlogging is a permanent feature of some soils, but in others it may be temporary, occurring only during a period of slow drainage following the input of excess water into the soil.

Water that saturates the deeper strata of soils is termed *ground water*. It often forms a continuous system with neighboring open bodies of water, such as lakes or rivers. Depending on local conditions, ground water may help sustain the level of water in these open bodies by flowing toward and draining into them, or it may be sustained by flow, or *seepage*, of water from the open bodies into the neighboring soil.

The second important mechanism of soil water retention is *capillarity*. This phenomenon is caused by two forces: (1) *adhesion*, in this instance, the attraction between soil particle surfaces and polar water molecules, and

(2) *cohesion*, here the attraction of water molecules for each other. Water is held in soil pores in films of varying thickness and against various forces of removal, including gravity, root absorptive forces, or those responsible for the evaporation of water into the atmosphere. If little water is present so that moisture films are spread thinly over soil particle surfaces, capillary forces hold the water very tightly and a strong force must be applied to remove it from the soil. With increasing water content and film thickness, however, the capillary force of attraction gradually declines until it becomes zero under conditions of *absolute saturation*. Absolute saturation is that moisture state where no more water can be drawn into soil pores by capillary forces even if they are unopposed by a force of removal.

Expressing the force of soil water retention

The pressure of water, ordinarily called *hydraulic pressure*, provides a practical means of expressing the force of soil water retention. Pressure is a direct measure of a force applied over a unit area, or:

$$\text{Pressure} = \text{Force}/\text{Area} \tag{8.5}$$

Thus, a change in the force holding water in the soil results in a comparable change in its hydraulic pressure, and it may be positive, negative, or zero. A positive hydraulic pressure is produced where water is subjected to a compressive force that tends to squeeze the water molecules into closer contact. In contrast, forces that tend to pull the water molecules apart produce a negative hydraulic pressure. If water is subjected to neither a compressing nor a pulling force, the hydraulic pressuree will be zero. Such a condition exists at a *free water surface*, that is, at the surface of an open body or container of water.

Pressures can be expressed in various ways. Typical examples are pounds per square inch and dynes per square centimeter, the dyne being an extremely small force corresponding to a weight of 0.00102 g. Other pressure terms have been more widely used for soil water, however. One is the *bar*, which is a fairly large unit equalling 10^6 dynes per square centimeter. If the hydraulic pressure is small, the term *centimeters of water* is commonly used, a practice that parallels the use of centimeters or millimeters of mercury to indicate barometric (atmospheric) pressures. A pressure of 1020 cm of water is equal to one bar; that is, a pressure of 1 bar is produced at the lower end of a water column 1020 cm high (33 ft., approximately) or at a depth of 1020 cm in an open body of water. A centimeter of water pressure is therefore equal to 0.00102 bar.[1]

Hydraulic pressure in an open body of water, such as a lake or river, is due to the weight of water induced by the pull of gravity. This weight, acting as a downward, compressive force, causes the hydraulic pressure of free water to be positive; it is zero at the free water surface and increases linearly by 1 bar for each 1020-cm increase in depth. In comparison with

[1]The SI unit of pressure is the pascal, abbreviated pa (see Appendix C). A bar is equal to 100,000 pa or 100 kpa (kilopascals). A bar is almost the same as the *standard atmosphere* of pressure, which is equal to 14.7 lbs per sq in, 760 mm of mercury, or 1033 cm of water.

Water relationships in soils

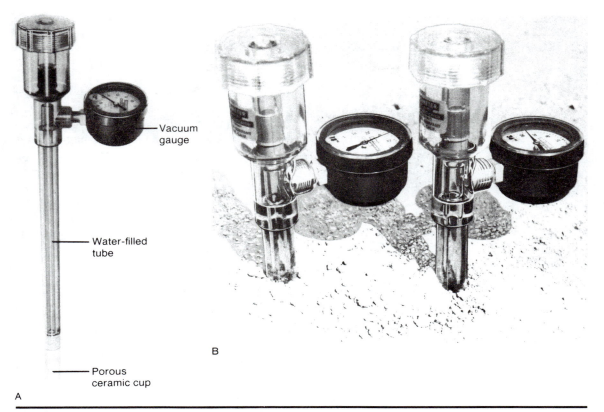

Vacuum
gauge

Water-filled
tube

Porous
ceramic cup

A

B

Figure 8.2.
(A) A tensiometer, which consists of a tube with a porous ceramic cup at the bottom and a vacuum gauge at the top. Water within the tensiometer forms a continuous system from the gauge through the pores in the ceramic cup to soil water outside the cup. (B) Two tensiometers installed in the soil. The one on the left is placed at a shallower depth and is in contact with drier soil, thus giving a higher tension reading than the more deeply placed tensiometer at the right. (Photos courtesy of the Irrometer Company, Riverside, CA.)

free water, that held by and subjected to the continuous pull of capillary forces in soils is under negative hydraulic pressure. The hydraulic pressure of capillary held water approaches zero in saturated soil but may drop to -1000 bars or less in soils that are air-dry. Water occurs in very thin films and is held with great tenacity in air-dry soils.

The use of negative values of hydraulic pressure is not particularly convenient and can be avoided through the use of positive terms such as *suction* or *tension*. In their simplest description, soil water suction or tension are a measure of the pulling force per unit area needed to produce a given negative hydraulic pressure in water. For example, if a suction of $+1$ bar were applied to a saturated soil at zero hydraulic pressure, water would flow from the soil causing the hydraulic pressure to drop until it reached -1 bar. At that point, water flow from the soil would stop, and the force of water retention by the soil, as indicated by a -1 bar hydraulic pressure, would have become equal to the applied suction force of removal.

The force of capillary retention of water in soils, as indicated by tension, can be rather easily measured by an instrument called the *tensiometer* (see fig. 8.2). This instrument consists of a porous ceramic cup connected through a water-filled tube to a vacuum gauge or other vacuum-measuring device. When the tensiometer is placed in the soil, water inside forms a continuous system with the soil water outside through the pores in the cup. Thus, any variation in the tension of soil water is transmitted throughout the system to ultimately register on the gauge. Tensiometers permit the continuous monitoring of soil water tension, but they are useful only in relatively moist soil, more specifically, where the tension remains below about 0.7 to 0.8 bar. At higher tensions, bubbles of water vapor form within the tensiometer. This interrupts the continuity of the liquid water system and renders the gauge reading invalid.

Water relationships in waterlogged soils

Waterlogged soils have a unique water-distribution pattern. Typically, the water occurs in two zones, one over the other, that differ with respect to both hydraulic pressure and water-behavioral characteristics. In the lower zone, the soil is absolutely saturated; thus, capillary forces have no effect on the water and its hydraulic pressure is positive. Above the zone of absolute saturation, water is held in the soil by capillarity and under negative hydraulic pressure. At the boundary between these two zones, the hydraulic pressure is zero. This boundary corresponds to the *water table* in waterlogged soils. Hydraulic pressure above the water table depends on the soil water content, but it tends to be most negative at or near the surface where the soil is usually the driest. Hydraulic pressure relationships below the water table are identical to those in a free body of water; the hydraulic pressure is zero at the water table and increases 1 bar for each 1020-cm increase in depth below.

As shown in figure 8.3, the water table in soil next to an open body of water is continuous with the free water surface outside. If there is no flow of water beneath the water table, the water table will be flat. Ordinarily, however, there is some movement of water beneath the water table, either toward or away from a natural drainageway or other body of water. Movement is indicated by a sloping water table, the direction of movement being in the direction of slope. Whether there is slope can be determined from the level of the water table at different locations over the landscape. The level of the water table at any one location is judged from the equilibrium level to which water rises in a well or hole that has been dug down into the saturated zone of a waterlogged soil (see fig. 8.3).

Capillarity not only contributes to the retention of water in soils but is also responsible for lifting water above the water table. This phenomenon is referred to as *capillary rise*, and it may take place in thin films pulled up

Figure 8.3.
Diagram of a waterlogged soil with a water table and an overlying capillary fringe. Shading indicates free water outside and saturation inside the soil. Note that the water table is continuous with the free water surfaces outside. It is also shown as being flat, which means that water is not flowing through the saturated soil below.

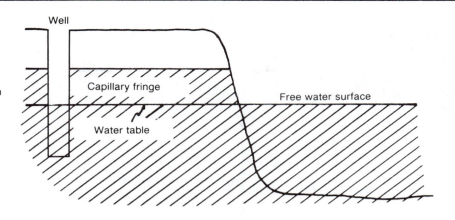

over pore walls or as continuous, pore-filling columns that rise at least a short distance above the water table. Adhesion is the prime cause of capillary rise as it pulls water upward to wet the pore wall. Cohesion holds the water molecules together so that they will rise as a continuous column filling the center of the pore. Upward movement of the column stops when its weight, acting as a downward force, becomes just equal to cohesional (surface tensional) forces among molecules forming the surface of the column that must support the entire weight of the column below.

The height to which a continuous capillary water column will rise depends on the diameter of the pore, and as shown in figure 8.4, the smaller the pore the greater the height. For water at 25°C and a wettable capillary material, such as glass, the relationship between height of rise and pore diameter is expressed by the equation:

$$\text{Height of rise } = 0.31/\text{Pore diameter} \tag{8.6}$$

where the height of rise and the pore diameter are in the same units, such as centimeters. According to the equation, a vertical capillary tube 1 mm (0.1 cm) in diameter will lift a water column to a height of 3.1 cm; if the diameter were 0.1 mm, the height would be 31 cm.

Because of capillary rise, most waterlogged soils have at least a thin zone of saturated soil immediately above the water table. This zone is called the *capillary fringe* (see fig. 8.3), and its thickness depends on how high the largest pores in a soil can lift water as a continuous column above the water table. A capillary fringe is likely to be absent in very coarse sandy or gravelly soils, but in fine, clayey soils it may extend for a meter or more above the water table. Because the roots of most plants will not penetrate saturated soil lacking oxygen, the presence of a capillary fringe can seriously reduce plant growth by limiting the depth of rooting in the soil.

Figure 8.4.
A model demonstrating the relationship between capillary tube size (pore diameter) and the height to which water is lifted as a continuous column above a free water surface.

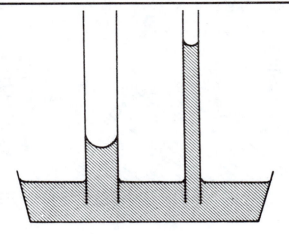

Energy relationships in soil water

As discussed above, the force of soil water retention can be expressed in pressure terms. However, forces that produce pressure in water also provide it with potential energy and the capacity to perform work.[2] It follows, therefore, that the force of soil water retention can be expressed by either the level of potential energy or hydraulic pressure. In most applications, the level of potential energy, or more simply, the *soil water potential*, is preferred.

The potential of soil water results from an interaction of the water with four different forces, namely: *matric* (capillary),[3] *osmotic* (ion-hydration), *gravitational*, and *weight*. The contribution of these forces is additive, their combined effect determining the *total potential* of soil water, or:

$$\psi_{tot} = \psi_g + \psi_p + \psi_m + \psi_o \qquad (8.7)$$

where ψ (Greek, *psi*) is the potential, and in combination with the subscripts, identifies the total potential or one of four *component potentials*, namely: gravitational (ψ_g), pressure (ψ_p), matric (ψ_m), and osmotic (ψ_o). However, there is no condition of water in soils where all four components act simultaneously as determinants of the total water potential. For example, the matric potential contributes to the total potential under unsaturated but not under conditions of absolute saturation. The reverse is true for the pressure potential.

[2]The potential energy of an object is a function of its position in a force field, the greater the distance from the center of force (center of attraction) the greater the potential energy. For example, if a stone is lifted above ground level, its potential energy increases because it is moved outward in the earth's gravitational field. The energy gained would come from the lifting force, and the amount would depend on the height to which the stone was raised. If the stone were then dropped, potential energy would decline, but it would not be lost. Instead, it would be converted to kinetic energy (energy of motion), and the conversion would be complete as the stone struck the ground. Any work performed by the stone, such as compressing the ground on impact, would depend on the kinetic energy content at the end of the fall.

[3]Matric is synonomous with capillary. It is derived from the word matrix, which here refers to the solid particles making up the porous body of the soil.

The soil water potential is important not only because of its relation to water retention but also because of its influence on the flow of water in the soil or between the soil and some external system. The one essential principle regarding flow is that it takes place only if there is a difference in total potential from place to place in the soil, the direction of flow being in the direction of decreasing potential.

Water potential values can be conveniently expressed in pressure terms such as the bar or centimeters of water. The reason for this is that pressure and potential are measures of the same thing, the interaction of water with one or more forces. Values assigned to water potentials are relative rather than absolute, however; they express the potential of the water in question relative to that of a *reference* state or condition where the potential is zero. As may be seen in the following discussions, the potential of soil water may be positive, negative, or zero depending on whether it is greater than, less than, or equal to the potential of water in the reference state.

Gravitational potential

This potential expresses the level of energy in water caused by the constant pull of gravity, and its magnitude is a function of the position, or elevation, of water in the earth's gravitational field. The gravity potential increases with increasing elevation, as when water is pulled upward by capillarity, and decreases with decreasing elevation, as when water is moved downward through soil by the pull of gravity. Water would lose all energy derived from gravity (the gravitational potential would drop to absolute zero) at the earth's center of gravity. However, since the location of the center of gravity is not known, the reference of the gravitational potential must be placed at an arbitrarily assigned elevation. In waterlogged soils, the water table is commonly used. If there is no water table, then the reference, or zero, gravitational potential can be assigned to any elevation that is convenient.

Once the reference of the gravitational potential has been established, the gravity potential at any point in the soil is determined by the vertical distance between that point and the reference. The relationship is such that the gravitational potential increases by 1 bar for each 1020-cm increase in elevation. Thus, water located 102 cm (about 3 ft.) above a water table assigned as the reference would have a gravitational potential of $+0.1$ bar, but if it were 102 cm below the water table, the gravitational potential would be -0.1 bar.

Pressure potential

This potential accounts for the influence of weight on the total potential of soil water. A build-up of pressure of entrapped air acting against the weight of overlying soil can be one cause of a pressure potential in soil water. Another is the positive hydraulic pressure beneath a water table. To distinguish it from pressure potentials produced by others causes, that due to positive hydraulic pressure beneath a water table is termed *submergence potential*.[4] The submergence potential is the only pressure potential considered here.

[4]See Rose, C. W., in the references at the end of this chapter.

The submergence potential, like positive hydraulic pressure, is zero at the level of the water table (or of a free water surface outside the soil) and increases by 1 bar for each 1020-cm increase in depth below. As will be shown later, a difference in submergence potential accounts for lateral water flow beneath a water table.

Matric potential

The matric potential of soil water is a function of the soil water content. In a soil pore at absolute saturation, matric (capillary) forces are completely satisfied, so they have no effect on the energy status of the water. The matric potential is zero under these circumstances. If the pore is partially emptied, however, the water remaining will be more tightly held, and its freedom of movement and level of energy, as indicated by the matric potential, will be reduced. This results in the matric potential dropping to a value less than that in saturated soil, where it is zero. Since a value less than zero is negative, matric potentials, if not zero, are negative, and the lower the water content the more negative the value.

Like negative hydraulic pressure, the matric potential of soil water can be judged from measured soil moisture tension or suction values. For example, if the suction or tension of soil water is +1 bar, the matric potential will be -1 bar. As with the negative hydraulic pressure, the matric potential may range from zero in soil that is absolutely saturated to -1000 bars or below in air-dry soil.

Osmotic potential

The osmotic potential expresses the difference between the potential of soil water and that of pure (ion-free) water occurring under otherwise identical gravitational and matric potentials. Since there are no ions in pure water, its osmotic potential is zero. Like matric forces, osmotic forces reduce the potential of water. Osmotic potentials are therefore negative in sign, and they vary between zero in pure water to very low (highly negative) values where the concentration of salts is high.

Unlike matric forces, osmotic forces associated with dissolved ions are unable to effectively move water from pore to pore in soils. Ions are not anchored like soil particles responsible for matric forces; rather, they move freely in the water or with it. However, ions isolated on one side of a semipermeable membrane, which is a boundary more permeable to water than to ions, do behave as though they were in a fixed position. As a result, variation in ion concentration through a semipermeable membrane produces a difference in potential capable of causing water flow. Flow occurs in the direction of highest concentration, since this corresponds to the direction of decreasing osmotic potential.

Osmotic effects on the soil water potential are important under two circumstances: in the evaporation of water from soil water films, the surface boundary of the films being a perfect semipermeable membrane, and in the transfer of water through the cell wall of living organisms, such as plant roots and soil microbes. Greatest concern over osmotic effects is in salt-laden soils where a high concentration of ions in the soil solution may limit the uptake of water by plants and germinating seeds.

Figure 8.5.
Diagram showing water under five different conditions, or energy levels, caused by differences in the matric, osmotic, and gravitational potentials. Changes from left to right represent differences in matric and osmotic potentials while the gravitational potential is held constant. Changes on the vertical are due to variation in the gravitational potential while the matric and osmotic potentials remain constant. For further details, see the text.

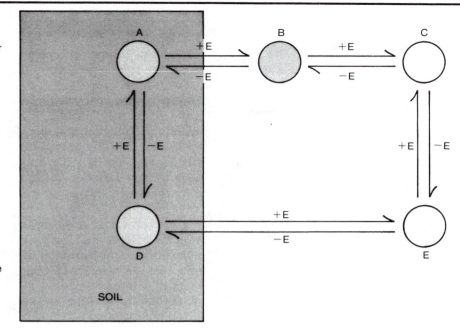

Total water potentials

In unsaturated soil the total potential of soil water is expressed relative to a reference where the matric, osmotic, and gravitational potentials are zero. Thus, the total potential is a function of the degree to which the component potentials vary from their reference, or zero, value. This relationship is demonstrated through use of figure 8.5.

In figure 8.5, A represents soil water with some soluble ions; both its matric and osmotic potentials are negative. If this water were pulled from the soil and converted to a free state without a change in ionic composition or elevation, the work performed on the water would increase its energy content ($+E$) and raise the matric potential from its original negative value to zero, as at B. If, in turn, the water at B were pulled away from the ions and converted to pure, free water without a change in elevation, the work performed would supply the energy needed to raise the osmotic potential from its original negative value to zero, as at C. If the reference, or zero value, of the gravitational potential were assigned to the same elevation as A, the gravity potential at C would also be zero. This would result in a total potential of zero at C, thus making C the reference of the total potential at A.

As may be judged from the above, the potential at A in figure 8.5 is expressed by the difference in potential between A and C, and it is equal to the sum of the differences in matric and osmotic potentials. The potential at A is less than at the reference C by the amount of energy lost ($-E$) if the pure, free water at C were reabsorbed by the soil and returned to the original condition at A. The energy lost would reappear as heat caused by the interaction of the water with the ions (heat of solution or ion-hydration) and with soil particle surfaces (heat of wetting). The effect of these reactions is easily recognized by the rise in temperature that occurs when dry soil is moistened with a small amount of water.

If the reference of the gravitational potential were moved from the level of A to D in figure 8.5, the reference of the total potential would shift from C to E, which represents pure, free water at the same elevation as D. This would result in an increase in the total potential at A by an amount equal to the increase in the gravitational potential at A. With the gravity potential reference at D, the gravitational potential at A would be greater than at D by the amount of energy gained in lifting water through the distance from D to A, or by the amount of energy lost by water draining back from A to D, either change taking place without variation in the matric and osmotic potentials. Energy gained by the water carried up from D to A would come from the force of lift; that lost during drainage would be the energy consumed in overcoming friction and would reappear in the system as heat.

The effect of moving the reference of the gravity potential from A to D on the total potential of water at A in figure 8.5 can be shown by a simple example. For this, assume that the matric and osmotic potentials at A are -1.0 and -0.1 bar, respectively. If the reference of the gravity potential, or zero, were assigned to A, the total potential there would be -1.1 bar $[-1.0 + (-0.1) + 0 = -1.1]$. If a gravitational potential of zero were assigned to D, and D was 102 cm below A, the gravitational potential at A would be $+0.1$ bar, and the total potential would be -1.0 bar $[-1.0 + (-0.1) + 0.1 = -1.0]$. In this example, then, the total potential at A is 0.1 bar greater when the reference of the gravitational potential is at D than when it is at A.

Very often, the gravitational and osmotic potentials can be ignored as determinants of the total soil water potential. Consider, for example, potentials that affect the absorption of water by plant roots. Absorption occurs if there is a decrease in total potential from outside to inside the root. The gravity potential does not contribute to this decrease because, on the average, it is the same in the root as in the water immediately surrounding the root. The gravity potential can therefore be eliminated from the determination of total water potential by setting it equal to zero at the elevation of the root.

The osmotic potential can be treated similarly. In the vast majority of soils, ions are present in such low concentrations that they have virtually no effect on the total soil water potential. When this applies, the osmotic potential may also be set equal to zero, and the total soil water potential thereupon judged solely from the matric potential. The osmotic potential will likely be a significant part of the root water potential, however.

Matric potential and the soil water content

The matric potential varies with and is therefore an index of the soil water content. However, the relationship between matric potential and water content differs among soils, primarily because of differences in texture and pore space configuration. This fact is indicated by curves in figure 8.6, which show water content as the water ratio plotted against the matric potential, which

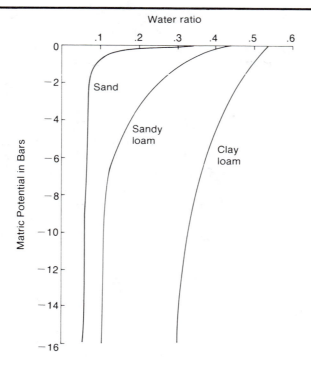

Figure 8.6.
Characteristic soil water curves for soils of three widely different textures.

starts with zero for saturated soil at the top of the graph and becomes increasingly negative in a downward direction. These curves, which are called *characteristic soil water*, or *water retention, curves*, are constructed using measured water contents of soils that have been adjusted to different matric potentials in the laboratory. Once constructed, these curves can be used to estimate water contents from known matric potentials, or matric potentials from known soil water contents.

According to the curves in figure 8.6, the finer the texture of the soil the higher the water content over a wide range of matric potential. This difference among soils reflects differences in both porosity (total pore space volume) and pore size. Fine soils have the highest porosity, so they retain the most water when saturated. They also contain a high proportion of small pores that remain filled, or nearly so, even with a fairly large decline in matric potential below zero. In contrast, coarse soils have a high proportion of large pores that are readily emptied by comparatively small removal forces. As a consequence, the curve for the coarse soil in figure 8.6 shows a sharp decline in water content with only a slight·decrease in the matric potential below zero. From that point on, however, small losses of water are accompanied by a rapid decline in the matric potential.

The range in matric potential shown in figure 8.6 corresponds closely to that measured in soils supporting plant growth. Generally speaking, the best plant growth occurs if the average matric potential in the root zone does not drop much below −2 to −3 bars. As the matric potential falls below this range, the availability of water declines and tends to become negligible at matric potentials of -15 bars or less. Thus, reductions in the matric potential below -15 bars usually result only from evaporation, as from the surface layer of soil. Air-drying may cause the matric potential of soil to drop -1000 bars and below.

Water flow in soils

Variation in the water potential is responsible for the movement of soil water in both liquid and vapor form. However, flow causes a reduction in the water content where the potential is high and increases the water content where the potential is low. Thus, the tendency in flow is to reduce the difference in potential and to establish an equilibrium characterized by the absence of water movement. Such equilibria are probably never attained, principally because the total quantity of soil water on which the equilibria depend undergoes continuous change, either because of additions of water to the soil, as by rain or irrigation, or because of water loss through plant use, evaporation, or drainage.

Although water flow in soils always depends on a difference in the total potential, this difference in potential is usually the result of variation in only one or two of the component potentials. For example, *saturated flow*, defined as flow through filled pores where the matric potential is zero, is due entirely to a difference in the submergence potential. In contrast, *unsaturated flow*, defined as flow where the matric potential is less than zero, even through water-filled pores, as in a capillary fringe, is due to a difference in the matric and gravitational potentials. Water transfer from soil moisture films into neighboring, empty pore spaces (evaporation), or from the films into roots is caused solely by a difference in matric and osmotic potentials. These relationships are summarized in table 8.1.

Rate of soil water flow

The rate of water flow in soils is measured as the volume of water delivered through a unit cross-sectional area of soil per unit of time, and it is a function of two variables: (1) the *potential gradient*, or the difference in potential per unit of flow distance, and (2) the *hydraulic conductivity*, or the ability of the soil to transmit water under an imposed potential gradient. More often than not, variation in the rate of water flow in soils is explained on the basis of variation in hydraulic conductivity rather than in the potential gradient. More attention is given to hydraulic conductivity for this reason.

Texture and hydraulic conductivity of soils

The ability of soils to conduct water is a function of pore size and the degree to which the pores are filled with water. The hydraulic conductivity is greatest in filled pores (saturated soil). It is less in partially filled pores because, as shown in figure 8.7, flow under these conditions is restricted to thin films over pore walls. This not only reduces the volume of water flowing through the pores at any one time, it increases the average length of the flow path through the pore and forces all flow to occur near the pore wall where resistance to flow (frictional loss) is the greatest.

Figure 8.7.
Diagram illustrating the effect of an air-filled void on water flow through a soil pore. The void not only reduces the total volume of water moving through the pore at any one time, it restricts flow to the relatively long path around the periphery of the pore where frictional loss is much higher than in the center of a filled pore.

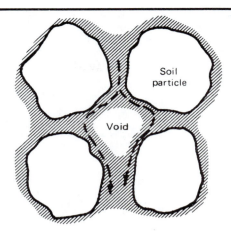

Table 8.1. The relationship between conditions of water flow in soils and the components of potential associated with the flow.

Condition of flow	Component of ψ_{tot}
Saturated flow[a]	ψ_s
Unsaturated flow, pore-to-pore	$\psi_m + \psi_g$
Unsaturated flow, pore-to-root	$\psi_m + \psi_o$
Evaporation	$\psi_m + \psi_o$

[a]Limited to conditions where the matric potential is zero.

In any pore, the friction encountered by moving water decreases with increasing distance from the pore wall and reaches a minimum at the center of the pore. In large soil pores, the distance from the wall to the pore center can be many times greater than in small pores. Because of this, the friction encountered in saturated flow through large pores is much less than in fine pores and accounts for a substantially higher hydraulic conductivity for saturated coarse- than for saturated fine-textured soils. This fact is indicated by the difference in hydraulic conductivity for coarse and fine soils at zero matric potential in figure 8.8.

In contrast to the above, the hydraulic conductivity of unsaturated coarse soils is usually less than that of unsaturated fine-textured soils (see fig. 8.8). This is explained in part by the characteristic soil water curves in figure 8.6, which show a much greater water loss from coarse than from fine soils as the matric potential declines from zero. As a result, the hydraulic conductivity of coarse soils drops below that of finer soils at matric potentials only slightly below zero. On the average, then, fine soils have the highest hydraulic conductivity when unsaturated. However, even when saturated, these soils tend not to be very good conductors of water.

Figure 8.8.
The relationship between hydraulic conductivity and matric potential for soil materials of widely different pore space configuration, as indicated by texture. The conductivity at zero matric potential is for saturated soil. The range in matric potential is from zero to about −10 bars.

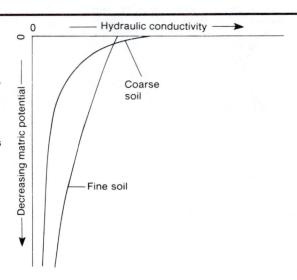

Water flow beneath a water table

Water flow beneath a water table is through saturated soil and is due solely to a difference in the submergence potential. This can be explained through use of figure 8.9 in which the sloping line shows the position of the water table. In this example, flow beneath the water table is from X toward Y, or in the direction of slope of the water table. Flow takes place in that direction because the submergence potential at X is greater than at Y.

The rate of water flow beneath a water table is a function of the saturated hydraulic conductivity and the gradient of the submergence potential. In uniform soil materials, the saturated hydraulic conductivity is essentially constant. Thus, flow beneath the water table will vary mainly with changes in the gradient.

In figure 8.9, the gradient of potential is equal to the difference in submergence potential from X to Y divided by the distance between these two points. Since Y is at the level of the water table, the submergence potential there is zero. The difference in potential between X and Y is therefore equal to the depth of X below the water table, as indicated by the letter s in the diagram. The distance of flow in this example is fixed and is equal to the distance from X to Y in the diagram. Thus, variation in the flow rate occurs whenever there is a change in the elevation of the water table above X, or in s. An increase in s increases the potential gradient and the rate of flow, a decrease in s having the reverse effect.

The distance s in figure 8.9 also determines the slope of the water table. Thus, the greater the value of s the steeper the slope of the water table and the faster the rate of saturated flow below. Flow will occur so long as s is greater than zero. Should the water table drop to the line XY, both the distance s and potential gradient would become zero and flow under the water table would stop.

Water relationships in soils

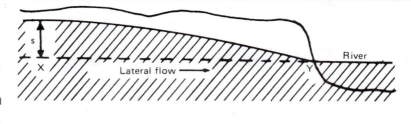

Figure 8.9.
Diagram illustrating conditions promoting saturated water flow beneath a water table. The vertical distance s represents the difference in submergence potential that occurs over the lateral distance from X to Y. This vertical distance, when divided by the distance from X to Y, gives the average potential gradient of flow between X and Y.

Saturated flow beneath a water table is due to a natural tendency for water to seek a uniform level. Except where the texture is very coarse, the rate of flow beneath a water table in most soils is comparatively slow, so that differences in elevation of the water table, once established, are not rapidly eliminated. Further, the input of water into a soil, as during a rainy period, limits the effect of drainage in lowering the water table. For these reasons, the water table in soils usually has some slope. Where the water table is flat, or nearly so, lateral flow keeps pace with input of water into the soil. This situation may be realized in coarse soils of high saturated conductivity, in soils where the distance of lateral flow is short, or where the input of surface-applied water is infrequent. Under the latter condition, the level of the water table may actually be maintained by seepage of water from a permanent waterway.

Unsaturated water flow in soils

The transfer of the great bulk of water in unsaturated soils is due to differences in the matric potential. Gravitational forces oppose or aid the transfer, provided there is a change in the elevation of the water as flow takes place. Gravity assists unsaturated flow if it is in a downward direction, but acts as an opposing force if flow is upward. Principally, unsaturated flow is of significance in: (1) the rise of water from a water table, (2) soil-wetting by surface-applied water, and (3) the transfer of stored water to plant roots.

Rise of water from a water table

The rise of water from a water table occurs whenever the total potential in the overlying soil decreases in an upward direction. The difference in potential responsible for flow reflects primarily a difference in the matric potential, which is zero at the water table but may fall to a negative value of several hundred bars in dry surface soil. Gravity opposes capillary rise, but its effect on the total potential is small; the gravity potential increases in an upward direction, but only by about 0.3 bar per meter change in elevation. Although gravity affects the total potential very little, it can have a profound effect on the rate of capillary rise by preventing the filling of large pores and the formation of a capillary fringe just above the water table. Recall that flow through saturated soil, as in a capillary fringe, is greater than through unsaturated soil.

On a sustained basis, the capillary rise of water in coarse soils that lack a capillary fringe is often too slow to be significant, especially if the water table is deep. The same is true for fine clays, even though they may have a thick capillary fringe. This is because clays tend to have such a low saturated hydraulic conductivity.

Soils of medium texture, especially those high in silt, allow the most rapid rise of water by capillarity. Pores in these soils are small enough to remain filled, or nearly so, for substantial distances above the water table, but they are not so small as to seriously limit the rate of saturated flow through frictional loss. The comparatively rapid rise of water in waterlogged, silty soils of colder regions often results in serious frost-heave problems as described on page 521.

The capillary rise of water is of greatest importance in drier regions where soil drying by evapotranspiration goes on most of the time. It is of less significance in humid regions where natural precipitation supplies more water than is lost by evapotranspiration. Under these conditions, the average direction of water flow in the profile is down.

Soil wetting

The entry, or *infiltration*, of surface-applied water into and its movement through the soil is in response to the combined influence of matric and gravitational forces. In wetting, pores at the surface are the first to receive water, and after partial or complete filling, they provide the pathway through which water flows to progressively greater depths in the soil. By this means a gradually expanded wetted zone develops. The water in this zone moves as a continuous system extending from the surface downward to the point of farthest water advance, or the *wetting front* (see fig. 8.10). If the soil being wetted is dry, the location of the wetting front is easily recognized as the boundary between the dark, wetted soil above and the light-colored, unwetted soil below.

Maximum infiltration and wetting occur where water is applied so rapidly that it accumulates on the surface and saturates the soil to at least a shallow depth. Under these conditions, infiltration is controlled by the rate at which water can be absorbed and transmitted through the saturated zone and on to greater depths in the profile. In general, this occurs most rapidly in coarse-textured soils, which have an inherently high saturated conductivity. However, infiltration into some highly weathered, fine-textured tropical soils can be substantial. Large pores associated with a well developed and stable structure accounts for the highly permeable nature of these soils.

The gradient of potential causing infiltration and wetting of soil is a function of the difference in matric and gravitational potentials from the soil surface to the wetting front. If the soil being wetted is dry, the matric potential at the wetting front may be several bars negative. The difference in matric potential across the wetted zone will be of equal magnitude, since the matric potential of the saturated soil at the surface will be zero. As with the

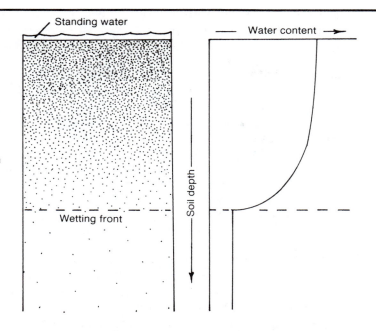

Figure 8.10.
The distribution of water in a soil profile during wetting as indicated by shading on the left and by a water content curve on the right. With increasing depth, the water content decreases from saturation at the surface to a minimum at the wetting front.

Standing water

Water content →

Soil depth

Wetting front

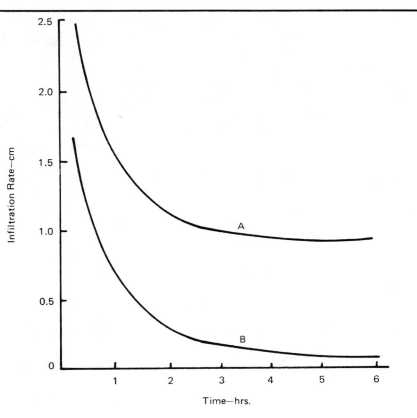

Figure 8.11.
Curves illustrating the change in the infiltration rate of soils with time. Curve A is an average for soils of inherently high infiltration capacity; curve B, for soils in which infiltration is minimal. The majority of soils lie between these extremes. (Adapted from B. W. Musgrave, in *Water, 1955 Yearbook of Agriculture,* pp. 151–59, Washington, D.C.: Government Printing Office, 1955.)

Infiltration Rate—cm

A

B

Time—hrs.

capillary rise of water, differences in the gravitational potential associated with soil wetting are usually quite small. Even so, in saturated soil at the surface, where the matric potential is zero, gravity is the only force causing the downward flow through the soil.

As shown in figure 8.11, infiltration decreases as the length of the wetting period increases. A principal reason for this is a decrease in the potential gradient as the distance of flow between the soil surface and the wetting front increases. The swelling of soil as it wets or the filling of pores in shallow soil also cause infiltration to decrease with time.

Curves in figure 8.11 show fairly typical infiltration rates encountered in the field. Many soils of medium texture, after an initial period of wetting, have infiltration rates of around 1 cm/hr, as shown by the upper curve in the figure. The lower curve is more characteristic of clayey soils, although where the clay is an expanding type, swelling may ultimately cause the infiltration rate to drop essentially to zero. The same is true for shallow soils of low total pore space volume.

Soil wetting from furrows

Irrigated soils are often wetted from furrows, or trenches, laid out in a parallel fashion to carry water across a field. The movement of water from a furrow into the soil involves lateral as well as downward flow and, as illustrated in figure 8.12, causes different wetting patterns depending on the texture of the soil.

Lateral flow from furrows into a soil is caused solely by matric forces. Wetting in a lateral direction is by unsaturated flow, therefore, and it tends to be slower in coarse- than in fine-textured soils. In comparison, downward flow, which is controlled by the saturated conductivity of the soil beneath the furrow, is faster in coarse- than in fine-textured soils. The net effect is that coarse soils wet more rapidly in a downward direction, which results in the oblong wetting pattern shown in figure 8.12. In finer soils, unsaturated flow to the sides is almost as fast as is flow in a downward direction beneath the furrow. As a result, the wetting pattern for these soils is almost semicircular.

Figure 8.12 indicates that different irrigation practices are required for soils of widely different textures. Sandy soils need not be irrigated for long periods, since they wet deeply and quickly. Where irrigation is from furrows, such as those shown in figure 8.12, the spacing between furrows should be narrower in sandy than in fine-textured soils. A narrow spacing is necessary so that complete wetting of the soil between furrows is assured. If the surface soil is not completely wetted, full advantage will not be taken of the total storage capacity of the soil for the water. Further, maximum root activity, including the uptake of nutrients, occurs in the surface soil. If the surface soil is not maintained in a moist state, plants are forced to exist on the less vigorous and less extensive system of roots in the subsoil.

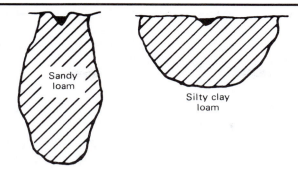

Figure 8.12.
Wetting patterns in coarse and fine soil materials after several hours of irrigation from a furrow.

Sandy loam

Silty clay loam

Drainage following soil wetting

Soil wetting is followed spontaneously by the drainage of excess water from the saturated pores at the surface and its redistribution into underlying soil layers. Drainage under these conditions is self-limiting, for it tends to equalize conditions responsible for the downward flow of water. Thus, drainage slows with time, eventually attaining a negligible rate. Once this point is reached, the amount of water left in the soil approaches a constant value referred to as the *field capacity*. In concept, the field capacity corresponds to the maximum amount of water a thoroughly (deeply) wetted soil will retain against normal drainage. Each time a soil is wetted and drained, it will return to this characteristic value.

In general, drainage to the field capacity takes place more rapidly in coarse- than in fine-textured soils (see fig. 8.13). This is due primarily to the ease, or speed, with which drainage forces can empty the predominantly large pores of coarse soils; drainage from them can come to a virtual standstill in a matter of only two or three days. In comparison, drainage to the field capacity may take up to a week in finer soil; not only is there more water to be drained in these soils, the rate of drainage is appreciably slower than in coarse-textured soils.

The slowing of drainage and the establishment of an almost constant water content characteristic of the field capacity is due principally to the emptying of pores and the consequent reduction in hydraulic conductivity to a limiting value. In sandy soils, conductivity becomes limiting to drainage while the films conducting water are still moderately thick; the matric potential may have dropped to only -0.1 to -0.15 bars by this time. In medium- to fine-textured soils, however, drainage goes on in a much larger number of pores, so films can become thinner before the drainage rate becomes critically low. The field capacity may not be reached in these soils until the matric potential drops to -0.3 to -0.5 bar, with an average of about $-1/3$rd bar for medium-textured soils. Although not particularly accurate for soils of extreme texture, an estimate of the field capacity is sometimes made by determining the water content of soil samples that have been adjusted to a matric potential of $-1/3$rd bar.

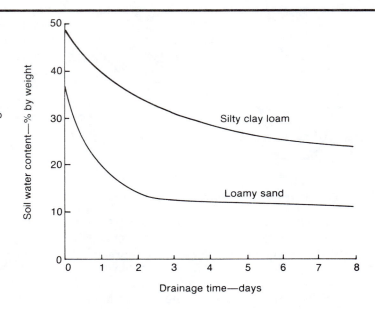

Figure 8.13.
Drainage following wetting as a function of soil texture and time. In general, coarse soils drain more quickly to an essentially constant water content than do soils of finer texture.

Soil water content—% by weight

Silty clay loam

Loamy sand

Drainage time—days

Textural stratification and drainage

After a thorough wetting, soils drain differently if they are texturally layered than if they are of reasonably uniform texture with depth. To illustrate, consider the effect of a sand layer in an otherwise uniform loam profile. If there were no sand layer, the loam, after wetting, would likely drain to a matric potential of about -0.3 bar. However, drainage through a sand layer inserted in the profile, and, therefore, from the overlying soil, would become limiting at a matric potential of no lower than about -0.1 to -0.15 bar. This would result in the loam retaining more water than had it drained to a matric potential of -0.3 bar. A similar effect would result from a compact clay layer in the profile, for such a layer would limit drainage from overlying soil even if the profile were saturated with water.

The incomplete drainage of texturally stratified soils may have either desirable or undesirable effects. One desirable consequence is the increased retention of water for plant use. Incomplete drainage is undesirable, however, where excess water causes serious aeration or plant-disease problems. An example of the latter effect is the frequent incidence of disease in grass on golf greens that drain incompletely because they consist of a few centimeters of soil placed over a layer of sand. Ironically, the purpose of the sand is to improve drainage, but unfortunately, this use overlooks the fact that sand conducts water well only when it is saturated or nearly so. The common practice of watering golf greens on a daily basis only adds to the problem.

A problem somewhat similar to that cited above often occurs with plants grown in pots or on greenhouse benches. Soil in pots or in shallow beds on benches is, in effect, suspended in air, that is, it lacks an underlying capillary system capable of extracting any excess water that may be present. Soils under these conditions tend to remain saturated unless they contain a system of large pores that can be largely emptied by the pull of gravity. A better pore-size distribution is provided in standard potting and greenhouse soil mixes prepared by combining rather large proportions of coarse materials, such as sand, peat, perlite, or vermiculite, with small amounts of finer soil materials.[5] The addition of the coarse materials more nearly assures a high content of large, easily drained pores in the soil mix.

Utilization of water by plants

Water is essential to all phases of plant growth. It is a reactant in the photosynthetic production of sugar, and it serves as the medium in which all biochemical reactions take place. The translocation of soluble materials, including nutrients and sugar, depends on water flow within plants. Water also plays an important role in plant-temperature control by carrying away excess heat during transpiration.

Water has an even more direct influence on growth because of its ability to exert a positive internal pressure (*turgor pressure*) on plant cells. This pressure is necessary if newly formed cells are to enlarge. Should a water shortage develop, turgor pressure is reduced and cell enlargement slowed. If the shortage becomes acute, turgidity is lost and wilting results. Normally, any sign of wilting in plants is an indication that active growth has stopped.

Plant growth can be restricted even if no wilting occurs. This fact is demonstrated by figure 8.14, which shows the effect of a reduction in soil water content on the rate of elongation of cotton leaves. The plants were grown in pots and the length of leaves measured daily as the plants gradually depleted the supply of water in the soil. At the beginning of the growth period, when the soil was moist, growth was rapid, as indicated by the sharp initial rise in the curve of figure 8.14. As the water content of the soil and the rate of water delivery to roots decreased, however, growth declined and eventually stopped altogether. When a second irrigation was made, the plants again started to grow normally. It is significant that at no time was there an indication of wilting in the cotton.

The use of soil water by plants

Under most circumstances the water content of soils undergoes an essentially continuous cyclic change. Increases in the water content occur each time the soil is wetted by a rain or irrigation. The water content is then reduced by the combined effects of drainage, evaporation, and plant use. Whereas significant water loss by drainage and evaporation continues for only a relatively short time following wetting, plant use extends over rather prolonged periods. In most instances, plant use accounts for the bulk of water loss from soils.

The absorption of water by plants is opposed by combined matric and osmotic forces and takes place so long as the total water potential is less inside than outside the root. Whereas both matric and osmotic effects are important determinants of the total potential within the root, it is usually only

[5]The perlite and vermiculite are commercially produced mineral particles that are highly porous and, therefore, of low density. The particles usually are about the size of fine gravel.

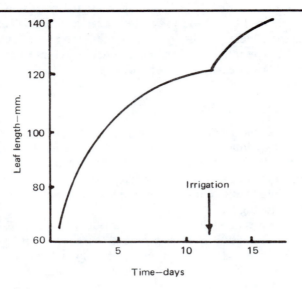

Figure 8.14.
The effect of soil drying and subsequent irrigation on the rate of cotton leaf elongation. (From C. H. Wadleigh, in *Water,* 1955 *Yearbook of Agriculture,* pp. 358–61. Washington, D.C.: Government Printing Office, 1955.)

the matric potential that is important on the outside. This is because of the very low concentration of soluble ions in most soils. Yet, osmotic effects can be highly significant in poorly drained, arid-region soils where substantial salts have accumulated. The origin, occurrence, and water-supplying characteristics of these soils are discussed in considerable detail in Chapter 18.

Soil water use and plant wilting

A soil that has been recently wetted and drained to the field capacity normally holds water rather weakly and offers little resistence to its uptake by roots. With time, however, a reduction in the soil water content and hydraulic conductivity slows the delivery of water to roots, thus slowing uptake. If the rate of uptake drops below the rate of water loss by transpiration, wilting results. If no more water is added to the soil, the severity of wilting will increase, and the plants may either go dormant, or they may ultimately die because of dessication.

The permanent wilting point

For plants subjected to normal daily fluctuations in temperature and humidity, the first signs of wilting are usually temporary, apparent only during mid-day when the temperature is high, the relative humidity low, and transpiration by the plant near maximum. With each passing day, however, the period of wilting increases until it eventually becomes continuous. When this point is reached, further removal of soil water drops to such a low level that the soil moisture content becomes essentially constant. The moisture content of the soil at this time is termed the *permanent wilting point,* and it corresponds to the lowest practical limit of water removal from soil by plants. Plants may reduce the water content to the permanent wilting point throughout a part of the root zone without harm, but removal beyond a certain level can lead to continuous wilting and ultimately to possible plant damage.

Water relationships in soils

A rather close relationship exists between the permanent wilting point and the average matric potential under a wide range of soil and plant conditions; more often than not, the average matric potential in the root zone at permanent wilting is near -15 bars. Because of this relationship, the water content of soil that has been adjusted to a matric potential of -15 bars is sometimes used as an estimate of the permanent wilting point. This procedure is quite satisfactory provided the matric potential gives a good estimate of the total potential of soil water. If salts are present in relative abundance, on the other hand, the total potential may be significantly influenced by the osmotic potential. This will cause the permanent wilting point to occur at a matric potential that is higher (less negative) than -15 bars.

Even in soils of negligible salt content, plants start to wilt at matric potentials considerably above -15 bars. However, matric potentials associated with the first signs of wilting are highly variable, for they depend not only on the rate at which water is being delivered to roots, but also on the immediate demand for water, as indicated by the transpiration rate, on the type of plant, and on the location within the root zone where the matric potential is measured.[6]

For some crop plants, allowing more than just the very surface of the soil to reach the permanent wilting point can have a harmful effect, either on yield or quality. Potatoes and certain other shallow-rooted vegetables are particularly sensitive to a moisture shortage, and where grown under irrigation, are often watered before the average matric potential in the upper root zone has dropped to -2 bars. Conversely, a crop like dryland wheat may reduce the soil water content to the permanent wilting point throughout most of the root zone and still yield a quality crop. Nonetheless, the yield of dryland wheat depends on the amount of water stored in the soil, and under any circumstance, there must be enough present to carry the plant to maturity if any grain is to be harvested.

Long-season crops, such as corn and sugar beets, must be grown under conditions where supplemental water is applied periodically during the growing season. If water is not added, crops of this type fail to reach a stage of maturity that makes a harvest worthwhile. Further, some plants demand easy access to water at certain critical periods during their life cycle. Corn, for instance, will not form seed properly if it is subjected to drouth at pollination time. Interestingly, however, many grasses tend to remain vegetative if the soil is kept continuously moist. When grown for seed, these plants require a period of moisture stress for the initiation of flowering.

Concept of available water in soils

The ability of soils to store water for plant use is one of their most important properties; it is a determinant of how well plants will survive long periods without rain or of the frequency with which water must be applied to irri-

---

[6]Wilting is not easy to recognize in all plants. Leaf tissue of grasses and small grains, for example, is often permanently damaged by dessication before other outward signs of severe moisture stress are clearly evident.

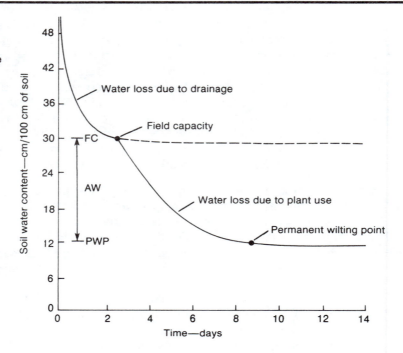

Figure 8.15.
Pattern of water loss from the root zone of a recently wetted soil as caused by drainage and plant use. The broken line projects continued drainage loss in the absence of plants. Abbreviations: FC, field capacity; PWP, permanent wilting point; and AW, available water storage capacity.

gated lands. The amount of water a soil can store for plant use is termed *available water*. This is an arbitrarily defined value computed as the difference in soil water contents at the field capacity and the permanent wilting point, or:

$$\text{Available water content} = \text{FC} - \text{PWP} \qquad (8.8)$$

where FC and PWP are the water contents, either gravimetric or volumetric, at the field capacity and the permanent wilting point.

The concept of available water storage and use is illustrated diagrammatically in figure 8.15, which refers to an arbitrarily selected rooting depth of 100 cm. As shown in the figure, the field capacity is taken as the upper limit of available water, for this represents the maximum amount of water that can be stored in the soil for any period of time. The amount of available water then becomes the amount plants remove as they reduce the water content to the lowest practical level, or the permanent wilting point.

The relationship between the theoretical and actual use of available soil water is illustrated for an eastern Washington, wheatland soil in figure 8.16. The soil is a deep silt loam formed in loess. The water supply comes mainly from winter precipitation, which is usually sufficient to re-establish the field capacity each year. Lack of rain during mid- and late summer allows wheat to use all of the available water throughout much of the root zone by harvest, which is usually in August.

Water relationships in soils

Figure 8.16.
The approximate distribution of water in the profile of a silt loam soil at the field capacity and permanent wilting point, and, as shown by the broken line, at the end of the growing season following a crop of wheat. Wilting-point values are based on laboratory data for soil samples from different depths in the profile. The field capacity curve represents the distribution of water present in the soil in early April. The available water is equal to the reduction in water content from the field capacity to the permanent wilting point.

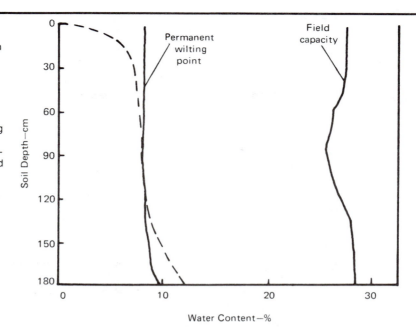

The solid lines in figure 8.16 are for the field capacity and permanent wilting point, expressed on a weight-percentage basis, at different depths in the profile. The field capacity is measured as the naturally established water content of the profile about April 1, which is before much growth and water use by plants have occurred. The permanent wilting percentages were determined in the laboratory on soil samples taken from throughout the profile. The broken line shows the water content of the profile at the time of harvest.

According to figure 8.16, wheat reduced the water content of the soil to the permanent wilting point to a depth of about 120 cm. The water content dropped below the permanent wilting point in the upper 60 cm of soil, but this was due mainly to evaporation from the soil surface and from cracks that form to depths of 30 cm or more. The plants did not remove all available water below a 120-cm depth before they matured. The water-content curve at harvest suggests that some water was taken up by the wheat even below a depth of 180 cm. This is not surprising, for wheat is known to use water from depth of 300 cm or more under some conditions.

Factors affecting storage and use of available water

The ability of soil to store water in available form depends mainly on the textural configuration, including textural stratification, and depth of the soil. Soil depth is important where shallowness, as over bedrock or other impermeable material, limits the volume of soil in which water can be stored. Texture is important because of its relationship to water retention and flow under soil water conditions at the field capacity and the permanent wilting point.

Table 8.2. Relationship between soil texture and estimated available-water content of soils as judged from differences in the soil water content at matric potentials of −1/3rd and −15 bars. (Data from various sources).

Textural class	Water content at ψ_m of −1/3rd bar	−15 bars	Estimated available water
	%	%	%
Silt loam	32	10	22
Silt loam	26	4	22
Loam	29	8	21
Silty clay	36	17	19
Silt loam	28	11	17
Sandy loam	20	4	16
Clay	40	25	15
Silty clay	37	23	14
Loam	21	9	12
Clay	42	31	11
Clay loam	22	12	10
Sandy loam	15	6	9
Sandy loam	14	6	8
Sandy loam	11	5	6
Clay loam	25	20	5
Sandy loam	7	3	4

The general relationship between soil texture and available water storage is indicated in table 8.2, where field capacity and permanent wilting point values are water contents of soil samples adjusted to matric potentials of −1/3rd and −15 bars, respectively. According to the table, soils with the highest capacity to store available water are medium-textured loams and silt loams. Pore space relations in these soils are such that they retain considerable water at the field capacity but give up much of it as the water content is reduced to the permanent wilting point. Fine soils also have comparatively high field capacities, as estimated at −1/3rd bar matric potential, but they give up less water as the matric potential is reduced to −15 bars. Coarse soils hold relatively little available water, simply because they retain so little water at the field capacity.

The ability of plants to utilize available water stored in the soil is a function of various factors. Among different plants, a principal determinant is their natural depth of rooting, with plants that root deeply being able to extract a greater quantity of stored water than can those that are shallow rooted. Within a given plant type, the depth of rooting will vary depending on growing conditions, including the supply of light and heat as well as the state of soil tilth and fertility. Of these factors, the fertility status of the soil is perhaps the most easily controlled. A principal reason for maintaining an adequate level of fertility is therefore to promote root growth, which allows greater utilization of available water stored in the soil. The importance of fertility to root growth is dramatically illustrated in figure 8.17.

Assessment of the available-water supply

The ability of a soil to store available water must be judged in a way that takes into account all the various factors affecting the retention of soil water and its ultimate delivery to plants. Inaccuracies in the judging center mostly

Figure 8.17.
The effect of fertility level on the growth of corn and wheat roots in two Illinois soils. Both the density and depth of rooting are increased at the higher level of fertility, effects that can be particularly important to the complete utilization of water stored in soil profiles. (Photo courtesy of J. B. Fehrenbacher, University of Illinois, Urbana.)

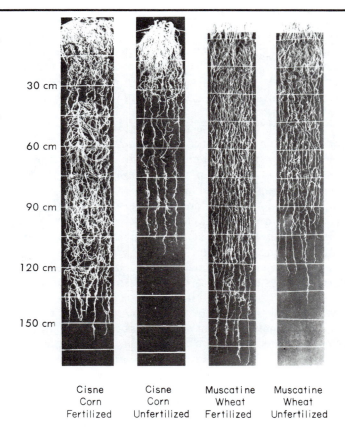

Cisne	Cisne	Muscatine	Muscatine
Corn	Corn	Wheat	Wheat
Fertilized	Unfertilized	Fertilized	Unfertilized

around the field-capacity determination; it should be assessed under field conditions and throughout the soil depth from which the bulk of water is taken up by plants. Permanent wilting points can be judged reasonably well in the laboratory, but like field-capacity measurements, they must account for variation in texture throughout the depth of rooting.

Even where available-water values are carefully determined, they may still fail to indicate the quantity of water actually taken up by plants under normal growing conditions. For example, these values do not account for the water used by plants during soil wetting, or that lost by evaporation or by continued slow drainage beyond the root zone during the period of active water use by plants. Further, the water content of the soil throughout the entire root zone is usually not reduced to the permanent wilting point under most cropping conditions. In fact, the use of available water between irrigations on irrigated land is often intentionally limited so as not to exceed one-half to two-thirds of the total supply. Any greater reduction could result in reduced water uptake and retarded growth for some crops (see Chapter 9).

Flow of water vapor in soils

The primary interest in the flow of water vapor in soils has to do with the loss of water by evaporation. This type of loss is important because it reduces the supply of available water and results in the accumulation of salts at the surface of soils in drier regions. Except for special situations, as with certain high-value crops, little that is economically practical can be done to eliminate evaporation.

As with liquid flow, the flow of water vapor in soils is in the direction of decreasing total water potential, this potential depending on combined matric and osmotic potentials. Generally, vapor flow in the soil is of consequence only over long periods of time. This is due for the most part to the very low density of the vapor, which may be only one-thousandth that of liquid water. As a mechanism of water transfer in soils, therefore, vapor flow is far less important than liquid flow. Vapor flow has its greatest impact on the evaporative loss of soil water into the atmosphere, especially where the soil is in contact with dry air.

In applying the above concepts to practical situations, we note that the loss of water by evaporation is maximum from soil that is moist to the surface and in contact with dry air. Such a condition favors the rapid escape of water vapor because the matric potential in pores at the surface of the soil is much higher than that of the air above. Since, under these circumstances, the vapor need not pass through soil to enter the atmosphere, neither the distance of flow nor the conductive capacity of the soil is significant in vapor transfer. The rate of evaporation thus depends on temperature (molecular activity) and on the rate at which moist air above the soil is removed, as by wind, so that a large difference in matric potential is continuously maintained.

So long as a soil remains moist to the surface, the loss of water by evaporation can be relatively rapid. Usually, however, the rate of evaporation exceeds the capacity of the soil to deliver water upward in liquid flow. Thus, where evaporation is rapid, exposed surface soil air-dries quickly. Continued delivery of soil water into the outer atmosphere then depends on vapor flow through the dry surface layer of soil. However, because of the low density of the vapor, the transfer is comparatively slow, and it becomes progressively slower as the thickness of dry layer increases. The formation of such a layer at the surface of a soil is one of the most effective means of limiting the continued loss of soil water by evaporation.

Summary

Water is important to essentially all physical, chemical, and biological reactions and processes in the soil, but its main role is the direct and indirect support of plant growth. A principal function of soils is therefore to store and deliver water to plant roots. This ability of soils at any given moment depends in part on their water content, which is a determinant of the force of water retention and the ease with which it moves from place to place in the soil.

Soil water contents are expressed quantitatively on either a gravimetric (weight) or volumetric basis, with the latter method having greater application in most practical situations. However, much can be judged about the behavior or status of soil water from knowledge of whether the soil is unsaturated or saturated with water. Under unsaturated conditions, pores are only partially filled with water that is retained in the soil by capillary (matric) forces. In saturated soils, pores are completely filled with water. Saturation normally results when water in excess of the amount capillary forces can retain drains downward to accumulate over an impermeable subsurface soil or rock layer.

The force with which water is retained by the soil is expressed either by the hydraulic pressure of the water or by its level of potential energy, or soil water potential. The total potential of soil water is determined as the sum of four component potentials: gravitational, pressure, matric, and osmotic. These result respectively from the interaction of water with one or more of four forces, namely: gravitational, weight, capillary (matric), and osmotic (ion-hydration). In unsaturated soils, the behavior of water is controlled largely by the matric potential; in saturated soils, it is controlled by a pressure (submergence) potential.

The water potential is an index of the ease with which water can be removed from the soil, as by plants, drainage, or evaporation. It is also a determinant of the rate and direction of water flow within the soil or between it and some external system. Flow occurs whenever there is a difference in total potential from one place to another, the direction of flow being in the direction of decreasing potential.

The rate of water flow in the soil system is a function of the gradient of the potential and the hydraulic conductivity of the soil. The conductivity of the soil depends on texture and the degree to which pores are filled with water. In general, water flows most rapidly in saturated soils and is greater in saturated coarse-textured than in saturated fine-textured soils. Under unsaturated conditions, fine-textured soils are the best conductors of water, but this capacity in fine soils is not very high even when they are saturated.

Water stored in soils for plant use is termed available water and is estimated by the difference in soil water contents at the field capacity and permanent wilting point. Available water storage depends on soil depth (depth of rooting) and texture, with soils of medium texture storing the most, fine-textured soils an intermediate amount, and coarse soils the least amount of water in available form.

The field capacity of soils designates the maximum amount of water a thoroughly wetted soil can store against normal drainage. Plant use ordinarily accounts for the greatest loss of water stored at the field capacity, but evaporation results in the loss of some. Over the short term, evaporation causes the loss of water from relatively shallow depths. A major concern with evaporation is its effect on salt accumulation at the surface of arid-region soils having a shallow water table.

Review questions

1. Explain why the equivalent depth of water per unit of soil depth can be used to express volumetric soil water contents?
2. Show that the water ratio of a soil with a bulk density of 1.5 g/cc and a 15 percent gravimetric water content is .225.
3. Why does the suction that must be applied to remove water from a soil increase as the water content decreases?
4. Explain the operation of a tensiometer.
5. Explain the relationship among the terms waterlogging, ground water, water table, and capillary fringe.
6. Why does the matric potential but not the gravitational potential at a given point in the soil change with a change in the soil water content at that point?
7. Why is lateral water flow along a horizontal plane beneath a water table not affected by differences in matric, osmotic, or gravitational potentials?
8. Why do osmotic forces influence the movement of water from soil moisture films into neighboring pore voids (in evaporation) or into roots but not its movement from one soil pore to the next through the soil moisture films?
9. Why does infiltration slow as the time of soil wetting increases?
10. Explain why coarse soils wet more rapidly in a downward direction than do fine-textured soils, but more slowly in a lateral direction, as from an irrigation furrow.
11. Explain the concept of available water in soils.
12. Why would the surface layer of a loam soil store more available water if underlain by a layer of coarse sand or gravel than if underlain by similar loam soil.
13. Why does the formation of a dry layer of soil at the surface slow the rate at which stored soil water is lost to the outer atmosphere by evaporation?

Selected references

Baver, L. D.; Gardner, W. H.; and Gardner, W. R. *Soil Physics.* New York: John Wiley & Sons, 1972.

Black, C. A., ed.-in-chief. *Methods of Soil Analysis*, pp. 82–298. Madison, Wis.: American Society of Agronomy, 1965. (Fundamental concepts of soil water and methods of measuring water-related soil properties are discussed.)

Hagan, R. M.; Haise, H. R.; and Edminister, T. W., eds. *Irrigation of Agricultural Lands.* Madison, Wis.: American Society of Agronomy, 1967.

Hanks, R. J. and Ashcroft, G. L. *Applied Soil Physics.* New York: Springer-Verlag, 1980.

Hillel, D. *Soil and Water: Physical Principles and Processes.* New York: Academic Press, 1971.

Parr, J. F., and Bertrand, A. R. Water Infiltration into Soils. *Advances in Agron.* 12:311-63, New York: Academic Press, Inc., 1960.

Richards, L. A., and Richards, S. J. *Soil Moisture.* In *Soil, 1957 Yearbook of Agriculture*, pp. 49-60. Washington, D.C.: Government Printing Office.

Rose, C. W. *Agricultural Physics.* New York: Pergamon Press, 1966.

Russell, M. B., coordinator. Water and Its Relation to Soils and Crops. *Advances in Agron.* 11:1-131. New York: Academic Press, Inc., 1959.

Stefferud, A. ed., *Water, 1955 Yearbook of Agriculture.* Washington, D.C.: Government Printing Office, 1955.

9

Soil water management

Soil water conditions are seldom ideal for plant growth. In many instances extreme conditions prevail, such as excessive wetness in waterlogged soils or the lack of water in well drained soils of dry regions. Even where conditions are more favorable, excesses or shortages of water sometimes place limitations on crop production. Under these circumstances, good management can often reduce the limitations, possibly to a point where they are of little consequence. As conditions tend toward the extremes, however, the satisfactory use of land for most agricultural purposes may ultimately depend on drastic changes such as the introduction of irrigation in arid regions or the artificial drainage of waterlogged soils.

A consideration of soil water management logically involves all aspects of water supply and use as they affect the growth of plants. Since problems associated with excess water differ markedly from those characterized by a water shortage, the two subjects are treated separately. The first to be dealt with is that where a lack of water functions as a limiting factor of plant growth.

Plant characteristics important to water use

Provided all other factors are optimum, plants exhibit their greatest potential for growth when continuously supplied with readily available water. The water supply determines not only the potential for growth but also the types of plants that can adapt successfully to the prevailing conditions. This applies to either native or domesticated plant species.

Plants differ in their ability to survive conditions of limited water supply depending on: (1) their capacity to extract stored water from the soil, and (2) their capacity to limit water loss by transpiration. For most common agricultural crops, the first of these characteristics is the more important; plants often differ materially in their ability to remove water from the soil but not in their capacity to control transpiration.

The utilization of soil water is largely a function of the kind of root system a plant has. One important feature is root density, or the concentration of roots per unit volume of soil. Water absorption from a volume of soil is potentially greater where the density is high, since increasing density shortens the average distance of water movement from sites of storage to roots. Because of the limited hydraulic conductivity of the soil as the wilting point is approached, stored water may never reach the plant roots by the time the plant is seriously wilted if the distance of flow to widely spaced roots is too great.

A second important characteristic is the depth of rooting. This feature determines the total volume of soil from which available water can be extracted. Generally speaking, the depth of rooting rather than density is the principal reason why plants differ in their ability to utilize water stored in soils.

The effective rooting depth of plants

The soil depth from which plants obtain the major part of their water supply is termed the *effective rooting depth*. This is a highly variable characteristic, for it differs among plants, with changing conditions of water supply in the soil, or with variation in atmospheric conditions outside the soil. The effective rooting depth should account for 80 percent or more of the total water removed from the soil by combined evaporation and transpiration. So long as this is true, relatively accurate judgments of the total water used in producing a crop plant can be made from measurements limited solely to the effective rooting depth.

When the supply of available water is adequate at all times, by far the greatest part of water removal by plants is from the surface layer of soil. However, of the total water extracted, the proportion taken up from the surface soil is less for deep-rooted than for shallow-rooted plants. This relationship is illustrated in figure 9.1, which shows highly generalized patterns of water extraction for three categories of effective rooting depth. Few plants extract much soil water beyond the maximum 180-cm depth shown in the diagram, provided they are adequately supplied with water most of the time.

If the soil is not kept continuously moist but goes instead for relatively long periods between water additions, the pattern of water removal is changed substantially. This fact is illustrated by figure 9.2, which shows the pattern of water extraction by sugar beets from a soil profile during a one-week and a four-week period following an irrigation. During the first week, 2.5 cm of water was removed from the upper 60-cm of soil, accounting for 78 percent of the total removed from a 120-cm depth. After four weeks, 6.6 cm of water had been taken up from the upper 60-cm, but it accounted for only 56 percent of the total extracted from the entire 120-cm depth. Although not clearly obvious in figure 9.2, the sugar beets extracted some water from below the 120-cm depth when they were forced to go four weeks without an irrigation. Such an effect would increase the effective rooting depth over that characteristic of these plants when watered at weekly intervals.

The effective rooting depths of several crop plants are listed in table 9.1. These values are averages and apply only to those conditions where the plants are generally well supplied with water. A number of these plants are commonly grown under conditions where a mild moisture stress develops at least occasionally, which has the effect of increasing the effective depth of rooting over that shown in the table. However, it also results in a reduction

Figure 9.1.
Generalized water-extraction patterns for shallow-, medium-, and deep-rooted plants kept continuously well supplied with water. Values show the water removed from each 30-cm depth of soil as a percentage of that removed throughout the total effective rooting depth.

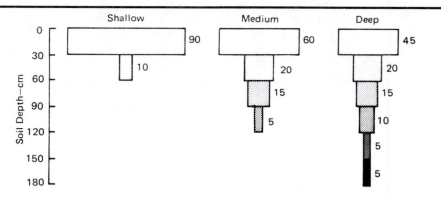

Figure 9.2.
Water use from successive 30-cm depths of soil by sugar beets during 1- and 4-week intervals following an irrigation. Percentages relate to the total water absorbed from a 120-cm depth of soil. (From S. A. Taylor, in *Soil, 1957 Yearbook of Agriculture*, pp. 61–66, Washington, D.C.: Government Printing Office, 1957.)

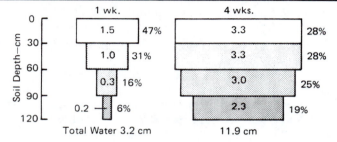

in yield. For example, small grains grown under dryland conditions may have an effective rooting depth of from 180 to 250 cm, as compared to 90 cm under irrigation, but the yield under irrigation can be substantially higher.

Rooting depth and the use of available water

The removal of water by plants becomes progressively more difficult as the water content of a soil is reduced. So long as water delivery to roots keeps pace with transpirational losses, decreased availability will have little adverse influence on overall plant functioning. However, once transpirational losses exceed the rate of absorption, a decrease in the internal (turgor) pressure of water in plant cells decreases, and the rate of growth is slowed. Yet, a limited water shortage is not harmful in all cases, as for example, where a period of moisture stress enhances flowering and seed formation in grasses. Similarly, experimental work has shown that sugar beets subjected to periods of temporary stress on a daily basis can accumulate more sugar than if continuously well supplied with water. Apparently, sugar that would be used in the synthesis of new vegetative (leaf) tissue is diverted to storage sites in the beet roots under stress conditions.

Table 9.1. Effective rooting depths of various plants.

Effective rooting depth cm	Plant type
30	Strawberries, tobacco, turf grasses
60	Beans, potatoes, Ladino clover
90	Corn, sorghum, soybeans, small grains
105	Sugar beets
120	Raspberries
180	Alfalfa, cotton, grapes, apples, peaches

The above examples notwithstanding, it is usually preferable to avoid moisture stress in most economic crop plants. This requires that the supply of available soil water, or that held between the field capacity and the permanent wilting point, not be excessively depleted before it is replenished by rain or irrigation. The extent to which the available water supply can be reduced without serious hindrance to growth varies among plants, but in general, it depends on the ability of the soil to deliver stored water to roots under an existing level of soil moisture and transpirational rate. For most plants grown under conditions of moderate to high rates of transpiration, it appears that growth starts to decline after about two-thirds of the available water in the effective rooting depth has been removed. Some plants do not tolerate this degree of water loss. Potatoes, for example, may suffer if the supply of available water in the effective depth of rooting is reduced by much more than one-third. Where transpiration is low, a higher proportion of the available water can be removed without seriously hindering plant growth.

The efficiency of water use by plants

The efficiency of water use by plants depends on the relationship between plant growth, here expressed by total yield, and the quantity of water consumed in achieving that yield. The efficient use of water by plants is most important where, due to low natural precipitation, the supply of available water is a growth-limiting factor, or where, though not limiting to growth, it must be supplied through irrigation. Under either of these circumstances, low water-use efficiency implies a loss in yield and income. However, the consequence may be most serious under irrigated conditions where production costs are high and include the cost of the water and its application. Regardless, the most efficient use of water should be the goal under any type of farming, and it can be most readily achieved by assuring that all factors of growth, including the availability of water, are as near optimal as possible.

Two measures have been used for the most part to express water-use efficiencies. One is the *transpiration ratio*, which is the amount of water transpired per unit of dry plant material produced. The second is the *consumptive use* of water, or the total evapotranspirational loss associated with the production of a crop. Both of these measures reflect the potential for the evaporative loss of water. Thus, when evaluated under comparable conditions, they tend to parallel each other. They express water use differently, however, since consumptive-use values are in absolute terms, such as centimeters of water, rather than as a ratio. Consumptive-use values are the most convenient, for they are in the same units used to indicate the amount of water added to or stored in the soil.

The transpiration ratio

The measurement of transpiration ratios is an attempt to detect differences among plants in their inherent ability to use soil water efficiently. Precise measurement of these ratios is possible only where the entire root system is enclosed so that all water loss takes place through the plant. The evaluation of transpiration ratios is therefore carried out with potted plants.

Numerous measurements of transpiration ratios have shown that different types of common agricultural crops, when grown under comparable conditons, exhibit relatively similar water-use efficiencies. Some differences are noted, however. For example, corn has a lower transpiration ratio than many other plants, small grains included.

Major differences in transpiration ratios occur with variation in the climatic factor. Under arid-region conditions they may be 800/1 or above. Under humid conditions, on the other hand, these values may be only one-third as high, with perhaps no more than 250 to 350 units of water being transpired for each unit of dry plant material produced.

Consumptive-use measurements

Contemporary evaluations of water-use efficiencies are most often by consumptive-use measurements. These are made under field conditions by determining the reduction in the soil water content over a period of time. Care must be taken to prevent or minimize water loss by deep percolation, for if it occurs the total water loss cannot be considered a measure of evapotranspiration.

Two aspects of consumptive use are important: the total evapotranspirational requirement for the production of a crop, and the rate of water use. Like transpiration ratios, both the total and the rate of consumptive use are climatically related. Total consumptive use also varies among crops, however, being greater for those having the longest growing season. The rate of consumptive use varies with the time of season as changes in atmospheric conditions increase or decrease the potential for evaporation.

Under comparable atmospheric and soil water conditions, consumptive-use rates for many common agricultural crops do not differ greatly. This fact is illustrated in figure 9.3, which shows that alfalfa and corn have comparable average daily rates of consumptive use when grown under similar conditions and for the same period of time. However, total consumptive use is greater for alfalfa, since it grows for a longer period during the year.

Factors affecting water-use efficiencies

As suggested above, the consumptive use of water does not vary greatly among widely different plants. It follows, therefore, that there should be only limited variation in the consumptive use of a single species when grown under different levels of productivity. More water is used when plant growth is increased by altering some factor of the environment, but the increase in water use is normally much less than the increase in productivity. In other words, water-use efficiency can be increased by improving conditions for plant growth.

Soil water management

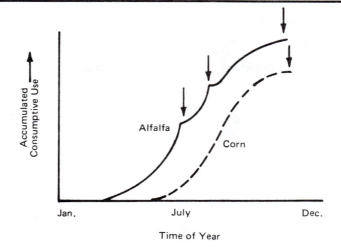

Figure 9.3. Relative accumulated consumptive use of water by alfalfa and corn. Arrows indicate times of harvest. (From D. E. Angus, *Advances in Agron.* 11:19–35. New York: Academic Press, Inc., 1959.)

Table 9.2. Effect of phosphorus fertility and moisture on yield and water-use efficiency of alfalfa. (From C. O. Stanberry et al., *Proc. Soil Sci. Soc. Amer.* 19:303–10, 1955).

Added P (kg/ha)	Irrigation Treatment					
	Dry			Wet		
	Yield (t/ha)	Y/ET[a] (kg/cm)	ET/Y (cm/t)	Yield (t/ha)	Y/ET (kg/cm)	ET/Y cm/t
50	19.3	42	24	24.9	46	22
100	21.3	47	21	28.2	53	19
200	25.8	57	18	33.4	62	16
400	26.0	58	17	35.2	66	15

[a]Y is yield; ET is evapotranspiration.

The data in table 9.2 show how variation in either of two growth factors, fertility and the supply of available water, can influence the water-use efficiency of plants. These data are for alfalfa grown over a three-year period at Yuma, Arizona. Differential treatments involved four levels of phosphorus (P) fertilization at two levels of available moisture, the latter being obtained by varying the frequency of irrigation. In the dry treatment, plants showed signs of wilting before irrigation water was applied; in the wet treatment, the frequency of irrigation was such that no significant moisture stress developed in the plants. On the average, 180 cm of water were applied each year to the dry treatment and 215 cm to the wet treatment. Drainage losses were small, as was rainfall, so that the consumptive use of water corresponded closely to that applied by irrigation.

The results of this experiment show clearly that increases in the level of either available phosphorus or water increased the efficiency of water use. These results are indicated in two ways in the table: the ratio of yield, in kilograms, to evapotranspiration in centimeters; or the centimeters of water required to produce a ton of hay. In the poorest treatment, 24 cm of water were needed to produce a 1-ton yield; in the best treatment, water use was reduced to 15 cm per metric ton.

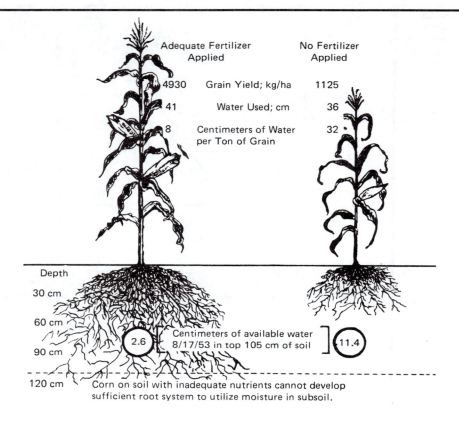

Figure 9.4.
The relationship between fertility level and the efficiency of water use by corn. (From G. E. Smith, Missouri Farmers Association Bulletin, 1953.)

Adequate Fertilizer Applied		No Fertilizer Applied
4930	Grain Yield; kg/ha	1125
41	Water Used; cm	36
8	Centimeters of Water per Ton of Grain	32

Depth
30 cm
60 cm
90 cm
120 cm

2.6   Centimeters of available water
8/17/53 in top 105 cm of soil   11.4

Corn on soil with inadequate nutrients cannot develop sufficient root system to utilize moisture in subsoil.

Plant nutrition-water supply interrelationships

The data in table 9.2 indicate that the efficient use of either fertility or available water depends on the degree to which both are adequately supplied to plants. Reasons for this interrelationship can be seen by considering how variation in each affects the availability of the other.

Water affects the nutrition of plants in a number of ways. It is essential for the release and transfer of nutrients to roots and their movement within the plant. It is also important to the proliferation of roots so that they continue to contact previously untapped sources of available nutrients in the soil. Root growth is limited if a water stress results in lowered cell expansion. Even in the absence of stress, root extension into dry soil takes place with difficulty. This fact is of great significance where drying prevents root exploration of the surface soil that normally contains the highest proportion of nutrients in readily available form.

When a low level of available nutrients limits root development, plants are unable to make complete use of available water and may suffer reduced growth for this reason. Data from experimental work in Missouri, shown in figure 9.4, indicate that a suitable level of nutrients greatly encourages the use of available water by corn. In this experiment, fertilized corn extracted all but 2.6 cm of the available water stored in the surface 120 cm of soil by the middle of August. Without fertilizer, 11.4 cm of available water remained

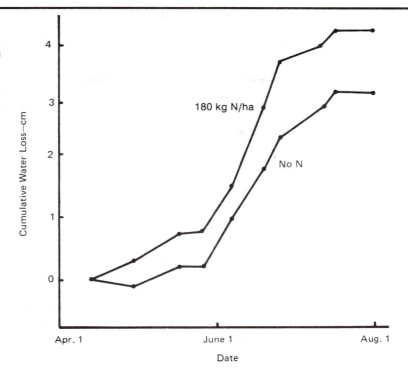

Figure 9.5.
The effect of fertilizer N on the utilization of water by wheat from a 30-cm depth of soil 180 cm below the surface. (Unpublished data of F. E. Koehler, Washington State University.)

unused within this soil depth. Although the most obvious benefit of the fertilizer was the increase in grain yield from 1125 to 4930 kg/ha, the greater efficiency of water use was of equal importance. With fertilization, the quantity of water required to produce each bushel of grain was only about one-fourth that needed where no fertilizer was applied.

Similar results have been noted many times with other crops. For example, fertilization results in the earlier and more extensive use of subsoil moisture by dryland wheat. The effect is shown in figure 9.5 by data for water extraction from a 30–cm layer of soil 180 cm below the surface and reflects a more extensive root system for the fertilized wheat. It may also be assumed that the total depth from which water intake occurred was greater in the fertilized than in the unfertilized soil. Where the water supply in a soil can be regenerated each year from natural precipitation, failure to utilize it completely results in an unnecessary reduction in yield.

Water conservation

Under conditions where the supply of water derived from precipitation is generally too small to allow maximum plant growth, special steps may be taken to maintain the best possible moisture relationships in the soil. Two approaches are usually followed: (1) increasing the potential for maximum build-up of water in the soil, and (2) assuring complete and efficient utilization of the stored water. For example, one of the best means of increasing the water supply in soils of drier regions is through use of standard erosion-control measures that slow runoff and thereby increase infiltration. Once extra water has been stored in the soil, its efficient use rests largely with sound crop and soil management practices.

Although the above approaches are general and therefore apply to most cropping situations, certain kinds of agricultural endeavors are amenable to special water-conservation measures. Some of these are considered in the following sections.

Water conservation under dryland conditions

One of the most practical means of water conservation for cropland of the drier regions is use of a two-year rotation in which land is alternated between small grain production and fallow. During the year when the soil is in the fallowed, or weed-free, state, water is collected and stored to supplement that received during the following year of cropping to small grain. Although the net increase in available water is not very large, usually no more than a 25-percent increase over that received on an annual basis, it is normally sufficient to prevent the total failure of a crop grown in an unusually dry year. The extra water also increases water-use efficiency by more nearly satisfying the total demand of the crop during the year it is grown. In addition to the above benefits, the two-year rotation reduces the costs associated with planting, growing, and harvesting the crop to only once every two years instead of every year.

A problem that besets many areas of dryland farming is the occurrence, at least occasionally, of torrential, highly erosive rainstorms. The control of rainfall erosion is therefore important to these areas, not only to protect the soil but also to recover as much of the water as possible for plant use. Increased growth caused by improved soil water relationships, in turn, benefits erosion control by supplying residues as a protective ground cover. Similar benefits accrue where more efficient water use and greater residue return results from the alternate grain—fallow system of dryland farming. However, these benefits may not materialize unless the residues are used to full advantage as a protective cover during the year of summer fallow.

Rangeland of drier areas also benefits from water-conservation practices that limit runoff and permit the efficient use of water by plants. Normally, however, special water-conservation practices are needed only where deterioration of the rangeland has set in, primarily because of overgrazing (see fig. 9.6). When a soil cover of well adapted, natural vegetation is destroyed, one of the best means of water conservation is lost, since runoff tends to increase and only limited use is made of the water that does infiltrate the soil. With high-quality range, it is recommended that no more than half, sometimes only one-third, of the vegetative cover be removed by grazing at any one time. When grazing is more extensive, the soil becomes unduly exposed to falling rain and is given only minimum protection from runoff. Overgrazing also increases soil compaction, which slows infiltration and increases the volume and rate of runoff.

Figure 9.6.
Good range at the left
compared to
overgrazed range at
the right. Loss of
cover due to
overgrazing is an
invitation to severe
soil erosion. (U.S.D.A.
Soil Conservation
Service photo by L. F.
Bredemeier.)

Deteriorated rangeland may be improved if all water normally lost as runoff can be retained for plant use. Two cultural practices are widely used to decrease the runoff: *contour-furrowing* and *soil-pitting*. For the first of these, a series of furrows is established by plowing at intervals across a slope as shown in figure 9.7. The spacing between furrows is determined by the potential for runoff, which varies with the climate and the soil, including its slope. Contour-furrowing is generally of little benefit on steep slopes or under any condition where it fails to prevent runoff. If, as the result of a heavy rain, the furrows are breached and the impounded water suddenly released, very serious erosion can result.

Soil-pitting is sometimes better than contour-furrowing for minimizing runoff from rangeland (see fig. 9.7). The pits are produced through use of a disc-plow with off-center discs. As the discs rotate they scoop out short pits about 5 cm deep at intervals of roughly 40 cm. Since the pits occur in a staggered pattern over the soil surface, they normally intercept all runoff before it moves very far downslope. A main advantage of soil pits is that they do not collect and concentrate large amounts of water in one location as contour furrows sometimes do. Since the water is distributed over a greater total area of soil surface, it is both more rapidly absorbed by the soil and more uniformly accessible to plants than is typical of water collected by contour furrows.

A   B

Figure 9.7.
Methods of reducing runoff from rangeland. (A) Contour furrows in
Lincoln County, Colorado, and (B) pitted rangeland in western South
Dakota. (Credits: (A) U.S.D.A. Soil Conservation Service; (B), J. T.
Nichols, South Dakota State University, Brookings.)

Other treatments that may be used to improve rangeland include the
seeding of desired grasses and the use of fertilizer, but neither of these is
practical in the absence of good moisture relationships in the soil. Sometimes
rangeland will recover naturally if nothing is done other than to control runoff, and even this is not necessary under all circumstances. However, regardless of the approach taken, little benefit will be realized if the initial
cause of deterioration, overgrazing, is allowed to continue.

**Water
conservation on
irrigated land**

Under irrigated conditions, water conservation depends mainly on avoiding
undue water loss by deep percolation, runoff, and evaporation. These losses
are a function of the specific kind of system used to irrigate the land, the
more important of which are described below.

Types of Irrigation Systems. There are four general methods of applying irrigation water to land: (1) surface irrigation, which includes the furrow
method and depends on gravity to distribute water over the land, (2) sprinkler
irrigation, (3) trickle, or drip, irrigation, and (4) subsurface irrigation. Of
these the last has perhaps the most stringent requirements. It utilizes ground
water maintained at an appropriate level for plant use through adjustment

Figure 9.8.
Furrow, or rill, type
irrigation. (U.S.D.A.
Soil Conservation
Service photo.)

of the rate of drainage from land naturally susceptible to waterlogging. For this the area must be esesntially flat, and it must contain permeable soil so that rapid adjustment of the subsurface water level is possible. Where subsurface irrigation is practiced, water is usually present in excess; thus, its conservation is of minor importance. Of greater concern may be the inability to maintain the water at an appropriate level for plant use, or where evaporation is high, the accumulation of salts carried upward by capillarity and concentrated in amounts that are damaging to plants.

Surface Irrigation. Most surface irrigation is carried out by the furrow method, but two others, *border* and *basin*, are sometimes used (see figs. 9.8, 9.9, and 9.10). For border irrigation, water is applied to broad strips bounded on each side by ridged soil that directs flow downslope. In furrow irrigation, parallel, V-shaped ditches (furrows) are worked into the soil in a downslope

Figure 9.9.
Orchard irrigation by the border method. (U.S.D.A. Soil Conservation Service photo.)

direction. Spacing between the furrows depends in part on the space that should be left between rows of the plants and in part on the soil, since lateral flow from irrigation furrows is better in fine- than in coarse-textured soils.

It is all but impossible to cause uniform soil wetting over the full length of a field by furrow or border irrigation. With either method, wetting at the upper end of the field starts well in advance of wetting at the lower end and continues so long as water is applied. Thus, adequate wetting at the lower end causes overirrigation at the upper end of the field, with the excess water being lost by deep drainage. The problem can be reduced by increasing the rate of water flow, which may increase erosion, or by shortening the length of the furrows or borders, which may not be practical.

Furrow and border irrigation do not work well if the slope is irregular; high spots tend to back up water in lower spots, which causes unequal wetting (see fig. 9.11). To overcome this problem requires leveling or smoothing of the field. The high cost of this procedure is one of the major disadvantages of furrow and border irrigation. Another disadvantage is that more water is ordinarily delivered into the furrows than can be absorbed by the soil down-slope. This is done to assure that adequate water reaches the lower end of the field, but it also assures that there will be water loss as runoff and, possibly, some erosion. Unless the runoff can be recovered for reuse, it results in a serious waste of water. If erosion occurs, there is an additional problem in dealing with the sediment.

The loss of runoff is avoided in the basin system of irrigation. Basin irrigation involves the flooding of an area enclosed in a border, usually of ridged soil. This system is most suitable for flat land, but it is even used on rather steep slopes that have been worked into level contour terraces or benches (see fig. 9.10). In addition to preventing water loss by runoff, basin irrigation allows precise control over the amount of water applied. It is also an effective means of flushing excess salts from the soil.

Figure 9.10.
Basin irrigation. This
method is well
adapted to flatland, as
in the upper photo, but
it has been used in
successful rice culture
on extremely steep
slopes (lower photo)
by the Ifugao people
in the Banawe Valley,
Philippines, for some
3000 years. (U.S.D.A.
Soil Conservation
Service and United
Nations photos.)

A

B

Sprinkler Irrigation. Sprinkler irrigation overcomes a number of problems inherent in the application of water by surface irrigation. An important advantage of sprinklers is that they allow the relatively uniform application of water across the full width of a field and to a depth that can be readily controlled. In addition, they require relatively little advance land preparation, such as the leveling or grading that is often necessary for surface systems. Sprinklers can be used on land that is either too uneven or too steep for surface irrigation.

Sprinkler systems may be either fixed or movable. Mobile systems may be moved by hand, or they may be wheel-mounted. Units of the latter type are shown in figures 9.12 *(wheel-line)* and 9.13 *(center-pivot).* Wheel-line systems are moved stepwise across a field; they remain in one position for several hours, during which time they supply enough water to cause deep wetting of the soil. In contrast, center-pivot systems move continuously in a circular pattern as illustrated in figure 9.14, and they normally make at least one revolution per day. Since water is applied daily, the soil need be wetted to a shallow depth only. This eliminates water loss by deep percolation and the removal of mobile nutrients by leaching. However, because of continual wetness, root rots, rusts, and other plant diseases may be strongly encouraged.

Figure 9.12.
A hydraulically driven wheel-line sprinkler system. (Courtesy of Wade Rain Sprinkler Irrigation Systems.)

Figure 9.13.
An electrically driven center-pivot sprinkler system (Courtesy of Gifford-Hill & Company, Inc.)

Although water loss by deep percolation or runoff can be avoided by sprinkler irrigation, it does increase water loss by evaporation. Also, the distribution of water may be uneven on windy days. Unless they are well-engineered, center-pivot systems may also have a distribution problem due to the faster rate of travel at the outer end of the line. To compensate for this, the volume of flow from the sprinklers is increased in an outward direction along the line (see fig. 9.14).

Figure 9.14.
The water distribution pattern of a center-pivot irrigation system. The water is distributed by individual sprinklers mounted at intervals along a pipe that moves continuously in a circular pattern during an irrigation. For uniform water application, the rate of delivery increases toward the outer end of the pipe and extends beyond the area planted to crops, which is the shaded area in the diagram. Conventional center-pivot sprinkler systems cover about 80 percent of the area in a square field.

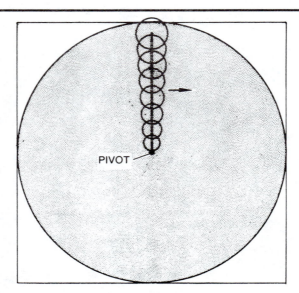

Figure 9.15.
An emitter on a trickle irrigation line. (Washington State University Engineering Extension Service photo.)

Trickle Irrigation. The trickle, or drip, system of irrigation is a fairly recent innovation. It is of advantage in that it can eliminate water loss by runoff and deep percolation while minimizing evaporative loss. In this system, water is delivered over the field through plastic tubing that is fitted at selected intervals with *emitters* (valves) that control the rate of water delivery into the soil (see fig. 9.15). In use, an attempt is made to match the rate of delivery to the evapotranspirational use of water. However, since evapotranspiration is not constant, periodic adjustment of the delivery rate is required.

There are two main limitations to the trickle system of irrigation. One is the high initial cost, which restricts its use mainly to widely spaced, high-value, vine, tree-fruit, or similar crops. Another is plugging of the small openings through which water is transferred to the soil. Plugging may be by mineral or organic particles or by microbial tissue forming at the outlet of the emitter. This type of problem can be controlled reasonably well through use of carefully filtered water that is periodically chlorinated to kill microorganisms.

Trickle systems are usually operated under pressure, with the rate of water delivery depending on the pressure and on the size of the emitter openings. Gravity flow can also be used, however. Since a gravity system would operate under very low pressure, it would allow the use of comparatively large lateral feeder tubes having only a limited tendency to clog.

Timing as a factor of irrigation efficiency

Where irrigation water is applied at infrequent intervals rather than on a daily or continuous (trickle) basis, there are two important considerations in the proper and efficient application of water. These are the determination of the time to irrigate and of the amount of water to apply. The time of irrigation depends on plant needs; it should avoid the development of a serious moisture stress in plants. The amount of water to apply depends on the quantity used since the last irrigation. Water applied in excess of this amount will be lost either as runoff or by deep percolation.

The time to irrigate can be judged in several ways. One is by measuring the soil water content throughout the effective rooting depth every few days after the soil has been irrigated. This may be done directly by sampling the soil and determining its water content, or it may be done indirectly through use of permanently installed, moisture-sensing equipment, such as *resistance blocks* or the *neutron probe*, as illustrated in figure 9.16. These are instrumented devices, which, after installation, require nothing more than taking electrical meter readings to determine the soil moisture content. However, these instruments must be calibrated before use; that is, meter readings must be related to actual soil moisture contents. The calibration consists of comparing meter readings with a range of soil moisture contents for the soil where the instruments are installed.

Resistance blocks are small units of porous material (nylon or fiberglass cloth, or gypsum blocks molded from Plaster of Paris) capable of absorbing water by capillary attraction. Each block contains two electrodes for measuring the flow (or resistance to flow) of electric current through water

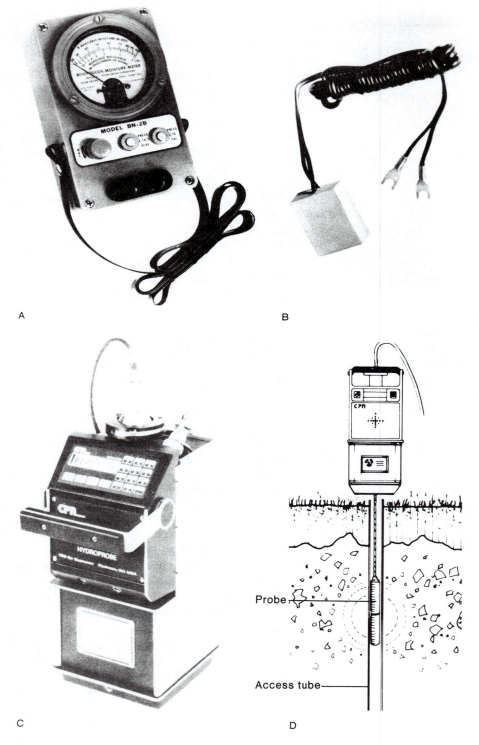

Figure 9.16.
(A) Bouyoucous moisture meter, which measures current flow through gypsum blocks (B) to determine the moisture content of soils. (C) A neutron meter with the probe stored in a shielded storage compartment, and (D) a diagram of the neutron meter and probe in use. (Photos (A) and (B) courtesy Beckman Instrument Company; (C) and (D), Campbell-Pacific Nuclear Corporation.)

A

B

C

D

Probe

Access tube

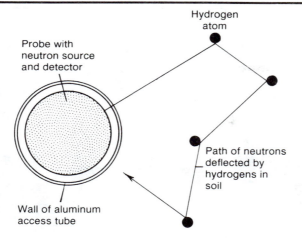

Figure 9.17.
Diagram illustrating the deflection path of neutrons in soil. Fast neutrons emitted from a source in the probe are slowed and deflected by hydrogen in soil water (circles). Neutrons that return to the probe are counted by a slow neutron detector.

Hydrogen atom

Probe with neutron source and detector

Path of neutrons deflected by hydrogens in soil

Wall of aluminum access tube

in the block. When buried in the soil, the blocks adjust to the water content of the soil, and the amount of water is judged by the measured current flow between the two electrodes. Because ions in water affect its ability to conduct an electrical current, resistance blocks are not too dependable in soils of high or variable salt content, nor do they function well in very wet soils.

The principle of the neutron probe is based on the ability of hydrogen ions in soil water to slow and deflect neutrons, which are electrically neutral, subatomic particles contained in the nucleus of many atoms. The neutrons are provided by a point source, usually by radioactive Americium, present in the probe. In use, the probe is lowered into the soil through an aluminum access tube. Neutrons emitted by the source enter the soil at high velocity, but on encountering hydrogen in water, they are both slowed and deflected, some returning to the probe (see fig. 9.17), where they are counted by a slow-neutron detector. Since the number of neutrons returning to the probe is a function of the amount of water (hydrogen) in the soil, the instrument can be calibrated to read the soil moisture content. The principal limitations of the neutron probe are its initial expense, the restrictions placed on its use because it contains highly radioactive material, and the need to calibrate it for each soil.

Whether one uses soil sampling or instrumentation for determining soil moisture contents, the procedure for judging the time to irrigate is the same. After a soil has been irrigated, its moisture content is periodically monitored, and it is irrigated again when the level of available water has been reduced to a minimum consistent with continued good plant growth. As indicated earlier, this point is reached for many crop plants when about two-thirds of the available water in the effective rooting depth has been removed.

Another way of determining the time to irrigate is through use of evaporation data that are systematically collected in some areas (see fig. 9.18). The amount of water lost from a standard evaporation pan, as shown in the figure, correlates closely with the consumptive use of water by plants grown

Figure 9.18.
A standard U.S. Weather Bureau evaporation pan installed for comparing evaporational water loss with consumptive use by crops grown in neighboring field plots.

in the same vicinity. The amount of evaporation, when multiplied by an appropriate factor, estimates the amount of water evapotranspired and, therefore, the extent to which the supply of available water has been depleted.

Data obtained by periodic soil sampling or from evaporation measurements provide a running record of the soil moisture status and a means for computing both the daily and the total consumptive use of water. With this kind of information, it is possible to predict when the supply of readily available water will have been exhausted and, also, how much water must be added when the soil is irrigated.

In some places, especially in the southwestern United States, tensiometers are used to determine the timing and length of irrigations. It may be recalled that tensiometers, by monitoring the matric potential, provide a measure of the momentary availability of water to plants. However, since tensiometers become inoperative at matric potentials below about -0.8 bar, they must be positioned at a depth where the matric potential is continuously above this value. On irrigated land, a 30-cm depth is usually recommended. For shallow-rooted crops, such as strawberries or beans, irrigation water may be applied when the matric potential has dropped to about -0.2 or -0.3 bar at the 30-cm depth. For more deeply rooted crops, such as alfalfa, cotton, or fruit trees, the matric potential at a 30-cm depth may be allowed to fall to -0.7 or -0.8 bar before irrigation is started. These values are only guidelines, however; the tensiometer readings that most accurately predict the time to irrigate must be determined by practical experience in the field.

Soil water management

Tensiometers used to judge when enough irrigation water has been applied are positioned to record a rise in the matric potential when the wetting front reaches specified depth. This may be at a 30–cm depth for shallow rooted plants and up to 90 cm for those that are deeply rooted. The tensiometers are placed in the field where wetting of the profile takes the longest time after the application of water has started.

Land drainage

In the United States, the use of artificial drainage to improve agricultural land dates back almost to the Revolutionary War. In some areas, particularly near the Great Lakes and along the lower Mississippi River, drainage has provided the only means whereby land could be brought under cultivation. In many other places drainage has been beneficial by increasing the productive potential of the soil and by extending the range of crops that can be grown on it.

In this country, cropland in excess of 40 million hectares, or about one-fourth of the total, is organized into district and county drainage enterprises. As shown in figure 9.19, most of these are concentrated in the eastern half of the United States. Those in the western states are primarily associated with irrigation. Improved drainage has also been established on many individual farms without their incorporation into large coordinated projects. With drainage being used on such a high proportion of the land, its importance can hardly be overemphasized.

The drainage problem in perspective

Water applied to land is lost by surface runoff, deep percolation, evaporation, and transpiration. Where the combination of these processes is inadequate to prevent crop damage from excess water, provisions may be required for its more rapid removal. This is done by the improvement of surface or subsurface (internal) drainage, or both. Often, the same facilities are used for both types of drainage, so the two are not wholly separable. Usually, both must be considered in the solution of a drainage problem.

Aside from plant response, the need for improved drainage is signaled in a number of ways, the most obvious of which is the collection of water over the surface of the ground for prolonged periods (see fig. 9.20). Less apparent, but still rather easily detected, is a high water table. The depth to the water table gives some indication of how far plant roots will extend into the soil. If a capillary fringe is present, however, roots will not even penetrate to the water table, for the fringe is saturated and lacks sufficient oxygen for root development. A capillary fringe may reduce the depth of root penetration by some 30 to 45 cm, or even more in some cases.

Climate and waterlogging

Except where irrigation is practiced, the extent and pattern of waterlogging, as judged by the depth to a water table, often relates to local climatic conditions. The general nature of the relationship is demonstrated in figure 9.21, which is for two widely different climatic conditions. Diagram A in the figure is for a humid region where a water table is maintained by the deep percolation of water through the soil. The root zone at the surface of this profile

Figure 9.19.
Drainage enterprises
on agricultural land in
the United States.
(From H. H. Wooten,
and L. A. Jones, in
*Water, 1955 Yearbook
of Agriculture,*
pp. 478–91.
Washington, D.C.,
Government Printing
Office, 1955.)

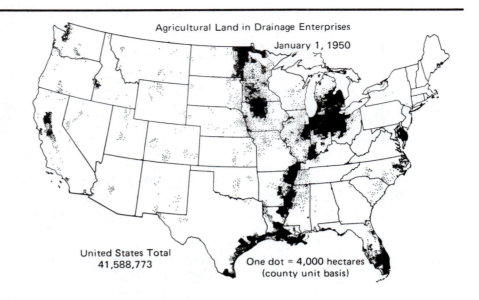

Agricultural Land in Drainage Enterprises

January 1, 1950

United States Total
41,588,773

One dot = 4,000 hectares
(county unit basis)

Figure 9.20.
Corn near Watertown,
Minnesota, seriously
damaged by water
standing in a
depressional area.
This problem could be
overcome by
underground drains.
(U.S.D.A. Soil
Conservation Service
photo by L. J. Eder.)

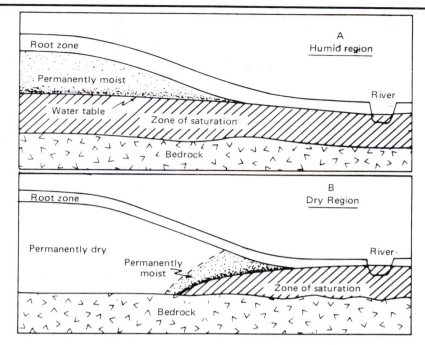

Figure 9.21.
The relationship between climate and topography as factors determining the presence and level of a permanent water table.

is subject to frequent and extensive changes in water content because of repeated wetting by rain and subsequent drying by plant use and evaporation. Water penetrating beyond the root zone maintains the substrata in a permanently moist state, with a part being continuously saturated. The water table, which here serves to indicate the upper boundary of the saturated zone, is far below the surface in upland areas, but is shallow near the river. The water content of the root zone in the soil adjacent to the river may be affected markedly by the nearness of the water table.

The slope of the water table in diagram A of figure 9.21 indicates that the movement of excess water is toward the river. During especially rainy periods the slope increases, as does the rate of internal drainage toward the river. If the increase is sufficient, it will eventually cause the level of the river to rise.

Conditions likely to occur in drier regions are depicted in diagram B of figure 9.21. Here only limited water from natural precipitation enters the soil, and in higher-lying positions, the thickness of the root zone depends largely on the depth of soil wetting in any particular season. Since surface-applied water does not move beyond this depth, the soil below is permanently dry.

A saturated zone may be maintained under conditions of low rainfall where, in low-lying positions, lateral seepage from a stream or river can occur. In this instance the slope of the water table is away from the river. Near the river the capillary rise of water affects not only the moisture status within the root zone but also contributes to the accumulation of salts at the surface of the soil.

Some fundamental aspects of drainage

The improvement of soil drainage involves one or both of two things: a reduction in the input of water into a soil, or an increase in the rate at which it is removed after entry has been made. Inputs can be reduced, depending on the cause of the excess water, by better irrigation management, by the rapid removal of surface runoff, or by the control of seepage from neighboring waterways or from wet soils on adjacent higher-lying slopes. Once in the soil, excess water is removed by installing subsurface outlets, or drains, that collect and transport water away by gravity flow. The success of the drains depends on how much they increase the rate of water flow to natural drainageways over that occurring by lateral transfer through the soil alone.

Factors important to drainage

Internal soil drainage depends on the movement of water through and its discharge from the soil by gravitational forces. A first and essential step is therefore to establish a drain below the level of the water table in the soil, that is, in a position where positive submergence potential in the soil water will force water into the drain.

The movement of drainage water through a soil is by saturated flow and occurs at a rate depending on the hyraulic conductivity and the lateral distance through which water must flow. The conductive capacity is determined largely by texture, with coarse materials having the highest conductivity. However, little can be done to change texture, so improvement of drainage rests primarily with reducing the distance through which lateral saturated flow in the soil takes place. The distance of flow, in turn, depends on how close to each other drains are installed in the soil. As shown in table 9.3, drains are normally placed at narrower intervals in fine- than in coarse-textured soils, the purpose of the narrower spacing being to compensate for the inherently low conductivity of the finer soil materials.

The texture of the main body of soil may not be an important factor in determining drain spacing if porous substrata can be utilized in the lateral conduction of excess water through the soil. For example, a field with a continuous gravelly subsurface layer may need only a single drain to correct a waterlogged condition associated with a rather large area. In this instance, the main requirements are that the drain be deep enough to intercept water as it flows out of the gravel layer and large enough to carry away all the water it receives.

In some soils, poor drainage is the result of an impermeable subsurface layer that prevents the rapid percolation of excess water to deeper strata. Here, drain placement should be at the upper boundary of the impermeable layer so that full advantage can be taken of the water-transmitting characteristics of the more permeable soil material above. Recognize, however that if the soil depth over an impermeable layer is shallow, the installation of a drainage system will not likely bring about much improvement in plant growth. The reason is that the effective depth of drainage can be no greater

Table 9.3. Ranges of drain spacing for soil materials of different permeability characteristics. (From K. H. Beauchamp, In *Water, the 1955 Yearbook of Agriculture*, pp. 508–20, 1955).

Soil	Permeability	Spacing
		m
Clays and clay loams	Very slow	10–20
Silts and silty clay loams	Slow to moderately slow	18–30
Sandy loams	Moderately slow to rapid	30–100
Mucks and peats	Variable	15–60

than the depth to the impermeable layer. Where soils with impermeable substrata occur on a slope, water may flow laterally over the impermeable layer to saturate a neighboring field at the foot of the slope. A single drain placed across the slope and in a position where it can intercept the water (interception drain) may be all that is necessary to protect the soil in the lower-lying position (see fig. 9.25).

The drainage requirement

Adequate drainage can be expected only from a well designed system, one that will maintain the water table below a level that might otherwise damage crops. The demand placed on a drainage system is expressed by the *drainage requirement*, or the quantity of water to be removed per unit of time. Factors important to the drainage requirement are: (1) the crops to be grown, and (2) the rate of water input into the soil. Where the drainage requirement cannot be met, an alternative is to grow crops that can best tolerate the adverse conditions associated with excess water.

The ability of a drainge system to take care of the drainage requirement depends upon how rapidly excess water can be collected and carried away. Important to this are such features of design as the spacing between drains and their size and grade, the latter being the downward pitch, or slope, of the drain toward the point where it discharges water. The size and grade of a drain determine the total amount of water it can transport per unit of time. Also of significance is the depth of drain placement, since this determines the maximum depth to which the level of the water table can be reduced. On irrigated land where the removal of salts is necessary, the depth may be 180 cm or more; in humid-region soils, it commonly ranges between 75 and 125 cm.

Figure 9.22 illustrates an instance where drainage provided by a system of buried tile pipes is inadequate for corn. The inadequacy of drainage is indicated by reduced growth between the drains, where, as is to be expected, the water table remains at its highest level. The fact that the roots of the corn were everywhere separated from the water table by a distance of at least 45 cm suggests the presence of a rather thick capillary fringe in this soil.

Figure 9.22.
The effect of variation
in soil drainage on the
rooting depth and top
growth of corn (From
C. R. B. Elliott,
Ecology 5:175–78,
1924.)

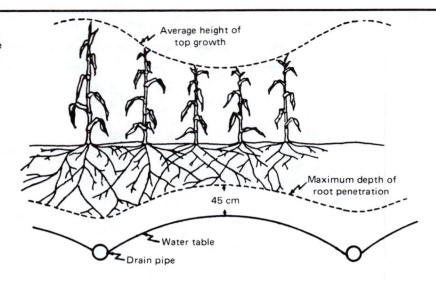

Average height of top growth

Maximum depth of root penetration

45 cm

Water table

Drain pipe

Systems of land drainage

The design of a drainage system should reflect the nature and magnitude of the problem as dictated by local conditions. Because of the complexity of many drainage problems, the design and installation of a drainage system should be under the guidance of an expert. More often than not, the solution rests with the development of an area-wide drainage network, if for no other reason than to provide facilities for the disposal of very large quantities of waste water.

The improvement of surface drainage

Where the origin of excess water is natural precipitation, often the most expedient control measure is the improvement of surface drainage. One reason is that it is much easier to collect and remove water before it has had a chance to infiltrate the soil. Another is that the removal of water by subsurface drainage results in the loss of soluble plant nutrients.

Provisions for bettering surface drainage sometimes involve little more than improving natural drainageways in low-lying positions in a field or by cleaning and supplying better outlets from poorly maintained roadside ditches. Other measures include such things as land-leveling or grading to eliminate low spots or ridges that impede the flow of water downslope, provided such changes do not encourage erosion.

Erosion-control systems that reduce runoff tend to increase the amount of water retained on a field for disposal by deep percolation. This not only increases the potential for nutrient loss by leaching but also, where the soil is slowly permeable, prolongs wetness until it may seriously interfere with aeration. Therefore, if the conservation of water for plant use is unimportant, well designed erosion-control systems should not eliminate runoff but, instead, should provide for its rapid disposal by orderly flow over the surface of the ground. Structures called *diversions* serve this purpose. A diversion is

Figure 9.23.
A diversion being checked for grade to assure proper flow of water across and off the side of the field. (U.S.D.A. Soil Conservation photo by James Hansen.)

a ditch dug across a slope to intercept runoff before it builds up to a highly erosive level (see fig. 9.23). Excess water that is collected is carried to the end of the diversion, where it is emptied out onto a protected surface, commonly a grass-covered waterway, for rapid, but nonerosive, flow downslope.

Subsurface drainage systems

The normal intent of subsurface drainage is to lower the level of groundwater in the soil or to prevent the development of a waterlogged condition by seepage. Two general types of systems are used: *open* and *closed*. An open system consists of one or more ditches that border or transect the land being drained and are continuous with a disposal system that carries the water to a natural drainageway. Closed systems, for the most part, consist of interconnected plastic, tile, or cement pipes located below the level of the water table, where they can collect the water and transport it to an open drainageway.

Open drains are used mostly where the rapid transport of large volumes of water is required (see fig. 9.24). For this reason they serve a principal function in the collection and disposal of water carried away from an area by other means, such as by surface flow or through a closed drainage system. Where soils are permeable, drainage may be sufficiently improved by the installation of a few open ditches scattered widely over an area.

Figure 9.24.
An open drain, which
has the advantage of
being able to transport
large volumes of
water. (U.S.D.A. Soil
Conservation Service
photo by Earl Baker.)

Figure 9.25.
Cross-sectional
diagram of an
interception drain
used to reduce or
prevent waterlogging
of adjacent, lower-
lying land.

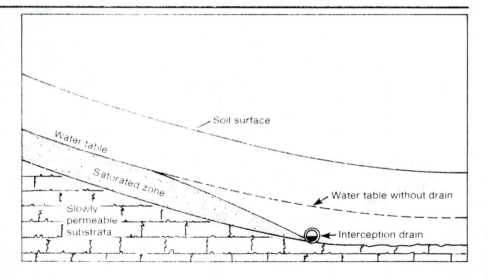

Closed systems vary from simple interception drains, which may consist of a single line of pipes located along the lower edge of a slope (see fig. 9.25), to the complex grids often needed to remove excess water from broad, flat areas as shown in figure 9.26. Interception drains are similar in effect to surface drainage; they remove water before it enters a soil to create a waterlogged condition. Interception drains are normally used to protect rather small land areas.

Figure 9.26.
Common patterns of
tile drainage systems.

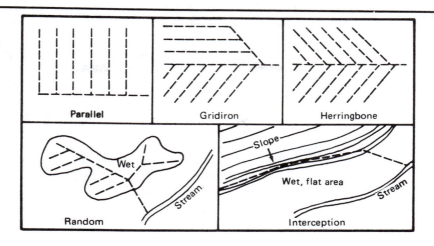

A

B

Figure 9.27.
Closed-drain installation: (A) cement pipe, and (B) flexible,
perforated plastic pipe. Where possible, installation is made during
the drier part of the year. (U.S.D.A. Soil Conservation Service photos
by A. F. Harms.)

Closed drains may be constructed of continuous, perforated plastic pipe, or of tile or cement pipes, which come in sections 60 to 90 cm long that are placed together in line but not sealed at the joints (see fig. 9.27). Water enters the line through the unsealed joints, so they must be protected against clogging, usually by a gravel covering. Special care is needed in drain construction through areas of unstable soil, for movement of the soil may throw a system of loosely joined pipes out of line. For such areas, long sections of perforated pipe with strong, permanently fixed joints are sometimes used.

Although initially more expensive, closed systems of drainge have the advantages of not taking land out of use and, if they are properly engineered and installed, provide trouble-free service for many years. Where the drainage requirement necessitates small intervals between drains, there is little choice but to use a closed system. Open drains are less expensive to install but not to maintain. They are the most practical to use where the disposal of large volumes of water is involved.

Drainage of irrigated land

Excess water in irrigated soils has the same adverse effect as it does under other conditions and, in arid regions, has the further disadvantage of encouraging salt accumulation. Irrigated agriculture often suffers because the land most suitable for water application is also the most difficult to drain. It is commonly located on flat areas where surface flow is relatively slow and where ground water derived by seepage or other means already exists at rather shallow depths. A problem may not arise until excess water supplied to irrigate the land causes the groundwater to rise dangerously near the surface. Much good land has been rendered useless because of the failure to provide drainage along with an irrigation system.

Drainage problems in irrigated areas are most common where the bulk of the water is applied in gravity systems, such as the furrow or border types. With these systems adequate wetting of the entire field ordinarily results in overirrigating upper slopes where water is first applied. The difficulty arises because water must be applied for a longer period on the upper than on the lower part of the slope. Such a problem is reduced with sprinklers that allow the same application of water across the full width of a field, and through careful control, both runoff and deep percolation can be avoided. Sprinkler irrigation systems are of particular value where, because of steep or complex topographic pattern, water cannot be applied properly under gravity flow.

In many locations throughout the world, the construction of irrigation canals through permeable soil materials has been the principal cause of rising ground water. In some of these systems there may be as much water lost by seepage as is finally delivered for use on the land. The control of seepage, which is usually achieved by lining canals with an impermeable material, provides the dual benefit of minimizing unwarranted deterioration of land by waterlogging and increasing the efficiency of the water-delivery system.

Benefits of drainage

The removal of excess water from soils is of both direct and indirect value to plants. Some of the benefits relate to improvement in soil physical, chemical, and biological properties that contribute to plant-growth functions. Usually, however, the change that most favors the plant is the increased availability of oxygen to roots.

A lack of oxygen reduces the uptake of nutrients and water by roots. Oddly, plants may wilt even when their roots are bathed in water if they do not obtain sufficient oxygen for respiration. Not only are root absorptive processes hindered by poor aeration, root exploration of the soil for stored nutrients will also be restricted.

Poor aeration also interferes with plant growth indirectly. It reduces organic matter decay and the release of nutrients through mineralization, and it tends to encourage the build-up of the reduced, more readily soluble form of manganese, which can be toxic to plants. It also allows common decay microbes to reduce nitrate (NO_3^-) ions to gases that can escape from the soil. Nitrate is an important and readily available form of nitrogen, and its transformation to gases and loss from the soil can have serious economic consequences where this nutrient is supplied in expensive fertilizers.

Excessive wetness limits the use that can be made of land. Slow drying in the spring delays cultivation and seeding. It also slows soil warming, so that seed germination and the subsequent development of plant roots and shoots are retarded. In combination, all of these effects delay harvest or may so shorten the length of the effective growing season that only a few kinds of plants can be grown.

Drainage increases the rate at which surface-applied water is taken in by soils. This is due in part to the reduction in the average water content of the soil, but it also reflects better structural conditions that result when drainage permits cyclic wetting and drying as well as an increase in the rate of microbiological activity. Increased intake of water by soils is of particular importance in flood and erosion control.

Summary

Soil water management involves the drainage of excessively wet soils or the use of measures, including proper irrigation techniques, that increase the efficiency of soil water use by plants. Maximum water use efficiency requires maintaining all factors of plant growth, including the supply of available water, as near optimum as possible. Where the water supply comes solely from natural precipitation, the water supply and the efficiency of its use can both be increased by erosion-control practices that reduce the loss of water as runoff.

The efficiency of water use by plants is customarily evaluatd by consumptive-use measurements, which determine evapotranspirational water loss. Such measurements are of particular value in irrigation control, since for maximum efficiency, irrigation water should be added in amounts just sufficient to replace that lost through evapotranspiration. The ability to control the amount of irrigation water added depends in part on the method used to apply the water.

Poor control over irrigation may cause excessive wetness in soils and the need for land drainage. However, this problem is more common to low-lying soils of the humid regions. Its correction may involve removal of excess water by encouraging rapid surface runoff, or by installing a system of drains, either open or closed, that allow the rapid transfer of groundwater to natural drainageways. The primary intent is to lower the water table. Benefits derived are increased aeration and depth of plant rooting, and improved soil temperature relationships that lengthen the growing season and increase the kinds of plants that can be grown on the soil.

Review questions

1. Why would the effective rooting depth of a plant likely be shallower if the plant were irrigated every week instead of every two weeks?
2. Why, on irrigated cropland, is it not wise to allow all available water in the effective rooting depth of a crop to be removed between successive irrigations?
3. Explain why improving the fertility status of soil increases the efficiency of water use by plants.
4. Why, under dryland farming conditions, do soil conservation practices tend to increase the amount of water made available to plants?
5. How can summer fallowing increase the efficiency of water use by plants in regions of low rainfall?
6. What are the advantages and disadvantages of the major methods of applying irrigation water, namely: rill, border, basin, sprinkler, and drip?
7. How are tensiometers, resistance blocks, and neutron probes used in irrigation scheduling and control?
8. Why is it necessary to place drains below the level of the water table if drainage of waterlogged soils is to be accomplished?
9. What is meant by improved surface drainage, and why should this approach to drainage be used wherever practical?
10. Name the advantages and disadvantages of closed and open drainage systems.
11. What are the major benefits derived from the drainage of waterlogged agricultural land?

Selected references

Edminster, T. W., and Reeves, R. C. Drainage Problems and Methods. In *Soil, 1957 Yearbook of Agriculture*, pp. 378–85. Washington, D.C.: Government Printing Office.

Hagan, R. M.; Haise, H. R.; and Edminster, T. W., eds. *Irrigation of Agricultural Lands*. Madison, Wis.: American Society of Agronomy, 1967.

Hansen, V. E., Israelsen, O. W., and Stringham, G. E. *Irrigation Principles and Practices*. New York: John Wiley & Sons, Inc., 1980.

Jensen, M. E., ed. *Design and Operation of Farm Irrigation Systems*. St. Joseph, MI.: American Society of Agricultural Engineers, 1980.

Officials of the Soil Conservation Service. *Drainage of Agricultural Land*. Port Washington, N.Y.: Water Information Center, Inc., 1973.

Stefferud, A. ed. *Water. 1955 Yearbook of Agriculture*. Washington, D.C.: Government Printing Office.

van Schilfgaarde, J., ed. *Drainage for Agriculture*. Madison, Wis.: American Society of Agronomy, 1974.

Viets, F. G., Jr. Fertilizers and the Efficient Use of Water. *Advances in Agron*. 14:223–64. New York: Academic Press, Inc., 1962.

10

Soil erosion: principles and control

With but few exceptions erosion under natural conditions is slight. As man has cut away the forests, allowed overgrazing of rangelands, and plowed the native sod, however, undue exposure has greatly accelerated soil destruction by erosion. A maximum price is paid for this folly if good land is converted to a state no longer suitable for the normal growth of plants.

The principal erosive agents with which we are concerned are wind and water. Although they tend to function under different circumstances, both lead to the loss of topsoil with its more abundant supply of available nutrients and preferred physical state. The deposition of the eroding material may, in turn, damage other land, choke roadside ditches, and fill streams and rivers with burdensome loads of sediment. This problem is not solely the concern of the individual whose land is being eroded, but must be dealt with by those who are affected by it indirectly as well.

The seriousness of erosion in the United States has been recognized for a long time. Expressions of concern at both the local and national level, following World War I, led to the formation in 1935 of a federal agency, the Soil Conservation Service, to work specifically with this problem. Today this organization is able to assist in erosion control throughout the nation.

The erosion problem in perspective

The extent to which land erodes is highly variable. Average annual soil losses from farmland may range from a small fraction to 300 metric tons per hectare, or more. The larger of these values, 300 tons, corresponds approximately to the weight of a layer of soil 2.5 cm deep covering an area of one hectare. At this rate the plow layer could be lost in only six or seven years. As a general rule, a long-term average annual loss of more than 10 tons of soil per hectare is considered serious.

The erosion problem is not limited to any particular region in this country, as may be seen from the map in figure 10.1; essentially all agricultural land has been affected to some extent. Those areas shown to be little affected

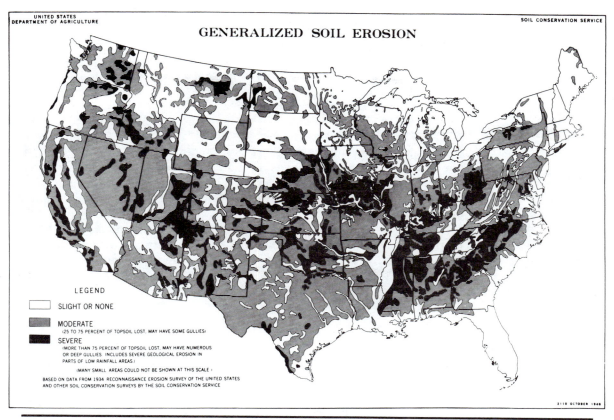

GENERALIZED SOIL EROSION

UNITED STATES
DEPARTMENT OF AGRICULTURE

SOIL CONSERVATION SERVICE

LEGEND

SLIGHT OR NONE

MODERATE
(25 TO 75 PERCENT OF TOPSOIL LOST. MAY HAVE SOME GULLIES)

SEVERE
(MORE THAN 75 PERCENT OF TOPSOIL LOST. MAY HAVE NUMEROUS
OR DEEP GULLIES. INCLUDES SEVERE GEOLOGICAL EROSION IN
PARTS OF LOW RAINFALL AREAS)

(MANY SMALL AREAS COULD NOT BE SHOWN AT THIS SCALE)

BASED ON DATA FROM 1934 RECONNAISSANCE EROSION SURVEY OF THE UNITED STATES
AND OTHER SOIL CONSERVATION SURVEYS BY THE SOIL CONSERVATION SERVICE

Figure 10.1.
Soil erosion in the United States. (U.S. Soil Conservation Service
map.)

by erosion represent mountainous terrain and other land not subjected to cultivation, or farmland on nearly level topography, such as the coastal plains of the East and South. The most severe erosion has occurred in the midsection of the Corn Belt and in the Southeast. Both of these areas have been utilized extensively for row-cropping without proper conservation practices. In the drier West and Midwest some places have been affected seriously by the erosive action of winds, and even by water, where rain, though limited in total quantity, often comes in torrents.

Some fundamental aspects of erosion

Erosion is a two-step process. The first of these is *detachment*, the breaking away of particles or small aggregates at the surface of the soil. The second step is *transportation*, which results in the actual loss of soil material.

Whether or not detachment will occur depends on the ability of erosive agents to overcome forces responsible for maintaining the soil in a coherent state. Thus well aggregated, fine-textured soils may erode less readily than non-coherent sandy soils, the latter requiring little energy for detachment. Once detached, however, small, lightweight particles are easier to transport than are those of larger size and greater weight.

Wind and moving water provide energy for particle detachment and transportation. The conditions determining the effectiveness of these two agencies in erosion differ in many respects, however. For example, a rain will not induce erosion unless there is a runoff, and this will occur only if the intensity of rainfall exceeds the infiltration capacity of the soil and if there is slope to the land. Wind, on the other hand, functions independently of slope and infiltration characteristics. For these reasons, wind and water erosion are treated separately.

Water erosion occurs under a variety of conditions and may be caused by rainfall, melting snow, irrigation water, or by stream or river flow. Rainfall erosion is by far the most widespread. Other types of water erosion tend to be of greater concern locally. Nonetheless, all may cause serious damage to the soil if allowed to go unchecked.

Rainfall erosion

The potential for rainfall erosion exists whenever there is a loss of water as runoff. This potential is generally greatest on cultivated land, which at times stands without a protective vegetative cover and, as often occurs, in a physical condition that limits the rate of water infiltration. Most erosion-control schemes include measures aimed at promoting maximum infiltration so as to minimize runoff. If runoff cannot be avoided, some means is usually provided to control its rate of flow and thereby reduce the tendency for erosion. All in all, a dense cover of plants is the best protection against rainfall erosion. It not only shelters the soil from the impact of falling rain but also slows runoff as it moves downslope.

Types of erosion caused by water

The removal of soil by flowing water follows two general erosional patterns. The first of these, *sheet* erosion, is the uniform detachment and removal of surface soil. It occurs where soil is exposed to the direct impact of intense rainfall, particularly if the surface has been previously pulverized by cultivation.

A second pattern of soil removal results from the channeling of runoff water and, depending upon the severity of process, causes either *rill* or *gully* erosion. Rills and gullies are distinguished by the size of channels left following erosion (see fig. 10.2). Rills are small and can be removed by normal soil cultivation; gullies are too large for this. Although gully erosion is highly important, sheet and rill erosion are more widespread and are of greater concern for this reason.

Figure 10.2.
(A) Relatively severe
rill erosion on
wheatland in eastern
Washington.
(B) Gullies in sandy
orchard soil in Florida.
These channels, which
were formed in a
single three-day
period, will greatly
hamper normal
cultural operations
and can be removed
only at great expense.
(U.S.D.A. Soil
Conservation Service
photos.)

A

B

Channeling markedly increases the destructive capacity of runoff, because water concentrated in deep channels flows more rapidly and more destructively than if spread out in a thin layer over the slope. Since channeled water also contacts less soil area and for a shorter period of time, infiltration from a given storm is reduced, and the lower the infiltration the greater the runoff and potential for erosion.

Factors of rainfall erosion

Five major factors contribute to the erosion of agricultural land by rain. These are: (1) the nature of the rainfall as determined by its frequency, intensity, and seasonal distribution, (2) the soil as it affects infiltration and susceptibility to detachment and transport, (3) the steepness and length of slope, (4) the nature of the cover provided by plants and their residues, and (5) cultural and soil-management practices that reduce runoff by modifying soil and cover conditions.

The factors of rainfall erosion have been known for a long time. However, it was not until 1961, when, following the detailed analysis of accumulated rainfall and erosion data, introduction of the *universal soil-loss equation* permitted the quantitative evaluation of rainfall erosion for a wide range of climatic, soil, slope, and cropping conditions. The equation is:

$$A = R \times K \times SL \times C \times P \tag{10.1}$$

where A is the predicted average soil loss in metric tons per hectare per year; R, the *rainfall factor* or *index*; K, the *soil erodibility factor*; SL, the *slope factor*, which is based on both the length and steepness of slope; C, the *cropping factor*; and P, the *management-* or *conservation-practices factor*. Of these factors, the first three are largely fixed by local conditions and together express the inherent potential for a soil to erode under a given set of rainfall, soil, and slope conditions. The last two factors take into account the potential for different crops or rotations, as well as specific management (conservation) practices to protect the soil against erosion. Where each of these factors has been suitably evaluated, their use in the soil-loss equation provides a basis for selecting cropping and management practices that will keep soil loss within acceptable limits. The evaluation and significance of these factors are considered in the following sections.

The rainfall factor

Runoff occurs whenever rain falls on sloping land at a rate the exceeds the rate of infiltration. As either the intensity or frequency of rains increases so does the potential for runoff and erosion. Particularly important is the number of heavy rains, for they cause most of the erosion even though they may represent only a small fraction of the total precipitation.

Much progress has been made in recent years in analyzing rainfall data and determining the relationship between rainfall patterns and soil erosion. Workers in the United States Runoff and Soil Loss Data Laboratory, at Lafayette, Indiana, through use of accumulated climatic records and associated erosion data, have established rainfall erosion indexes that express the relative erosion potential of rains at any location in the lower forty-eight United States and Hawaii. These values take into account two characteristics of individual rainstorms that correlate well with the potential for erosion, namely: (1) the *total energy* of the storm, which depends on the size and number of raindrops as well as the total amount (weight) of rain that falls, and (2) the *maximum 30–minute intensity*, or the average rate of fall during the 30–minute period of greatest intensity. Since rains of low volume have little effect on erosion, only the data for storms exceeding 1.25 cm of rainfall are considered. The annual rainfall index is then computed as the sum of

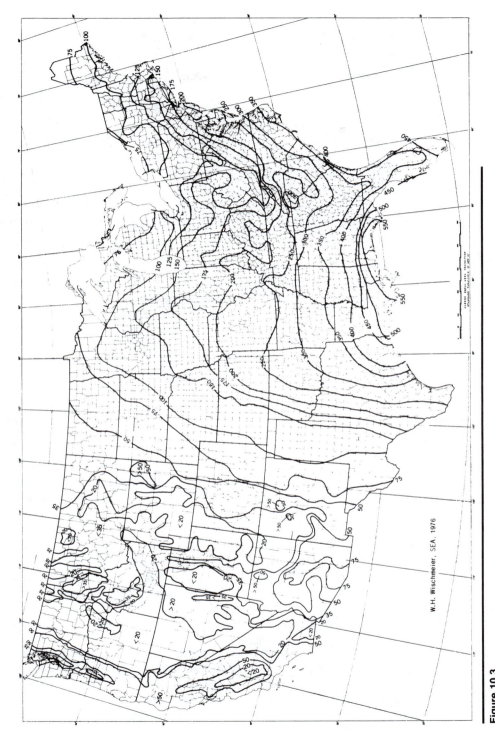

Figure 10.3.
Rainfall erosion indexes for the United States. (From W. H. Wischmeier, and D. D. Smith, *U.S.D.A. Handbook 537*, 1978.)

the indexes for all storms occurring during the year, with the index for use in the soil-loss equation being an average for many years. Since the rainfall index tends to vary markedly from one year to another, it cannot be used to predict soil loss on a year-by-year basis.

When sites of like rainfall-index are plotted on a map and joined by a continuous line, the pattern shown in figure 10.3 evolves. Interpretation of these values is as follows: an index of 600 indicates that the erosion potential identifiable with rainfall alone is 6 times greater than where the index is 100. A comparison with figure 10.1 shows, however, that areas having the highest indexes are not necessarily plagued with the most serious erosion problem. This points up the importance of the several other factors that influence rainfall erosion. The low incidence of severe erosion in the southern Atlantic and Gulf Coast lowlands, in spite of the high rainfall index for the area, results largely from relatively favorable topographic and soil physical conditions.

The slope factor

The ability of rainfall to erode soils increases with an increase in either the length or steepness of slope. The principal reason for this is that changes in these slope characteristics alter both the volume of runoff and the velocity of flow downslope. Slope length is a factor because it determines the total area from which runoff accumulates. The primary effect of steepness is on runoff velocity, for the steeper the slope the higher the velocity. In addition, as the rate of water flow off a slope increases, less time is available for infiltration into the soil. As a consequence, the greater the steepness of slope the larger the total volume of water lost as runoff.

The general relationship between slope characteristics and erosion can be determined from a chart such as the one in figure 10.4. This figure shows the ratio of soil loss for any of a wide range of slope conditions. A value of 1.0 on this chart has been arbitrarily assigned to the slope conditions of standard experimental plots used over many years in the United States for the collection of runoff and erosion data. These plots have a length of 22 m and are located on a 9 percent slope. Because they have a soil-loss ratio of 1.0, these plots are commonly referred to as *unity* plots.

Data obtained for soils studied on unity plots can be applied to other slope conditions through use of figure 10.4. For example, an 8-percent slope, 120 m long, or a 12-percent slope, 38 m long would, on the average, be associated with twice the soil loss as occurs on a unity plot under comparable soil and climatic conditions. Similarly, a 4-percent slope, 300 m long should experience about the same loss as a unity plot, whereas the loss from a 2-percent slope, 300 m long should be only 0.4 of that amount.

The soil-erodibility factor

For the most part, texture, structure, and organic matter content, which are properties that strongly influence infiltration and the ease of particle detachment and transport, determine the erodibility of a soil. Subsoil conditions that affect internal soil drainage are also important, for where internal drainage is poor, infiltration may be retarded and the potential for erosion thereby increased.

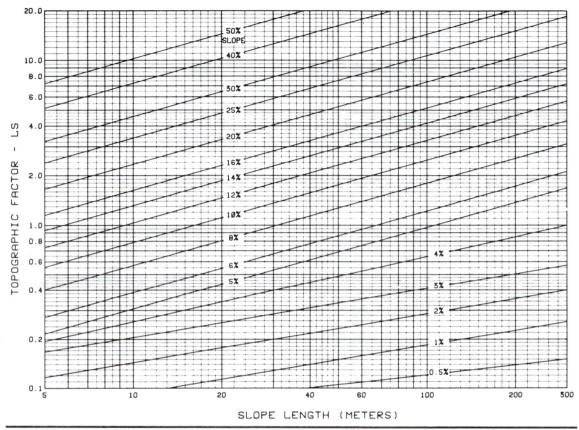

Figure 10.4.
Chart for determining the slope (topographic) factor for soils under
varying conditions of slope length and steepness. (From W. H.
Wischmeier and D. D. Smith, U.S.D.A. Agriculture Handbook 537,
1978.)

The soil-erodibility (K) factor, or index, is a measure of the inherent
tendency for a soil to erode irrespective of other erosion factors. It is expressed as soil loss in metric tons per hectare per unit of rainfall erosion
index. The erodibility factor is measured on unity plots, where the soil-loss
ratio is 1.0 and under clean-tilled fallow so plants do not affect runoff and
erosion. In the determination, annual soil loss is measured for a period of
years. The erodibility (K) value is then computed by dividing the average
annual soil loss by the local rainfall erosion index (R). Thus, if the average
loss of a soil were 60 metric tons per hectare per year where the rainfall
erosion index was 100, the K value for the soil would be:

$$K = \text{Soil loss}/R = 60/100 = 0.6 \qquad (10.2)$$

Example erodibility indexes determined in this way for a number of soils
are shown in table 10.1.

Table 10.1. Soil-erodibility factors for soils on erosion-research stations in the eastern United States. (From W. H. Wischmeier, and D. D. Smith, *U.S.D.A. Handbook* 537, 1978).

Soil type	Location	Erodibility factor[a]
Dunkirk silt loam	Geneva, N.Y.	1.54
Fayette silt loam	La Cross, Wis.	.85
Hagerstown silty clay loam	State College, Pa.	.69
Cecil sandy loam	Clemson, S.C.	.63
Tifton loamy sand	Tifton, Ga.	.22
Albia gravelly loam	Beemerville, N.J.	.07

[a]These values are 2.24 times those for soil loss expressed in English tons per acre.

The direct measurement of erodibility factors on unity plots has been possible with only relatively few soils. However, additional studies have been made with other soils using small plots and simulated rainfall. From these studies it has been found possible to estimate the erodibility factor of any soil from five measured characteristics that have dominant control over erodibility. These are: (1) percentage content of silt plus very fine sand (0.002–0.1 mm), (2) percent total sand exclusive of the very fine fraction (0.1–2.0 mm), (3) percent organic matter, (4) structural class, and (5) permeability class (see page 519). Values for these characteristics are normally available from standard soil surveys published by the U.S. Department of Agriculture. They are used in estimating erodibility factors by means of the nomograph in figure 10.5.

The use of the nomograph for estimating erodibility factors is illustrated by the broken line in the diagram. The starting point in the determination is with the percentage content of silt plus very fine sand, as shown on the left-hand axis of the nomograph. A line is then drawn to intersect, in order, curves for percent sand (0.1–2.0 mm), percent organic matter, structural class, and permeability class. From the last intersect, the plotted line is then projected horizontally until it meets the erodibility-factor axis. In the example plot, an erodibility value of slightly above 0.3 is obtained for a soil containing 65 percent silt plus very fine sand, 5 percent sand coarser than very fine sand, 3 percent organic matter, structural class 2 (fine granular), and permeability class 4 (slow to moderate).

The cropping factor

Plants vary in the protection they offer against the hazard of rainfall erosion. Differences relate primarily to (1) the amount and duration of the cover they provide, (2) the quantity and type of residue left upon or incorporated into the soil, and (3) the nature of tillage practices where cultivated crops are grown. The last factor is particularly important if the soil is cultivated excessively, as in weed control, and thereby reduced to a highly erodible state because of granule destruction.

Plants can be grouped according to their relative ability to protect the soil against rainfall erosion, as in table 10.2. The greatest protection is afforded by noncultivated plants that provide a good ground cover. Similarly, hay crops are highly effective in erosion control, for they are not often disturbed by cultivation. Well established small grains provide almost as good

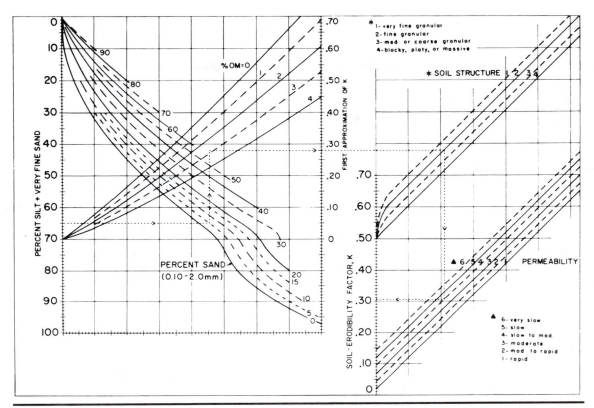

Figure 10.5.
Nomograph for estimating the erodibility (K) factor of a soil. For use,
see the text. (From W. H. Wischmeier, and D. D. Smith, U.S.D.A.
Agriculture Handbook 537, 1978.)

Table 10.2. The relative value of different plant covers in erosion control.

Type of cover	Degree of protection
Permanent woods or grass sod	Complete
Legumes or grasses for hay	High
Small grains	Medium
Row crops	Low
None (fallow)	None

protection as hay crops. However, land devoted to the production of small
grains is usually cultivated at least once a year and often remains exposed
for substantial periods of time. For these reasons, small grains are not rated
as high as hay crops so far as erosion control is concerned.

Except for fallowland, soils planted to row crops such as cotton, corn,
and potatoes are most susceptible to erosive influences. Usually, plants of
this type do not provide a canopy that shields the soil from rain until comparatively late in the season. Also, land preparation for seeding and culti-

vation for weed control may break down soil structure so that the soil absorbs water slowly and erodes easily. The plants, being widely spaced, offer little resistance to water flow over the surface of the soil and therefore do not contribute much to the reduction of runoff.

Many common cropping systems have been evaluated to show their relative ability to reduce soil erosion. In general, such evaluations take into account the sequence of cropping, which may consist of one crop plant grown continuously or a variety of crops grown in rotation. It is important that a sequence be considered because of differences in carry-over effects from one season to the next. For example, corn following corn results in a higher susceptibility to erosion than when it follows a sod crop. The benefit of the sod crop results mostly from its enhancement of soil structure, an effect that may be evident for several years after the sod has been plowed under.

The level of crop production must also be taken into account in the evaluation of cropping systems, the reason being that variation in the level of productivity has a profound effect on the potential for erosion. Vigorous plants associated with high productivity offer both earlier and more extensive protection during the growing season. They also supply larger quantities of residues for use as a post-harvest cover or as a source of soil organic matter. Such benefits are often overlooked in assessing the gains that result from maintaining a soil in a highly productive state.

Because of the large number of cropping systems employed in the United States, it is impractical to consider the specific effect of each on erosion. However, some of the general relationships involved can be demonstrated through use of an example. For this, assume that the soil loss associated with continuous corn at one level of soil and residue management averages about half that occurring under clean-tilled fallow. If, instead of every year, corn were grown only one year out of three, with the land being used for small grain and hay meadow during the other two, erosion during the three-year period might be reduced to about 10 percent of that under clean-tilled fallow, or to about one-fifth of that associated with continuous corn. If, as an alternative to the above rotation, the land were used for the production of hay or small grain only, the average soil loss might be further reduced to about 2.5 percent of that experienced with continuous fallow, or 5 percent of that associated with corn grown year after year.

The conservation-practices factor

Although any plant cover provides some protection to the land and thereby reduces erosion below that associated with clean-tilled fallow, the degree of protection can be enhanced where the cover is grown in association with well devised conservation practices. For example, continuous corn planted up and down the slope may reduce erosion to about 75 percent of that under clean-tilled fallow, but by means of the relatively simple practice of tilling and planting across the slope (contour tillage), the loss may be reduced to as little as one-third that under clean-tilled fallow (see fig. 10.6). The effect of such practices is accounted for through use of the conservation practices (P) factor in the soil-loss equation. The factor is assigned a value of 1.0 where no conservation practices are applied, as when a row crop is planted up and

Figure 10.6.
The distinction
between contour
cultivation and
cultivation up and
down the slope.

down the slope, but where soil loss is reduced by an applied conservation practice, the P factor for use in the soil-loss equation will be less than 1.0.

In addition to contour tillage, there are several other conservation practices in common use on cultivated land. These include: (1) *minimum tillage*, (2) *contour stripcropping*, and (3) *terracing*. Where the potential for erosion is great, the application of more than one of these practices may be necessary for adequate control of soil loss.

As the name implies, *minimum tillage* requires the least possible mechanical disturbance of the land, either during seedbed preparation or by cultivation for weed control. A standard practice in preparing a seedbed is to turn the entire field with a plow and then reduce cloddiness by discing and harrowing. Two purposes are involved: to bury plant residues, and to produce a granular state favorable for planting and seed germination. Unfortunately, however, cultivation in seedbed preparation is often excessive and may create conditions that encourage erosion. To overcome this disadvantage several minimum-tillage techniques have been proposed. These go by such names as *plow-plant, wheel-track planting, strip-planting,* and *zero tillage*. In plow-plant, seeds are drilled directly into plowed soil, whereas in track-planting, the seeds are placed in plowed soil that has been compressed and smoothed by the tractor wheels. In strip-planting, only narrow strips of soil are tilled, and planting is done in the tilled strips. In zero-tillage, illustrated in figure 10.7, planting is directly into untilled ground. Either strip-planting or zero-tillage takes advantage of undisturbed residues from a previous crop to protect the soil against erosion. In plow-plant and wheel-track planting, erosion is reduced by the capture of water in the loosened, plowed soil remaining between the planted rows.

Minimum-tillage procedures require the control of weeds by chemical means. In strip-planting and zero tillage, this control may also include killing of the previous crop prior to planting. Because of the repeated and extensive use of herbicides under these circumstances, the potential for serious environmental pollution or damage to the soil or subsequent crops should not be overlooked.

Contour tillage, a widely used practice for the control of rainfall erosion, consists of carrying out all cultural operations across the slope. Furrows and ridges thus formed obstruct the flow of water downgrade. In areas of intense rainfall, the soil may be worked into tall ridges and deep furrows that retain much larger quantities of water than do depressions left by normal planting operations. This type of soil preparation is known as *listing.* Here, planting is on top of the ridges.[1]

[1]Listing is used also for planting where the surface soil is too dry for seed germination. In this instance the seeds are placed in moist soil at the bottom of the furrow.

In *contour stripcropping*, narrow belts of different crops are planted across the slope as shown in figure 10.8. Usually the strips alternate between thickly seeded meadow and row crops, or, in dry-land areas where the wheat-fallow system is used, the strips may consist alternately of cropped and fallow ground. Stripcropping is beneficial in that it limits the length of exposed slope. Also, runoff lost from a tilled strip can be slowed by a meadow immediately below so that suspended soil is deposited instead of being carried off the field. It is a standard practice to subject the strips to normal crop rotation.

The most drastic contour operation is *land-terracing*, which involves forming a series of low ridges across the slope to intercept runoff (see fig. 10.9). Water that collects behind the terraces may be diverted to the edges of the field, where it can flow downslope on protected (grassed) waterways. However, if the amount of water is limited, it may be allowed to accumulate and soak into the soil.

If terraces running across the slope are held close to the contour, the space between them varies, and often considerably, as may be seen in figure 10.10. This makes planting and harvesting with machinery difficult. Such a problem can be overcome in part by constructing terraces parallel to each other. When this is done, however, the terraces do not follow the contour but, instead, are lower in some places than in others. As a result, water flowing

Soil erosion: principles and control

Figure 10.8.
Contour stripcropping. Row crops (dark areas) planted on the contour are effective in limiting soil erosion during light rains. The more thickly planted buffer strips (light areas) retard runoff during severe rainstorms. (Photo by U.S. Soil Conservation Service—U.S.D.A.)

Figure 10.9.
Forming soil into a terrace to control rainfall erosion on a slope. (U.S.D.A. Soil Conservation Service photo by Earl Baker.)

Figure 10.10.
Land terraces. Note the permanent grass cover on the terraces and in the channels behind the terraces. (Photo by U.S. Soil Conservation Service—U.S.D.A.)

laterally from either side may accumulate in sufficient amounts to overflow and erode the terrace at the lowest points. To avoid this problem, provisions are made to remove the excess water, either through underground drains connected to slotted-pipe intakes placed in the channel behind the terrace (see fig. 10.11), or by means of adequately protected, surface drainageways.

Steep slopes do not lend themselves well to terracing. Where steepness is a problem, runoff may be reduced by constructing one or two diversions across the slope. Diversions are like terraces, but they are usually higher and have a deeper channel, which permits the collection of larger volumes of water (see fig. 9.23). Some grade, or slope, is built into the channels in order to move water across and off the field.

The importance of stripcropping and contour tillage in erosion control can be judged from table 10.3. The factors listed indicate the probable erosion associated with these practices relative to the erosion expected with rows planted up and down the slope. Stripcropping combined with contour tillage is twice as effective as contour tillage alone, and either practice attains maximum efficiency on moderate slopes ranging in grade between 2 and 7 percent. This arises from the fact that runoff is difficult to stop by any measure on very steep slopes. On nearly level land, however, one cultural method is as effective as any other in erosion control.

It is difficult to index the average effect of terracing in erosion control; thus, factors for this practice are not listed in table 10.3. The efficiency of a terrace depends on other protective measures used in conjunction with it. Under appropriate management, such as the maintenance of a sod crop on steep terrace faces and the use of contour tillage between terraces, this practice is considered to be equal in effect to contour stripcropping. Without proper management, substantial local erosion of the terrace can occur and

Soil erosion: principles and control

Figure 10.11.
(A) Examining a slotted-pipe intake on a parallel terrace. (B) Diagram of a drain installed in parallel terraces. (Photo courtesy of U.S.D.A. Soil Conservation Service; diagram adapted from a U.S.D.A. Soil Conservation Service drawing.)

A

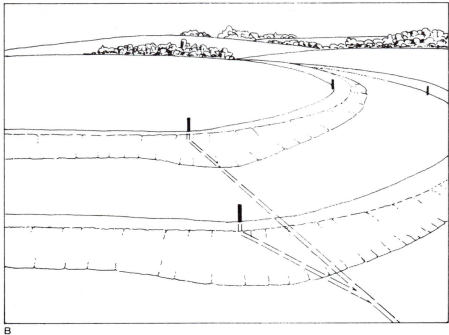

B

Table 10.3. Erosion-control practice factors for contouring and contour stripcropping. (From W. H. Wischmeier, and D. D. Smith. *U.S.D.A. Agriculture Handbook* 537, 1978).

Percent slope	Contour tillage	Contour stripcropping
1–2	0.60	0.30
2–7	0.50	0.25
7–12	0.60	0.30
12–18	0.80	0.40
18–24	0.90	0.45

causes sediment to accumulate in the terrace channels. Even so, terraces can reduce markedly the amount of soil lost from the field. Like terracing, contour listing is considered to be about equal in value to stripcropping in erosion control.

A practical example

Where factors of soil erosion have been quantitatively assessed, judgments of the potential soil loss under different cropping and management systems can be made. A system that provides satisfactory return from the land without permitting serious erosion can then be selected. Steps involved in the selection are illustrated by an example.[2] For this we will judge the effects of different combinations of cropping and management practices on reducing erosion below that indicated for clean-tilled fallow in the example on page 233. In this earlier example, the specified soil loss was 60 tons per hectare per year where the rainfall index was 100, the soil erodibility index was 0.6, and the soil-loss ratio was 1.0. Through the selection of a system of cropping and management, our objective would be to reduce potential soil loss to less than 10 tons per hectare per year.

Assume that for the climatic and soil conditions above a common cropping system is a four-year rotation including two years in row crops, one year in wheat, and one year in a grass meadow cut for hay. Assume further that this cropping system yields a cropping (C) factor of 0.25 and is therefore capable of reducing soil loss to one-fourth that for clean-tilled fallow, or to 15 tons per hectare per year. This is determined through use of the soil-loss equation:

$$R \times K \times SL \times C \times P = 100 \times 0.6 \times 1 \times 0.25 \times 1 = 15 \text{ t/ha/yr} \quad (10.3)$$

Since this loss would exceed the limit of 10 tons per hectare per year, use of the above rotation with a special conservation practice, such as contour tillage, would logically be considered. The effectiveness of this treatment in further reducing erosion would be determined by applying the conservation-practice (P) factor for contour tillage in the soil-loss equation.

The P factor for contour tillage varies with the degree of slope. According to table 10.3, contour tillage on a 9 percent slope, which applies to our example, has a P factor of 0.6 and therefore has the potential for re-

[2]For an in-depth treatment of crop and management-practice selection for erosion control, see W. H. Wischmeier, and D. D. Smith, *U.S.D.A. Handbook* 537, 1978.

Soil erosion: principles and control

ducing erosion to 60 percent of that for tillage and planting up and down slope. Inserting this factor into the soil-loss equation we find that:

$$R \times K \times SL \times C \times P = 100 \times 0.6 \times 1.0 \times 0.25 \times 0.6$$
$$= 9 \text{ t/ha/yr} \qquad (10.4)$$

Since this level of soil loss is less than 10 tons per hectare per year, the use of the above rotation with contour tillage should provide adequate protection of the soil.

For conditions more critical than in the foregoing example, measures of greater stringency would be required to keep erosion in check. Additional steps that might be taken would include raising the level of productivity, incorporation of all residues into the soil, and use of minimum-tillage techniques, such as plow-plant. If further restrictive measures were required, changes in the rotation might be necessary, the principal one being a reduction in the frequency with which row crops were planted. Under extreme conditions the use of any cultivated crop would likely be eliminated, and the soil maintained permanently under pasture or as hay meadow or woodlot.

Some special considerations in water-erosion control

Previous discussions of erosion have dealt largely with conditions common to the majority of agricultural lands in regions of moderate to high rainfall. However, certain important aspects of erosion control fall outside these more general considerations. They relate to the control of gully erosion and to the nature of erosion problems on timber and irrigated land.

Soil loss under irrigation

Soil loss due to erosion can be a very serious problem on irrigated land, particularly where the water is applied by the furrow method. The greatest erosion occurs at the upper end of the furrow (see fig. 10.12), for this is where water is first applied and where it flows in greater volume than elsewhere down the furrow. Water is added to the furrows in relatively large volume in an effort to move it down the full length of furrow as quickly as possible. The greater the time required to move water down a furrow the greater the disparity between the depth of soil wetting at the upper and lower ends of the field.

It is usually difficult to move water across a furrow-irrigated field fast enough to ensure uniform soil wetting from one end of the furrow to the other. This is demonstrated by the data in table 10.4, which shows that it took more than 7 hours to move water down furrows through a distance of 270 m. If, in this instance, irrigation had been stopped after 12 hours, the lower end of the furrow would have received water for only 5 hours, while the upper end would have received it for the full 12 hours.

Erosion was measured in the experiment from which the above data were obtained, and as indicated in table 10.4, it was greatest at the upper end of the furrows. According to the table, an average of 62 kg of soil was carried away from the first 90 m of furrow.[3] Yet, only 1 kg of soil moved

[3]If these furrows were at a common 60–cm spacing, the rate of soil loss would have been 11 metric tons per hectare for the single irrigation, approximately.

Figure 10.12.
Erosion of irrigated
land caused by the
application of water to
furrows at an
excessive rate. (Photo
by U.S. Soil
Conservation
Service—U.S.D.A.).

Table 10.4. Flow rates, soil loss, and the travel time of water along a 270-m irrigation furrow in a fine sandy loam. (After S. J. Mech, *Agri. Eng.* 30:379–83, 289, 1949).

Distance along furrow	Flow rate[a]	Soil loss[b]	Travel time[c]
m	l/min	kg/furrow	hrs:min
90	20.8	62	0:24
180	11.8	17	1:38
270	5.3	1	7:16

[a]The rate of flow at the specified distances after establishment of constant rate of runoff.
[b]Soil carried past the specified distance during the entire irrigation period.
[c]Total time from initial input of water.

beyond 270 m. Thus, a measurement of the total loss of soil from the lower end of the furrows would give no indication of the erosion pattern. It is not uncommon for large quantities of soil to be carried off in irrigation water, however, the amount sometimes exceeding 100 tons per hectare per year.

To obtain complete soil wetting, the rate at which irrigation water is applied to a field must be in excess of the total infiltration capacity along the full length of the furrow. Thus, long furrows, espcially when they have been laid out on soils of high infiltration capacity, may require an exceptionally heavy and erosive input to transport the water along the full length of the furrows in a reasonable period. The solution of such a problem may have to be shortening the length of furrows. This may require splitting a field into two parts, with one set of furrows irrigating the upper part and a second set irrigating the lower part of the field.

Figure 10.13.
Irrigated lettuce planted on the contour to allow more nearly uniform soil wetting and to reduce the loss of soil by erosion. (Photo by U.S. Soil Conservation Service—U.S.D.A.)

Steep slopes that cause the rapid downhill flow of irrigation water can also result in excessive erosion of irrigation furrows. This problem can often be countered by angling the furrows across the slope so that they fall more nearly on the contour and, therefore, have a more gentle slope (see fig. 10.13). An alternative is sprinkler irrigation, which permits water to be applied at the same rate and for the same period of time across the full width of a field.

After irrigation furrows have first been formed, or if they have been reshaped between irrigations, they usually contain a residue of loose soil that not only interferes with water flow but also is subject to ready movement downslope. For this reason soil losses generally are considerably greater during the first irrigation following ditch preparation than at any other time. Measurements have shown that the bulk of soil movement in rill-irrigated fields occurs during the first four hours of the initial irrigation and sometimes may take place primarily during the first hour.

Control of gully erosion

Most farmland is on topography containing a network of drainageways that carry away runoff water. Since water flow in these drainageways is often great, removal of natural vegetation from them through normal farming operations may lead to extensive erosion and the development of gullies (see fig. 10.14). As may be seen in figure 10.15, the damage caused by gully erosion may be difficult, if not impossible, to control.

The best solution to the problem of gully erosion is prevention, usually by maintenance of natural drainageways under permanent cover, as shown in figure 10.16. Where gullies have already formed, however, steps should be taken to prevent further soil deterioration and to reduce the damage already done. Establishment of a vegetative cover is essential, and this may or may not be preceded by some reshaping and smoothing of the eroded area. Often the cover is allowed to develop naturally, although a special seed-

Figure 10.14.
Channeled runoff, as
shown on this Brown
County, Wisconsin,
farm, can lead to
serious gully erosion if
not brought under
control. (U.S.D.A. Soil
Conservation Service
photo by F. M. Stone.)

Figure 10.15.
Severely eroded
loessial plain on the
Potwar Plateau, in
north-central Pakistan.
This virtually total
destruction of once-
excellent farmland is
blamed on
uncontrolled grazing
by sheep over a
period of some 400
years.

Soil erosion: principles and control

A   B

Figure 10.16.
(A) A gully that could have been avoided by maintaining the natural
drainageway under a permanent grass cover. as shown in (B).
(U.S.D.A. Soil Conservation Service photos.)

ing of a rapidly growing grass or legume may be required. After the eroded
area has been stabilized, it may be profitably used as hayland if the slope is
not too steep.

Where no other control measure is immediately practical, it is helpful
to construct one or more dams across the bottom of a gully to slow runoff
and entrap sediment (see fig. 10.17). The dams may be temporary or permanent depending upon the demands of the particular case involved.

**Erosion in
forested areas**

Wooded areas preserved in their natural state can withstand severe punish-
ment from the elements, evident where forests maintain stable conditions
in areas of exceptionally high precipitation. Once the cover is removed, however, serious erosion may follow. Such effects are most critical on mountain
slopes, where the soils tend often to be shallow over hard rock and are essentially irreplaceable by normal soil-forming processes.

Erosion problems of major proportions often develop on cutover timberland. For the most part the difficulty arises when the plant cover is destroyed by the moving of logs from place to place or by the uncontrolled
burning of slash. Logs are normally transported by skidding, either along
graded skid roads by tractor or down skid trails by cable line (see fig. 10.18).
In either instance the vegetation and protective surface litter are removed

Figure 10.17.
A steel dam in a smoothed, seriously eroded drainageway installed to slow runoff and reduce sediment loss. The drainageway will be seeded to a permanent grass cover. (U.S.D.A. Soil Conservation Service photo by R. W. Lowery.)

Figure 10.18.
Logs that have been skidded to a loading site. Note the essentially total destruction of protective cover and pulverization of the soil, making it highly susceptible to erosion. (U.S.D.A. Forest Service photo by Frank Flack.)

and the mineral soil exposed. Repeated use of the trails and roads compacts the soil and reduces the potential for infiltration. Where the roads or trails lie on a long, smooth grade, they provide an ideal pathway for the flow of water downhill. It is usually here that water erosion makes its first serious inroads in forests used for commercial wood production (see fig. 10.19).

Logs are commonly delivered to a central point for loading. All trails and roads converge on this point, as does the runoff they carry during rainy

Figure 10.19.
Serious erosion of a skid trail left unprotected after abandonment. (U.S.D.A. Forest Service photo.)

periods. For this reason loading sites are often subjected to extreme damage from erosion. The careful selection and protection of these sites is essential if serious erosion is to be avoided.

Only a few simple suggestions are given to aid in minimizing the erosion of commercial forestlands. First and foremost is the need for careful planning, an essential prerequisite to all lumbering operations. The laying out of trails and roads should be preceded by a preliminary survey of soil and slope conditions so that the most satisfactory routes can be established. Areas of unstable soil should be avoided and the number of trails and roads held to a minimum. A preferred grade of skid roads is less than 7 or 8 percent wherever possible, and long grades should be interrupted periodically with diversions that prevent runoff water from developing into erosive streams. Water diverted from the roadway, when fed onto the natural soil cover, will seep slowly into the ground. It is important also to protect the beds of streams by preventing the skidding of logs either along or across them.

When logging in an area is completed, exposed soil should not be left unprotected. Sometimes natural vegetation reestablishes itself quickly, but if it fails to do so, the seeding of an adapted, rapidly growing grass may be desirable. When appropriate precautions are taken, even the clearcutting of forestland can be carried out without particular fear of serious damage to the watershed. At the same time, pollution of streams and rivers with eroded materials can be largely avoided.

Fire can set the stage for serious erosion of forestland (see fig. 10.20). Fire often destroys not only a protective plant cover but also any organic litter layers that might occur on the surface of the mineral soil. These layers are storage sites for nutrients, both in organic combination and in exchangeable form, and their loss can have a serious impact on the plant nutrient supply. This is especially true for organic nitrogen, which can be totally lost

Figure 10.20.
The total loss of soil
cover from a forest
fire. The soil, left bare,
thus becomes
susceptible to erosion
by water. (U.S.D.A.
photo.)

on burning. Thus, fire may not only destroy an existing protective cover of organic litter and vegetation, it may make the re-establishment of a new vegetative cover more difficult.

Wind erosion

Erosion of soil by wind is a problem primarily of dry regions. There are a number of reasons why this is so: high winds occur frequently in these areas; soil texture and moisture conditions are often such that easy detachment and transport of soil particles is possible; and the vegetation tends to be sparse, which limits the amount of cover available for protecting the soil against damaging effects of the wind. Within the United States the wind-erosion hazard is greatest where the average annual precipitation is less than 50 cm. There are exceptions, however. Beaches or other areas of accumulated sands, or organic soils in humid regions are often subject to wind erosion where wind velocities are at least occasionally high.

Mechanics of wind erosion

Wind erosion is usually initiated when soil particles or aggregates lying loose at the soil surface are set in motion. These particles, in turn, assist in detaching others as they bounce along and strike the soil surface repeatedly. Particle bounce, or *saltation*, is probably the most important mechanism of wind erosion. A less pronounced effect results from *surface creep*, a sliding or rolling motion limited principally to the larger, heavier soil particles or aggregates.

In addition to saltation and soil creep, a third type of motion, particle *suspension*, is normally involved in wind erosion. Suspension occurs when air flow is sufficiently rapid to prevent particles from dropping back to the ground once they have been lifted into the main air current. Particles most susceptible to the suspension process are those of small size, but they include very fine sands. Coarser particles, when picked up by the wind, tend to settle back to the earth rapidly, whereupon they contribute to the saltation process.

Factors of wind erosion

As with rainfall erosion, erosion of soil by wind is a function of several factors that can be evaluated and related mathematically in order to predict potential soil erosion under a given set of climatic, soil, and cropping conditions. However, due to interaction among these factors, where a variation in one causes one or more of the other factors to also vary, only a highly generalized prediction equation can be given; and it serves mainly to identify the factors individually. The equation is:

$$E = f(I', K', C', L', V') \tag{10.5}$$

where E is the average annual soil loss in metric tons per hectare; I', the soil-erodibility factor in metric tons per hectare per year; K', a surface-roughness factor associated with cloddiness or ridging; C', the climatic factor; L', the width of unprotected soil surface exposed to the wind and expressed in meters; and V', a vegetative factor expressing the amount of surface cover provided by growing plants or their residues. Because of the complexity of the prediction equation, estimates of the potential loss of soil by wind erosion are made from graphs covering different sets of climate, soil, and cover conditions. The use of these graphs is illustrated following an initial discussion of the factors of wind erosion.

Climatic factor of wind erosion

The relationship between climate and wind erosion depends upon the intensity and frequency of windstorms and the contribution of rainfall and evaporation to the general moisture status of the surface soil during periods of high winds. Moisture conditions are important because soil is less easily eroded by wind when moist than when dry.

The climatic effect in wind erosion has been determined experimentally under a wide range of conditions. Some of this work has been conducted in the field, but because of the need for rigid control, much of the experimentation has been under simulated conditions, such as in wind tunnels. On the basis of the correlations established in these studies, the potential climatic influence on wind erosion has been computed for many locations throughout the continental United States. The values obtained, which define the *wind-erosion climatic factor*, are relative; they express the potential for erosion associated with the climate at a specified location as a percentage of the potential at Garden City, Kansas, where extensive wind-erosion studies have been conducted. The climatic factor at the latter location has been arbitrarily assigned a value of 100 percent. Soil loss potentials are greater than those at Garden City when the climatic factor exceeds 100 percent and are less if it is under 100 percent.

Figure 10.21.
The wind-erosion
climatic factor for the
plains and western
regions of the United
States. (Adapted from
U.S. Soil Conservation
Service maps.)

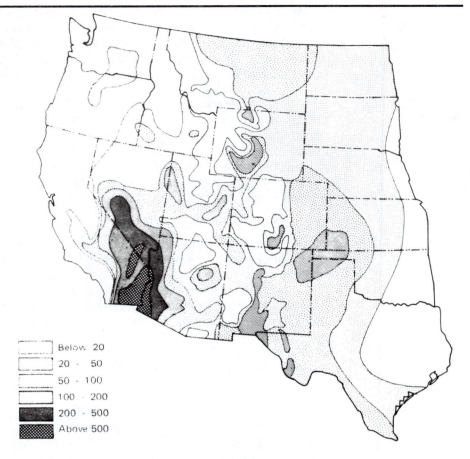

Below 20
20 - 50
50 - 100
100 - 200
200 - 500
Above 500

Figure 10.21 shows the general distribution of the wind-erosion climatic factor for the western part of the United States. The highest potential for wind erosion is in the Southwest, where it is encouraged by both a lack of soil moisture and frequent strong wind storms. An area centered on the Oklahoma Panhandle, where the climatic factor exceeds 100 percent, identifies the Dust Bowl that was so devastated by wind erosion in the early 1930s (see fig. 10.22).

Soil-erodibility factor

Soil properties important to erosion by wind relate primarily to (1) the ease with which individual particles or aggregates may be detached from the main body of soil, and (2) the size of these units once they have been detached. The latter characteristic is significant because, on the average, the smaller the particle the greater the ease of movement by the wind.

Detachment is not a restrictive factor in the erodibility of sands if they occur naturally in a single-grained state. Particle size, as it affects the ease

Figure 10.22.
Dust storms such as this, in Prowers County of southeastern Colorado, devastated much farmland in the Dust Bowl during the 1930s. (U.S.D.A. Soil Conservation Service photo by T. G. Meier.)

of transport, thus exerts a controlling influence under these circumstances. However, coarse sand grains are not moved readily by winds because of their weight. With a decrease in average particle size, erodibility increases until it attains a maximum with particles about 0.1 mm in diameter. With a further decrease in size the property of cohesion, or the ability of particles to cling to each other, becomes important. Cohesion reduces erosion because it makes particle detachment more difficult. For this reason, primarily, soils containing particles predominantly of small size tend not to be easily eroded by wind.

Soil-erodibility values express in tons per hectare per year the potential loss of soil from a wide, unprotected, smooth surface where the wind-erosion climatic factor is 100 percent. The specification of a wide, unprotected, smooth surface is important because this combination of conditions permits maximum soil erosion by wind. Thus, the measured values may be treated as constants that reflect solely the relationship between other soil characteristics and erodibility. Soil-erodibility values, when multiplied by the climatic factor, yield an estimate of the potential soil loss from a given wide, smooth, unprotected field.

The general relationship between soil texture and measured erodibility values is shown in table 10.5. For convenience, soils are placed in seven *wind-erodibility groups* as listed in the table. Each group is identified by a number and consists of soils of similar texture and erodibility. The erodibility values in the table are for a climatic factor of 100 percent.

Table 10.5. Wind-erodibility groups of soils defined on the basis of textural class and typical soil-erodibility values. (From D. G. Craig, and J. W. Turelle, *Guide for Wind Erosion Control on Cropland in the Great Plains States.* Soil Conservation Service, U.S.D.A., Washington, D.C. 1964).

Erodibility group	Predominant textural classes	Erodibility value[a]
		t/ha/yr
1	Very fine to medium sands	490
2	Loamy very fine to medium sands	300
3	Very fine to coarse sandy loams	195
4	Clays and silty clays[b]	195
5	Loams, sandy clay loams, and sandy clays	125
6	Silt loams and clay loams	105
7	Silty clay loams	85

[a]Representative values for the erodibility groups.
[b]Fine-textured soils that separate readily into very small aggregates.

From the table it may be seen that noncoherent soils containing particles predominantly within the smaller size range of sand are by far the most susceptible to wind erosion. Erodibility Group 4 embraces those fine-textured soils that contain a high proportion of expanding clay, such as montmorillonite. Even though these soils are cohesive, repeated wetting and drying causes them to separate naturally into tiny aggregates that are rather easily eroded by wind. In contrast, soils of wind-erodibility groups 5 through 7 are medium- to fine-textured materials that tend to remain in larger aggregates or masses that resist being moved by the wind.

Determination of the soil erodibility factor is by a simple screening procedure using a screen with openings 0.84 mm in diameter. By means of this test, which uses surface soil as it comes from the field, a soil is placed in one of the seven erodibility groups depending on the amount of material that passes the screen; the greater the amount the lower the erodibility group number. Coarse sand grains or stable soil aggregates, either of which are normally associated with a low erosion potential, tend not to pass through the screen under the gentle shaking prescribed for the test. In contrast, soils containing high quantities of very fine to medium sand pass largely through the screen and therefore fall naturally into groups identified with high erodibility.

Surface soil roughness

A rough soil surface aids in the control of wind erosion by particle entrapment. Surface roughness may result from a cloddy condition produced by plowing fine-textured soils or by working loose, coarse-textured soils into ridges with cultivation or planting implements. Particle entrapment occurs behind the clods or ridges where the velocity of wind is greatly reduced and is therefore unable to keep particles in suspension. Clods entrap particles regardless of the direction of the wind; ridges do so only if they lie approximately at right angles to the wind, as illustrated in figure 10.23.

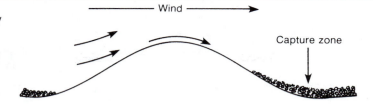

Figure 10.23.
Particle entrapment by soil ridges or other obstructions is due to settlement of particles into low spots where they are protected from the wind.

Wind →

Capture zone

One of the more critical times for wind erosion is during the period immediately following planting of a crop. Standard practices of seedbed preparation and planting usually leave the soil in a somewhat pulverized, smooth, and therefore easily eroded state. However, planting can be in deep to moderately deep furrows, which leaves the soil in a ridged condition (see fig. 10.24). If the planting is at right angles to the prevailing wind, the ridges formed during planting provide substantial protection against soil loss. Although the overall effect of ridging is variable, it can be expected to reduce soil loss from a broad sweep of land to 50 percent or less of that occurring where the soil is left in a comparatively smooth state by conventional planting equipment.

Plant and residue covers in wind-erosion control

Plants or their residues function in much the same way as ridges or other surface irregularities in minimizing soil loss by wind erosion; they retard the velocity of the wind at ground level and provide sheltered sites for the collection of loose soil grains. In areas of severe wind erosion it is essential that a cover be maintained as nearly continuously as possible for suitable control of erosion.

The protection offered by plants or their residues increases with the amount of cover present, sharply at first and then more gradually until maximum protection is attained with an essentially complete cover. Standing plants or stubble provide the best protection; they are from 2 to 2.5 times more effective in erosion control than are flattened residues. Residues offer little protection to the soil if they are not anchored; thus, stubble or other residues left standing in the field following harvest should not be disturbed any sooner or to any greater extent than necessary.

Field width as a factor in wind erosion

The more times moving particles strike the surface of the ground the greater the potential for detachment and erosion of other particles by wind. For this reason the total quantity of soil set in motion by wind is dependent upon the unobstructed distance over which soil particles can move across a field. Although little if any erosion will occur on the windward side of a field sheltered by a barrier of plants, large quantities of soil may be carried off the opposite side if the field contains no protective cover.

Even though the potential for wind erosion increases with field width, there is for each set of climatic and soil conditions a maximum width beyond which no further increase in erosion takes place. For highly erodible soils,

Figure 10.24.
The relatively smooth soil surface in the upper photograph offers little protection against wind erosion. Maximum benefit from ridging is obtained from a condition like that shown below. (U.S. Soil Conservation Service photos.)

the amount of material set into motion by wind builds to a maximum over relatively short distances. Areas of unprotected, easily eroded soil should therefore be kept as narrow as possible.

Stripcropping is one of the more suitable means of limiting the distance of unobstructed wind and soil flow across a field. In this instance the strips are oriented across the path of prevailing winds. Most commonly, stripcropping is incorporated into the system of dryland farming where small grains,

Soil erosion: principles and control

Figure 10.25.
Stripcropping for
wind-erosion control in
Montana. The strips
are alternated each
year between wheat
and summer fallow.
Note the planting of
trees, which provides
a windbreak around
the farmstead in the
foreground. (Photo by
U.S. Soil Conservation
Service—U.S.D.A.)

wheat especially, are grown only every other year. In the years when it is not cropped, the land is held in weed-free fallow as a means of increasing the supply of available water for the succeeding small grain crop. In practice, planted strips are alternated with fallowland as shown in figure 10.25.

Predicting soil loss caused by wind erosion

The kind of erosion-control measures undertaken in areas of frequent high winds depends largely on prevailing soil and climatic conditions. Where the potential for erosion is slight, suitable control may be realized from the application of but a single erosion-control measure. As conditions increase in severity, however, more than one practice may be essential, and each such practice may require rigorous application. In most instances the intent is to reduce the potential for soil loss to below 10 tons per hectare per year. For soils of high erodibility existing under extreme climatic conditions, it may be possible to keep wind erosion within limits only by maintaining the land under permanent cover.

Curves such as those shown in figure 10.26 are used for determining cropping and management needs for cultivated land subject to wind erosion. Their fundamental purpose is to identify the field width and ridging requirements necessary to maintain erosion within acceptable limits where the quantity of residue cover, if any, is known. Adjustments in field width are usually made by varying the distance across planted and fallow strips on stripcropped land. The curves in the figure are for only one set of soil and climatic conditions. Different curves are used where other conditions prevail.

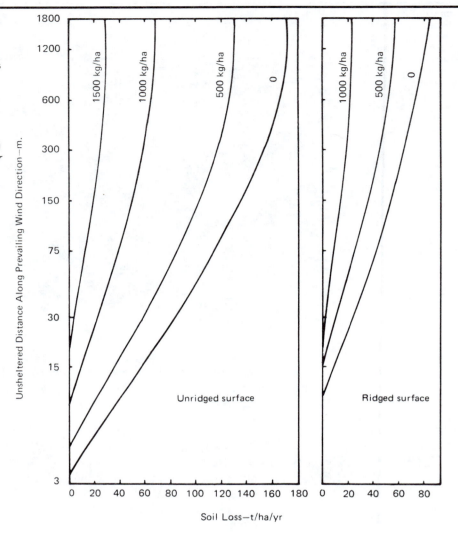

Figure 10.26.
Curves for predicting the soil loss by wind erosion for soils of wind-erodibility groups 3 and 4 occurring where the climatic factor is 100 percent. Values on curves are for quantities of flat, small grain residues. (From D. G. Craig, and J. W. Turelle, *Guide for Wind Erosion Control on Cropland in the Great Plains States.* Soil Conservation Service, U.S.D.A., Washington, D.C., 1964.)

The curves in figure 10.26 are of direct use only with anchored small grain residues that lie flat on the ground. The quantity of this kind of residue may be judged either by an experienced person or through use of guides such as the photographs in figure 10.27. The effect of standing stubble or growing plants can also be determined from the curves, provided the quantity of either is first converted to an equivalent quantity of flat residue. The conversion can be carried out readily through the use of appropriate charts.

The curves in figure 10.26 are based on average data. They do not provide accurate estimates of soil loss during any one period, for conditions affecting wind erosion are so highly variable. Over long periods, however, the predicted and actual loss of soil should agree reasonably well. If they do not, and if it is found that soil losses remain excessive, adjustments in the system of farming will be necessary.

Soil erosion: principles and control

Figure 10.27.
Guides for judging the density of flat wheat straw residues: left photograph, 1700 kg per hectare; right, 3400 kg per hectare. (Photo by U.S. Soil Conservation Service—U.S.D.A.)

Summer-fallowing and wind erosion

Summer-fallowing for water conservation is a practice common to many of the wheat-growing areas of the United States and Canada where annual cropping is risky because of potential water shortages during years of low precipitation. In the Great Plains east of the Rocky Mountains, summer-fallowing is used most extensively in areas receiving less than 60 cm of precipitation annually. In the Pacific Northwest, where comparatively low temperatures during the growing season reduce water loss by evapotranspiration, summer-fallowing is not an important practice unless the average annual precipitation is below 45 cm.

Fallowing to conserve soil water necessitates the control of weeds. Herein lies the most critical aspect of the fallow system in areas of high wind-erosion potential. If cultivation is used for weed control, undue destruction and flattening of residues may occur and crushing or other disturbance may leave the soil particles in a loose, detached state. Either of these effects is an open invitation to wind erosion. For this reason, it has been necessary to develop mechanical weed-control techniques that result in minimum disturbance of both the soil and residue cover.

The most satisfactory mechanical control of weeds in areas of high wind-erosion potential is achieved with subsurface tillage implements. These devices consist of either rods or wide blades that are drawn through the soil just beneath the surface (see fig. 10.28). They destroy weeds by shearing roots while causing minimum disturbance of surface residues. The standing residue that is left by this type of tillage is referred to as a *stubble* or *trash mulch*.

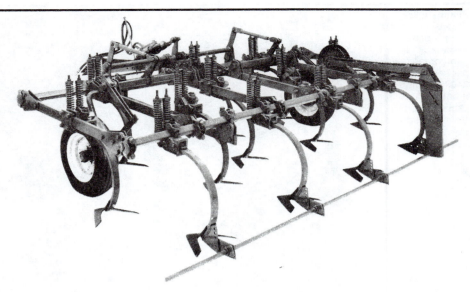

Figure 10.28.
A rod weeder with sweep cultivators. The cultivators loosen the soil and reduce drag on the rod, which is shown at the rear. The rod rotates in a clockwise direction as it advances through the soil. The rotation helps shear roots without burying the stubble. (Photo courtesy of Deere and Company.)

Where wind erosion is a problem, land destined for summer fallow should not be cultivated until the spring following harvest. This delay allows stubble from the previous crop to protect the soil throughout the winter. Cultivation is initiated in the spring before the weeds affect the soil moisture supply significantly. Since even the best tillage implements tend to bury stubble, the number of weeding operations should be held to a minimum. In some places seeding directly into the stubble is recommended, this practice assuring a continuous protective cover for the soil.

Windbreaks for erosion control

In areas of frequent high winds it is sometimes desirable to provide protection to fields by means of *shelterbelts* or *windbreaks* consisting of dense row-plantings of trees or shrubs as shown in figure 10.29. These are positioned along the windward side of the field. One function of windbreaks is to cause the deposition of wind-borne soil grains before they are blown onto a field. Windbreaks also reduce the velocity of the wind that might otherwise initiate erosion.

By reducing wind velocities, windbreaks shorten the effective distance of unobstructed wind flow across a field. In practice it is assumed that the protection is complete over a distance equal to ten times the height of the windbreak. This effect is taken into account in judgment of the probable effect of field width on potential soil loss. Though the protection offered decreases with distance, some reduction in wind velocity can be detected as far away as twenty times the height of the windbreak.

It is essential for a windbreak or shelterbelt to provide a continuous barrier to the wind. Openings beneath or between trees tend to funnel the air and can increase the velocity of flow well above that of unobstructed wind. More harm than good can come from a windbreak under these circumstances.

Figure 10.29.
Willow windbreaks cast a picturesque pattern of shadows while protecting highly erosive organic soil in Michigan against movement by wind. (U.S.D.A. Soil Conservation Service photo by F. J. Stock.)

Summary

Soil erosion is caused by either wind or water. It is most serious if the soil is unprotected and in a loose, pulverized condition, as often occurs on fallow-land or at planting time. Timber-harvest procedures that leave the soil exposed are the main cause of forestland erosion. On rangeland, it is the removal of vegetative cover by overgrazing.

Factors of erosion have been identified and can be quantitatively expressed for use in predicting erosion, either by rainfall or wind. For rainfall erosion, these factors are climate, soil, slope, the kind of crop, and the cultural (erosion-control) practices used in producing the crop. For wind erosion, the factors are climate, soil, soil surface roughness, the kind and amount of plant or residue cover, and unprotected field width.

Protection of agricultural land against erosion rests mainly with the selection of crops and cropping practices that hold erosion to a minimum. Preferred crops are those that provide a ground cover for the greatest length of time during the year. Erosion-control practices include minimum tillage, maintaining exposed ground in a roughened condition, and planting crops perpendicularly to the prevailing direction of water or wind flow. Erosion control under irrigation normally requires procedures that reduce the rate at which irrigation water is applied to the land.

Review questions

1. What soil characteristics are important to both the detachment and transport of soil particles during erosion?
2. Distinguish among sheet, rill, and gully erosion.
3. Why does the control of rainfall erosion rest largely with the cropping and cropping (conservation) practices factors?
4. Show that the predicted soil loss from a 5–percent slope 100 m long, where the rainfall erosion factor is 100 and the soil with an erodibility factor of 0.05 is held under clean-tilled fallow, would average 5 t/ha, approximately.
5. Explain the function of contour stripcropping in rainfall-erosion control.
6. Why does lengthening irrigation furrows tend to increase soil erosion?
7. Why is erosion of irrigated land subject to better control where sprinklers rather than gravity irrigation systems are used?
8. What can be done to reduce erosion of roads and trails constructed for timber harvest?
9. How does saltation increase the potential for wind erosion?
10. Why are rainfall data included in evaluating the climatic factor of wind erosion?
11. Describe the screening method of determining the wind erodibility of soil, and explain how data obtained by this technique are used to rate a soil.
12. Explain why soil ridging, stripcropping, stubble mulches, and windbreaks are able to reduce soil erosion by wind.

Selected references

Barrows, H. L., and Kilmer, V. J. Plant Nutrient Losses from Soils by Water Erosion. *Advances in Agron.* 15:303–16. New York: Academic Press, Inc., 1963.

Bennett, H. H. *Elements of Soil Conservation.* New York: McGraw-Hill Book Company, Inc., 1955.

Chepil, W. S., and Woodruff, N. P. The Physics of Wind Erosion and Its Control. *Advances in Agron.* 15:211–302. New York: Academic Press, Inc., 1963.

Cook, R. L. *Soil Management for Conservation and Production.* New York: John Wiley & Sons, Inc., 1962.

Hudson, N. *Soil Conservation.* Ithaca, N.Y.: Cornell University Press, 1981.

Stefferud, A. ed. In *Soil, 1957 Yearbook of Agriculture*, pp. 290–395. Washington, D.C.: Government Printing Office.

Troeh, F. R.; Hobbs, J. A.; and Donahue, R. L. *Soil and Water Conservation.* Englewood Cliffs, N.J.: Prentice-Hall, Inc., 1980.

Wischmeier, W. H. and Smith, D. D. *Predicting Rainfall Erosion Losses.* U.S.D.A. Agriculture Handbook 537. Washington, D.C.: Government Printing Office, 1978.

Woodruff, N. P.; Lyles, L.; Siddoway, F. H.; and Fryrear, D. W. *How to Control Wind Erosion.* U.S.D.A. Agriculture Information Bulletin 354, 1972.

11

Soil aeration and temperature

There are two properties of soils that are subject to continuous variation and must therefore be characterized by momentary measurements or by average values obtained over an interval of time. These are *aeration* and *temperature*. Each reflects an interaction of soil properties, largely physical, with the external environment. Both are strongly influenced by and vary with changes in the soil-water conditions. Aeration and temperature are important factors in the development of soils. They are also factors that may partially or wholly determine the suitability of soil as a medium for plant growth.

Soil aeration

Aeration is the process responsible for maintaining the supply of gaseous oxygen (O_2) in the soil. Oxygen is important because of its direct involvement in respiration and other biologically induced reactions. In respiration, organic carbon, as in plant sugar or in soil organic matter undergoing decay, is combined with O_2 and given off as carbon dioxide. The process supplies energy for the proper functioning of living organisms. For example, if O_2 is lacking, both the exploration of the soil and the uptake of nutrients and water by plant roots may be limited, for the energy required by these processes is supplied by respiration that takes place in the roots. A lack of O_2 retards organic matter decay and the associated release, or mineralization, of organically bound nutrients. The transformation of elements such as manganese and sulfur to forms that are toxic to plants may also take place if the O_2 level is seriously low. These transformations are carried out by anaerobic organisms that operate in a low-O_2 environment. For various reasons, then, the supply of O_2 in the soil is critical to plant and microbial growth. Our immediate goal is to consider some of the more important soil factors that control this supply. A major factor is the rate of gaseous interchange between the soil and outer atmosphere.

Principles of gaseous interchange in soils

The primary process of gaseous interchange in soils is *diffusion*. Diffusion is the result of random, heat-induced motion of ions, molecules, or other very finely divided particles that tends to cause their uniform distribution within a continuous system. For example, if a drop of dye were carefully placed in a glass of water, it would gradually expand as the particles or molecules of dye moved, or diffused, outward into the water. Ultimately, the dye would become uniformly distributed throughout the water; that is, its concentration in the water would become the same everywhere.

Diffusion of a gas occurs whenever its concentration in one part of a system differs from that in another part. For example, O_2 diffuses into the soil so long as its concentration there is less than that in the outer atmosphere. Similarly, carbon dioxide (CO_2) diffuses out of the soil because its release by root and microbial respiration keeps the concentration in soil air higher than that in the outer atmosphere. In soil aeration, therefore, diffusion tends to maintain a uniform concentration of CO_2 and O_2 throughout the continuous system between the soil and the air above. However, a uniform distribution of these gases in a soil will not occur so long as the use of O_2 and release of CO_2 by respiration in the soil continues, and it is likely to continue, at least feebly, in all soils that are reasonably well drained.

Composition of soil air

The composition of air expresses the concentrations of gases as percentages of all gas molecules present. On this basis, the earth's atmosphere is mainly a mixture of nitrogen (N_2) and O_2, for they comprise, respectively, about 78 and 21 percent of the air. The remaining 1 percent is made up of water vapor, CO_2, and a number of inert gases. The concentration of CO_2 in the earth's atmosphere is very low, averaging only 0.03 percent.

In soil air the concentration of N_2 is virtually always the same as in the outer atmosphere (78 percent). This is not true for O_2 and CO_2 because of respiration. In the soil, O_2 is continuously below 21 percent and the CO_2 continuously above 0.03 percent, with the variation from these norms being an index of the state of aeration. The magnitude of deviation from the norm for either gas is the same, although in opposite directions, since for each O_2 molecule taken out of soil air by respiration, one molecule of CO_2 is released back into the air. This is demonstrated in a simplified equation for respiration:

$$C + O_2 \rightarrow CO_2 + \text{Energy} \qquad (11.1)$$

Thus, the combined concentration of O_2 and CO_2 is relatively constant, remaining near 21 percent, or:

$$\text{Percent } O_2 + \text{Percent } CO_2 = 21 \qquad (11.2)$$

As a consequence of this relationship, one can estimate the concentration of either gas in soil air if the concentration of the other is known. For example, if the concentration of O_2 in soil air were 14 percent, the approximate concentration of CO_2 would be 7 percent.

The most important use of O_2 in soils is in respiration. However, it is just this use that results in the reduced O_2 content associated with poor aeration, and it is also the reason why O_2 flow into the soil is initiated. A declining O_2 content in the soil slows respiration, but at the same time, it causes an increased inflow of O_2 from the outer atmosphere. The rates of O_2 consumption by respiration and of replenishment by flow into the soil therefore tend toward the same limiting value. Should the two rates become equal, the O_2 concentration would become fixed. However, because factors affecting the flow rate and the consumption of O_2 undergo continuous change, the O_2 content of soil air seldom remains steady for long.

Factors of gaseous diffusion in soils

The diffusion of gases requires a difference in gas concentration from one place to another, and in part, the rate at which it takes place is a function of how large the difference in concentration is. However, the principal factor controlling diffusion in soils is the conductive capacity of the soil for gases. If the conductivity for gases is low, the replenishment of O_2 from the outer atmosphere will not keep pace with its consumption in respiration, and the concentration of O_2 may drop markedly below 21 percent. If the conduction is rapid, on the other hand, the concentration of O_2 may be held nearer to 21 percent. The soil that conducts well is therefore more likely to be well aerated. The main cause of low conductivity and poor aeration is excess water that fills soil pores and blocks the flow of gas. The situation may be further compounded where, due to a fine texture or a lack of well developed soil structure, the average size of pores is small. It follows, therefore, that soil characteristics most important to aeration are texture, structure, and water content.

In numerous soils, a factor determining the state of aeration is the rate of soil drainage after wetting, and this can be generally related to soil textural and structural relationships. Often, a condition of poor aeration exists because it takes so long for a soil to drain to the field capacity. In some soils, aeration remains suboptimal even after the field capacity has been attained. This is indicated by data in table 11.1, which shows the aeration porosity (percent of the total soil volume occupied by air space) for several soils and quartz sand at their respective field capacities. In considering these data, recall that aeration may be inadequate if the aeration porosity falls much below about 10 percent.

Several of the soils listed in table 11.1 show aeration porosities at the field capacity that are close to, if not below, the limiting value of 10 percent, which suggests a potential aeration problem for some of them. For example, computation would show that the Davidson clay, when at the field capacity, could have 90 percent of its total pore space filled with water, thus leaving only 10 percent for the circulation of air. In comparison, only about half of the pore space in quartz sand would likely be water-filled at the field capacity. This reflects the inability of the large pores in the sand to retain much water against normal drainage. Indeed, aeration in well drained, coarse sandy soils is usually more than adequate, their principal limitation being an inability to store sufficient water for plant growth except where the water supply can be frequently replenished from external sources.

Structure is important to aeration principally as a means of introducing large pore spaces into fine-textured soils. Large pores provide a good avenue for gaseous interchange because they tend to drain rapidly after soil wetting.

Table 11.1. The relationship of soil texture to the air content of soils at their field capacity. Air contents expressed as a percent of the total soil volume. (Modified from L. D. Baver. *Proc. Soil Sci. Soc. Amer.* 3:52–56. 1938).

Soil name and texture	Air content
	%
Cecil clay	13
Davidson clay	6
Chenango loam	11
Iredell sandy clay loam	9
Genesee silt loam	15
Paulding clay	11
Wooster silt loam	10
Quartz sand	22

However, if drainage is impeded by a factor such as an impervious subsoil layer, a waterlogged condition may develop regardless of pore space relations in the soil and so reduce aeration that the growth of many agricultural crops becomes all but impossible.

Aeration and plant growth

Although it is decidedly important to stress the need for proper aeration in soils, it does not necessarily follow that plant growth is impaired when the concentration of O_2 in the root zone drops below 21 percent. As shown by the data in figure 11.1, even crops such as corn and cotton, both of which are relatively sensitive to poor aeration, can perform well when the O_2 content of the soil air is no higher than 10 percent by volume. Marked reductions in the growth of these plants do occur at O_2 concentrations of five percent or less. That a low O_2 concentration can seriously impair the growth of cotton and corn is verified by observations of the poor performance of these two plants on inadequately drained land.

Very few crop plants can long withstand the adverse conditions associated with waterlogging. Many woody species are particularly susceptible to damage when soil aeration is restricted. Peach trees, for instance, can be killed if the soil in the root zone is saturated for only a few days. The death of most plants subjected suddenly to waterlogged conditions can generally be traced to deterioration of the root system and an attendant decrease in the uptake of nutrients and water.

Some plant species that normally inhabit well drained soils can survive submersion for considerable periods and may often produce measureable growth under these circumstances. Oxygen appears to be absorbed by the leaves of these plants and is then transmitted to the roots through a system of large pores in the stem. Irrigated rice is the only important agricultural crop that grows well in waterlogged soil.

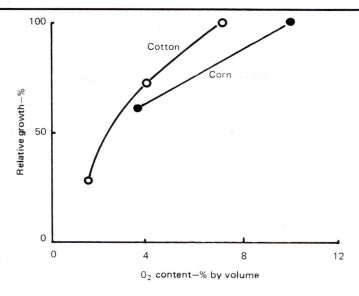

Figure 11.1.
The effect of varying O_2 concentration in soil air on the growth of cotton and corn. (After W. A. Canon, 1925, *Carnegie Institute Pub.* 368, Washington, D.C.: Carnegie Institution.)

Aeration in field soils

As implied above, aeration problems in soils assume their most serious proportions under waterlogged conditions. There are two practical solutions to such a problem. One is to increase the rate of water removal through improved underdrainage. The second is to limit the use of land to plants capable of withstanding waterlogged conditions. Normally, the second alternative is the least desirable since there are few agricultural crops of importance that grow satisfactorily in waterlogged soils. Drained land, on the other hand, may be used for any crop adapted to the area.

Waterlogging, as an enduring problem, is most prevalent in flat, low-lying soils adjacent to rivers or other permanent bodies of water, or in depressional areas where runoff and seepage from the surrounding land can accumulate. Soils in other topographic positions, such as the sloping uplands, are less often affected by excess water, at least for prolonged periods. When they are, the wetness is usually associated with a high water input, either by rain or irrigation, and with such soil characteristics as shallowness or slow permeability. Except where better control over irrigation is possible, it is often more practical to live with a problem of this type than to attempt either to correct or improve it.

Aeration in well drained soils is better at the surface than in the subsoil, with the differences between the two profile positions being greatest in soils that are slowly permeable to air. If texture, structure, and water relations are not highly variable with depth, the decrease in O_2 down through the profile is approximately linear (see fig. 11.2). However, most soils are neither texturally nor structurally uniform with depth; often they are higher in clay and have a less favorable structure in the subsoil. Under these circumstances, subsoil aeration may be exceedingly poor even though that of the surface soil is good (see curve A in fig 11.2).

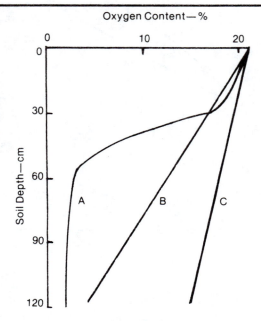

Figure 11.2.
Curves illustrating the distribution of O_2 with depth in soils of three different profile characteristics: A, subsoil higher in clay or less well structured than the surface, or both; B, and C, soils of uniform texture and structure, with B having a finer texture than C.

Coarse-textured soils are seldom poorly aerated, unless they are inadequately drained. In fact, some soils that are very coarse may be excessively aerated. This is not to imply that O_2 is present in damaging quantities; only that the potential for gas flow may be undesirably high. The most undesirable aspect of excessive aeration is the loss of soil water by evaporation.

Soil temperature

Temperature is important to a wide range of chemical, physical, and biological functions in soils. Of greatest immediate concern is the relationship of temperature to the growth of plants and soil microorganisms. Soil temperature affects plants directly as it influences seed germination and root growth, and indirectly as it influences the microbial population of the soil, the soil water supply, and the temperature of the air above the ground. Over long periods, temperature has a critical influence on mineral weathering and organic matter accumulation in soils, either process being of great importance to the fertility status of soils.

The temperature of the soil depends on the energy balance resulting from a continuous, two-way exchange of energy across the boundary at the soil surface. Incoming energy is mostly radiation from the sun and the atmosphere. The outflow of energy from the soil results largely from radiation to outer space, the transfer of heat to air, and the evaporation of water. Of these processes, radiation is the most important to temperature relationships in soils.

Radiation, which is the transfer of energy through space in wave form, is characterized by two properties: (1) *intensity*, or amount of energy per unit of time, and (2) *wavelength*, or the distance between wave nodes or crests. Any object above absolute zero ($-273°C$) gives off radiation continuously, but both the intensity and wavelength of the radiation vary depending on the temperature. For example, radiation from the sun, which is at about $6000°C$, is far more intense than that given off by the earth, where the average temperature is about $5°C$. The wavelengths also differ, with that of solar radiation being the shortest, on the average. The wavelength of radiation is important to the way it interacts with other matter.

Composition and behavior of solar radiation

Three general types of radiation are present in the sun's rays: *ultraviolet*, *visible light*, and *infrared*. Each of these subclasses of radiation covers a different range in wavelength and displays different behavioral traits. Ultraviolet has the shortest wavelength, infrared has the longest, with visible light covering the intermediate range between the other two. Ultraviolet is the component of solar radiation that causes sunburn. Unlike visible light, it cannot be seen. Nor can infrared be detected visually, although it can be sensed by the skin because of its warming effect. About 90 percent of the energy in solar radiation comes from visible light and infrared radiation, and it is about equally divided between them. The remaining 10 percent is ultraviolet.

Both visible light and infrared radiation show variable properties depending on wavelength. For example, variation in wavelength is responsible for the different colors of visible light. Similarly, infrared radiation behaves differently depending on whether it is of short wavelength *(near infrared)* or of long wavelength *(far infrared)*. A principal difference between these two types of infrared is their ability to penetrate the atmosphere. This capacity is greater for near than for far infrared, although the intensity of either can be reduced by absorption in the atmosphere.

The absorption of infrared radiation in the atmosphere is principally by water, either in vapor or droplet (cloud) form. Even on clear days, the intensity of all infrared from the sun is reduced substantially as it enters and passes through the atmosphere. On cloudy days, essentially all of the far and most of the near infrared are screened out by the atmosphere before reaching ground level. When a cloud cover is present, therefore, the major part of solar radiation reaching the ground consists of visible light (see fig. 11.3).

Radiation from the earth

Heat is lost continuously from the ground as radiation toward space. The type of radiation is totally long-wave, or far, infrared, which can be intercepted by the atmosphere. Complete interception occurs if there is a cloud cover. However, energy that is intercepted by the atmosphere is again lost as infrared radiation, roughly half being directed on toward space and the remainder back to earth. The return of radiant energy to earth can greatly reduce the net loss of heat from the soil and from the air immediately above. The effect is particularly noticeable at night, which is the time when soil and

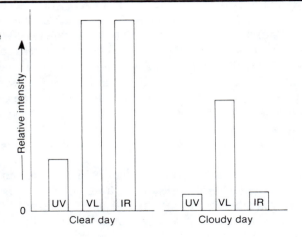

Figure 11.3.
Relative intensities of ultra violet (UV), visible light (VL), and infrared (IR) components of solar radiation at ground level on clear and cloudy days. Clouds cause the most marked reduction in the intensity of infrared.

the air near the ground undergo a natural cooling trend. Because of this effect, temperatures at or near the ground normally tend to remain higher, on the average, on cloudy than on clear nights.

Factors affecting solar intensity at ground level

The intensity of all components of solar radiation is reduced as they pass through the atmosphere. As indicated above, this is in part due to water in the air. Another important factor, however, is the length of the pathway the radiation must follow as it passes through the atmosphere and on to the earth's surface, for the greater the length the lower the intensity at ground level. The length of the pathway changes with both time of day and season of the year, and at any given moment, differs from one geographic location to another.

As shown in figure 11.4, the distance of travel through the atmosphere is shortest when incoming radiation approaches the earth perpendicularly. Thus, the intensity is maximum when the sun is directly overhead, and it decreases as the sun approaches the horizon. This relationship explains why the average intensity of sunlight is greater in equatorial regions than at higher latitudes, and why, as the higher latitudes are tilted toward the sun in the summertime, the intensity of radiation they receive increases. It also explains why the sun's rays are more intense at midday than either during the early morning or late afternoon hours.

Once solar radiation reaches the earth's surface, its effectiveness in warming the soil depends on the angle at which it strikes the soil surface. Greatest warming occurs where the rays approach the soil surface perpendicularly, or at an angle of 90°. The reason for this is that the farther the angle from 90°, the greater the area over which a unit of incoming radiation is distributed (see fig. 11.5). Because of the slope effect, soil-warming by solar radiation is greatest on level land in equatorial regions; but in a poleward direction, it is greatest on slopes that face the equator. Similarly, slopes with an easterly exposure are warmed most rapidly by the morning sun, and those facing west are more strongly affected by the afternoon sun.

Soil aeration and temperature

Figure 11.4.
The effect of the angle of incidence of incoming solar radiation on the distance of travel through the earth's atmosphere. The rays of the sun travel a shorter distance (d_1) where they arrive perpendicularly to the outer surface of the atmosphere than where they arrive at other than the perpendicular (d_2). Sunlight follows path d_1 when the sun is directly overhead, and d_2 when it is nearer the horizon.

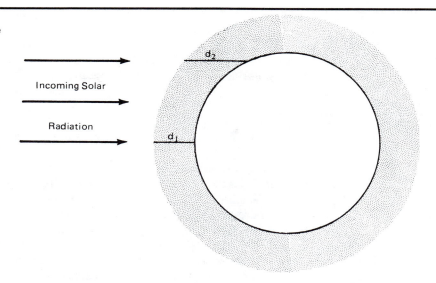

Incoming Solar

Radiation

Figure 11.5.
The effect of angle of incidence on the intensity of solar radiation at the soil surface. A shaft of light of width d is spread over a surface of equal width when the angle of incidence is 90° (left diagram). Increasing the slope to that shown at the right increases the width of the absorbing surface by 25 percent, or to 1.25d, and reduces the intensity of the intercepted energy accordingly.

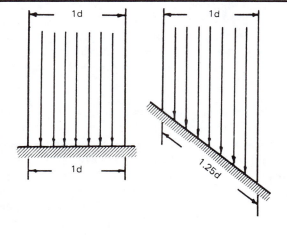

Energy exchange by soils

No object has the capacity to retain thermal energy. Instead, the energy is lost by radiation or by direct transfer to some other object of lower temperature. For the earth, energy loss is by radiation to outer space. At a specific location, any excess of heat accumulated during the day is largely lost at night, and the build-up of heat during the warmer parts of the year is countered by losses during cooler seasons. However, on a global basis, radiation losses just equal the solar radiation received. The average temperature of the earth is therefore constant.

Solar radiation received at the earth's surface may be either absorbed or reflected. Radiation that is absorbed is converted immediately to thermal energy, which causes a rise in temperature at the point of absorption. Reflected radiation is lost as a source of heat, however. Thus, the warming effect of solar radiation depends on the absorptive and reflective properties of the surface receiving it. For example, solar radiation tends to warm dark-colored soils more readily than it warms soils of light color. This relates to the greater reflection, especially of visible light, by light-colored soils. Infrared tends not to be reflected much, for it is effectively absorbed by soils regardless of their color.

Energy absorbed at the soil surface results in soil warming if the amount absorbed exceeds the amount being lost from the surface at that particular moment. This is normally the case during daylight hours when a high level of radiant energy is received from the sun. The net result is a rise in temperature at the soil surface above the temperature in the subsoil, which causes the downward flow of heat and its accumulation in the profile. At night, however, in the absence of direct radiation from the sun, more energy is lost than is received at the soil surface. As a consequence, the surface of the soil cools during the night, its temperature drops below that of the subsoil, and heat accumulated in the profile during the day flows back toward the surface to be lost mainly as infrared radiation. The net heat transfer between the surface and subsoil during any 24-hour period is in part a function of the amount of solar energy received during the daytime and in part dependent on the ability of the soil to transmit the heat downward during the day and upward during the night.

Heat transfer in soils

Heat transfer through soils is largely by *conduction*. The conduction of heat results from the exchange of thermal energy (motion) between adjacent atoms or molecules in matter. In soils, conduction of heat takes place through both the solid particles, principally mineral grains, and the pore spaces between the grains. Heat flow through the mineral particles occurs more readily than through the pores, and the rate of heat conduction through pores depends on whether they contain water or air. Water, which is some 25 times better than air at conducting heat, provides the best conduction of heat through the pores (see fig. 11.6). Where other factors are the same, a soil that is moist may conduct heat ten times more readily than when it is dry.

Since pore spaces place the greatest limitation on heat transfer through soils, anything that increases porosity tends to reduce thermal conductivity. Thus, soils in which the size of pores has been reduced by compaction are better conductors of heat than when they are in a loose, highly porous state. Because of the latter factor, cultivation can have a marked influence on heat conduction and overall temperature relationships in soils.

Variation in the energy balance at the soil surface

Energy exchange at the surface of the soil is largely a function of the amount of radiant energy received. This fact is illustrated diagrammatically in figure 11.7, which shows that energy exchange is much greater during the day than at night. This difference results from the higher total input of energy during the day, which includes solar as well as atmospheric radiation.

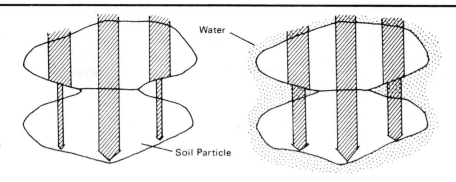

Figure 11.6.
The pathway of heat transfer from particle to particle in dry soil (left) is restricted largely to contact points between soil grains. Water films between adjacent particles (right) increase the rate of transfer, since heat is conducted more readily by water than by air. Arrow widths indicate relative heat transfer through soil mineral particles, air, and water but are not to scale.

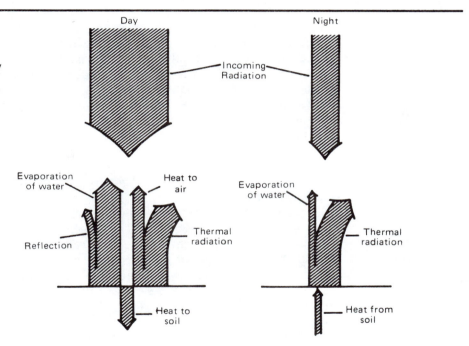

Figure 11.7.
A diagram illustrating the relative influence of different mechanisms of energy exchange at the surface of the earth during the day and at night. Incoming radiation during daylight hours is from the sun and the atmosphere; at night, it is only from the atmosphere. During the day, incoming energy exceeds losses by the amount retained to warm the soil. At night, losses exceed input, which results in soil cooling. Arrow widths in the diagrams indicate the relative amounts of heat exchanged by each mechanism.

An important fact illustrated in figure 11.7 is the rather inefficient retention of energy received by the soil. During the day by far the greater part of incoming energy is lost, the remainder being retained by the soil. At night, more energy is lost than is received, which results in a drop in soil temperature. Most of the nocturnal heat loss is by radiation, since evaporation proceeds very slowly and negligible heat is transferred to the air. In fact, air near the ground normally loses heat to the soil during the night, for the temperature of the air at that time tends to be higher than the temperature at the surface of the soil.

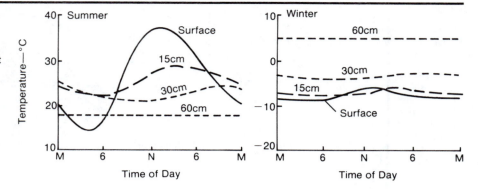

Figure 11.8.
Curves typifying the daily variation in soil temperature at four different depths and at two times of the year.

Patterns of soil temperature change

Diagrams in figure 11.8 show the pattern of soil temperature change on a daily basis; they depict temperature variation at four soil depths and at two different times of the year. The diagram on the left is typical of the early summer when the soil is undergoing a gradual warming trend. The diagram on the right is for early winter, at which time the soil profile is losing heat. A warming trend is indicated where, on the average, the soil is warmest near the surface; a cooling trend is indicated where the subsoil is the warmest.

As indicated by figure 11.8, the temperature of the soil varies continuously. At the surface, nighttime heat loss results in a drop in temperature until a minimum is reached at about sunrise. With the appearance of the sun, the soil begins to receive more energy than it loses, which causes the temperature to rise. This trend continues into the early afternoon, or until a declining intensity of incoming radiation causes the amount of energy received to fall below the amount lost from the soil surface. From this time on, the temperature at the surface of the soil declines until it again attains a minimum just before sunrise.

Because heat is transferred slowly through soils, variation in temperature below the surface is both later and less pronounced than at the surface. As shown in figure 11.8, changes in temperature at a 15-cm depth lag some 3 hours behind changes at the surface; at a 30-cm depth, the temperature response is delayed some 8 to 10 hours. No change in temperature is indicated at a soil depth of 60 cm. As a general rule, variation in soil temperature at a depth of 60 cm or more is rarely if ever significant during a single 24-hour period.

Daily variations in soil temperature are comparatively small in the winter, the result of solar input that is of both low intensity and short duration. Under these conditions, the average loss of heat is greater than the gains, which causes a gradual decline in overall soil temperature. Since the subsoil is warmer than the surface, heat is transferred upward through the profile. The greater the amount of heat moving up from the subsoil the warmer the surface soil during the winter.

The daily variation in surface soil temperature shown for summer conditions in figure 11.8 is fairly typical of humid, temperate regions where water

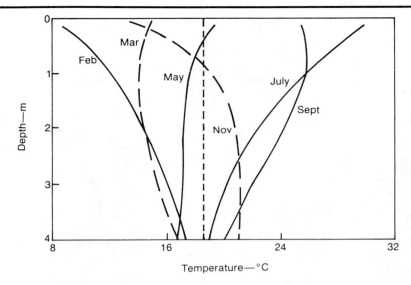

Figure 11.9.
Variation in soil temperature with depth at different times of the year. The broken line is an estimate of the average temperature. Measurements at Davis, California. (Modified from A. Smith, 1932, *Jour. Ag. Res.* 44:421–28).

vapor greatly limits the transfer of infrared radiation through the atmosphere. A much wider temperature swing would be noted for the surface of a dry, desert soil. Relatively large amounts of infrared radiation are transmitted by the dry air of desert regions, with the result the desert soils normally show a strong warming trend during the day and an equally strong cooling trend at night. Further, desert soils usually lack the water that places a distinct restraint on temperature variation in the soils of the more humid regions.

A pattern typical of soil-temperature variation on an annual basis is shown in figure 11.9. Compared to daily soil-temperature changes, annual variations cover a wider range and are apparent to much greater depths in the soil. Further, with increasing depth, there is a greater lag in temperature response. For example, the soil considered in figure 11.9 attains a maximum temperature at the surface in July, but at a 4-meter depth, the maximum does not occur until about November. Once it has reached maximum temperature, the subsoil then starts to lose heat in an upward direction and continues to do so until a minimum temperature is attained in the late spring.

Over the long term, annual gains and loses of heat by a soil are equal, so that the average temperature of the soil is constant and uniform with depth, as shown by the straight, vertical line in figure 11.9. This average also applies to the zone of constant temperature in the deep subsoil as well as to the temperature of the air above. Where the average temperature is below 0°C, as is generally the case in soils of arctic and subarctic regions, the profile will be completely frozen during the winter and will thaw to only a limited depth during the summer. Below the depth of thawing, the soil remains permanently frozen, a condition referred to as *permafrost*. Soils with permafrost tend to be very fragile, and anything that results in unusually deep thawing can cause rapid and serious soil deterioration (see fig. 11.10).

Figure 11.10.
Instability of ground with permafrost. (A) A thermokarst hole near
Fairbanks. Alaska, caused by subsidence due to localized melting of
permafrost. (B) The effects of differential subsidence of ground with
permafrost on a railway in the Copper River Valley northeast of
Valdez. Alaska. The railway was abandoned in 1938. (Credits:
(A) U.S.D.A. Soil Conservation Service photo by T. H. Day; (B) U.S.
Geological Survey photo by L. A. Yehle.)

Soil warming and plant growth

Once a soil has reached its minimum temperature in the winter, it undergoes a gradual warming trend until mid- or late summer. The rate of warming depends on several factors, including soil color, degree of compaction, water content, and the extent to which it is shaded by plants or their residues. Where land is used to produce common agricultural crops, rapid soil warming in the spring is usually of advantage; it not only allows earlier seed germination and harvest, it also lengthens the growing season, which is sometimes critically short in cooler areas.

Slow soil warming in the spring may have several causes, including a cool climate or an unfavorable topographic position, as on cooler, north-facing slopes. Standing plant stubble, which is often maintained as an erosion-control measure, also slows warming by shading the soil from the sun's rays. In general, however, the most frequent cause of slow soil warming in the spring is excess water. Excess water limits the rate of soil warming by encouraging heat loss through evaporation, by conducting heat to the deeper subsoil, and by increasing the total mass of material—water as well as soil minerals—that must be warmed by a fixed amount of incoming energy. This effect of water is of concern primarily where the soil is waterlogged, and correction of the problem can be brought about only by improved internal soil drainage.

Relationships between air and soil temperature

Because of the ready exchange of energy between the earth and the atmosphere, there is a rather close relationship between the temperature of the soil and the overlying air. This relationship applies mainly to the air near the ground. At high elevations, the temperature of the air is independent of the temperature of the earth below.

The pattern of air temperature variation has a number of similarities to the soil temperature pattern. At ground level, the temperature of the air is the same as the temperature at the soil surface. Thus, it reaches a maximum in midafternoon and a minimum around sunrise. In the afternoon, on still, or windless days, the temperature is highest at ground level and decreases gradually in an upward direction as illustrated in figure 11.11. Convection, or the rise of warm air into the cooler air above, occurs under these conditions. Convection is usually desirable, for it maintains lower temperatures at ground level on very hot days, and it also helps cleanse the air by carrying away unwanted smoke and fumes.

In the early morning hours, especially just before sunrise, air temperature relationships are just the reverse of those prevalent in the midafternoon (see fig. 11.11). At this time, air near the ground is the coldest and will not rise into the warmer, less dense air above. This pattern is referred to as *temperature inversion*. When a temperature inversion exists, colder air tends to settle into the lowest positions on the landscape, such as valleys or depressional areas. On chilly, windless nights, the lowest temperatures are experienced in these positions. Thus, the wise orchardist or horticulturist will likely avoid these locations in areas where late spring frosts are common. Instead, they would tend to locate frost-sensitive plants on the surrounding slopes that remain warmer during the early morning hours.

The effect of wind

Winds greatly affect temperature relationships at or near the surface of the ground. They do so by increasing the rate of energy exchange between soil and air and by the turbulent mixing of air, which results in more uniform air temperatures from ground level up. The rate of energy exchange between air and soils depends on the degree to which they differ in temperature and on the velocity of the wind. Where the velocity is moderate to strong, the effects of wind may completely override those due to the normal day-to-day energy exchange at the soil surface. Even in the absence of strong winds, large air masses of different temperature that move in succession through an area can cause marked changes in temperature at the surface of the soil from one day to the next.

The effect that moving air can have on temperature is evident in the way warm winds hasten the melting of snow or thawing of frozen ground. Winds also increase the rate of water loss from soils and plants, particularly

Figure 11.11.
Air-temperature profiles in the early morning and midafternoon on a still summer day. The average air temperature decreases in an upward direction. The upper limit of the diagram is about 10,000 m.

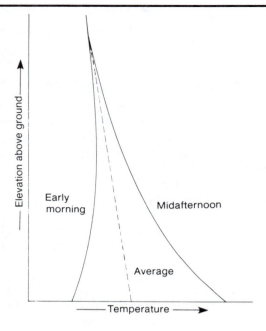

if the air is dry. The rate of water loss is increased in part because of the high capacity of moving air to supply energy for evaporation and in part from the removal of water vapor as rapidly as it is released by evaporation.

The effect of vegetation

Vegetation tends to reduce the variation in soil temperature. Where plants occur as a complete canopy, they act as a barrier to radiant energy coming either from above or below. As a result, the net exchange of energy at the surface of the soil is reduced, as is the daily variation in temperature. Commonly, maximum daytime temperatures of the soil under a complete plant cover may be 15°C or more below those at the surface of a bare soil under otherwise similar conditions. Differences observed at night tend not to be so great, but soils under a plant cover are usually significantly warmer than where they are bare.

A plant cover also affects the temperature of the overlying air; it usually lowers the temperature relative to that over bare ground. During the day, plants maintain lower air temperatures principally by absorbing heat in transpiration. At night, more energy is absorbed from the air by plants than by bare soil. The reason is that plants contain relatively little stored heat, and they are cooled to a lower temperature by radiation loss than are soils of higher heat content. Because of the effect of plants on air temperature, tender flowers of fruit trees are sometimes damaged by frost where the soil below is covered by a mat of grass or weeds, whereas they are not if the soil is bare. The benefit derived from bare soil is due to the free transfer of heat from the soil to the air and trees above. Such transfer is greatly restricted by an insulating vegetative cover over the soil surface.

Figure 11.12.
Experimental plots designed to test the efficiency of plastic strips for soil temperature and moisture control in the winter production of strawberries. (Photo courtesy of Victor Voth, University of California, South Coast Field Station, Santa Ana.)

Control of soil and air temperatures

In the production of agricultural crops, it is often desirable to adjust the temperature of the plant environment to improve average conditions for growth or to prevent plant damage from temperature extremes. The most frequent limitations to plant development occur in the spring when low soil temperatures may retard seed germination and seedling growth, or when frost may damage plants on abnormally cool nights.

There are several techniques of temperature control used to hasten seed germination. A standard treatment is to compact the soil after it has been loosened by the planting operation. Compaction reduces pore size and the circulation of air that might otherwise cause soil around the seed to dry excessively. Because compaction reduces pore size and maintains a higher water content in the conducting pores, heat received at the surface is more quickly transferred downward to warm the seed and thereby speed its germination.

For high value crops, such as vegetables destined for an early market, soil warming is sometimes hastened by covering the ground with plastic. As shown in figure 11.12, the plants grow through holes in the plastic. This approach increases soil warming primarily by limiting heat loss through evaporation.

Frost damage to plants is a common problem on clear, cool spring nights when heat loss by radiation is rapid. Low temperatures that harm plants are most likely to occur near the ground, since with a normal nighttime temperature inversion, the farther above the ground the higher the temperature of the air.

Figure 11.13.
Methods of controlling
air temperature to
minimize frost damage
in orchards.
(A) Overhead sprinkler,
(B) fan for mixing
warm upper air with
cooler air at ground
level, and (C) an
orchard heater. (Photo
(B) courtesy of The
Goodfruit Grower,
Yakima, Washington.)

A

B

C

Soil aeration and temperature

Most agricultural plants can be given some protection against frost. The kind of protection depends in part on plant size. Small plants can be easily protected by water-proofed paper covers (hot caps) that capture and hold heat rising from the ground. Plants that are too large for this can be protected by heating or mixing air near the ground, or by sprinkling the plants with water during periods of low temperature (see fig. 11.13). Water sprinkled on plants minimizes frost damage because it serves as a continuous source of heat.

The heating of air as a frost preventative is used extensively by orchardists. When air near the ground is warmed to a limited degree, the natural temperature inversion is maintained and heat loss by convection largely avoided.

In the artificial mixing of air, warm air from above is drawn downward to displace the cooler air near the ground. Winds produce this effect naturally, with the result that frost damage rarely occurs on windy nights. Frost damage is also rare on cloudy nights, when the net loss of radiant energy to space is small.

Often desirable air temperatures in orchards can be maintained during the spring by careful management of conditions near the ground. The reduction in air temperatures by a plant cover has been previously mentioned; better air-temperature relationships are maintained on cool nights where the ground is bare. However, this benefit is largely lost if the soil porosity is increased by cultivation. Much as they affect the downward transfer of heat by day, loose, porous soils limit the return flow of heat to the surface at night and thereby cause lower air temperatures above. Cultivation can reduce nighttime air temperatures 2 to 3 meters above the ground by 3°C or more.

Soil in a bare, compact state usually warms up more rapidly in the spring than if it is cultivated or insulated by a vegetative ground cover. Since air temperatures near the ground tend to follow those of the soil, treatments that speed the warming of orchard soils in the spring will likely encourage earlier flowering in the fruit trees. This may be of advantage in frost-free areas, but not where it increases the potential for frost damage. Indeed, where late frosts are to be expected, it may be preferable to discourage early flowering in fruit trees through soil- and air-temperature control. For this, one may decide to cultivate an orchard soil, or to shade it from the sun by a cover of low-growing plants. An alternative is the frequent application of water by sprinkling, which will hold average temperatures down by evaporative cooling. This latter technique has been used successfully to delay flowering in orchards by as much as two weeks in some instances.

Attempts to minimize temperature extremes in soils are usually to prevent damage to perennial plant roots during severe winters or to limit the evaporative loss of water in the summer. A mulch, such as a layer of straw, serves either purpose, since it functions much like a complete plant cover in minimizing soil-temperature variations. Mulches also limit air circulation that hastens soil-drying.

Where irrigation is practiced, temperature relationships in the surface soil can often be improved by irrigating the soil late in the fall. Moisture in the soil increases the upward conduction of stored subsoil heat and maintains a warmer surface soil during the winter.

Summary

Aeration and temperature are highly variable soil properties. Aeration depends on the balance between the rate of oxygen consumption (respiration) in the soil and the rate at which the oxygen is replenished from the outer atmosphere. The most common cause of poor aeration is the blockage of pores by excess soil water.

The temperature of the soil depends on the balance between incoming energy from the sun and energy loss from the soil through evaporation, heating of air, and radiation toward outer space. The temperature of the soil varies continuously, principally because of variation in incoming energy from above. On a daily basis, the soil undergoes a warming trend in the morning and early afternoon, but then undergoes a cooling trend until sunrise the next day. On an annual basis, soil warming occurs during the spring and early summer, and cooling during the rest of the year. Since all energy gained by the soil is eventually lost, its temperature is constant on the average.

By reducing solar radiation during the day and heat loss from the soil during the night, the atmosphere moderates swings in soil temperature, over both daily and annual time periods. A similar effect is produced by a plant cover or even cultivated surface soil, either of which reduces the flow of energy into and out of the soil.

Radiation from the soil warms overlying vegetation and can help protect it against frost damage on cold spring nights. This effect is limited, however, and if air temperatures drop too low, supplemental heating or mixing of air, or sprinkling the plants with water, may be required to limit the damage.

Review questions

1. What role does root respiration play in plant growth?
2. Why do the concentrations of oxygen and carbon dioxide in the soil atmosphere vary inversely to each other?
3. Why is root respiration the basic cause of the diffusion of oxygen into and the diffusion of carbon dioxide out of the soil?
4. Why does excess water interfere with soil aeration?
5. Why does moisture in the atmosphere reduce the intensity of solar radiation during the day but increase the amount of radiation returned from the atmosphere to the ground during the night?
6. Explain why variation in the position of the sun relative to the earth's horizon influences the intensity of solar radiation received at ground level.

7. Why is soil warming the greatest when solar radiation strikes the ground on the perpendicular, or at a 90–degree angle?
8. Explain why relatively little of the solar energy received during the daytime goes to warm the soil.
9. Why does cultivating a soil reduce its ability to conduct heat?
10. Describe the similarities and differences in the pattern of temperature change with depth in the soil on a daily and on an annual basis.
11. Why does water in the soil slow soil warming?
12. What is an atmospheric temperature inversion, and how does this condition aid in the protection of plants against frost damage?
13. Why does a vegetative cover reduce the daily variation in soil temperature?

Selected references

Baver, L. D.; Gardner, W. H.; and Gardner, W. R. *Soil Physics.* New York: John Wiley & Sons, Inc., 1972.

Campbell, G. S. *An Introduction to Environmental Biophysics.* New York: Springer-Verlag New York, Inc., 1977.

Geiger, R. *The Climate Near the Ground.* Cambridge, Mass.: Harvard University Press, 1965.

Lowry, W. P. *Weather and Life.* New York: Academic Press, 1969.

Monteith, J. L. *Principles of Environmental Physics.* New York: American Elsevier Publishing Co., Inc., 1973.

Rose, C. W. *Agricultural Physics.* New York: Pergamon Press, 1966.

Russell, M. B. Physical Properties. *Soil, 1957 Yearbook of Agriculture*, pp. 31–38. Superintendent of Documents. Washington, D.C.: Government Printing Office.

Waggoner, P. E. *et al. Agricultural Meterology.* Meterological Monographs, Vol. 6, No. 28. Boston, Mass.: American Meterological Society, 1965.

12

Soil classification and survey

Soil classification and survey involve the collection, organization, and interpretation of information on the kinds and distribution of soils over the surface of the earth. The collection of this information is a function of soil survey; its organization and interpretation are within the province of soil classification.

When carried to a practical end, soil classification serves two primary objectives. One is to group soils on the basis of their *morphology*, or form, as judged from the nature of the profile. A classification developed on this basis is a *taxonomic classification*, or simply, *taxonomy*. The second objective in classification is to group soils according to their suitability for one or more uses. This is the *utilitarian* or *interpretive classification* of soils. The term interpretive is used since the determination of suitability requires an interpretation, or judgment, of how closely the properties of a soil match those needed for a particular use. Since many uses, both agricultural and nonagricultural, are made of the soil, and since each use requires a separate interpretation of the basic information collected on a soil, there are many different systems of utilitarian classification. Several of these are discussed in this chapter.

The United States system of soil taxonomy[1]

Classification, whether of soils or of other objects, involves two steps: (1) the recognition of individuals on the basis of sets of identifying properties, and (2) the organization of classes containing individuals alike in many though not all properties. The greater the similarity among all members of a class

[1]The soil classification system described herein is from *Soil Taxonomy: A Basic System of Soil Classification for Making and Interpreting Soil Surveys*, prepared by the Soil Survey Staff, Soil Conservation Service, U.S. Department of Agriculture, and published in 1975. All information in tables 9.1 through 9.8 is derived from this publication. An earlier system of soil classification, published in 1938 in *Soils and Men*, the *Yearbook of Agriculture* for that year and modified in 1949 (see Thorp, J., and Smith, G. D., 1949, *Soil Sci.* 67: 117–26) was used until replaced by the current system in 1965.

the greater the likelihood of uniformity in behavior and response to treatment. Thus, a fundamental purpose in classification is established; it permits the application of knowledge gained from the study of one member of a class to other members of the same class.

Concepts of soil individuals

Soils differ from members of the plant and animal kingdoms in that they do not occur as distinct, easily separable entities. Instead, they form a continuum that grades in properties and behavior from one position on the landscape to the next. Such changes can normally be related to variation in one or more of the five soil-forming factors,[2] and along a lateral dimension, they may be abrupt or gradual depending on the manner in which the soil-forming factors change individually or collectively. The most common cause of variation among soils, either on a worldwide or a local basis, is variation in the effective climate for soil formation.

Three things are required of soil individuals for their placement in a practical system of soil classification. First, each individual must cover a large enough area to make its delineation on the landscape worthwhile. Second, individuals of like character must occur in sufficient number to justify their inclusion as a class in a broadly based classification system. Finally, each individual must be sufficiently uniform throughout its length and breadth so that it can be described in its entirety by a fairly narrow range of profile properties. If a soil body is too variable in its properties and behavior, then a specific recommendation for its use and management cannot be made and the purpose for its classification thereby nullified.

As a guide to the establishment of classifiable soil units, two types of individuals are recognized: the *pedon* and the *polypedon*. The pedon (Gr. *pedon*, ground) is defined as the smallest volume that can be called *a soil*. A pedon should be no larger than is necessary to show the normal variation allowed in an individual soil along its lateral dimensions. The area of a pedon is restricted to 1 sq m if the profile is uniform laterally, or to 10 sq m if the profile shows considerable lateral variation. For example, a soil may show a natural wavelike variation in the thickness of one or more horizons. The size of the pedon should be adequate to show the full range in this type of variation.

A polypedon is a soil volume consisting of two or more pedons sufficiently alike to be called the same soil (see fig. 12.1). There is no set limit to the size of a polypedon, the only restriction being that it show relative uniformity within its boundaries. The polypedon conceptualizes the soil individual used in practical soil classification, with groups of like polypedons being placed together as a single class in the soil taxonomic system. The type of class formed by combining like polypedons is called a *soil series*. Soil series are the basic units of classification in the U.S. system of soil taxonomy. They also provide a basis for distinguishing among soils during soil mapping.

---

[2]The factors of soil formation were shown in Chapter 3 to be parent material, climate, living matter, topography, and time. The first four of these determine the nature of the soil-forming environment.

Figure 12.1.
Conceptualization of a
polypedon and one of
its many similar
pedons. The size of
the polypedon
depends on the
number of like pedons
coexisting in a
continuous soil
volume.

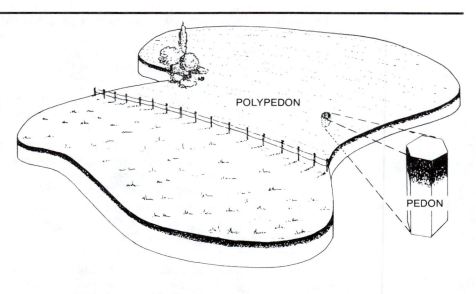

Figure 12.2.
Changes in profile
properties with
changing topographic
positions and effective
precipitation. Two
soils (polypedons A
and B) are recognized,
and they are
connected by a
transitional zone.
Vertical scale is
greatly exaggerated.

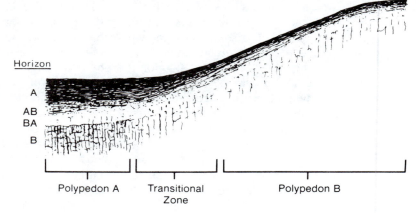

Typically, most landscapes consist of scattered polypedons of two or
more soils separated by transitional zones as illustrated in figure 12.2. This
figure shows, in profile, two polypedons of contrasting properties. As polypedons, each may be assumed to consist of many similar pedons. The transitional zone connecting the two polypedons also consists of many pedons,
but they occur in a graded series in which no one pedon is sufficiently like
its neighbor to be called the same soil. Because of this condition, transitional
zones surrounding polypedons do not meet the requirements necessary for
their classification as individuals in the taxonomic system; they show too
much variation to be described by a specific set of profile properties.

In soil mapping, all soil on the landscape, both that in polypedons and
in the transitional zones between, must be recognized and displayed on the
map. Ordinarily, this is done by combining a polypedon with a part of the

transitional zone surrounding it and showing the combined area as a single unit (mapping unit) on the map. The mapped area is then named for the soil series identified with the polypedon, a step that places the mapping unit into the taxonomic system of soil classification. Obviously, the properties of the mapping unit will not correspond totally to those of the polypedon it contains. Even so, in usual agricultural applications, the behavior of the entire mapping unit will not differ greatly from that of the polypedon for which the unit is named.

The soil series   Within an area identified by a single topographic pattern, slopes of essentially identical character occur repeatedly over the landscape. Where other factors of soil formation remain relatively unchanged, the recurrence of similar slope conditions results in the development of scattered bodies (polypedons) of soil displaying approximately the same kind and range of profile characteristics. All such occurrences of like soil may be considered together and treated as a single class, namely, a soil series.

All members of a given soil series normally occur in the same general geographic area or region, although this is not absolutely necessary. In reality, members of the same series can occur anywhere provided they meet the criteria established for defining the series. For this, they must have formed from the same specific kind of parent material, and they must have the same kinds and sequence of horizons. Yet, some variability is allowed in horizon characteristics, although the range of variability is arbitrarily limited. Thus, members of the same series may differ slightly in color, horizon thickness, pH, and other features. Subsoil texture must be very nearly the same in all members of a series, but the surface soil may vary among two or three different textural classes, provided they do not represent large differences in particle-size analyses.

Different soil series can result from independent variation in any one of the soil-forming factors. For example, the soils depicted in figure 12.2 may be considered as members of two series that have acquired different characteristics because of different topographic influences. Similarly, figure 12.3 shows soils of four series that have developed under conditions in which texture of the parent material has been the principal variable in the soil-forming environment. The profile features apparent in the figure are acquired traits that reflect the influence of varying texture on the processes of soil formation. The differences noted either in texture or in the profile features shown in figure 12.3 provide an ample basis for separation of the four soils into different series.

Sometimes, simultaneous variation in two or more factors can take place without causing significant change in the soil-forming environment. For instance, different combinations of slope and external climate can result in similar conditions for soil development, as can variations in other factors that tend to compensate for each other. For this reason the occurrence of the same kind of soil in widely scattered locations and in different topographic positions is not unusual.

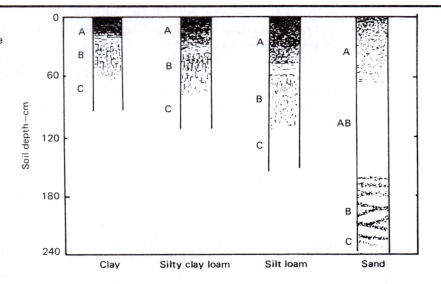

Figure 12.3.
The relationship between texture of the parent material and the depth of profile development for four soils in Iowa. (From G. D. Smith, W. H. Allaway, and F. F. Riecken, *Advances in Agron.* 2:157–205, 1950.)

Tentatively, new series are described during the conduct of a soil survey. As the survey progresses, the soils being examined are compared with those of previously surveyed areas. When a soil unlike any other is discovered and found to be of sufficient extent to warrant separate recognition, it is studied, described in detail, and assigned a series name. The name is taken from some local landmark, often that of a town or city. Examples are the Memphis series along the Mississippi River, the Fort Collins series in Colorado, and the Mohave series in California. Thousands of series are currently recognized in the United States, and many more will be discovered as soil surveys are extended into unsurveyed areas.

Categories of the taxonomic system

The U.S. system of soil taxonomy provides for the classification of soils at six different levels of generalization, with each level being called a *category* of classification. *Soil orders* comprise the highest, or most generalized, category, and moving downward through the system, are followed in succession by *suborders, great groups, subgroups, families,* and finally, *series.* Series are combined to make up soil classes within each of the higher categories. For this reason, they are termed the fundamental units of the taxonomic system.

The level of generalization in soil taxonomy relates to the range in properties allowed within the different classes making up a category. The most highly generalized classes, the soil orders, vary widely in properties because they are made up of many different soil series. These series must be alike in only those few properties that define the order. Down through the system, classes are defined by an increasing number of properties. This limits the number of series assigned to a class as well as the range in overall properties of the class. Thus, the lower the level of classification the greater the amount of information required to classify a soil.

Soil formation and classification

The development of a classification system is facilitated if properties defining a soil class can be related to a causal agency. This is reasonable, for soils subjected to the same kinds of influences during formation will likely have many of the same properties and should therefore fit naturally into a common class. Although this concept has been applied throughout the history of modern soil classification, some of the earlier systems failed because the principal factors that determine soil characteristics were not fully recognized. One such system was based primarily on the geology of the parent material. It proved unsatisfactory under test, since it overlooked the important effect of acquired characteristics on soil behavior.

A major advance came in soil classification with the recognition of soils as natural bodies produced in time by definable environmental influences. As a result of this discovery, earlier systems of classification were modified to account for the effect of the several soil-forming factors on soil properties. In this country a classification scheme based primarily on the concept of soils as natural bodies was adopted officially in 1938. In some instances classification units within this system were inadequately defined, a fact that became more apparent as new knowledge about soils was acquired. The 1938 system was used with considerable success for about 25 years.

The current system of classification in the United States was adopted officially in January, 1965. In it an effort was made to define units of classification as rigidly as possible. However, no definition is better than the knowledge upon which it is based. Thus, we may conclude that even the present system is hampered by our limited understanding of soils and will undoubtedly undergo continuing modification in the future.

Diagnostic properties

This term applies to the properties used for classifying soils into the taxonomic system. Either a single property or a combination of properties can be diagnostic of a soil class. If it is the latter, the properties usually appear as one or more specific kinds of *diagnostic horizons*. To the extent possible, diagnostic properties are selected on the basis of how well they reflect the influence of the soil-forming factors, especially climate and vegetation, on soil development.

Diagnostic horizons may occur in either a surface or a subsurface position. Diagnostic surface horizons are called *epipedons*, and they are defined by properties that are not altered by normal cultivation. Definitions of epipedons and diagnostic subsurface horizons are given in tables 12.1 and 12.2, respectively. Descriptions of diagnostic horizons that are not restricted to a specific position in the profile, along with descriptions of certain other diagnostic features, including those of horizons used in classifying organic soils, are listed in table 12.3. Our discussion of soil classification will deal directly with relatively few of the diagnostic properties, the remainder being listed primarily for reference.

Classification within the taxonomic system is sometimes based on the *temperature* or *moisture regime* of the soil. Soil temperature regimes are

Table 12.1. Generalized descriptions of epipedons.

Epipedon	Description
Mollic	With some exceptions, as for very shallow soils, a surface layer at least 25 cm thick that contains more than 1 percent organic matter, but not sufficient to make it an organic layer. It must be more than 50 percent saturated with bases. Typical of many grassland soils.
Histic	A thin organic layer overlying mineral soil, or if mixed with mineral soil, as by plowing, a layer containing at least 20 percent organic matter if sandy or up to 30 percent organic matter if high in clay. Seasonally saturated with water. Should be no thicker than 30 cm if dry, or 45 cm if wet.[a]
Anthrophic	A surface horizon having the properties of a mollic epipedon, but in addition, contains a high content of extractable phosphorus normally attributable to heavy fertilization.
Plaggen	A surface horizon greater than 50 cm thick produced by heavy manuring.
Umbric	A surface horizon of mineral soil darkened by accumulated organic matter but too low in bases or extractable phosphorus to be classed as mollic or anthrophic, or too thin to be classed as plaggen. Typical of certain soils of the temperate forested region.
Ochric	Essentially, any surface horizon not meeting the requirements of the above epipedons. Typically, ochric epipedons have a light color due to a lack of accumulated organic matter.

[a]Histic epipedons should also be wet for at least 30 consecutive days in most years. This is to distinguish histic epipedons formed under swampy conditions from O horizons formed in well drained, forested soils.

Table 12.2. Generalized descriptions of diagnostic subsurface horizons.

Horizon	Description
Argillic	A subsurface horizon formed in part by the illuviation of crystalline, layer clays. Certain limits in both clay content and thickness must be attained before an illuvial horizon can be classed as argillic.
Natric	An argillic horizon in which the content of exchangeable Na exceeds 15 percent of the cation-exchange capacity. Natric horizons have either a prismatic or a columnar structure.
Spodic	A subsurface horizon containing illuvial humus and amorphous oxides of Fe and Al in excess of certain specified minima. Typical of many soils forming under strongly acid leaching.
Cambic	A subsurface horizon containing illuviated clay, humus, or sesquioxides, but insufficient to be classed either as argillic or spodic.
Oxic	A highly weathered horizon, high in 1:1 and sesquioxides clays, sometimes in combination with other highly resistant minerals such as quartz. Must not exceed certain limits in cation-exchange capacity and content of exchangeable bases.
Agric	A horizon formed immediately beneath the plow layer, presumably through eluviation and illuviation of clay and humus.

defined ranges in mean annual soil temperature, as modified in some instances by the magnitude of the difference between the winter and summer mean soil temperatures. Soil moisture regimes refer to defined moisture levels of wet, moist, or dry in relation to the periods of the year when these levels are maintained in the soil. Highly generalized descriptions of soil temperature and moisture regimes are given in tables 12.4 and 12.5.

Table 12.3. Generalized descriptions of other diagnostic horizons or features used in taxonomic soil classification.

Horizon or feature	Description
	Horizons in mineral soils
Albic	Typically, a light-colored E horizon.
Calcic	Lime-enriched layer due to limited leaching.
Duripan	Layer cemented by precipitated silica.
Fragipan	Layer of brittle soil, commonly of a loam texture.
Gypsic	Gypsum-enriched layer due to limited leaching.
Petrocalcic	Lime-cemented layer. Caliche is an example.
Placic	Layer cemented by iron (ironpan).
Plinthite	Material cemented by precipitated sesquioxides, often in a continuous layer. Soft while moist, but hardens irreversibly on drying (laterite).
Salic	Layer high in salts more soluble than gypsum.
Sombric	A subsurface horizon high in illuvial humus, but with base saturation below 50 percent. Restricted to well drained tropical and subtropical soils.
Sulfuric	A horizon in any position that has a pH below 3.5 produced by the oxidation of sulfur in minerals or organic matter to sulfuric acid, usually when a soil is drained.
	Horizons of organic soils[a]
Fibric	Fibers compose over two-thirds of the mass, suggesting very limited decomposition.
Hemic	Fibers compose between one-third and two-thirds of the mass.
Sapric	Fibers compose less than one-third of the mass.
	Additional soil features
Lithic contact	Boundary between soil and unweathered bedrock. Used mainly in specifying soil depth.
Paralithic contact	Similar to lithic contact, except it is between soil and rock softened by weathering.
Permafrost	Permanently frozen soil.

[a]Organic deposits that exceed the limiting thickness of a histic epipedon.

Table 12.4. Soil temperature regimes.

Regime	Mean annual temperature
	°C
Pergelic	<0
Cryic	0–8
Frigid[a]	<8
Mesic	8–15
Thermic	15–22
Hyperthermic	>22

[a]Temperature variation from winter to summer is greater for frigid than for cryic or pergelic, the difference resulting from a higher average summer temperature for the frigid regime.

Table 12.5. Soil moisture regimes.

Regime	Characteristics[a]
Aquic	Saturated at least part of the time; reducing conditions in the soil prevail.
Udic	Moist, but not wet, most of the time.
Ustic	Intermediate between udic and aridic.
Aridic (or Torric)[b]	Dry most of the time.
Xeric	Mediterranean: wet winters, dry summers.

[a]Wet implies saturation or near saturation; moist means above and dry means at or below the permanent wilting point, or a matric potential of −15 bars.
[b]Aridic and torric are identical but are used at different levels in the classification system.

Table 12.6. Names, central concepts, and diagnostic properties of soil orders in the U.S. system of soil taxonomy.

Order name	Formative element of the name			Diagnostic property
	Element	Derivation	Connotation	
Entisol	ent	Meaningless syllable	Soils of very limited development; may contain horizons that form quickly.	None specific to all members of the order.
Inceptisol	ept	L. *inceptum*, beginning	Soils of humid regions showing an intermediate level of development.	None specific to all members of the order.
Aridisol	id	L. *aridus*, dry	Soils of arid regions showing limited change because of a low effective precipitation.	An ochric epipedon with >50 percent base saturation.
Mollisol	oll	L. *mollis*, soft	Soils of the subhumid to semiarid grasslands with deep, dark, friable surface horizons.	A mollic epipedon.
Spodosol	od	L. *spodos*, wood ash	Highly leached, strongly acid, coarse-textured soils of the humid forests.	A spodic horizon.
Alfisol	alf	Meaningless syllable	Soils of temperate-region forests showing moderate effects of weathering and leaching, but strong eluviation and illuviation.	An argillic horizon dominated by 2:1 clays; >35 percent base saturation.
Ultisol	ult	L. *ultimus*, last	Soils of subtropical forests showing effects of strong weathering, leaching, eluviation and illuviation.	An argillic horizon dominated by 1:1 and sesquioxide clays; <35 percent base saturation.
Oxisol	ox	Fr. *oxide*	Soils of tropical forests showing maximum effects of weathering and leaching.	An oxic horizon or a continuous plinthite layer.
Histosol	ist	Gr. *histos*, tissue	Soils of wet places high in organic matter.	One or more soil layers containing 20–30 percent organic matter, depending on clay content, with a combined thickness of >40 cm.
Vertisol	ert	L. *verto*, to turn	Soils high in expanding clay that form large cracks on drying; self-mixing.	An orchric or mollic epipedon containing >30 percent expanding clay.

The soil orders

A consideration of the ten soil orders in the U. S. System of Soil Taxonomy provides a relatively simple means of viewing the full spectrum of soils as they occur in nature, for definitions of the orders take into account only a few major properties, and they theoretically encompass all naturally occurring soils. In the discussions that follow, the orders are characterized by their principal diagnostic features and are related to the general conditions under which they form. In addition, some indication of their general suitability for the production of economic plants is given.

A list of the ten orders, along with their diagnostic properties, appears in table 12.6. Also shown is information pertaining to the derivation of the order names. These names are coined words, all of which end in *sol*, the Latin word for soil. With but two exceptions, the order names are derived from roots, most often Latin, selected to infer or imply the central concept of the order. Suitable roots were not found for Entisols and Alfisols, due mainly to the lack of a single set of properties that clearly distinguishes these orders from the other eight.

Except for Entisols and Alfisols, the roots of the order names, along with their connation and an identifying syllable, known as a *formative element*, are shown in table 12.6. The formative elements appear in both the order names and their roots. Except for the soil series, these elements also appear in class names within each category below the order. Thus, you can determine the order of a soil from its suborder, great group, subgroup, or family name.

Where possible, diagnostic properties of soil orders reflect conditions under which soils of the order have formed. For six of the orders, these properties are specific kinds of diagnostic horizons that show the dominating influence of climate and vegetation on soil development. All soils in these six orders are viewed as mature; that is, they have been subjected to development sufficiently long so as to show the full impact of environmental conditions on profile properties. Two of the orders, Entisols and Inceptisols, contain soils of limited development and, therefore, are not recognized on the basis of climatically- and vegetationally-related diagnostic properties. In the two remaining orders, Vertisols and Histosols, diagnostic properties recognize unique parent materials rather than climate and vegetation as the major determinant of soil properties.

The general relationship between soil orders and climatic-vegetational zones, as expressed by the vegetational pattern in the United States, can be judged from a comparison between figures 12.4 and 12.5. Note particularly the distribution of Aridisols and Mollisols within the desert and grassland regions of the Continental United States, with Spodosols, Alfisols, and Ultisols being limited to the forested areas. Note also that the Entisols, which are soils with little or no development, are widely scattered and show no relation to the vegetational pattern. Inceptisols, which show greater development than do the Entisols, are also widely scattered, but they are limited to wetter regions where climatic influences are able to bring about the degree of development required for this order. Implications of these distribution patterns are considered in greater detail in the following discussions of the ten soil orders.

Entisols

This order includes soils that display no more than a minimum expression of properties acquired under the current cycle of soil formation. Entisols occur in parent materials that are of recent origin or are highly resistant to change under the imposed environment. They may also be observed on sloping land where erosion prevents the development of significant profile features or erases those acquired during an earlier phase of soil formation.

Though Entisol profiles are like the parent material in most respects, they may have weakly to moderately well developed horizons. For example, dark surface horizons may be artificially induced in Entisols by heavy manuring. Gleyed layers may be present, for they form quite rapidly, and textural horizons may occur if the soil is forming from stratified parent material.

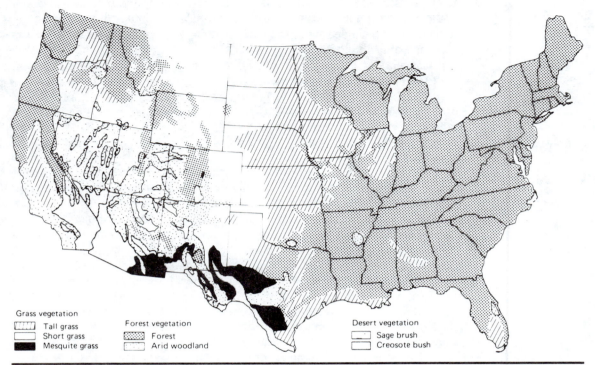

Figure 12.4.
Native vegetation zones in the conterminous United States. In
Alaska, the vegetation ranges from Pacific rainforest in the south,
through boreal forest, to tundra in the north. Tropical rainforests are
dominant in Hawaii. (U.S.D.A. Forest Service-Bureau of Plant
Industry map by Rapheal Zon and H. L. Shantz.)

Legend:

Grass vegetation
- Tall grass
- Short grass
- Mesquite grass

Forest vegetation
- Forest
- Arid woodland

Desert vegetation
- Sage brush
- Creosote bush

Typical Entisols are shifting sand dunes and soils forming from alluvial
or glacial deposits that have been in place for only a short time. Coarse quartz
sands that undergo little change even in humid regions also provide one kind
of Entisol.

Entisols occur in any geographic or climatic area (see fig. 12.5). Since
this order is not defined by genetically derived properties, its subdivision
into lower classes is based on such things as differences in parent material
or in temperature or moisture regimes.

Inceptisols

Inceptisols are soils of the humid regions that are at an intermediate stage
of development. They may have histic, plaggen, anthropic, or umbric epipedons, for these horizons can form without other major changes occurring
in the profile. They may also have a mollic epipedon, as where a soil with
an umbric horizon has been limed. However, the most common diagnostic
horizons in Inceptisols are ochric or umbric epipedons, cambic subsurface
horizons, and fragipans or duripans. Inceptisols occur under any condition,
except an arid climate, and form from either transported or residual parent

materials. Those forming in residual parent material are usually shallow. One class of Inceptisols includes soils forming in volcanic ash or coarser tephra. These may have a dark umbric epipedon, or an O horizon if under forest, and amorphous materials, which have a high pH-dependent charge, are the main contributors to cation-exchange characteristics in these ash-rich soils.

In the United States, the largest body of Inceptisols is in the East extending southward from New York, where they have formed mostly in glacial material, through the Appalachian Mountains, where the parent materials are dominantly weathered sedimentary and metamorphic rocks (see fig. 12.5). Inceptisols in Michigan occur in lacustrine deposits, and those paralleling the Mississippi River, in old lacustrine and alluvial materials. Inceptisols in the Pacific Northwest, Alaska, and Hawaii are strongly influenced by volcanic ash, or locally, coarser tephra.

Because of their diversity, Inceptisols are used for a wide range of crops, but predominantly for timber. They are often inherently fertile, although many respond to lime and fertilizer treatment.

Aridisols

Soils of this order have an aridic soil moisture regime typical of dry climatic conditions. Under these conditions vegetation ranges from typical desert species to short grasses. Total growth of the vegetation is never great, with the result that Aridisols are little darkened by accumulating organic matter. Because of a lack of organic matter in dry-region soils, an ochric epipedon has been selected as diagnostic of the Aridisol order. However, some soils of humid regions also have light-colored ochric epipedons. Thus, to distinguish between soils occurring under these two different conditions, a second requirement is placed on the diagnostic epipedons of Aridisols; they must be more than 50–percent saturated with bases, which is essentially a universal condition in arid-region soils. In contrast, more highly leached soils of humid regions normally have surface horizons that are below 50 percent base saturation.

The weathering of silicate rocks is negligible under arid climatic conditions; thus, Aridisols form in transported parent materials. During soil development, weathering is limited to the solution and downward translocation of lime and other soluble compounds. Calcic horizons formed by this means are frequent components of the B or C horizons of Aridisols. Gypsic or salic horizons may also be present at times.

In spite of the limited precipitation of dry regions, eluviation and illuviation of silicate clays is an important process in the formation of some Aridisols. The translocation of clay is most extensive in either highly permeable, coarse-textured soil materials that contain some clay or in soils containing a high level of exchangeable sodium. Exchangeable sodium encourages the breakdown of structure and the separation, or dispersion, of clay particles so that they can move independently in the soil. When dispersed, clay can be more readily translocated to produce an argillic horizon in the subsoil. Many Aridisols contain well developed argillic horizons, and some have sufficient exchangeable sodium to make them natric horizons.

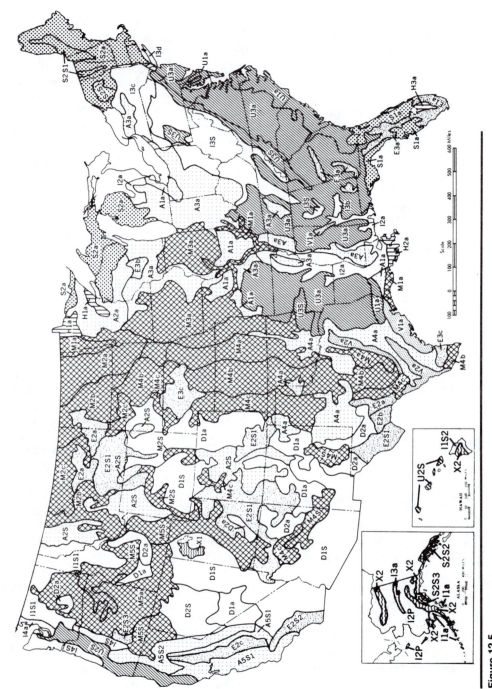

Figure 12.5.
The distribution of soil orders and suborders in the United States. (U.S.D.A. Soil Conservation Service map.)

(Names listed first are of the dominant suborder or great group; names in parentheses are for the principal associated soils.)

Alfisols

A1a Aqualfs (Udalfs, Haplaquepts, Udolls)
A2a Boralfs (Udipsamments, Histosols)
A2S Cryoboralfs (Borolls, Cryochrepts, Cryorthods, Rock outcrops)
A3a Udalfs (Aqualfs, Aquolls, Rendolls, Udolls, Udults)
A4a Ustalfs (Ustochrepts, Ustolls, Usterts, Ustorthents)
A5S1 Xeralfs (Xerolls, Xerorthents, Xererts)
A5S2 Haploxeralfs (Andepts, Xerults, Xerolls, Xerochrepts)

Aridisols

D1a Argids (Orthids, Orthents, Psamments, Ustolls)
D1S Argids (Orthids, Torriorthents)
D2a Orthids (Argids, Orthents, Xerolls)
D2s Orthids (Argids, Torriorthents, Xerorthents)

Entisols

E1a Aquents (Quartzipsamments, Aquepts, Aquolls, Aquods)
E2a Torriorthents (Aridisols, Usterts, Borolls)
E2b Torriorthents (Torrerts)
E2c Xerorthents (Xeralfs, Orthids, Argids)
E2S1 Torriorthents (Argids, Torrifluvents, Ustolls, Borolls)
E2S2 Xerorthents (Xeralfs, Xerolls)
E2S3 Cryorthents (Cryopsamments, Cryandepts)
E3a Quartzipsamments (Aquults, Udults)
E3b Udipsamments (Aquolls, Udalfs)
E3c Ustipsamments (Ustalfs, Aquolls)

Histosols

H1a Hemists (Psammaquents, Udipsamments)

H2a Hemists and Saprists (Fluvaquents, Haplaquents)
H3a Fibrists, Hemists, and Saprists (Psammaquents)

Inceptisols

I1a Cryandepts (Cryaquepts, Histosols, Rock land)
I1S1 Cryandepts (Cryochrepts, Cryumbrepts, Cryorthods)
I1S2 Andepts (Tropepts, Ustolls, Tropofolists)
I2a Haplaquepts (Aqualfs, Aquolls, Udalfs, Fluvaquents)
I2P Cryaquepts (Orthents, Histosols, Ochrepts)
I3a Cryochrepts (Aquepts, Histosols, Orthods)
I3b Eutrochrepts (Uderts)
I3c Fragiochrepts (Fragiaquepts)
I3d Dystrochrepts (Udipsamments, Haplorthods)
I3S Dystrochrepts (Udalfs and Udults)
I4a Haplumbrepts (Aquepts, Orthods)
I4S Haplumbrepts (Orthods, Xerolls, Andepts)

Mollisols

M1a Aquolls (Udalfs, Fluvents, Udipsamments, Ustipsamments, Aquepts, Eutrochrepts, Borolls)
M2a Borolls (Aquolls, Ustorthents)
M2b Borolls (Ustipsamments, Ustorthents, Boralfs)
M2c Borolls (Argids, Orthids, Torriorthents)
M2S Borolls (Boralfs, Argids, Torriorthents, Ustolls)
M3a Udolls (Aquolls, Udalfs, Aqualfs, Fluvents, Psamments, Ustorthents, Aquepts, Albolls)
M4a Ustolls (Orthents, Ustochrepts, Usterts, Aquents, Fluvents, Udolls)
M4b Ustolls (Ustalfs, Ustipsamments, Ustorthents, Ustochrepts, Aquolls, Usterts)

M4c Ustolls (Ustalfs, Orthids, Ustipsamments, Ustorthents, Ustochrepts, Torriorthents, Borolls, Ustolls, Usterts)
M4S Ustolls (Argids, Torriorthents)
M5a Xerolls (Argids, Orthids, Fluvents, Cryoboralfs, Cryoborolls, Xerorthents)
M5S Xerolls (Cryoboralfs, Xeralfs, Xerorthents, Xererts)

Spodosols

S1a Aquods (Psammaquents, Aquolls, Humods, Aquults)
S2a Orthods (Boralfs, Aquents, Orthents, Psamments, Histosols, Aquepts, Fragiochrepts, Dystrochrepts)
S2S1 Orthods (Histosols, Aquents, Aquepts)
S2S2 Cryorthods (Histosols)
S2S3 Cryorthods (Histosols, Andepts, Aquepts)

Ultisols

U1a Aquults (Aquents, Histosols, Quartzisamments, Udults)
U2S Humults (Andepts, Tropepts, Xerolls, Ustolls, Orthox, Torrox)
U3a Udults (Udalfs, Fluvents, Aquents, Quartzipsamments, Aquepts, Dystrochrepts, Aquults)
U3S Udults (Dystrochrepts)

Vertisols

V1a Uderts (Aqualfs, Eutrochrepts, Aquolls, Ustolls)
V2a Usterts (Aqualfs, Orthids, Udifluvents, Aquolls, Ustolls, Torrerts)

Areas With Little Soil

X1 Salt flats
X2 Rock land (plus permanent snow fields and glaciers)

The presence of argillic, natric, calcic, gypsic and similar horizons is the mark of soil development in arid regions and is taken as signs of maturity required to classify a soil as an Aridisol. Soils lacking such signs of maturity are placed in the Entisol order.

In their natural state, Aridisols are used primarily for grazing, but they are also used for the production of cultivated crops when irrigated. However, irrigation often raises the level of the ground water, which may allow soluble salts to rise in the profile and accumulate in the surface horizon to the point of damaging plants. Such problems normally require the installation of drainage systems for the removal of excess water from the land.

Mollisols

The central concept of a Mollisol is a soil with a deep, dark surface horizon produced by the accumulation of organic matter under steppe or prairie vegetation. In drier areas, Mollisols grade toward Aridisols; in wetter areas, they grade toward soils of the forested-regions (see fig. 12.5).

Because of certain similarities in the conditions under which they develop, Aridisols and Mollisols occur in a continuum in which the effective moisture for soil development grades from very low (arid) to moderate (subhumid). The effect of this gradation in climate on certain soil properties is illustrated diagrammatically in figure 12.6. As shown in the figure, increasing precipitation causes a decrease in the average pH of the surface soil, an increase in the amount and depth of accumulated organic matter, and in the depth to free lime. Though not shown in the figure, the eluviation and illuviation of clay also increase with increasing precipitation. Thus, argillic horizons are more common in Mollisols than in Aridisols. As illustrated by the well drained soils in figure 12.7, argillic horizons in Mollisols tend to be deeper, thicker, and more well developed. The A horizon also appears darker and thicker in the Mollisols than in the Aridisol.

Not only is precipitation higher in Mollisol than in Aridisol regions, it also covers a broader range. This is indicated by the fact that Mollisols occur under ustic, udic, aquic, and xeric soil moisture regimes. Mollisols with a relatively dry ustic moisture regime are the most similar to Aridisols. For example, Mollisols retaining lime in the profile, as shown in the diagram of figure 12.6, generally have a ustic soil moisture regime. Mollisols with a ustic moisture regime may also have natric horizons. In contrast, leaching is sufficient in soils with udic or aquic soil moisture regimes so as to completely remove both soluble salts and lime from the profile. Thus, calcic or natric horizons that may occur in Mollisols with a ustic soil moisture regime are not found in those with udic or aquic moisture regimes. Mollisols with a xeric moisture regime (wet winters, dry summers) may or may not be swept free of lime and soluble salts depending on the total average precipitation they receive each year.

Mollisols with aquic moisture regimes typically show evidence of strong eluviation and illuviation of clays. This frequently produces not only intensely developed argillic (B) horizons, but also easily recognizable light-colored albic (E) horizons of very low clay content. A profile with these characteristics is shown for the imperfectly drained Mollisol in figure 12.7. Soils

Figure 12.6.
Changes in some important profile characteristics with variation in average annual precipitation for well drained soils of the Aridisol and Mollisol orders. Shading indicates the depth and relative intensity of organic matter accumulation. The pH values are for the surface soil.

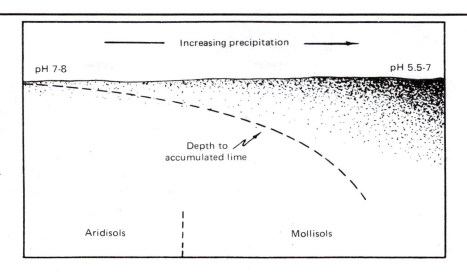

Figure 12.7.
Profiles typical of the Aridisol and Mollisol soil orders formed under different drainage conditions. (U.S. Soil Conservation Service photographs.)

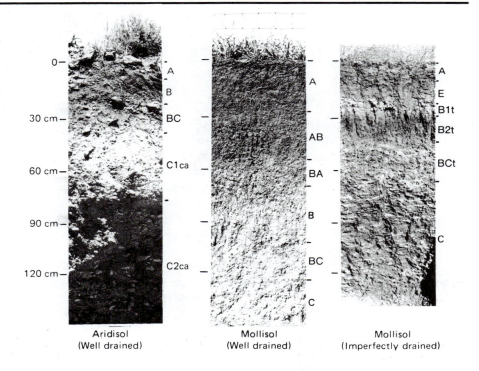

of this type tend to occur on flat surfaces that lose little water as runoff, or they may be found at the foot of slopes or in depressional areas that receive runoff from surrounding slopes.

Climatic conditions associated with Mollisol development, like those producing Aridisols, are also incapable of producing residual parent material by the weathering of hard, silicate rocks. Thus, Mollisols ordinarily develop in transported parent materials. One exception is the formation of Mollisols in residual parent material produced by the weathering of limestone in place. In the United States, these Mollisols occur mainly in association with Alfisols in Kentucky and Tennessee.

Mollisols are the dominant soils of the Great Plains east of the Rocky Mountain region. Those in the Northern Great Plains occur principally under either udic or aquic moisture regimes. In the south, where high temperatures reduce the effectiveness of precipitation for soil formation, Mollisols typically have a ustic moisture regime. In the southwestern part of the continental United States, they also have a ustic moisture regime, whereas in the Pacific Northwest, it is xeric.

Many Mollisols have formed from parent materials derived either directly or indirectly from glacial action. These include glacial till in the northern sections of the continental United States, and glacial outwash or wind-transported loess farther to the south. These soils are inherently fertile, except where dry climates have limited the accumulation of organic matter and nitrogen. Principal uses of Mollisols are in the production of corn, soybeans, and similar crops in the more humid regions, and small grains or range in drier areas.

Spodosols

Although Spodosols occur under varying conditions, they form most often in coarse, quartz-bearing parent materials, usually of glacial origin, in association with boreal forests. Strongly acid weathering and leaching are major factors in the formation of these soils, the acidity being attributable to a low level of base cycling and to the formation of acids, largely organic, in the litter layer (O horizon) common to boreal forest vegetation. Due to the strong acidity, rapid decomposition of nonquartz minerals, including layer clays, results in the formation of a whitish E (albic) horizon high in quartz and positioned immediately beneath the O horizon (see fig. 12.8). Iron and aluminum derived from the weathering minerals, in combination with soluble organic material from the O horizon, form a soluble complex that can then be moved downward by percolating water. In the subsoil, these materials separate and precipitate as coatings over other mineral grains, presumably because of somewhat lower acidity in the subsoil. The humic material tends to precipitate first, forming a dark-colored layer just below the E horizon. The iron and aluminum may precipitate as amorphous oxides in the same zone with the humic material or in a somewhat wider band that extends below the humic layer. The horizon formed by the precipitation of these materials is the spodic horizon, and it is diagnostic of Spodosols. Although O and E horizons are common to Spodosols, they are not diagnostic of this order. They are not diagnostic because they are often too thin to survive cultivation. For any horizon to be diagnostic requires it have the same general appearance after as before cultivation.

Figure 12.8.
Profile of an imperfectly drained Spodosol developed in silty and sandy lacustrine material. The symbol Bhs indicates an accumulation of both humus and iron oxide. Note the variability in the thickness of the E and B horizons. (U.S. Soil Conservation Service photograph.)

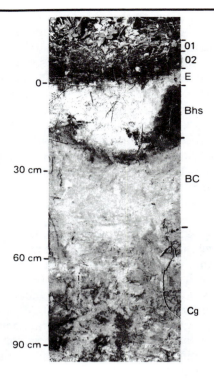

In addition to O, E, and B (spodic) horizons, Spodosols may contain duripans, fragipans, and placic (ironpan) horizons. Duripans result from cementation in the E horizon by silica. Placic horizons usually form in the upper part of the spodic horizon, and are sometimes essentially impermeable to water. Strong acidity is normal to Spodosols; the pH of the O and E horizons is frequently as low as 3.5. In many of these soils, the pH may be below 4.5 throughout the profile.

Spodosols occur most widely in spruce-fir woodlands, but other species of trees, both coniferous and deciduous, and even heath and tundra vegetation, appear capable of supporting spodic horizon formation by encouraging the translocation of sesquioxides. This is thought to result from specific kinds of organic substances that form soluble complexes capable of moving Fe and Al through the soil. These organic materials also provide an unusually high cation-exchange capacity to the layer in which the humus accumulates. However, unless the acidity of this layer is neutralized and the pH-dependent charge of the organic fraction activated the ion retained will be H^+.

The principal use of Spodosols in their natural state is for timber. Where the land has been cleared, these soils are used for various types of crops, but most satisfactorily if they have been limed to overcome the strong acidity. Unlimed Spodosols are used rather extensively for blueberries, which require strongly acid conditions for satisfactory growth.

Alfisols

This order occurs under the same moisture regimes as Mollisols (*i.e.*, ustic, udic, aquic, and xeric). However, Alfisols differ from Mollisols in that they lack a mollic epipedon. Under a ustic moisture regime, the vegetation is grass, but Alfisols lack a mollic epipedon because they form under comparatively high temperatures that prevent much build-up of organic matter in the surface horizon. A light-colored ochric epipedon is characteristic of these Alfisols. Alfisols of wetter regions fail to develop a mollic epipedon because they occur under deciduous forests where organic matter accumulates in mineral horizons that are either too thin or too low in exchangeable bases for a mollic epipedon. Alfisols of the forested regions have either an ochric or an umbric epipedon. These soils occur generally in a belt between Mollisols of the grassland regions and forested soils of the cool, humid regions (Spodosols), or of the warm, humid regions (Ultisols). Alfisols under ustic moisture regimes lie between Aridisols and forested soils of the warm, humid regions (Ultisols) or of the hot, humid regions (Oxisols) (see fig. 12.5).

The moisture regimes of Alfisols encourage moderate to extensive eluviation and illuviation, but without excessive leaching. An illuvial horizon, either argillic or natric, has therefore been selected as the prime diagnostic feature of this order. The base saturation of the diagnostic horizon must exceed 35 percent, even though the surface horizon of wetter Alfisols may be strongly acid. Acidity in Alfisols is less than that in Spodosols, however, the principal reasons being that either grass or deciduous forest vegetation cycles more bases than do conifers, and there is less percolation of water through the Alfisol profile. Both a somewhat lower effective precipitation and a restriction on water flow due to the argillic horizon may explain the limited leaching. Interference with water flow by the argillic horizon helps maintain an aquic moisture regime in some Alfisols.

The clays in Alfisols are predominantly 2:1 layer types of moderate to high cation-exchange capacity. This type of clay is comparatively stable at the pH of argillic horizons in Alfisols. Most Alfisols also contain easily weathered primary minerals, which, as a source of nutrients, contribute to a relatively high fertility level. These soils are used extensively for corn, soybeans, small grains, and forages.

Ultisols

Ultisols, like Alfisols, show signs of clay translocation, but it is generally under more intense weathering and leaching than is characteristic of Alfisols. Together, these two orders make up a continuum in which properties of the Ultisols show the influence of a warmer, wetter climate. The average temperature of Ultisols must be above 8°C (mesic or warmer temperature regime). Where Alfisols and Ultisols occur in the same vicinity, Ultisols are found on older land surfaces where time has been sufficient for strong profile development.

An argillic horizon low in exchangeable bases (<35 percent base saturation) is diagnostic of Ultisols. The clays in this horizon show the effects of extensive weathering and are predominantly kaolinite, sesquioxides, and Al-chlorite. Having a low negative charge, these clays contribute only modestly to the cation-exchange capacity of the soil. The level of base saturation is higher near the surface than in the subsoil, presumably the result of base cycling by plants. The level of bases in the surface horizon has been increased substantially in many cultivated soils through treatment with lime and fertilizer.

Most Ultisols have a characteristic red or yellow color derived from accumulated Fe oxides. Organic matter may darken the surface soil, but more so in poorly drained than in well drained profiles. The normal epipedon is ochric, although in some wetter Ultisols, it is umbric. Albic horizons are of frequent occurrence in Ultisols derived from quartz-bearing parent materials. Some Ultisols have fragipans, and some may have plinthite in or below the argillic horizon.

Except for micas and quartz, most primary minerals have been lost from Ultisols through weathering. As a result, the fertility of these soils is low on the average, a condition that is accentuated if organic matter is also low. Where adequately fertilized, however, Ultisols can be highly productive. They are used for a wide range of crop plants, some of the more typical being corn, tobacco, cotton, small grains, and forages.

Oxisols

The unique properties of Oxisols relate to extreme weathering, for their principal occurrence is on old landscapes of the tropical regions. Mineralogically, they consist chiefly of kaolinite and sesquioxide clays, with or without other resistant minerals such as quartz. They have a low cation-exchange capacity and a low content of exchangeable bases. Because of the low cation-exchange capacity, they are weakly buffered, with the pH of the soil often being controlled by the ionic composition of a very dilute soil solution. Many of these soils are only slightly to moderately acid.

Due to their high content of Fe oxides, Oxisols normally have an intense red or yellow color. Because of cementation by the oxides, these soils also tend to be strongly aggregated. As a consequence, they have a high infiltration rate, they are able to resist erosion, and they display limited stickiness and plasticity. Unlike temperate-region soils that contain swelling, sticky clays, Oxisols can be worked when quite wet without fear of serious effects on the physical state. This is fortunate, for many times the farmer of the tropics is hard pressed to find a dry period when he can attend to all cultural operations on the land.

Cementation by Fe is also responsible for the formation of plinthite in some Oxisols. In various places, plinthite is cut into a brick shape, dried, and used as a building material.

Because of their low nutrient reserves, Oxisols are often infertile, so infertile, in fact, that they occasionally support no vegetation in their natural state. A common practice in some underdeveloped tropical areas is to cut the forest and farm the land until depleted of its fertility. The nutrients are supplied primarily by the organic fraction of the soil. Through rapid decay, this source of nutrients is soon lost; it may sustain worthwhile plant growth for as little as two or three years.

Oxisols must contain either an oxic horizon or a plinthite layer. Beyond this, umbric or ochric epipedons and colors due to Fe oxides, sometimes modified by organic matter, are the only notable profile features. Many Oxisols are so uniform in appearance that it is difficult to recognize A, B, or C horizons.

In the United States, Oxisols, other than relics, are limited to Hawaii and Puerto Rico.

Vertisols

Vertisols comprise a unique class of soils identified by properties attributable to the influence of expanding clay, primarily montmorillonite. A fundamental characteristic of these soils is extensive swelling and shrinking on wetting and drying. Shrinkage on drying causes the formation of wide, deep cracks (see fig. 12.9), and sloughing or washing of surface materials into the cracks results in the continual inversion, or *self-plowing*, of the soil, commonly to depths of 75 to 100 cm. Because of mixing, the character of the surface layer of soil may change little throughout its entire thickness.

The easy entry of water into the cracks of Vertisols causes variation in the rate and extent of wetting and swelling with depth. As a consequence, the profile is often distorted by unequal mass-shifting of the soil material. Evidence of this effect is the tilt and loss of alignment often seen in fences and utility poles after they have been in these soils for a period of time.

Most Vertisols occur under grassland vegetation in generally warm, dry-to-moist climates. They must be seasonally dry in order for cracking to occur. They have either an ochric or a mollic epipedon, they must contain more than 30 percent clay, which distinguishes them from other orders with ochric or mollic epipedons, and they must be at least 50 cm deep. As shown in figure 12.5, the principal areas of Vertisols in the United States are in Texas, Mississippi, and Alabama, but they also occur in association with Mollisols in northern California, Nevada, and southern Oregon, with Alfisols in the coastal mountains of California, and with Entisols across central Montana and western North Dakota.

Vertisols may form residually from weathered limestone or basalt. Often they develop in fine-textured alluvial or marine deposits. Vertisols occurring under relatively moist conditions have a very dark, if not black, color. Because of their slow permeability to water, Vertisols are little affected by leaching and may remain calcareous indefinitely if formed in calcareous parent materials. Slow permeability also increases runoff and the potential for erosion, which can be severe at times. Vertisols are used for both irrigated and unirrigated agriculture, but their main use is for pasture.

Histosols

Soils of this order show a dominating influence of organic matter on properties. The usual concept of a Histosol is a deposit of partially or extensively decomposed aquatic plant residues, commonly of *Sphagnum* moss, that have accumulated under water in low-lying ponds or swamps. However, some organic soils form on sloping surfaces where water can be readily lost as runoff (see fig. 12.10). These occur under climates that maintain extremely wet conditions at the soil surface most of the time.

Organic deposits frequently contain considerable mineral matter. This can result from the accumulation of root tissue within the surface layer of mineral soils, but, more likely, it will be caused by mechanical mixing or by

Figure 12.9.
Cracking in a Vertisol.
(From J. R. Johnson,
and H. O. Hill, *Proc.
Soil Sci. Soc. Amer.*
9:24–29, 1945. By
permission of the Soil
Society of America.)

Figure 12.10.
A peat soil formed
under heather on a
relatively steep slope
in the Scottish
Highlands. The trench
exposing the peat is
for drainage.

the periodic deposition of mineral matter in an accumulating organic deposit. Examples of such mineral matter are diatomaceous earth, a siliceous product of water-inhabiting diatoms, volcanic dusts, loess, or erosional debris washed into a pond from the neighboring watershed. To be called organic, mixtures of this type must have a minimum of from 20 to 30 percent organic matter depending on the amount of clay present; 20 percent if clay is absent, and increasing amounts up to 30 percent as the clay content increases from zero to 60 percent. To be called a Histosol, an organic deposit must be at least 40 cm thick. If the organic deposit is interbedded with layers of mineral matter, the combined thickness of all organic layers must be 40 cm or more. Thinner layers on the surface of mineral soil material are classed as histic epipedons rather than as Histosols.

As with mineral soils, Histosol profiles can be described by means of a series of letter-number symbols that identify specific kinds of accumulating materials or layers. These symbols are:

Oi	Fibric material.
Oe	Hemic material.
Oa	Sapric material.
C	Mineral layer of eroded soil, loess, etc.

If organic horizons occur interbedded with two or more mineral layers, special horizon notations may be needed. For example, if different kinds of mineral layers occur, then they are identified by Arabic numerals, starting with 2, placed before the horizon symbols in increasing sequence down through the soil. Multiple occurrences of the same kind of horizon, mineral or organic, are indicated by prime marks following the capital letter of the horizon symbol. To illustrate, the symbol sequence for a Histosol containing two Oe horizons separated by a C (mineral) horizon would be Oe-C-O'e. If one more C and Oe horizon were present, the sequence would be Oe-C-O'e-C'-O''e. However, if the second mineral layer were not the same as the first, the sequence would become Oe-C-O'e-2C-O''e.

Organic soils are highly prized for agricultural use, particularly for vegetable farming. They are desirable because of their good physical character and a lack of serious nutritional problems. These soils, when drained, are usually highly permeable, have a high water-holding capacity, and are easily cultivated. They do have some undesirable characteristics. Following drainage, which increases aeration, they subside as decomposition sets in, and they may develop layers that restrict the downward flow of water. They are also easily eroded by wind when dry. Because of low strength, their tendency to subside, and the ease with which they can be compressed under load, organic soils are unsuitable for most engineering uses.

Although Histosols are more prevalent in humid regions, they can occur under any climatic condition so long as water is available. As a result, Histosols lack the kinds of climatically related properties that characterize most other orders.

Table 12.7. Formative elements of suborder names.

Formative element	Derivation	Connotation
Alb	L. *albus*, white	Presence of an albic horizon
And	Modified from ando	Andolike
Aqu	L. *aqua*, water	Aquic moisture regime
Ar	L. *arare*, to plow	Mixed horizons
Arg	L. *argilla*, white clay	Presence of an argillic horizon
Bor	Gr. *boreas*, northern	Cool
Ferr	L. *ferrum*, iron	Presence of iron
Fibr	L. *fibra*, fiber	Least decomposed stage
Fluv	L. *fluvius*, river	Flood plain
Fol	L. *folia*, leaf	Mass of leaves
Hem	Gr. *hemi*, half	Intermediate stage of decomposition
Hum	L. *humus*, earth	Presence of organic matter
Ochr	Gr. *ochros*, pale	Presence of an ochric epipedon
Orth	Gr. *orthos*, true	The common ones
Plagg	Ger. *plaggen*, sod	Presence of a plaggen epipedon
Psamm	Gr. *psammos*, sand	Sand texture
Rend	From Rendzina	High carbonate content
Sapr	Gr. *sapros*, rotten	Most decomposed state
Torr	L. *torridus*, hot and dry	Torric moisture regime
Trop	Gr. *tropikas*	Tropical (humid, warm)
Ud	L. *udus*, humid	Udic moisture regime
Umbr	L. *umbra*, shade	Presence of an umbric epipedon
Ust	L. *ustus*, burnt	Ustic moisture regime
Xer	Gr. *xeros*, dry	Xeric moisture regime

Soil suborders and great groups

The classification of soils into the lower categories of suborders through families provides more information about a soil than does its classification at the order level. This information is supplied by class names, which, like order names, are coined words based on roots identified with certain specific soil conditions or characteristics. Since our purpose here is mainly to just illustrate the general nature of soil taxonomy, the discussion will consider only the naming of suborders and great groups of soils.

Names of suborders and great groups are based on two sets of formative elements, one for suborders and one for great groups. These are listed in tables 12.7 and 12.8, along with the roots from which the formative elements are derived and their intended connotation. Certain of the formative elements appear in both lists. The reason for this is that the properties used to identify or separate suborders within some orders are used to identify great groups in other orders.

Names of suborders consist of a suborder formative element plus the formative element of the order to which the soil belongs. Names of great groups consist of a formative element for great groups plus the suborder name. Since the formative element of the order appears in both the suborder and great group names, one can immediately determine the order of a soil from either of these names. This is also the case for subgroup and family names, for they consist of great group names plus additional descriptive terms that supply increasing amounts of information about the soil being classified.

Table 12.8. Formative elements of great group names.

Formative element	Derivation	Connotation
Acr	Gr. *akros*, at the end	Extreme weathering
Agr	L. *ager*, field	An agric horizon
Alb	L. *albus*, white	An albic horizon
And	From ando	Andolike
Arg	L. *argilla*, white clay	An argillic horizon
Bor	Gr. *boreas*, northern	Cool
Calc	L. *calcis*, lime,	A calcic horizon
Camb	L. *cambiare*, to exchange	A cambic horizon
Chrom	Gr. *chroma*, color	High chroma (distinct color)
Cry	Gr. *kryos*, icy cold	Cold
Dur	L. *durus*, hard	A duripan
Dystr, dys	Gr. *dys*, ill; infertile	Low base saturation
Eutr, eu	Gr. *eu*, good; fertile	High base saturation
Ferr	L. *ferrum*, iron	Presence of iron
Fluv	L. *fluvus*, river	Flood plain
Frag	L. *fragilis*, brittle	Presence of fragipan
Fragloss	Compound of fra (g) and gloss	See frag and gloss
Gibbs	Gibbsite	Presence of gibbsite in sheets or nodules
Gyps	L. *gypsum*, gypsum	Presence of a gypsic horizon
Gloss	Gr. *glossa*, tongue	Tongued
Hal	Gr. *hals*, salt	Salty
Hapl	Gr. *haplous*, simple	Minimum horizon
Hum	L. *humus*, earth	Presence of humus
Hydr	Gr. *hydor*, water	Presence of water
Luv	Gr. *louo*, to wash	Illuvial
Med	L. *media*, middle	Of temperate climates
Nadur	Compound of na(tr) and dur	See natr and dur
Natr	L. *natrium*, sodium	Presence of natric horizon
Ochr	Gr. *ochros*, pale	Presence of ochric epipedon
Pale	Gr. *paleos*, old	Excessive development
Pell	Gr. *pellos*, dusky	Low chroma
Plac	Gr. *plax*, flat stone	Presence of a thin pan
Plagg	Ger. *plaggen*, sod	Presence of plaggen epipedon
Plinth	Gr. *plinthos*, brick	Presence of plinthite
Psamm	Gr. *psammos*, sand	Sand texture
Quartz	Ger. *quarz*, quartz	High quartz content
Rhod	Gr. *rhodon*, rose	Dark red color
Sal	L. *sal*, salt	Presence of salic horizon
Sider	Gr. *sideros*, iron	Presence of free iron oxides
Sombr	F. *sombre*, dark	A dark horizon
Sphagn	Gr. *sphagnos*, bog	Presence of Sphagnum
Sulf	L. *sulfur*, sulfur	Presence of sulfides or sulfates
Torr	L. *torridus*, hot and dry	Torric moisture regime
Trop	Gr. *tropikos*, of the solstice	Humid and continually warm
Ud	L. *udus*, humid	Udic moisture regime
Umbr	L. *umbra*, shade	Presence of umbric epipedon
Ust	L. *ustus*, burnt	Ustic moisture regime
Verm	L. *vermes*, worm	Wormy, or mixed by animals
Vitr	L., *vitrum*, glass	Presence of glass
Xer	Gr. *xeros*, dry	A xeric moisture regime

Soil classification and survey

Table 12.9. Names of suborders within the U. S. system of soil taxonomy.

Order	Suborder	Order	Suborder
Aridisol	Argid	Oxisol	Aquox
	Orthid		Humox
Mollisol	Alboll		Orthox
	Aquoll		Torrox
	Boroll		Ustox
	Rendoll	Entisol	Aquent
	Udoll		Arent
	Ustoll		Fluvent
	Xeroll		Orthent
			Psamment
Spodosol	Aquod		
	Ferrod	Inceptisol	Andept
	Humod		Aquept
	Orthod		Ochrept
Alfisol	Aqualf		Plaggept
	Boralf		Tropept
	Udalf		Umbrept
	Ustalf	Vertisol	Torrert
	Xeralf		Udert
Ultisol	Aquult		Ustert
	Humult		Xerert
	Udult	Histosol	Fibrist
	Ustult		Hemist
	Xerult		Saprist
			Folist

All suborders of soils currently recognized by the U. S. Department of Agriculture are shown in table 12.9. Some notion of the general nature of these groups can be obtained from the connotations of the formative elements for orders and suborders shown in tables 12.6 and 12.7. For example, consider the two suborders of Aridisols: Argids and Orthids. As may be determined from table 12.6, the last syllable, id, in each suborder name is the order formative element that associates these soils with Aridisols. The first syllables are the formative elements for the suborders. From the connotations in table 12.7, it may be judged that Argids have an argillic horizon that is diagnostic of this suborder, and by implication, that Orthids do not. An Orthid may show the effect of eluviation and illuviation, but not enough to produce an argillic horizon. According to the formative element, orth, Orthids are the more common of the two Aridisol suborders.

The method of subdividing suborders into great groups is illustrated through the use of six great groups, as shown in table 12.10. These great groups belong to two suborders of the Mollisol order: Ustolls and Udolls, the former having a drier moisture regime (ustic) than the latter (udic). Each of these suborders contain one great group in which an argillic horizon is diagnostic, and one in which the diagnostic properties have been produced by biological mixing. Soils corresponding to two of the great groups shown for Ustolls, Natrustolls and Calciustolls, are not found in Udolls, however. These two great groups have, respectively, a natric horizon (argillic horizon with more than

Table 12.10. Names of selected suborders and great groups of Mollisols, with the formative elements of the names and their connotation of soil characteristics.

| Order | Suborder | Great group | Formative element | | |
			Element	Derivation	Connotation
Mollisol			oll	L. *mollis*, soft	Surface horizon high in organic matter and in degree of base saturation
	Udoll		ud	L. *udus*, humid	Of humid climates
		Argiudoll	arg	L. *argilla*, white clay	Horizon high in clay (Bt)
		Vermudoll	verm	L. *vermes*, worm	Horizons mixed by worms or animals
	Ustoll		ust	L. *ustus*, burn	Hot, dry summers
		Argiustoll	arg	L. *argilla*, white clay	Horizon high in clay (Bt)
		Vermustoll	verm	L. *vermes*, worm	Horizons mixed by worms or animals
		Natrustoll	natr	L. *natrium*, sodium	Horizon high in exchangeable sodium
		Calciustoll	calc	L. *calcis*, lime	Horizon high in lime

15 percent exchangeable sodium) and a calcic horizon formed by the translocation of lime within the profile. The drier moisture regime of Ustolls permits the development of natric and calcic horizons but not the wetter moisture regime of Udolls. Under the latter conditions, leaching is sufficient to remove both readily soluble sodium salts and free lime from the profile. The sodium salts are necessary for natric-horizon formation.

Soil survey

Soil survey is the collection of information about soils with two primary objectives in mind: one is to show the distribution of soils on a soil map; the other is to provide information about the soils in the survey area that will benefit land-use planning. Our main concern is with soil surveys carried out by or under the sponsorship of the U. S. Department of Agriculture. These surveys have a strong agricultural bias, but they supply much information that is of value to nonagricultural soil use.

The soil survey report

Soil survey reports consist of four principal parts: (1) a set of maps, (2) a map legend, (3) a description of the soils in the surveyed area, and (4) a use and management report. The maps, in conjunction with the legend and soil descriptions, permits one to determine the distribution of different kinds of soil within the surveyed area. The use and management report serves as a guide to efficient land utilization by describing the capabilities and limitations of the various soils considered. It also contains information concerning crop adaptation, yield potential, erodibility, and the nutritional status of soils.

Intensity of surveys

Intensity refers to the detail of a survey. The amount of detail collected depends in part on the predicted use that will be made of the survey. In general, land that is used intensively, irrigated land for example, is characterized in considerable detail. Conversely, surveys of low intensity are suitable where land use is for such things as dryland wheat, range, or timber.

The mapping carried out in high-intensity survey work is with maps of large scale, often 12.5 cm per kilometer (8 inches to the mile).[3] An area of uniform dimensions as small as 1.2 hectares (about 3 acres) can be shown at this scale, which permits the delineation of changes in soils that occur frequently over the landscape. Mapping for low-intensity surveys may be at a scale no larger than 3 cm per kilometer (about 2 inches to the mile). Areas smaller than 8 hectares (about 20 acres) cannot be shown conveniently at this scale, which limits the separation of soils except on the basis of major changes in properties that occur relatively infrequently. For this reason low-intensity surveys provide only generalized information about the distribution of specific kinds of soils within an area.

The field mapping of soils is based on and carried out primarily with aerial photographs. This is of particular advantage because photographs can be prepared at any desired scale and show clearly many of the identifying features of the landscape. Frequently, changes in landscape features coincide with the boundaries between different kinds of soil, which is an important aid to the mapping operation.

Mapping units for soil survey

Whereas soil survey is largely devoted to the collection of information about soils, it is not without intimate reference to soil classification. This is because the field identification of soils is based on taxonomic units. For example, the different soil areas shown in the United States map in figure 12.5 are identified with soil suborders and great groups. These areas are not uniform, however, but contain a variety of soils in addition to the suborder or great group for which they are named. Such areas shown on soil maps are called *mapping units*. They are named for a single taxonomic unit, but they show a much wider range in properties than is implied by the name.

The nature of mapping units used in survey depends on the intensity of survey. For example, the information required to produce the map of figure 12.5 could be obtained from a survey of very low intensity; that is, it could be carried out by examining soils at relatively infrequent intervals over the landscape. As a consequence, such a survey would supply little information about the soil at any particular location. If more information is needed, the intensity of survey must be increased. This requires maps of larger scale and a reduction in the amount of variation permitted in the mapping units. Thus, as the need for detail in information varies, so will the intensity of survey. A well designed survey will collect and present only as much information as is needed to satisfy the purpose of the survey.

There are several types of mapping units used in soil survey, but those of greatest importance are *soil phases*, *soil complexes*, and *soil associations*. These, along with a separate category called *miscellaneous land types*, are discussed in order.

[3]English units of measure are shown in parenthesis. These units are used in essentially all soil surveys published in the United States.

Soil phases

The soil phase is the principal mapping unit used in the standard soil surveys published by the U.S. Department of Agriculture. Phases are separations within soil series to show differences important to soil use and management. Most phase separations are based on the actual or predicted principal use of the land, but more often than not, it is for the production of agricultural crops. Ideally, each phase separated in mapping consists of an area that is uniform in profile properties as well as in other properties important to the production of plants. Substantial information about a phase can be determined from its name.

Phase names identify the soil series as well as the soil properties or conditions used to distinguish among the phases within the series. Phases are distinguished from each other for various reasons, but most often it is because of differences in texture or coarse-fragment content of the surface soil, degree of slope,[4] degree of past erosion, depth over bedrock, drainage state, or salt content. Some series may not be subdivided into phases because they occur under a relatively uniform set of conditions, but this is rare. Others may occur under such a variable set of conditions that they may be separated into a dozen or more phases, each theoretically with its own unique use and management requirements.

Occasionally, a soil with a distinctive profile (pedon) is too limited in extent to be recognized as a separate soil series and is mapped as a *variant* of another series of somewhat similar characteristics. The difference between the profile of the variant and the series to which it is assigned is indicated by the name of the variant phase. An example of a variant is included in the following list, which contains ten different soil phases:

1. Ellisforde silt loam, 3 to 8 percent slopes
2. Ellisforde silt loam, hardpan variant, 3 to 8 percent slopes
3. Couse silt loam, 3 to 8 percent slopes
4. Couse silt loam, 8 to 15 percent slopes
5. Couse silt loam, 8 to 15 percent slopes, eroded
6. Gilpin silt loam, 0 to 5 percent slopes
7. Gilpin channery silt loam, 0 to 5 percent slopes
8. Semiahmoo muck
9. Semiahmoo muck, drained
10. Semiahmoo muck, moderately shallow, drained

The first two soils in the above list show two phases of the Ellisforde series that occur on the same slope but differ because one, a variant, has a hardpan that is absent in the normal Ellisforde soil. The first and third soils differ because they belong to different soil series. The three phases of Couse soil recognize differences in slope or in the degree of erosion. This particular soil is mapped as 10 different phases altogether. For the last two soils, phase separations indicate the presence or absence of coarse limestone fragments, which produce a channery condition, or they indicate differences in soil depth or drainage state.

[4]Slope is indicated by percentage values, which are equal to the number of units rise in elevation per 100 units of horizontal distance.

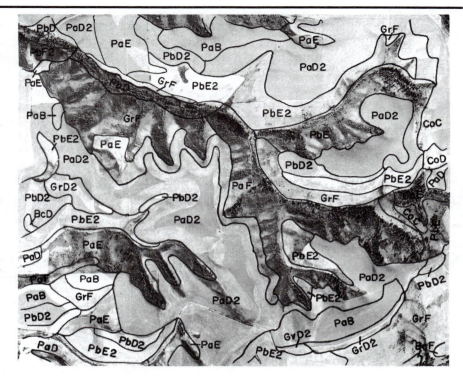

Figure 12.11.
The method of illustrating the distribution of soils on a modern soil map. The first capital letter of the mapping unit symbol is the first letter of the soil name. If the name identifies a soil series, the lower case letter indicates soil texture. Slope classes established for this survey are shown by the second capital letter. The final number indicates defined erosion classes, 2 for an eroded, 3 for a severely eroded. (Adapted from the *Soil Survey of Walla Walla County, Washington,* 1964.)

Most of the soils in surveys published by the U.S. Department of Agriculture are mapped as phases. Currently, maps developed from these surveys are superimposed directly over aerial photographs similar to the one shown in figure 12.11. The aerial photograph, when used as an integral part of the survey map, is as valuable for use in interpreting the mapped data as it is for use in doing the original mapping.

It is impractical to separate all soils in an area strictly on the basis of phase definitions, even in high-intensity surveys where large-scale maps are used. This is because some soils in an area are invariably too small to be shown on the map. When this occurs, it is necessary to include these small bodies with others of larger size and map the combination as a single unit. Small bodies of this type are called *inclusions.* Unless they greatly alter the nature of the mapping unit, they are ignored, the mapping unit name identifying only the dominant soil present.

If a mapping unit is to represent a uniform area of soil, it should contain only a limited proportion of inclusions that differ greatly from the principal soil in the unit. However, if the inclusions are similar to the principal soil, they may make up the majority of the mapping unit. Where this occurs, and it usually involves only a few, rather isolated cases, the name may actually identify only a small part of the soil in the mapping unit. Even so, since the unidentified soil is like the principal soil, the name still provides an adequate basis for judging the behavior of the unit as a whole.

Soil complexes   It is sometimes necessary to combine bodies of highly contrasting soils that are too small to be shown individually on large-scale maps. Combinations of this type are called soil complexes, and to indicate that they are variable in nature, they are named to show each of the principal soils present. Some example soil complex names are Athena-Lance silt loams, 0 to 30 percent slopes; Bath, Valois, and Lansing soils, 35 to 60 percent slopes; and Hesseltine very rocky complex, 0 to 30 percent slopes. The last name suggests soil areas intermixed with periodic rock outcrops. Rock outcrops are not considered to be soil, but rather one of a wide range of miscellaneous land types (see below).

Soil associations   Like soil complexes, soil associations are also mapping units formed by combining different soils, but the combinations are made intentionally, usually so that the distribution of soils can be shown on small-scale maps. Associations also have compound names that identify the principal soils present. In mapping, associations are separated from each other along natural lines, the more important reasons being to show differences in topographic position, parent material, or even climate, if it undergoes significant change across the surveyed area. In this sense, figure 12.5 may be viewed as one type of association map.

Most modern soil surveys published by the U.S. Department of Agriculture contain an association map as illustrated in figure 12.12. The map in this figure shows roughly one half of the association map for Ford County, Kansas. Seven associations are recognized in Ford County, with highly generalized descriptions of each appearing in the legend of the map. The topographic distribution of five of the associations is shown on the block diagram in figure 12.13.

Soil catenas (from *catenary,* meaning chain) are a special type of association consisting of soils forming as a continuum in the same parent material but differing in profile characteristics because of topographically-induced differences in effective climate. Catenas are not often used as mapping units, but they are important because they so clearly demonstrate the significance of topographic-climatic interactions in soil formation. An example that illustrates this is given in figure 12.14; it is a catena consisting of four soils that have developed under prairie vegetation in deep, calcareous loess in the Palouse Region of eastern Washington.

Most of the precipitation in eastern Washington comes during the cooler half of the year, with a substantial part consisting of wind-driven snow. The snow is largely swept from the hilltops, but it accumulates in deep drifts on the northeasterly slopes. This unique pattern of moisture distribution results in four distinctly different soils. Excess moisture from runoff and melting, drifted snow produces Latah and Thatuna soils, which have deep, dark A horizons and well developed B horizons that show substantial accumulations of illuvial clay. Of these two soils, the most intensely formed horizons are in the Latah series. In contrast, the Staley soil shows minimal development; it not only has a shallow A and a weakly developed B horizon, it retains free lime in the subsoil even though all other members of the catena are lime-free. Compared to the other soils, the Palouse is intermediate in both its effective moisture supply and the degree of horizon development in the profile.

Miscellaneous land types

Some areas shown on soil maps contain no soil. A part may be covered by water, or it may be by one of a number of miscellaneous land types. Some examples are made land; strip mines; rock land; steep, broken land; badlands; and riverwash. These terms refer to specific kinds of conditions, some of which are implicit in the names.

The utilitarian classification of soils

The purpose of utilitarian classification is to organize soils into classes to show their relative suitability for a particular use. Soil units combined into utilitarian classes are phases for the most part, with those placed together in a class being equally suitable for the use in question. Four utilitarian systems in current use classify soils according to their suitability for producing a wide range of plants. One of these is a general system termed *land use capability classification*. It rates all soils according to their suitability and limitations for the production of four categories of plants: cultivated crops, pasture and range, woodland, and plants of wildlife habitats. The other three systems group soils to show their ability to produce range, woodland, or wildlife habitat plants. Other utilitarian classification systems rate soils for recreational as well as for a wide range of engineering uses. With the exception of soil classification for engineering use, the above types of utilitarian systems are described briefly in the sections that follow. The classification of soils for engineering use is discussed in greater detail in Chapter 20.

Land use capability classification

The classification of soils into land use capability classes, or, more simply, capability classes, has been carried out for many years by the United States Department of Agriculture. The system is applied to all soils and miscellaneous land types and embraces three levels of generalization, the most general being the capability classes. There are eight such classes, numbered I through VIII, and they are defined so that the larger the class number the greater the limitation for use in the production of economic plants. The relationship between capability class number and the kinds of plants considered suitable for each class is shown in table 12.11. A photograph showing the distribution of these classes over a single landscape appears in figure 12.15.

As may be judged from table 12.11, only classes I through IV are deemed suitable for use with cultivated crops, the limitation for such use increasing with increasing class number. There are no limitations on the use of class I land for cultivated crops beyond reasonable management, which includes the use of well adapted plant varieties and, perhaps, fertilizers. The same is true for classes I through IV when they are used less intensively for the production of pasture or woodland plants. There are restrictions on soil use for the latter kinds of plants in classes V through VIII, however. Indeed, soils of

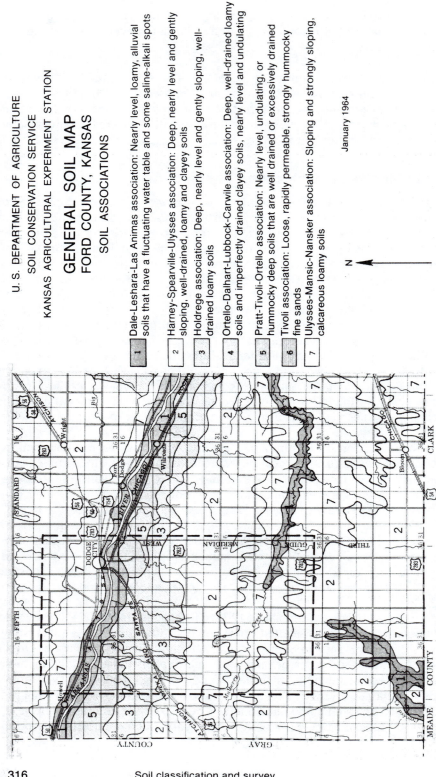

U. S. DEPARTMENT OF AGRICULTURE
SOIL CONSERVATION SERVICE
KANSAS AGRICULTURAL EXPERIMENT STATION

GENERAL SOIL MAP
FORD COUNTY, KANSAS
SOIL ASSOCIATIONS

1 Dale-Leshara-Las Animas association: Nearly level, loamy, alluvial soils that have a fluctuating water table and some saline-alkali spots

2 Harney-Spearville-Ulysses association: Deep, nearly level and gently sloping, well-drained, loamy and clayey soils

3 Holdrege association: Deep, nearly level and gently sloping, well-drained loamy soils

4 Ortello-Dalhart-Lubbock-Carwile association: Deep, well-drained loamy soils and imperfectly drained clayey soils, nearly level and undulating

5 Pratt-Tivoli-Ortello association: Nearly level, undulating, or hummocky deep soils that are well drained or excessively drained

6 Tivoli association: Loose, rapidly permeable, strongly hummocky fine sands

7 Ulysses-Mansic-Nansker association: Sloping and strongly sloping, calcareous loamy soils

January 1964

N

Figure 12.12.
Association map (western half) of Ford County, Kansas. The rectangle formed by broken lines, when viewed from the left, corresponds approximately to the area of the topographic diagram in figure 12.13. (From Soil Survey of Ford County, Kansas, U.S.D.A.)

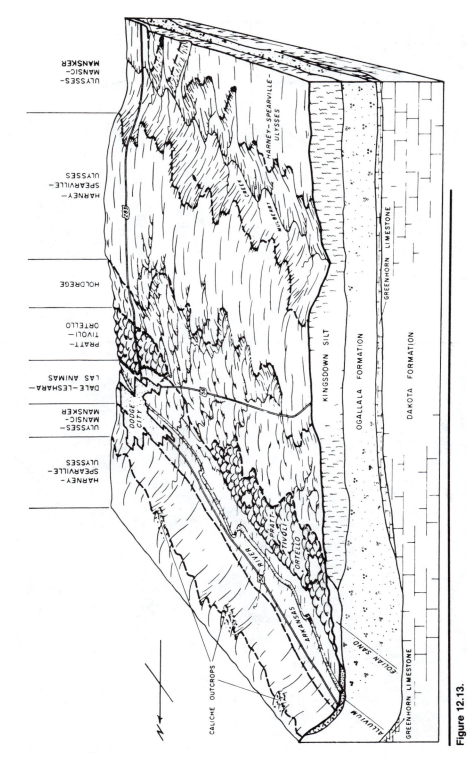

Figure 12.13.
Block diagram showing the topographic relationship among soil associations within the broken-line rectangle of figure 12.12. (From Soil Survey of Ford County, Kansas, U.S.D.A.)

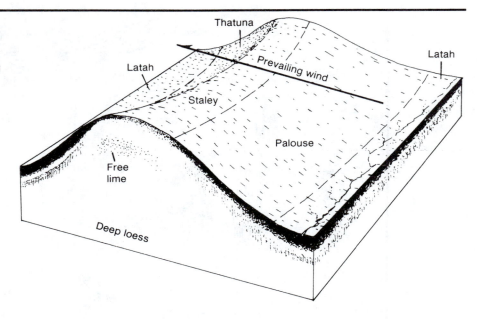

Figure 12.14.
A catena of four soil series in the Palouse Region of eastern Washington reflecting a variable effective climate for soil formation.

Figure 12.15.
A landscape showing the eight land-use capability classes. (Soil Conservation Service, U.S.D.A., photo.)

Soil classification and survey

Table 12.11. The degree of limitation in the production of cultivated crops, pasture, woodland, or wildlife as indicated by the capability class of the soil.

Use	Capability class							
	I	II	III	IV	V	VI	VII	VIII
Cultivated crops	none	mod	sev	v sev	←——————ns——————→			
Pasture	←————none————→				mod	sev	v sev	ns
Woodland	←————none————→				mod	sev	v sev	ns
Wildlife	←——————————————none——————————————→							

Degree of limitation: mod = moderate; sev = severe; v sev = very severe; and ns = not suitable for the specified use.

class VIII should not be used for any economic plant production. These soils are so susceptible to deterioration when disturbed in any way, or they are otherwise so poorly endowed that their use for plant production would be economically unsound. For wildlife production, on the other hand, even class VIII land has no restrictions in use.

Limitations to use that result in the placement of soils in classes above class I are varied and are indicated by appending one of four lower-case letter symbols to the capability class number, as in IIw, IIe, IVe, etc. By this means capability classes are subdivided into *capability subclasses*. Limitations indicated by the subclass symbols are:

e— erosion, either past or potential
w—wetness, as in poorly drained soils
c— climate, such as a growth-limiting temperature or moisture factor
s— soil properties, such as a high salt content, coarseness of texture or shallowness that limits water storage, or a stony or rocky condition that interferes with cultivation.

With but one exception, any of the above factors may be the reason a soil is placed in capability classes II or above. The exception is class V; soils are not placed in this class because of a problem due to erosion. Instead, class V is reserved for soils with a limitation that is difficult to remove, such as stoniness, rockiness, or a suboptimal climatic factor, which, if absent, would result in that soil being placed in a lower class, possibly even class I. Although class V land may be restricted in the kinds of plants it will produce, maximum production of these plants is often possible with only a minimum of management input. In general, management requirements for each class increases with class number, which says that the higher the class number the more difficult it is to maintain a desired level of productivity while avoiding serious deterioration of the land.

The final subdivision within the system of land use capability classification is into *capability units*. All soils within a capability unit belong to the same capability subclass and therefore have the same general limitation for use in plant production. However, soils making up a given capability subclass will vary in other ways important to plant production. Separation into capability units is on the basis of these characteristics. All soils placed in a single capability unit are judged to have the same potential productivity, crop adaptability, and special management requirements. Since these characteristics depend on local conditions, capability units are established on a local basis. Each is assigned a symbol consisting of a subclass symbol and an Arabic numeral, as IIe-1, IIe-2, etc. The number of units established for a given locality will depend on the need; in general, the greater the diversity of soils the greater the number of capability units.

Descriptions of capability units are a standard part of the soil surveys published by the U.S. Department of Agriculture. These descriptions provide the names the soils contained in the unit, identify the potential productivity of these soils for different kinds of plants, and provide guidelines for their management under sustained use.

Soil classification for woodland use

Woodland suitability groups, or *sites,* consist of soils, usually phases, that are similar in potential productivity and in limitations and management requirements for sustained timber production. The productivity of a woodland site depends on any soil or environmental factor important to tree growth. Whether or not the productivity of a site can be sustained over long periods depends on the severity of one or more of five limitations or hazards associated with the reestablishment, maintenance, and harvest of timber. These are (a) seedling mortality, (b) plant competition, (c) equipment limitations, (d) erosion hazard, and (e) windthrow hazard. The first two items in this list relate to the reestablishment of a timber stand. Equipment limitations apply to harvest. An erosion hazard is most evident following harvest, and the windthrow hazard most important to trees along the unprotected border of a timber stand.

The potential timber yield is a key factor in determining placement of soils in woodland suitability groups. The determination of yield is based on a standard unit of measure known as the *site index.* A site index is the average height of a species at a specified age, such as 50 or 100 years. Through use of the site index and appropriate conversion tables included in soil survey reports for forested areas, it is possible to estimate the probable yield of lumber a harvestable, uniform stand of trees will supply.

Woodland suitability groups separate soils not only on the basis of productivity but also with respect to the five limitations or hazards specified above. The degree of limitation is expressed by relative terms such as slight, moderate, or severe. Soils of the same productivity range and that show the same kind and degree of limitations under use are then placed together in the same woodland suitability group. The groups are identified in different ways; sometimes by a number and sometimes by special symbols devised to identify certain major features of the group.

Soil classification for range use

To show differences in their ability to provide forage for grazing, soils are placed into classes known as *range sites* or *range suitability groups.* All soils within the same range site produce about the same kind and amount of veg-

Figure 12.16.
A Colorado landscape showing six range sites. (Soil Conservation Service, U.S.D.A., photo.)

etation when in excellent condition, but different range sites may differ greatly in the production of forage under sustained use. For example, range sites made up of deep soils with a high water-storage capacity have a greater potential for producing more total forage during a single grazing season than do shallow soils within the same climatic zone. Soil evaluation for placement in a range site or suitability group involves an estimate or measurement of the kinds and proportions of plant species in the natural climax vegetation of the site, and in the potential productivity of the site.

Usually, soil phases falling into the same range suitability groups occur under similar conditions or are similar in a property highly important to forage production. As a result, range suitability groups are named in a way that identifies this particular condition or property (see figure 12.16). Typical names are bottomland site, shallow site, loamy site, rockland site, forested upland-pine site, and forested upland-fir site. These names recognize local conditions and may therefore vary greatly from one area to another.

The descriptions of range sites given in soil survey reports include information on the names, usually phase names, and general nature of the soils present, vegetation, potential productivity for both dry and moist years, and management procedures for reseeding, fertilization, moisture conservation, and erosion control. Range sites are not established for all surveyed areas; only for those in which soils are used for this purpose. These are limited predominantly to the drier regions.

Soil classification for wildlife production

One of the major effects of the encroachment of man on wildlands has been to change the nature of the habitat for wildlife. This is not to say that the effect is wholly detrimental to wildlife, although it has often caused significant shifts in the established wildlife species of an area. For example, the conversion of forest to crop or pastureland may eliminate forest species such as deer and elk, but it may greatly increase the kinds and numbers of seed-eating birds and animals. Similarly, dense forest untouched by man may support a minimum of wildlife because of the lack of grass, shrubs, and other low-growing plants that provide food for browsing animals. Yet, when such forests are cut, the early natural development of grasses and shrubs preceding the regrowth of dense forest may cause a marked increase in the number of browsing animals and other wildlife.

To show their suitability for wildlife production, soils are grouped into *wildlife suitability groups* or *wildlife sites*. Basically, all soils within each such group have about the same ability to supply food, water, and cover for wildlife; that is, they provide similar habitats and will support similar populations of particular kinds of wildlife. However, different kinds of wildlife have different habitat requirements. As a consequence, a given soil area may be suitable for only one or two general kinds of wildlife. Based on habitat requirements, wildlife can be classified into five categories as follows:

Wildlife category	Typical members
Farmland (Openland)	Pheasant, quail, partridge
Wetland	Ducks, geese, terns
Brushland (Rangeland)	Rabbit, squirrel, grouse
Forest (Woodland)	Deer, beaver, bear
Pond	Fish, wetland fowl

Whereas the above examples are limited mainly to game, other nongame animals fit into these categories as well.

The suitability of land for wildlife depends on its ability to supply the right kinds and amount of vegetation under a given level of management. Both food and cover must be adequate, otherwise an area will not be suitable for a particular wildlife category. As the level of adequacy varies so will the rating assigned to the soil to show its suitability for use in wildlife production. Suitability ratings are expressed relatively by terms such as excellent, good, fair, poor, and unsuitable.

The suitability of a soil for wildlife depends on its ability to produce various combinations of eight different *habitat elements*. The elements are as follows:

1. Grain and seed crops
2. Domestic grasses and legumes
3. Wild herbaceous plants
4. Hardwood trees
5. Coniferous plants
6. Shrubs
7. Wetland plants
8. Shallow water areas

A soil need provide only certain of these elements for a particular category of wildlife. For example, if an area is rated high for elements 1 through 3, it will also be rated high as a site for farmland wildlife. Similarly, a high rating for elements 2, 3, and 5 would result in a high rating for forest wildlife. Exceptions do occur, however. Land well suited for the production of coniferous forests (element 5) may produce stands so dense that understory plants

Figure 12.17.
An excellent site for wildlife on the Valentine-Loup soil association bordering the Dismal River in west-central Nebraska. (From Soil Survey of Hooker County, Nebraska. U.S.D.A. Soil Conservation Service photo by C. Bohart.)

(elements 2 and 3) needed as a source of food are crowded out. Such land would be rated unsuitable for forest wildlife. For a similar reason land rated excellent for cultivated crops may have to be given a low or unsuitable rating for farmland wildlife if the land is likely to be totally cultivated and all plants providing a permanent, protective cover thereby eliminated.

Whereas soil ratings for most wildlife involve judgments of the potential productivity for elements providing food and cover, the suitability of a site for pond wildlife depends only on whether or not the site will support element 8, shallow water areas. Soils suitable for ponding occur on essentially level land or in drainageways that can be easily dammed. They should also be relatively impermeable.

Figure 12.17 shows a wildlife site along the Dismal River in Hooker County of west-central Nebraska. The site is rated good to excellent for various types of wildlife. Two soil series are included in this site: the Valentine on the slopes and the Loup on the valley floor. The area provides good cover as well as adequate food for game such as grouse, pheasant, and deer. The river is also suitable for fishing.

Soil classification for recreational use

There are four recreational uses of soil for which ratings are supplied in the current soil surveys published by the U.S. Department of Agriculture. These are for soil use as (1) camp sites, (2) picnic areas, (3) paths and trails, and

Figure 12.18.
Foot traffic at recreational sites may lead to destruction of the natural plant cover and serious erosion, as shown at this canoe landing on the Crow River in Minnesota. (U.S.D.A. Soil Conservation Service photo by E. Schober.)

(4) playgrounds. Soil evaluation for each of these uses is very similar. It takes into account soil properties important to trafficability, workability, drainage, and moisture retention. The first of these concerns foot traffic for the most part, although automobile traffic is also a significant factor in camp-site evaluation. Ease with which the soil can be worked is of greatest concern where land-leveling or forming is necessary. Drainage and moisture relationships are of major importance to soil stability under traffic, although in some cases these properties determine the suitability of the soil for a vegetative cover. In general, the best soils for recreation sites are well drained and permeable, they occur on level to only slightly sloping surfaces that are not subject to flooding, and they have a very low content of coarse fragments at or near the surface. Recreation normally concentrates relatively large amounts of traffic on the land, and where it destroys the plant cover, can lead to serious soil erosion (see fig. 12.18).

Soil classification and survey

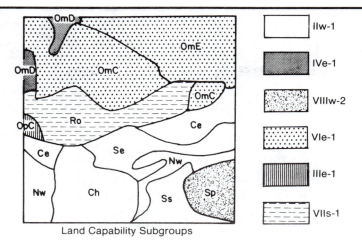

Figure 12.19.
An interpretive map showing the land capability units of several different soil phases.

Legend:
- IIw-1
- IVe-1
- VIIIw-2
- VIe-1
- IIIe-1
- VIIs-1

Land Capability Subgroups

Interpretive maps for agricultural and recreational land use planning

Utilitarian, or interpretive, classifications organize soil mapping units, usually phases, into groups of like use and management requirements. Maps showing the distribution of utilitarian classes of soils therefore provide a basis for land use planning, for they delineate areas differing in potential use and in inputs required to maintain productivity or to prevent deterioration under use. Examples of these types of maps are shown in figures 12.19 through 12.21. Figures 12.19 and 12.20 relate to land use for plant production, whereas figure 12.21 is for three types of recreational use. All three figures apply to the same area, which covers a section of land approximately of 260 ha, or 640 ac, in area.

Figure 12.19 shows the distribution of soil phases and one miscellaneous land type (rockland, symbolized Ro) in the area under consideration. Also shown is the utilitarian classification of these soils into capability units. Phases with symbols OmC, OmD, OmE, and OpC, as well as rockland, occur in upland positions, some of them moderately to steeply sloping. The remainder of the phases, which occur on the lower half of the map, are nearly level, bottomland soils that are generally poorly to very poorly drained (see Appendix B, Soil Drainage Classes). One of these soils is a perennially wet peat (Sp).

As may be judged from the map in figure 12.19, soils suitable for cultivation (capability classes II and III in this example) are concentrated largely in the lowlands. For these, the principal limitation for use in growing agricultural crops is wetness. This will restrict the kinds of crops that can be grown satisfactorily unless the soils are improved through artificial drainage. So long as the soils remain undrained, their use should be restricted to grasses or legume forages that are tolerant of wetness. If drainage is installed, on the other hand, a wide range of vegetable and other crops adaptable to the general area could be grown.

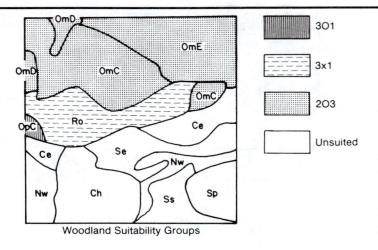

Figure 12.20.
An interpretive map showing the classification of several soil phases into woodland suitability groups.

Woodland Suitability Groups

▨	3O1
▤	3x1
▦	2O3
☐	Unsuited

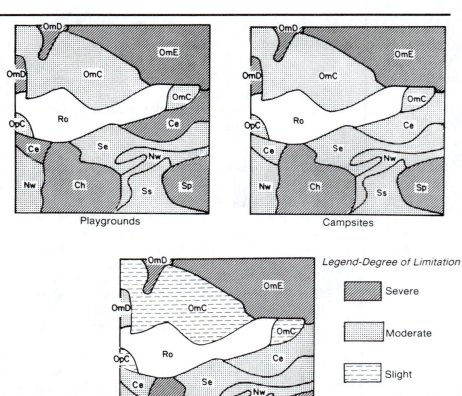

Figure 12.21.
An interpretive map showing the degree of limitation for several soil phases for use as sites for playgrounds, camps, and trails and secondary roads.

Playgrounds

Campsites

Trails and Secondary Roads

Legend-Degree of Limitation

▨	Severe
▦	Moderate
▤	Slight
☐	Not rated

The upland soils in figure 12.19 are either unsuited for cultivation (capability classes VI and VII), or where they are suited, as is the case for capability subclass IIIe, they must be treated with great care because of an erosion hazard. Rockland can, of course, not be cultivated by conventional means, and is classed VIIs for this reason. All of the soils in upland positions are suited for woodland use, however.

An interpretive map of the area in figure 12.19 for woodland use is shown in figure 12.20. Because of wetness, the lowland soils are not suited for commercial timber production. Those in the uplands fall into three woodland suitability groups, symbolized 2o3, 3o1, and 3x1, which differ in one or more of six basic characteristics: productivity, and the degree of limitation for use due to seedling mortality, erosion hazard, windthrow hazard, plant competition, and equipment limitations. Of the three groups, 3x1 has the greatest overall limitations for woodland use, and they reflect a comparatively low productivity as well as severe limitations in equipment use and in seedling mortality. These limitations are due primarily to the rocky nature of the land comprising woodland group 3x1.

Interpretive maps for three recreational uses are shown in figure 12.21: playgrounds, trails and secondary roads, and campsites. For such uses, in only one instance is the limitation slight, and that is the rating for use of phase OmC for trails and secondary roads. In all other cases, limitations are either moderate or severe, and they are due primarily to steepness of slope for the upland soils and wetness in the lowland soils.

Summary

Soil classification involves the grouping of soil individuals into classes according to properties that reflect conditions of soil development (natural, or taxonomic, classification) or that are important to soil use (utilitarian, or interpretive, classification). The basic unit of taxonomic classification is the soil series, a collection of soil bodies (polypedons) similar in profile properties. Similar series are, in turn, grouped into higher, more generalized classes of families, subgroups, great groups, suborders, and orders, the last being the broadest, or most highly generalized, grouping.

There are ten orders in the U. S. system of soil taxonomy. Six consist of mature soils grouped according to similarities in specific acquired properties. Two consist of immature soils, one showing minimal and the other showing an intermediate degree of development. The last two orders contain soils formed in unique parent materials: organic parent materials and parent materials high in expanding clay.

Soil phases, which are subdivisions of soil series, are the basic units of utilitarian classification systems. In utilitarian classification, phases are grouped to show their suitability for the production of cultivated crops, range, or pasture, and timber, or as sites for wildlife or recreation. With some limitations, phases can be characterized to show their suitability for numerous engineering uses.

Review questions

1. Explain the advantage of classifying soils on the basis of properties that reflect the conditions under which they have formed.
2. Define polypedon, and explain its use in the taxonomic system of soil classification.
3. Explain why soil series are the basic units of the taxonomic system of soil classification.
4. Why should diagnostic properties not be altered by soil cultivation?
5. How are soil order names identified?
6. Show how the diagnostic properties of Aridisols, Mollisols, and Spodosols relate to different climatic-vegetational regimes for soil development.
7. As may be judged from the connotations in table 12.7, indicate the primary differences among the four suborders of Histosols shown in table 12.9. Do the same for the two suborders of Aridisols.
8. Point out the similarities and dissimilarities between the two great groups of Mollisols, Argiudolls and Argiustolls (see table 12.10).
9. Distinguish among the following types of soil mapping units: phases, complexes, associations, and miscellaneous land types.
10. What property of the variant shown in the list on page 312 is not typical of the Ellisforde soil series?
11. Why are fewer capability classes suitable for use with cultivated crops than are suitable for use as pasture, woodland, or wildlife habitats?
12. What information is given by the site index of a woodland suitability group?
13. What factors are taken into account when assigning soil phases to woodland suitability groups? to range suitability groups? to wildlife suitability groups?

Selected references

Birkeland, P. W. *Soils and Geomorphology*. New York: Oxford University Press, 1984.

Buol, S. W., Hole, F. D., and McCracken, R. J. *Soil Genesis and Classification*. Ames, IA: The Iowa State University Press, 1980.

Committee on Tropical Soils, National Research Council. *Soils of the Humid Tropics*. Washington, D.C.: National Academy of Science, 1972.

Dregne, H. E. *Soils of Arid Regions*. New York: Elsevier Scientific Publishing Company, 1976.

Foth, H. D. and Schafer, J. W. *Soil Geography and Land Use*. New York: John Wiley & Sons, 1980.

Staff, Soil Conservation Service, U.S.D.A. *Soil Taxonomy*. Washington, D.C.: U.S. Government Printing Office, 1975.

13

Concepts of soil productivity and fertility

One of the more critical aspects of soil use and management is the maintenance of the capacity to produce plants that supply man with essential food and fiber. This capacity, referred to as *soil productivity*, is measured in terms of yield and is a function of all factors of plant growth, mainly light, heat, water, and nutrients. Productivity of soils is also affected by negative factors such as disease and insect pests, salts and other toxic substances, and adverse climatic conditions. Maximum growth occurs only where all factors are as near the optimum as possible, and failure of any one of them to measure up to the optimum reduces the effectiveness of all other factors on plant growth.

For the most part, the supply of light, heat, and water for plant growth is a climatic function and, therefore, subject to only limited control by man. True, the water supply can be increased through irrigation, but irrigated land makes up only a relatively minor part of the total utilized in the production of all agricultural, range, and woodland crops. A factor much more amenable to man's control is the nutritional status, or *fertility*, of the soil. Indeed, the ease with which many nutritional problems can be corrected and the spectacular results that often follow such correction have made fertility one of the more readily accepted aspects of soil management. In spite of its widespread popularity, however, the fundamentals of fertility control are not well understood, a fact that has frequently led to the improper use of fertilizers, sometimes to the serious detriment of the productive capacity of the soil.

The nature of soil fertility

Fertility is the potential of a soil to supply nutrient elements in amounts, forms, and proportions required for maximum plant growth. It is measured directly in terms of ions or compounds important to plant nutrition and indirectly in terms of the productive capacity of the soil. However, it cannot be assumed that a fertile soil is also productive, for the fertility status gives no indication of the adequacy of the other essential growth factors, nor does

Table 13.1. The essential plant nutrients by name and by chemical symbol.

Macronutrients		Micronutrients	
Carbon (C)	Potassium (K)	Iron (Fe)	Copper (Cu)
Hydrogen (H)	Calcium (Ca)	Manganese (Mn)	Zinc (Zn)
Oxygen (O)	Magnesium (Mg)	Molybdenum (Mo)	Chlorine (Cl)
Nitrogen (N)	Sulfur (S)	Boron (B)	
Phosphorus (P)			

it provide any basis for judging harmful influences that might have accrued from waterlogging, insects, diseases, and other undesirable features of the environment.

The fundamental components of fertility are the essential nutrients absorbed by plants and utilized for various growth processes (see table 13.1). The nutrients occur in the soil predominantly as constituents of minerals and organic matter and in lesser amounts as exchangeable ions, which are held loosely at the surfaces of clay or organic particles, or as free (soluble) ions in the soil solution. Nutrients in either soluble or exchangeable form are extracted readily when contacted by plant roots and provide the *active* fertility of the soil. Nutrient elements that are not immediately available to plants, such as those in primary and secondary minerals and in semiresistant organic combination, comprise the *potential* fertility of the soil. Often, crop production is highly dependent upon the rate of transfer from the potential to the active forms.

The essential elements

As shown in table 13.1, there are sixteen known essential elements, or nutrients, necessary for the successful growth of plants. They are classified into two groups, *macronutrients* and *micronutrients*, depending on the quantity needed by plants for satisfactory growth. Macronutrients are used in comparatively large amounts, for they comprise the bulk of structural and protoplasmic tissue in plants. Micronutrients are used in small quantities, since they are associated with components or systems that comprise very little of the plant. Micronutrients are important to enzymatic, oxidation-reduction, and similar biochemical reactions in plants.

Most of the essential elements used by plants are derived from the mineral and organic fractions of the soil. Exceptions are carbon, hydrogen, oxygen, and in a few instances, nitrogen. Nitrogen is an exception because it is sometimes derived from the atmosphere by plants associated with root-inhabiting microorganisms that convert inert atmospheric nitrogen (N_2) gas into a plant-usable form. These plants are legumes for the most part, but the phenomenon occurs with nonlegumes as well. Other plants obtain nitrogen from the soil.

The carbon, hydrogen, and oxygen used by plants come from air and water. One need for all three of these elements is in the photosynthetic production of sugars. The carbon and oxygen used in photosynthesis are supplied by atmospheric carbon dioxide, and the hydrogen by water. Oxygen is also necessary for respiration, the process that provides energy to all living organisms. For higher plants, oxygen used in respiration must be in free (O_2)

Table 13.2. Typical ranges in the content of nutrients in plant tissue.[a] (Data from various sources.)

Micronutrients		Macronutrients	
	ppm		%
Molybdenum	1–5	Sulfur	0.1–0.2
Copper	5–15	Phosphorus	0.1–0.4
Boron	5–50	Magnesium	0.1–1.0
Zinc	10–50	Calcium	0.2–3.0
Manganese	15–100	Potassium	1.0–3.0
Iron	30–150	Nitrogen	0.5–5.0
Chlorine	50–200	Hydrogen	5.0–6.0
		Carbon	40 – 45
		Oxygen	45 – 50

[a]Values exceeding these ranges are commonly encountered.

form. Respiration occurs in all parts of the plant, with that taking place in roots being especially important because it is essential to the uptake of nutrients and water.

In addition to the essential nutrients, many other elements are absorbed by plants. The latter elements seem to have no important physiological function after absorption, although sodium has been observed to enhance the growth of some plants. Further, some consider cobalt to be a plant nutrient because it is necessary for the functioning of legume-root organisms that convert atmospheric nitrogen to a form the legumes can use. In certain instances, elements harmful to man and livestock may accumulate in plants. Among these are radioactive isotopes residual from nuclear explosions, selenium, fluorine, molybdenum, and nitrite-nitrogen.

Table 13.2 shows typical ranges of nutrient contents in plant tissues. The concentrations of the micronutrients are given in parts per million (ppm), whereas for the macronutrients, they are percentages.[1] Ranges in concentration are shown, since the nutrient content varies widely depending on the nutrient supply in the soil and on the kind, age, and part of the plant used for the chemical analysis. According to the table, there is a clear distinction between macro- and micronutrients; the maximum value shown for the micronutrients (200 ppm) is only one-fiftieth of that for the lowest macronutrient value (0.1 percent). Particularly noteworthy are the high contents of carbon and oxygen, which, along with lesser amounts of hydrogen and nitrogen, are concentrated in structural and protoplasmic tissue.

Three principal approaches are used to determine if an element is necessary for plants: (1) measurement of growth response when the element is applied to the plant, (2) isolation of an essential plant compound of which

[1]Conventionally, nutrient contents are based on the weights of the nutrients expressed relative to the oven-dry weight of a sample of plant tissue, usually as percentages or parts per million. Percentages state the contents in units per 100 units of dry-tissue weight. Thus, a content of 1 percent corresponds to 1 g of nutrient per 100 g, or 10 g of nutrient per kilogram of plant tissue. Parts per million state nutrient contents in units per 1 million units of oven-dry tissue weight. A concentration of 1 ppm would therefore correspond to 1 mg (milligram) of nutrient per kilogram of dry tissue. Parts per million are used for micronutrients to avoid unwieldy decimal fractions that would result if their concentrations were expressed on a percentage basis.

the element is a part, and (3) measurement of a change in a specific plant process, other than growth, that results when the element is added to or removed from the growth medium. Examples of the latter processes are respiration, photosynthesis, and the absorption of water and nutrients. The essential nature of most nutrients listed has been demonstrated through use of all three approaches, many times over for some of them.

Factors of plant-nutrient supply

Thirteen of the sixteen essential elements needed by land plants must be supplied by the soil. They must be supplied not only in the amounts and proportions necessary to satisfy plant requirements, but also in plant-usable, or *available*, form. Some of the principal factors that control nutrient availability and supply to plants are discussed in the following sections.

A concept of nutrient availability

The availability of a nutrient to plants is a function of its chemical form and its position relative to actively absorbing roots. Nutrients are in *chemically available* form if they are present as either soluble or exchangeable ions; they are *positionally available* when in direct contact with a root surface. Both conditions of availability must be satisfied before a nutrient can be absorbed by plant roots.

Exchangeable ions are included with soluble ions as chemically available forms of nutrients even though nutrient absorption by roots appears to be almost entirely from the soil solution. Exchangeable ions are classed as chemically available because of their continuous and rapid interchange with ions in solution. However, for a given complement of exchangeable ions to be made totally available for plant uptake requires that they be displaced by other ions. For nutrient cations, such as calcium (Ca^{2+}), magnesium (Mg^{2+}), potassium (K^+), and nitrogen in ammonium (NH_4^+) form, the ion most involved in this displacement is H^+. The main source of H^+ ions is acids produced by biological reactions in plant roots or the soil.

A prime and continuing source of soil acids is root and microbial respiration. Through respiration, carbon used for energy by living organisms is released as carbon dioxide (CO_2). This gas, on dissolving in the soil solution, combines with water to form carbonic acid, H_2CO_3:

$$CO_2 + H_2O \rightarrow H_2CO_3 \tag{13.1}$$

On dissociation, the carbonic acid then releases free H^+ ions, along with an equal number of bicarbonate (HCO_3^-) ions:

$$H_2CO_3 \rightleftharpoons H^+ + HCO_3^- \tag{13.2}$$

Additional H^+ is also provided by the dissociation of organic acids secreted in small quantities by both roots and microorganisms, as well as from inorganic acids, such as sulfuric or nitric, produced by the microbial oxidation of sulfur and nitrogen compounds in the soil.

Any nutrient present in a soil in either exchangeable or soluble form is considered to be available to plants. All other nutrients, such as those occurring in mineral or organic solids, are termed *unavailable*. Ordinarily, the

Table 13.3. The principal forms of nutrients utilized by plants.

Element	Cations	Anions
	Macronutrients	
Nitrogen	NH_4^+	NO_3
Calcium	Ca^{2+}	
Magnesium	Mg^{2+}	
Potassium	K^+	
Phosphorus		HPO_4^2 , H_2PO_4
Sulfur		SO_4^2
	Micronutrients	
Copper	Cu^{2+}	
Iron	Fe^{2+}, Fe^3	
Manganese	Mn^{2+}, Mn^{4+}	
Zinc	Zn^{2+}	
Boron		H_3BO_3
Molybdenum		MoO_4^2
Chlorine		Cl

fertility status of soils is judged from the quantity of nutrients present in available form at a particular moment. However, it also depends on the release of nutrients stored in mineral and organic combinations. Thus, fertile soils are capable not only of quickly supplying large quantities of nutrients already in available form but, through release of unavailable nutrients, can meet a sustained demand over comparatively long periods. Conversely, infertile soils can meet a large immediate demand for nutrients only if they are fertilized, and they can satisfy a long-term demand only if they are fertilized repeatedly.

Nutrient forms used by plants

Nutrients are absorbed by plants primarily as ions,[2] the more important of which are listed in table 13.3. Some nutrients are absorbed only as positively charged cations, others only as negatively charged anions. Nitrogen is an exception, for this element occurs in both cationic and anionic forms that are equally utilizable by plants. The catonic form of nitrogen is the ammonium (NH_4^+) ion, whereas the anion is nitrate (NO_3^-).

Nutrients that are cations occur simultaneously in both exchangeable and soluble form, and in general, the greater the quantity present in the soil solution the greater the proportionate concentration in exchangeable form. Nutrients retained in highest concentrations as exchangeable ions are the macronutrients calcium, magnesium, and potassium. This is consistent with the relatively high average concentration of these three nutrients in the soil solution. In contrast, micronutrient cations seldom occur in more than trace amounts in either exchangeable or soluble form. To a degree, the same is true for nitrogen in the ammonium form, unless ammonium fertilizers have been recently applied. Even when added in fertilizer, nitrogen does not per-

[2]There are exceptions to this in that some nutrients can be absorbed when in combined, or molecular, form. For example, urea, $CO(NH_2)_2$, is a compound that does not dissociate into ions, although it dissolves readily in water and can be taken up by plants as a source of nitrogen. Similarly, boron is apparently absorbed primarily as undissociated, or molecular, boric acid (H_3BO_3). Boric acid does not dissociate because of the extremely strong bond between H^+ and BO_3^{3-} ions when in compound form.

sist in the ammonium form for long. Instead, it is usually rather rapidly converted to nitrate by a specialized group of microorganisms common to all soils.

Nutrient anions are much less susceptible to retention in exchangeable form than are cations. Such retention does occur, but mainly in strongly acid soils high in kaolinite or sesquioxide clays. Retention of sulfur as sulfate (SO_4^{2-}) ions is believed to be important to the availability of this element in some soils, but apparently, the adsorption of phosphate, primarily as the $H_2PO_4^-$ ion, is with such tenacity that it is likely to reduce rather than preserve the availability of the phosphorus to plants. Because of these relationships, the availability of anionic nutrients is determined largely by the quantities present in soluble form in soils. Ions in this form may be easily leached from soils of humid regions, with those most susceptible being nitrate, sulfate, and borate.

Soluble anions can accumulate in dry-region soils where leaching is limited. Those that accumulate most often are chloride, sulfate, borate, and nitrate. Of these, the first three are usually derived from naturally occurring salts of moderate to high solubility. Any build-up of nitrate is ordinarily the consequence of soil treatment with fertilizer.

Soil characteristics important to fertility

As implied by the foregoing discussions, the inherent fertility of soils is a function of their mineral and organic make-up. Mineral and organic components determine the quantity of reserve nutrients stored in the soil as well as the rate at which these nutrients are released in available form. Minerals (clays) and organic matter also provide sites for the retention of exchangeable nutrients. Exchange retention is of particular value since it permits the accumulation of nutrients in readily available form while greatly limiting their potential loss through leaching.

Minerals are the principal storage site for most plant nutrients in soils. However, whether the nutrient supply is adequate depends on the kinds and amounts of minerals present. It may not be adequate if the principal nutrient-bearing minerals have been largely lost from the soil by weathering or if they were initially absent from the soil parent material.

Although nutrient deficiencies that relate back to the mineralogy of the parent material are frequently observed, they tend to be of localized occurrence. Deficiencies associated with mineral depletion by weathering are more widespread and often more intense, but even these are primarily limited to humid regions. The fertility of Oxisols and Ultisols in tropical and subtropical regions is the most severely affected by weathering; that of Spodosols and Alfisols of the temperate-forested regions, both less often and less severely. Mollisols and Aridisols of the grassland and desert regions seldom show an inferior nutrient status that can be blamed on weathering. Shortages of mineral-derived nutrients in the latter soils are more often than not a function of the mineralogy of the soil parent material.

Relatively few nutrients are stored in soil organic matter. Even so, this soil component still makes a critical contribution to fertility, for it is often the chief source of some nutrients. For example, organic matter contains 95

Concepts of soil productivity and fertility

percent or more of the total nitrogen present in most soils, and in some, it may contain as much as 60 percent of the total phosphorus and 80 percent of the total sulfur.

Because of the way organic matter accumulates in soil profiles, the build-up of organic nitrogen, phosphorus, and sulfur is greater in the surface than in the subsoil. However, this effect is not limited solely to these three elements; it applies to any nutrient that is absorbed in the subsoil and returned to the surface in plant residues. By means of this nutrient-cycling process, the total concentration of nutrients in the surface soil is increased. Perhaps of equal, if not greater, importance is the fact that the forms in which nutrients accumulate after being recycled to the surface soil permit a comparatively high level of availability to plants. These forms include exchangeable and soluble ions as well as organic and inorganic compounds from which the nutrients can be released with relative ease.

The accumulation of available nutrients in the surface soil is especially beneficial to plants because it concentrates them where maximum root growth and absorption occur. However, for most elements, the gain in surface soil fertility is at the expense of the subsoil. Loss of the surface soil, as by erosion, may therefore expose subsoils that are often quite infertile. They may also be in a poor physical state, which in combination with low fertility, can provide a condition that makes economic plant production most difficult. It is for reasons such as these that conservationists have long encouraged the application of rigid erosion-control practices to our cultivated lands.

Plants and nutrient availability

Plants have two direct effects on nutrient availability, both of which relate to soil-plant root interactions. One has to do with the growth and exploration of the soil by roots as a means of assuring the positional availability of nutrient ions. The second has to do with the capacity of roots to increase the chemical availability of some nutrients. They do so as they encourage nutrient release from combined mineral form or from exchange sites in the soil.

Roots promote the release of nutrients from minerals in two ways. One is by increasing the acidity of the soil and, as a consequence, the solubility of minerals in which the nutrients are stored. The second way in which roots promote nutrient release from minerals is through the absorption of ions from the soil solution, for the lower the concentration of ions in the soil solution the greater the rate at which they are released from weathering minerals.

The transfer of soluble nutrients to plant roots

Soluble nutrients are present in soil water distributed throughout the network of pore spaces in soils. Since roots make direct contact with only a small part of these pores (see fig. 13.1), the transfer of nutrients from more distant pores to root surfaces becomes highly important to their continuous availability to plants. Two mechanisms are largely responsible for the transfer of soluble nutrients to roots. These are *mass flow* and *diffusion*.

Mass flow is the movement of ions with soil water. It occurs when water moves downward in drainage or upward in capillary rise. However, the principal transfer of nutrients by mass flow is in the *transpiration stream*, that

Figure 13.1.
Cross-section of a soybean root in soil. Note how little of the soil in the upper-left third of the photo is in direct contact with the root. (From Tan, K. H. and Napamornbodi, O. *Soil Sci.* 131:100–06. © 1981, Williams & Wilkins, Baltimore.)

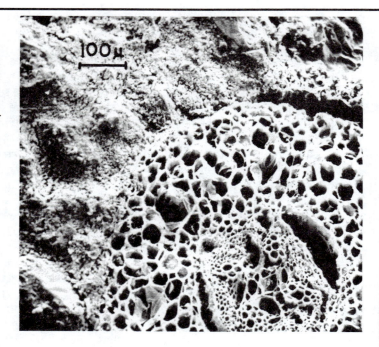

is, in water flowing through the soil and into roots to replace that lost from plant leaves by transpiration. The amount of nutrient moved in mass flow depends on its concentration in the soil solution and the rate of water flow toward roots. If the concentration is low, the amount delivered may be less than can be absorbed. However, if the concentration is high, the amount delivered may exceed the absorptive capacity of the plant, which causes the nutrient to accumulate outside the root (see fig. 13.2).

Diffusion of nutrients is the result of random ionic motion that tends to maintain a uniform concentration of ions throughout the soil solution. In diffusion, average movement is from regions of high to regions of lower ion concentration (see fig. 13.2). Net diffusion is toward roots so long as the absorption of ions keeps their concentration near roots below that in the soil solution more distant from the roots.

As a mechanism of nutrient supply, diffusion is of greatest importance to nutrients that occur in low concentration in the soil solution. Principally, it applies to nutrients derived from slowly weathering minerals and decaying organic matter. As opposed to diffusion, mass flow is the chief means of transfer for nutrients that occur in relatively high concentration in the soil solution. These nutrients come mostly from readily soluble compounds, such as natural salts or fertilizers, or by the displacement of exchangeable ions, especially Ca^{2+}, Mg^{2+}, and even K^+ in potassium-rich soils. Mass flow is also the cause of transfer of nitrate, sulfate, chloride, or borate ions added to soils in fertilizers. Normally, these ions, when added in fertilizer, dissolve completely in the soil solution and are, theoretically, totally transferable to roots by mass flow.

Concepts of soil productivity and fertility

Figure 13.2.
Diagrams illustrating mass flow and diffusion of ions to plant roots. In these examples, mass flow delivers more ions and diffusion less ions than can be absorbed by the roots.

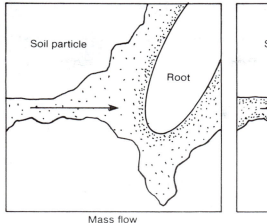

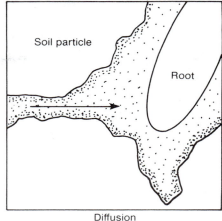

Mass flow   Diffusion

Nutrient requirement and fertility

Whether a soil is fertile depends on its ability to satisfy the *nutrient requirement* of plants. By definition, the nutrient requirement of a plant, when grown under a given set of environmental conditions, is the total quantity of nutrients the plant contains when each nutrient has been supplied in exactly the optimum amount. Since it is essentially impossible to determine when the supply of every nutrient is at the optimum, nutrient requirements are ordinarily judged from measured nutrient contents of plants grown at a high level of productivity. Such is the basis for the values in table 13.4, which shows nutrient utilization by a diverse group of plants. The five nutrients in the table are the ones supplied in greatest amounts through fertilization.

When productivity is high, some nutrients may be supplied and absorbed in greater amounts than is actually required for optimum growth. This reaction to an excess of nutrient is called *luxury consumption*, and it may result from either a high inherent fertility level in a soil or from overfertilization. Luxury consumption caused by overfertilization is undesirable, for the excess nutrient will be carried away in harvested produce, or, if some of the nutrient remains in the soil, it may eventually be lost through leaching. Some nutrients have a toxic effect on plants when present in excessive amounts.

Plant type and the nutrient requirement

Plants differ with respect to both the total amount and the proportions of nutrients they can effectively utilize. As a generalization, plants that produce large yields during a season tend to place the greatest total demand on the soil for nutrients. Recognize, however, that some plants with a comparatively small total requirement may still have a large immediate demand for nutrients during periods of peak growth. For this reason, the rate of nutrient supply is often equally as important a factor of fertility as is the total amount of nutrients a soil can supply during the full growing season.

Table 13.4. Annual uptake of five nutrients, in kilograms per hectare, by selected plants grown at high levels of productivity. (Modified from *Better Crops with Plant Food,* 1972, Vol. 1, and *Ohio Agric. Res. Bull.* 1090, 1976.)

Plant	Yield[a]	N	P	K	Mg	S
				kg/ha		
Corn	12T grain	190	35	45	18	16
	9T stover	78	15	178	38	18
Grain sorghum	9T grain	135	29	28	16	25
	9T stover	145	15	158	34	118
Barley	6T grain	123	20	32	9	11
	— straw	45	8	108	10	11
Oats	6T grain	90	12	19	6	—
	— straw	40	8	116	17	—
Wheat	5T grain	161	21	25	13	6
	7T straw	47	4	125	13	17
Rice	8T grain	86	26	26	9	6
	8T straw	39	7	130	7	8
Soybeans[b]	4T grain	282	25	81	19	13
	8T stalks, leaves, pods	94	8	54	10	15
Peanuts[b]	5T nuts	174	12	36	7	12
	6T vines	120	9	149	24	13
Cotton	1.7T lint, 2.5T seed	105	19	41	12	8
	Stalks, leaves, burrs	96	12	26	27	26
Tobacco (Burley)	4.5T leaves	162	7	139	20	27
	4T stalks, tops, suckers	106	8	106	10	24
Sugar beets	67T roots	140	8	232	30	11
	36T tops	146	12	280	60	39
Sugar cane	224T stalks	180	45	310	45	60
	Leaves, tops	224	33	255	67	35
Potatoes	56T tubers	168	39	245	13	13
	Vines	114	17	84	22	13
Apples	600 boxes (47kg)	22	5	46	2	—
	Blossoms, leaves, new wood	90	19	121	24	—
Grapes	27T fruit	74	11	112	—	—
	Vines	40	6	34	—	—
Peas (fresh)[b]	6T shelled peas	50	4	16	9	—
	Pods, vines	118	8	57	16	—
Tomatoes	90T fruit	160	34	263	11	31
	5T vines	100	10	162	29	29
Bermuda grass[c]	25T	560	60	440	—	—
Johnson grass[c]	27T	1000	94	584	67	56
Orchard grass	13T	335	50	350	—	—
Alfalfa[bc]	8T	500	40	450	—	—
Hardwood forest	—	95	9	30	—	—

[a]Yield per hectare. Metric tons are indicated.
[b]Can obtain nitrogen from the atmosphere.
[c]Yield and nutrient uptake are for warm regions with a long growing season.

Concepts of soil productivity and fertility

Variation in the proportionate amounts of nutrients required by plants often relates to differences in vital plant functions or processes. For example, potassium is critical to carbohydrate production and storage in plants. Thus, plants capable of producing large amounts of carbohydrates, such as sugar or starch, frequently have a comparatively high requirement for potassium. Potatoes, sugar crops, and some grain crops are typical. Similarly, nitrogen and sulfur, as components of protein, are normally needed in greatest amounts by plants of high protein content. Legumes, which are noted for their high protein content, are representative of plants with a relatively large requirement for nitrogen and sulfur.

Environment and the nutrient requirement

An important aspect of plant growth and nutrient requirement is the condition under which growth takes place. Since growth and the demand for nutrients vary under different environmental conditions, the nutrient requirement of any plant species must be expressed by a range rather than by a single value. Nutrient requirements are low where environmental conditions greatly limit growth and are high where all growth factors, including soil fertility, approach the ideal.

Aside from the fertility status of the soil, environmental influences on plant growth relate mainly to soil and climatic conditions that control the amount of heat, light, and water made available to plants. The availability of water, characterized by the total quantity and continuity of supply, is probably the most important single environmental factor responsible for variation in plant growth from one place to another. The length of the growing season is also of great importance, as implied by data in table 13.4.

Yield and quality as determinants of the nutrient requirement

Under most circumstances, the production of economic crop plants is with the intent of obtaining maximum yields. For this, attempts are made to select types of plants and fertility programs that make the most efficient use of existing environmental growth factors. There are instances where fertilization for maximum plant yield is not the most satisfactory approach, however. A case in point is in the fertilization of sugar beets, for the yield of sugar is often greatest where the growth of beet roots is held somewhat below the maximum by restricting the amount of nitrogen made available to the plant.

When well supplied with nitrogen, sugar beets tend to remain vegetative, whereupon photosynthesized sugar is consumed in the continuing production of leaf and root tissue instead of being stored in the root. As a result, if nitrogen is in excess, total plant growth (tops plus roots) increases without a corresponding increase in sugar content.

The effect of variable nitrogen availability on total sugar yield is illustrated in figure 13.3, where, on the left, representative trends in yield and sugar content of sugar beet roots are plotted against the level of available nitrogen in the soil. Multiplying the yield of beets by the percent sugar content and then dividing by 100 give the yield of sugar. Sugar yields computed in this way are plotted against the yield of beet roots in the right-hand diagram of figure 13.3. The results show the greatest return of sugar when the

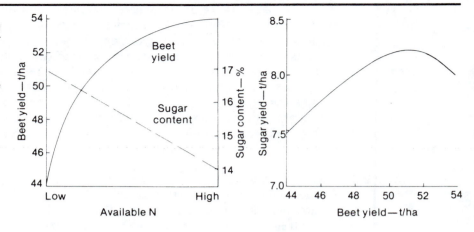

Figure 13.3.
Trends in the yield and sugar content of sugar beet roots and in the total yield of sugar as influenced by variation in the level of available nitrogen in the soil.

yield of beets is between 51 and 52 metric tons per hectare. It follows, therefore, that the amount of nitrogen required for maximum sugar yield is slightly less than that needed to produce the maximum quantity of beet roots.

Nutrient balance and the nutrient requirement

The need for each plant nutrient is dependent upon the availability of all other nutrients in the soil. In general, this relates to the fact that maximum yields result where all growing conditions, including the nutrient supply, are optimal. Optimal nutrition, in turn, suggests that all nutrients should be supplied in exactly the right amounts, since an excess or deficiency of one might reduce the efficiency with which all others are utilized. Such a point of view implies a need for proper *balance* among plant nutrients.

There are circumstances where ionic interactions affect the uptake and utilization of one or more nutrients by plants. For example, plants appear to have a rather fixed capacity for absorbing nutrient bases, but it makes little difference which of the bases are absorbed to satisfy this capacity. Thus, if an overabundance of calcium is supplied, the absorption of potassium and magnesium may be depressed, possibly even to the point of deficiency. An interaction such as this does not alter the total requirement for bases, but it is a partial determinant of whether or not the soil is able to satisfy the requirement for all of them.

Sometimes a nutrient may interact with another ion, often another nutrient, in a way that reduces the efficiency with which it is utilized after it has been absorbed by a plant. For example, a nonfunctional ion, when present in an excessive amount, may displace a nutrient ion from its normal position in a physiological system or process, or it may convert the nutrient to an unusable form. Such interactions usually require an above-normal concentration of the affected nutrient in the plant if proper growth is to occur.

How efficiently a nutrient is utilized after being absorbed is sometimes referred to as *physiological availability*, and an ionic interaction that reduces the physiological availability of a nutrient is called *antagonism*. For the most part, harmful antagonistic effects involve micronutrients rather than macronutrients.

Concepts of soil productivity and fertility

Evaluation of soil fertility

Soil fertility, briefly defined, is the ability of the soil to supply the nutrients required for maximum plant growth. Implicit in this definition are the kinds of measurements that can be used in soil fertility evaluation. One is plant growth, which is expressed by yield. The yield may be of the entire plant or of some harvestable portion, such as grain. Since yield is the normal objective in any soil management program, including fertility control, it serves as the fundamental reference for fertility evaluation.

Under a fixed set of plant and environmental conditions, a positive relationship can be established between yield and fertility, so that one can be inferred from the measurement of the other. If yield is the measure, then a maximum value is indicative of a high level of fertility, and any lesser yield is suggestive of an inadequate nutrient supply. However, merely knowing that yield is less than maximum gives no explanation of why; it could just as easily be due to a non-nutritional as to a nutritional factor. Determination of the specific cause customarily requires a second means of fertility evaluation, chemical analysis.

The evaluation of fertility by chemical means involves either the soil or the plant, but its goal is to measure the nutrient-supplying power of the soil. Whether the soil or the plant is used in the analysis depends on the particular advantages and limitations inherent in either approach. An advantage of soil analysis is that it can be carried out prior to the growing season, which allows nutrient deficiencies to be corrected before they reduce plant growth. Plant analysis, on the other hand, is not possible until plants have become well established, with the result that a revealed nutritional disorder may not be corrected until it is too late to be of maximum benefit. In spite of this limitation, plant analysis is an important adjunct to fertility control, since it provides the only reliable chemical means for judging the availability of some nutrients.

Soil tests and plant tissue tests are widely used as an aid to maintaining soil productivity at its most desirable level. These tests are helpful not only in detecting shortages of essential plant nutrients but also excesses of nutrients or other elements that can be harmful to either the quality or yield of plant produce. Because of their importance, tests of this type need to be given rather detailed consideration. This will be done after we have explored some of the more basic relationships between soil characteristics and nutrient availability (see Chapter 16).

Nutrient availability and plant growth

Nutrient availability-plant growth relationships follow an intricate pattern depending on the type of plant and the conditions under which it is grown. One important factor is whether the plant is an annual, a biennial, or a perennial. Annuals, for example, complete their life cycle in a single season, with the amount of fruit, seed, or other harvestable produce depending on the abundance of vegetative growth earlier in the season. For most annuals, the nutrient supply during the early phases of growth is critical to overall

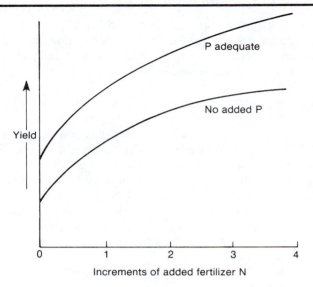

Figure 13.4.
Plant response to nitrogen and phosphorus applied to the soil in fertilizers. The upper curve shows, in effect, the increase in efficiency of fertilizer-nitrogen use where a phosphorus deficiency is corrected through fertilization.

P adequate

No added P

Yield

0   1   2   3   4

Increments of added fertilizer N

production, therefore. This need is widely recognized, and it accounts for the standard practice of applying fertilizers early to annual crops, usually either just prior to or at the same time as planting.

Unlike annuals, biennials have a two-year life cycle, and perennials, an indefinite life span. For these plants, fertilization will not only affect the immediate growth of the plant, but also that in subsequent years. For example, the development of buds and flowers in fruit trees or the formation of asparagus and rhubarb shoots in the very early part of the growing season are strongly influenced by fertility and vegetative growth the previous year.

The way in which plants react to fertilization is usually shown by *response curves*, such as those appearing in figure 13.4. These curves relate the yield of the plant, or of some harvestable part, to different levels of nutrition established by adding increasing amounts of fertilizer to the soil. One such set of curves was given for sugar beets in figure 13.3. The pair of curves in figure 13.4 illustrate the general nature of plant response to increases in the level of two growth-limiting factors; in this instance, available nitrogen and phosphorus.

The lower curve in figure 13.4 shows the effect of adding fertilizer nitrogen alone to a soil that is deficient in both nitrogen and phosphorus. The upper curve also shows plant response to increasing levels of nitrogen, but in this instance, where the phosphorus deficiency has been corrected through fertilization. Obviously, the greater response is obtained where both nutrients are added at the same time. Phosphorus alone causes a relatively small increase in yield, which is expressed by the difference in the two curves where they intersect the vertical axis. The effect of nitrogen alone is, of course, shown by the lower curve, the yield increase here depending on the amount of nitrogen added to the soil.

General considerations in fertility management

The maintenance or improvement of soil productivity requires attention to various aspects of plant growth, but the main emphasis is usually on the fertility status of the soil. The reason for this is the widespread shortage of plant nutrients in soils, as well as the relative ease with which these shortages can ordinarily be overcome. Often, infertility in soils is an inherited characteristic, as with soils that lack certain minerals needed as a source of nutrients. The more likely cause, however, is faulty husbandry that has allowed soil deterioration, particularly the loss of topsoil through erosion. For soils subject to this type of loss, the first step in proper management should be the initiation of a soil-conserving program that will hold further erosion to a minimum. In this regard, conservation of the soil and fertility maintenance become mutually beneficial, for a high level of fertility is essential to providing a good cover of plants as an aid to erosion control, while erosion control, by preserving the topsoil, aids in the maintenance of fertility.

Several things are done to sustain or improve the fertility of soils. Principal among them are the addition of fertilizers and soil amendments, the application of organic supplements, such as animal manures, packing-house and food-processing wastes, or sewage, or the production on the land of green-manure crops grown specifically for later incorporation into the soil (see fig. 13.5). Green-manure crops are beneficial because they return accumulated nutrients to the soil in a form that is quickly released to succeeding crops. The principal limitation to the use of green manures is that they remove land from the production of other high-value crops.

Organic supplements, such as animal manures, food-industry wastes, and sewage, are not used widely in fertility programs because they are not readily available to the vast majority of farms. The problem is the isolation of centers where these materials are generated in large quantities. Transportation costs thus limit their use primarily on land that lies close to the source of supply. As a consequence of the abundance of supply on a local basis, use of these materials becomes as much a matter of waste disposal as of fertility maintenance. Benefits to soil fertility are realized, of course, but only on a very small part of all agricultural land.

In view of the foregoing, most fertility programs are based on the use of chemical fertilizers and amendments capable of increasing the level of available nutrients in the soil. Fertilizers supply the nutrients directly, usually in highly purified compounds that have only a modest effect on the status of exchangeable and soluble ions in the soil. Amendments, which are, in concept, additives capable of improving various plant-growth conditions in soils, may also be direct sources of plant nutrients. Examples are crushed limestone, which usually adds both calcium and magnesium to the soil, agricultural sulfur, which is a nutrient in itself, and animal manure, a source of all nutrients plants ordinarily obtain from the soil. As an amendment, lime is applied primarily to neutralize excess acidity in the soil, and in so doing, it may increase the availability of native phosphorus, molybdenum, and other nutrients to plants. Sulfur is added to soils as an acidifying agent, the purpose

Figure 13.5.
Organic materials
such as barnyard
manure (A) and green
manures (B), when
worked into the soil,
benefit both the
fertility and tilth of
agricultural land.
(U.S.D.A. Soil
Conservation Service
photos.)

A

B

Concepts of soil productivity and fertility

sometimes being to increase the solubility and availability of iron. Animal manures are valued as amendments because they not only add a wide range of nutrients but can greatly benefit the physical state of the soil.

Fertility programs, if well devised, are ordinarily based on various analytical tools available for determining nutrient and other plant needs. Soil and plant-tissue tests, along with visual symptoms of abnormal growth or development, provide the principal means of determining whether a nutrient is adequately supplied by the soil. In addition, these tests may also identify nutrient excesses. Excesses of some nutrients have toxic effects on plants, and if the nutrients occur in mobile form in the soil, they can contaminate groundwater if leached from the profile. Fertility control means more than just the correction of deficiencies, therefore; it means the avoidance of unnecessary, and often harmful, excesses as well.

Nutritional disorders in plants can be dealt with most effectively if the relationship between nutrients on the one hand and the soil and plant on the other are well understood. For example, knowledge of the role of a nutrient in plants provides an understanding of its effect on growth and why it may produce a distinct abnormality, or deficiency symptom, in plants when inadequately supplied by the soil. Knowledge of reactions between a nutrient and the soil helps explain the cause and correction of problems that may result from either an inadequate or an excessive nutrient supply. The nature of the more important of these relationships are discussed in the next few chapters of the text.

Summary

The productivity of soils is a function of all factors of plant growth, both beneficial and harmful. Among the beneficial factors, the one most subject to control is fertility, or the ability of the soil to supply nutrients to plants. Fertility problems usually stem from nutrient shortages, but they can also arise from excesses that result in plant toxicities or a nutrient imbalance.

Sixteen nutrients are known to be essential for plants. Thirteen of these are derived from the soil. The ability of the soil to supply nutrients depends on the quantities present in chemically available (exchangeable + soluble) form and on the rate of release from combined (unavailable) forms in minerals and organic matter. Infertile soils normally lack one or more nutrients in both chemically available and in reserve, or unavailable, form.

The total amount of nutrient needed by plants (nutrient requirement) is difficult to measure, but it is often judged from the nutrient content of plants grown under a high level of productivity. Where the soil is unable to satisfy the nutrient requirement, supplemental nutrients can be added in fertilizers. Care is needed in soil fertilization, however, for excesses of some nutrients can reduce either yield or quality of plant produce. Valuable aids in estimating the nutrient status or fertilizer needs of soils are soil and plant-tissue tests.

Soil productivity can be enhanced through the use of manures and amendments that supply nutrients while promoting more nearly ideal chemical and physical conditions in the soil. Also important is proper crop and residue management designed to limit soil erosion and the loss of fertile topsoil.

Review questions

1. Explain why productive soils are fertile but fertile soils are not necessarily productive.
2. In general, how do the roles or functions of macro- and micronutrients explain differences in the amounts of these two groups of elements required for satisfactory plant growth?
3. Why are nutrients held in exchangeable form considered to be chemically available to plants?
4. Explain why increasing the concentration of a nutrient in the soil solution increases the potential for its delivery to roots by mass flow.
5. The supply of which plant nutrients depends in part on the organic matter content of the soil, and why?
6. Organize the five nutrients in table 13.4 in an order of average decreasing total uptake by plants.
7. What is luxury consumption, and why is it undesirable?
8. Why is fertility management sometimes tied to the quality of plant produce rather than to total yield?
9. What is meant by physiological nutrient availability?
10. Explain the benefits of green-manuring, the use of organic wastes as soil supplements, and the prevention of soil erosion to soil fertility.

Selected references

Bear, F. E. *Soils and Fertilizers.* New York: John Wiley and Sons, Inc., 1953.
Tisdale, S. L., Nelson, W. L., and Beaton, J. D. *Soil Fertility and Fertilizers.* New York: The Macmillan Company, 1984.

14

Nutrient relationships in soils and plants

This chapter discusses the 13 nutrients plants derived from the soil; it considers their behavior in the soil and their principal roles, or functions, in plants. The behavior of nutrients in the soil determines their availability to plants, not only from naturally occurring minerals and organic matter but also from fertilizers and other materials added to the soil. A discussion of the role of nutrients in plants provides a clue to their need as well as to how they affect processes of plant growth and maturation. The more that is known about plant-soil-nutrient relationships the greater the likelihood of success in the diagnosis and correction of nutritional disorders in plants.

An advantage in the diagnosis of nutrient problems in plants is that they are more often than not caused by only three elements: nitrogen, phosphorus, and potassium, either alone or in combination. These three nutrients are added to soils far more often and in considerably greater amounts in fertilizers than are any of the other nutrients. This fact accounts for the occasional reference to nitrogen, phosphorus, and potassium as the *fertilizer nutrients*. Of the three, nitrogen is the one most often found deficient. It is also needed in greater amounts by plants than is either phosphorus or potassium. The nutrient first to be discussed in the sections that follow is nitrogen, therefore. Sulfur is considered next because of its close relationship to nitrogen, both in plants and in the soil.

Nitrogen

By and large, current levels of nitrogen in soils reflect its accumulation in the organic fraction over long periods of time. As with the build-up of organic matter, soil nitrogen accumulation tends toward an equilibrium wherein additions made each year are essentially in balance with losses, these losses resulting principally from leaching, erosion, and the escape of nitrogen in gaseous form back into the atmosphere. Soils of highest total-nitrogen content are also highest in organic matter; they occur where conditions favor the addition of large amounts of plant residues while limiting residue decay.

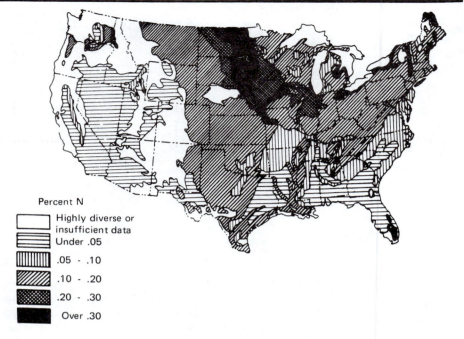

Percent N

☐	Highly diverse or insufficient data
▤	Under .05
▥	.05 - .10
▨	.10 - .20
▦	.20 - .30
■	Over .30

Except for swampy areas where organic soils accumulate, these conditions are most prevalent in the cooler, wetter grassland regions, as in the north-central part of the continental United States (see fig. 14.1).[1]

The nitrogen in soils came originally from the atmosphere; the build-up has resulted from the conversion of atmospheric nitrogen, or N_2 gas, to forms capable of being both used by plants and accumulating in the soil. The process responsible for the conversion is termed *nitrogen fixation*, and it goes on continuously. Some fixation takes place in the atmosphere, either through photochemical reactions or by lightning discharge, with the products being carried to the earth in precipitation. However, much more extensive natural fixation of nitrogen is implemented by various soil- and root-inhabiting microorganisms. These organisms release nitrogen as ammonia, as simple amino compounds, or on death, as microbial protein. Regardless of how the nitrogen is fixed, it ultimately finds its way into the organic fraction of soils, where it is retained in semistable organic combination.

Nitrogen uptake and utilization by plants

Nitrogen is absorbed from the soil and utilized by plants as either the ammonium or nitrate ion. Both of these ions occur naturally in soils and are common components of inorganic nitrogen fertilizers. Ammonium-nitrogen in soils is

[1]The percent-nitrogen content of organic matter in different soils is fairly uniform, so the amount present in soils can be estimated from their organic matter content, and vice versa. The average nitrogen content of soil organic matter is usually about 5 percent, although it may be somewhat less in soils of forested regions. Using the 5–percent value, an estimate of the organic matter content in soils of the continental United States can be obtained by multiplying the percent-nitrogen values in figure 14.1 by 20.

Figure 14.2.
Chlorosis due to nitrogen deficiency results in an overall yellowing of leaf tissue, as suggested by the two light-colored leaves in the photo. A normal leaf is at the left. The tip of the leaf at the extreme right has died because of extreme nitrogen shortage.

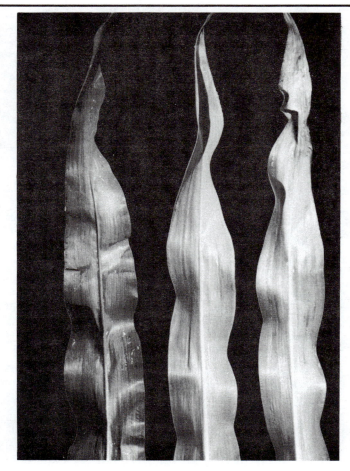

retained mostly in the immobile, exchangeable form, and to become positionally available, it must be sought out by plant roots. Nitrate is mobile in the soil, so it moves rather easily to roots, either with or through soil water. Because of its mobile nature, and because it is usually more abundant than ammonium, nitrate is the principal form in which nitrogen is taken up by plants.

Nitrogen in plants occurs primarily in protein, the center of life processes. It is also contained in chlorophyll, the green pigment responsible for the photosynthetic production of sugar. Where nitrogen is in short supply, the synthesis of both protein and chlorophyll is limited. This results in reduced plant growth and in *chlorosis* (chlorophyll deficiency), the latter being recognized by a yellowing of leaf tissue (see fig. 14.2). Since nitrogen, even when deficient, tends to move to newly forming tissue, chlorosis appears first in the older leaves of plants. If the deficiency is severe, the entire plant may turn yellow and the older leaves may die. A shortage of nitrogen is not so limiting to the formation of structural cellulose and lignin as it is to the for-

Figure 14.3.
The accumulation of nitrogen in corn as relates to height and age of the plant. The dates indicated by T and S identify the initiation of tasseling and silking. The data are from an experiment in Ohio. (After J. D. Sayre, *Plant Physiol.* 23:267–81, 1948.)

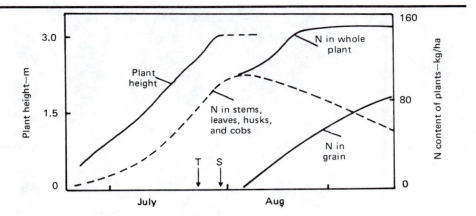

mation of protein. As a consequence, tissue that is normally soft and succulent may be stiff and woody when a nitrogen deficiency occurs.

As opposed to a shortage, nitrogen excesses favor the production of protoplasmic (proteinaceous) over structural tissue. Plants with excess nitrogen may grow rapidly, but they tend to be weak-stemmed and may lodge, or fall over, easily. By promoting vegetative growth, nitrogen excesses also delay maturity. Thus, plants may come up at harvest time in an overly succulent condition; sugar crops and fruit may be low in stored sugar, and the accumulation of starch in potatoes and other root crops or in the seed of grain crops may be below normal. This problem is most serious where the growing season is short.

Nitrogen is absorbed in relatively large amounts by plants during earlier stages of lush vegetative growth, but the uptake falls off sharply as maturity is approached. As maturity approaches, much of the nitrogen in stems and leaves of the plant is transferred to seed and fruit (see fig. 14.3). This causes the nitrogen content of the vegetative tissue to decline as the growing season progresses. In young vegetative plant parts, nitrogen may make up as much as 5 percent of the dry-tissue weight, but it may fall to a half percent or less as this tissue matures. Woody tissue, at maturity, is particularly low in nitrogen. In contrast, legumes, such as alfalfa and clover, may contain 2 to 3 percent nitrogen at maturity. These plants receive fixed nitrogen supplied by root-inhabiting bacteria, and are used widely as animal forage because of their high nitrogen (protein) content.

Transformations affecting soil-nitrogen availability

Nitrogen in humus is said to occur in a semistable form because it is gradually released through mineralization. In mineralization, organic nitrogen is transformed to inorganic ammonium ions. Once released, however, these ions are usually rather quickly converted to nitrate by a process called *nitrification*. Both mineralization and nitrification affect the availability of nitrogen to plants. Through mineralization, unavailable organic nitrogen is changed to the available, though relatively immobile, ammonium ions; nitrification transforms the immobile ammonium ion to mobile nitrate.

Mineralization of organic nitrogen is carried out by common decay organisms seeking carbon for energy and growth. Since humus lacks sugars, starch, and other easily degraded sources of carbon, this nutrient must be obtained from more complex compounds, including proteinaceous substances that comprise a substantial part of the humus. When these compounds are attacked just for the carbon they contain, the nitrogen present is released as a by-product and appears in the soil solution as ammonium ions. The ammonium can be taken up by plants, or it may be adsorbed to the exchange complex of the soil. However, under most soil conditions, ammonium is quickly converted to nitrate through nitrification.

Nitrification is carried out in a stepwise manner by two microorganisms, *Nitrosomonas* and *Nitrobacter*, each obtaining energy in the process. In the first step, *Nitrosomonas* oxidizes ammonium ions to nitrite (NO_2^-):

$$2NH_4^+ + 3O_2 \rightarrow 2NO_2^- + 4H^+ + 2H_2O + \text{Energy} \qquad (14.1)$$

In the second, nitrite is oxidized to nitrate by the *Nitrobacter*:

$$2NO_2^- + O_2 \rightarrow 2NO_3^- + \text{Energy} \qquad (14.2)$$

Under favorable conditions, as much as half of the ammonium-nitrogen present in the soil at any one time may be transformed to nitrate in one-week's time. At this rate, essentially all the ammonium will convert to nitrate in about one month (see fig. 14.4). Because of the speed of this reaction, it is usually safe to assume that inorganic nitrogen in a soil is or soon will be nitrate and, therefore occurs in a relatively mobile form.

Rapid nitrification requires good aeration, adequate moisture, a pH between about 5 and 8, and a temperature between 25 and 40°C. Nitrification slows as the temperature drops below 25°C, and all but stops as freezing is approached. For this reason, ammonium fertilizer applied to cold soils in the fall may not undergo much nitrification until the following spring. This may be of advantage, since the nitrogen held as exchangeable ammonium will not be subject to loss by winter leaching.

Nitrification proceeds very slowly, if at all, in poorly aerated, water-logged soils. It is also impeded by strong soil acidity, say below a pH of about 4.5. Since nitrification produces 2 H^+ ions for each NH_4^+ oxidized [see equation (14.1)], the process tends to be self-limiting because of acid production. For soils of low buffering capacity, repeated and heavy applications of ammonium fertilizers can cause a marked increase in soil acidity.

Nitrogen fixation

Nitrogen fixation, or the conversion of gaseous nitrogen to a plant-usable form, is carried out in several ways: (1) symbiotically, that is, by organisms that live in a mutually beneficial relationship with plants, mainly witih legumes, (2) nonsymbiotically, or by free-living organisms in the soil, (3) through lightning discharge or photochemical reactions in the atmosphere, and (4) synthetically, as in the production of commercial fertilizer compounds. Of these processes, symbiotic nitrogen fixation or that associated with fertilizer production are of greatest immediate importance to agricultural crops. The quantities of nitrogen fixed in the atmosphere or

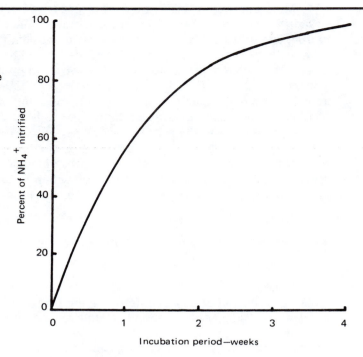

Figure 14.4.
The conversion of NH_4-N to NO_3-N in a loam soil under laboratory conditions. Nitrogen applied at the rate of 200 kg per 2 million kg of soil (After F. E. Broadbent et al., *Hilgardia* 27:247–67, 1957.)

nonsymbiotically in the soil are generally, though not always, too small to be of much immediate significance to plants. They are nonetheless highly important to the long-term accumulation of nitrogen in the organic fraction of soils.

Nonsymbiotic nitrogen fixation

Nonsymbiotic, or free-fixation, of nitrogen is carried out by numerous organisms, but two that have received the most attention are *Azotobacter*, an aerobe, and *Clostridium*, an anaerobe. Both are heterotropic bacteria that obtain energy by oxidizing organic carbon in the soil. When functioning strictly as independent organisms in the soil, neither fixes a large amount of nitrogen, probably no more than a few kilograms per hectare per year, on the average. The limitation is attributed to the lack of abundant, readily available carbon in most soils. However, certain free-fixing organisms seem able to establish a close relationship with the roots of selected plant species that increases the amount of nitrogen fixed substantially. In these instances, the microbes obtain energy from organic substances exuded by the roots. Such relationships have been observed on a number of important crop plants, including corn, rice, millet, sorghum, and various other grasses.

Another group of organisms capable of nonsymbiotic nitrogen fixation is the blue-green algae. These organisms supply their own energy carbon through photosynthesis. Thus, they function most satisfactorily in the presence of sunlight, as at the very surface of the soil or in ponded water. Because of their need for moisture and sunlight, blue-green algae will fix only limited

Nutrient relationships in soils and plants

amounts of nitrogen if isolated at the surface of a soil that is either dry or excessively shaded by higher plants. Greater amounts of nitrogen are fixed where the soil remains moist at the surface and the penetration of light is good.

Blue-green algae appear to have an important impact on the nitrogen regime of paddy rice, which is grown in diked, flooded fields. In the warmer climates where this crop is commonly produced, paddies often are covered with thick, floating algal mats consisting principally of blue-green algae. As the algae die and decompose, they contribute substantial amounts of fixed nitrogen in the soil. In some places in the world, this source of nitrogen has apparently sustained good rice production for centuries. In many cases, however, blue-green algae are not able to supply much nitrogen because their population is kept low by other water-inhabiting organisms, particularly protozoa, that use them as food.

Symbiotic nitrogen fixation in legumes

The fixation of nitrogen in association with legumes is of great significance to agriculture. These plants include forages, such as alfalfa, clover, and legume grains, as exemplified by soybeans, field beans, and peas. Each of these plants has long been used as a source of high-protein food for man and animals. Legumes, especially the forages, have also been used extensively as green manures, mainly because of their ability to collect nitrogen for return to the soil.

A single genus of bacteria, *Rhizobium*, is responsible for nitrogen fixation in legumes. These organisms invade young roots from the soil and, upon entry, cause the formation of swollen structures, called *root nodules*, in which nitrogen fixation takes place (see fig. 14.5). The *Rhizobium* obtains energy compounds from the plant and, in turn, provide the plant with nitrogen.

The amount of nitrogen fixed by legumes can be appreciable, but it varies depending on the type of plant, the length of its growing period, and the conditions under which it is grown. Perennial legumes regularly fix the most nitrogen. Annuals, mainly those with a short growing season, fix the least. Typical ranges of nitrogen fixed on an annual basis are:

Legume	Nitrogen fixed kg/ha/yr
Alfalfa, Ladino clover	150–250
Sweet, red, and white clover, Kudzu	100–150
Peas, soybeans, vetch, lespedeza	50–100
Peanuts, field beans	Less than 50

Actual amounts of nitrogen fixed may exceed these values. For example, in warm regions where long growing seasons permit abundant and continuous plant growth, the perennial alfalfa has been observed to fix 500 or more kilograms of nitrogen per hectare per year.

There are a number of species of *Rhizobium*, but each is compatible with only one, or at most, only a very few different species of legumes. However, even incompatible *Rhizobia* can invade legume roots and cause no-

Figure 14.5.
Well-nodulated root of
a pea plant. (Photo
courtesy of the
Nitragin Co.,
Milwaukee, Wis.)

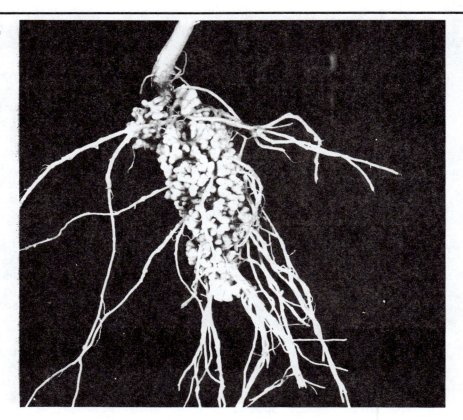

dulation, but they fix little, if any, nitrogen. A test for compatibility is the internal color of the nodules. Nodules that are active in nitrogen fixation show a red color when cut open; inactive nodules do not. The red color is caused by an iron-bearing compound, leghemoglobin, which controls the supply of oxygen to the *Rhizobia*.

To assure the presence of the correct nitrogen-fixing organisms in legumes, seed should be treated, or *inoculated*, with a specially prepared *Rhizobium* culture, preferably just prior to planting. The culture material is commercially produced and widely available. Preinoculated seed is also marketed, but the inoculum is sometimes ineffective because too little adheres to the seed or its viability has been lost during prolonged storage. Once established, *Rhizobium* persists in a soil for long periods, even in the absence of the proper host plant.

The amount of nitrogen fixed by legumes depends on the amount of available nitrogen in the soil, for any fixing organism will use available nitrogen in preference to obtaining it from the atmosphere. The general nature of this relationship is illustrated in figure 14.6, which shows the proportionate quantities of nitrogen in alfalfa obtained respectively from the soil, from added fertilizer, and from fixation. In the experiment to which these data apply, about 55 percent of the nitrogen in the alfalfa came from the soil

Nutrient relationships in soils and plants

Figure 14.6.
The effect of fertilization with nitrogen on the fixation of nitrogen by alfalfa. (Reproduced from Johnson, J. W. et al., *Jour. Environ. Qual.* 4:303–06, 1975. By permission of the American Society of Agronomy, Crop Science Society of America, and Soil Science Society of America.)

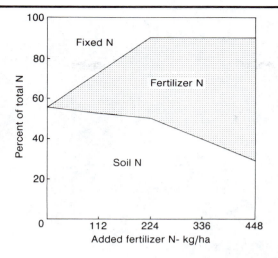

and about 45 percent from fixation where no fertilizer was added to the soil. As the amount of added fertilizer nitrogen was increased, however, the amount derived from the soil or through fixation decreased.

Symbiotic nitrogen fixation in nonlegumes

It has been long known that various nonlegumes have a tightly knit, mutually beneficial relationship with nitrogen-fixing microorganisms. None of the nonlegumes are associated with *Rhizobium*, however, but rather with two other groups, namely: actinomycetes and blue-green algae. Fixation by actinomycetes is in root nodules on various shrubs and trees. Blue-green algae occur in stem and leaf structures of smaller, non-woody plants.

Probably, the most widely distributed woody plant that fixes nitrogen is the alder, a member of the birch family (see fig. 14.7). These trees are early invaders during forest regeneration and, through nitrogen fixation and accumulation, are highly beneficial to later generations of shrubs and trees that do not fix nitrogen. Many shrubs are equally as efficient as alder in fixing nitrogen.[2] Estimates of annual nitrogen fixation by these plants, when occurring as a complete ground cover, are frequently from 100 to 200 kg/ha, but may run as high as 300 kg/ha or more.

Whereas blue-green algae can photosynthesize sugar for energy, they depend to a large part on the host plant for energy compounds when in a symbiotic relationship. In these associations, the algae are usually located in above-ground plant parts. For example, in the tropical plant, *Gunnera* (see fig. 14.8a), they occur in special glands located in the stems. In *Azolla*, a diminutive, floating fern, the algae are found in pores on the underside of leaves. Because of its ability to grow rapidly in water (see fig. 14.8b), *Azolla* has been extensively used in the culture of paddy rice as a source of ni-

[2]In addition to alder (genus, *Alnus*), genera of woody species that fix nitrogen include *Casuarina, Coriaria, Dryas,* and *Hippophae.* For a discussion of these and other plants, see Sylvester, W. B., 1974, *Proceedings of the 1st International Symposium on Nitrogen Fixation.* (Washington State University, Pullman, WA 99164).

Figure 14.7.
A stand of Alder, a tree capable of fixing atmospheric nitrogen, in Cascade Head Experimental Forest, Oregon. (U.S. Forest Service photo.)

trogen. The plant is allowed to grow either before or after planting the rice, and is then intentionally killed by temporarily draining the paddy. Nitrogen yields of up to 300 kg/ha/yr have been reported for *Azolla*.

Nitrogen losses from soils

The nitrogen in soils is subject to continual loss, the principal causes being: (1) removal in harvested plant material, (2) erosion, (3) leaching, and (4) volatilization, or loss to the atmosphere in gaseous form. The removal of nitrogen in harvested produce is, of course, unavoidable, although it can be held to a minimum if as much residue as possible is returned to the soil. Losses caused by volatilization, erosion, and leaching constitute a serious, and often avoidable, waste of nitrogen.

Volatilization loss of nitrogen

The nitrogen in either of the two most important inorganic forms, ammonium and nitrate, can be lost from the soil as gases. The loss from nitrate is caused by common anaerobic or facultative heterotrophs that use nitrate as a source of oxygen. The process is termed *denitrification*. In denitrification, oxygen is either partially or completely removed from the NO_3^- ion, which

Nutrient relationships in soils and plants

Figure 14.8.
Plants that fix atmospheric nitrogen in association with blue-green algae. (A) Gunnera, a broadleafed tropical plant, and (B) Azolla, a floating fern found commonly in rice paddies. Insets show a close-up of Azolla and of a leaf cavity lined with chains of spheroidal algal cells. (Photo (A), J. H. Becking, Institute for Atomic Science in Agriculture, Wageningen, The Netherlands; (B), T. H. Lumpkin, University of Hawaii, Honolulu.)

A

B

yields one of three gases capable of escaping readily from the soil: nitric oxide (NO), nitrous oxide (N_2O), and molecular nitrogen (N_2). Denitrification occurs most readily in poorly aerated soils and is most serious where nitrogen has been added to the soil in nitrate fertilizers.

The loss of nitrogen in ammonium form results from transformation of ammonium ions to ammonia (NH_3) gas, primarily in soils where the pH and the concentration of OH^- ions are high. The reaction involves two steps. In the first, the ammonium ions combine with hydroxyl ions to form ammonium hydroxide:

$$NH_4^+ + OH^- \rightarrow NH_4OH \tag{14.3}$$

In the second, the ammonium hydroxide decomposes to water and ammonia gas;

$$NH_4OH \rightarrow H_2O + NH_3\uparrow \tag{14.4}$$

So long as the NH_3 continues to escape from the soil so will the preceding steps in the reaction. Theoretically, all of the ammonium ions in a soil, even those held in exchangeable form, can be lost under the right conditions. Fortunately, there are restraints that prevent this. For example, the transformation of ammonium to gaseous NH_3 is not likely to occur in acid soils where a low OH^--ion concentration limits the formation of NH_4OH. Another is nitrification that converts ammonium to nonvolatile nitrate ions.

Nitrogen loss through erosion and leaching

Among the several ways in which nitrogen is lost from the soil, those of greatest consequence are leaching and erosion. Erosion is especially serious because it carries away organic matter that is a major source of soil nitrogen. This problem plagues those areas lacking a protective plant cover and where the rainfall tends to be torrential. Maintenance of a continuous cover, especially of close-growing plants, is the best safeguard against erosional losses.

Leaching losses of nitrogen, which are common where excess water percolates through the soil, involve nitrate-nitrogen for the most part. So long as nitrogen remains as ammonium and is retained by exchange adsorption, its leaching is largely avoided. Yet, where the potential for rapid nitrification exists, the presence of nitrogen in the ammonium form is not a guarantee against eventual loss by leaching. The nitrogen in soil organic matter is much less subject to leaching, for it seldom converts to nitrate at a faster rate than it can be absorbed by plants. A plant cover is therefore a means of limiting nitrogen loss either by erosion or by leaching.

Microbial competition for nitrogen

Nonlegume plants, as typified by corn and small grains, often suffer reduced growth where they must compete with common decay organisms capable of absorbing and immobilizing available nitrogen in the soil. This problem arises primarily where residues of wide C/N ratio are incorporated into the soil just prior to seeding the nonlegume. Residues of this type often contain an abundance of readily oxidizable carbon in relation to nitrogen. In utilizing this carbon, the decay organisms tie up available soil nitrogen and thereby limit the amount plants can obtain. Under these conditions, suitable growth

Nutrient relationships in soils and plants

of plants can be expected only if fertilizer nitrogen is applied in an amount adequate to meet the needs of both the plant and the decay organisms.

Added residues are not the sole cause of the immobilization of soil nitrogen by decay organisms. For example, soluble fertilizer nitrogen applied to a well-established stand of grass tends to disappear more rapidly than can be accounted for by plant absorption alone. A possible contributor to the disappearance is the myriad of organisms active in the decay of organic root excretions or dead tissue left in the soil by the roots themselves. The fact that pasture and grassland vegetation sometimes seems to require excessive amounts of nitrogen fertilizer for maximum production may be explained by this complex interrelationship involving both roots and microorganisms in the soil.

Ammonium and potassium entrapment

Ammonium and potassium ions are the same, or nearly so, in valence and size. For this reason, both behave similarly in their reaction with the clay mineral vermiculite. When their concentration in vermiculitic soils is raised sharply, as by fertilization, these ions become entrapped in the interlayer space of the vermiculite. Entrapment is accompanied by the collapse, or contraction, of the vermiculite from an expanded to a nonexpanded state and the transformation of the ammonium and potassium ions to a nonexchangeable form. This process is termed ammonium or potassium entrapment (sometimes *fixation*) and can materially reduce the availability of added fertilizer ammonium or potassium to plants. Entrapment is not permanent, however, for with depletion of these ions in the external soil solution, those in entrapped form are slowly released and again made available to plants. The significance of this is indicated later in a consideration of the release of nonexchangeable potassium from soils (see page 371).

The nitrogen cycle

Nitrogen in the earth environs undergoes a cyclic existence as illustrated in figure 14.9. The cycle is multifaceted, but it follows two pathways for the most part. One involves the interchange of gaseous forms of nitrogen between the soil and the atmosphere, with the loss from the soil being due to volatilization, and the return to the soil resulting from fixation. Fixation may occur in the atmosphere, with the fixed nitrogen being carried to the soil in precipitation, it may be by biological reactions in the soil, or it may be in the preparation of fertilizers for application to the land.

The second pathway followed in nitrogen cycling involves biochemical transformations that convert inorganic forms of nitrogen to immobilized organic forms in plants, animals, or soil microorganisms followed by the eventual return of the nitrogen to an inorganic-nitrogen pool through mineralization and nitrification. On a long-term basis, neither the transformation of the latter types nor the exchange of nitrogen between the soil and the outer atmosphere have a siginificant effect on the average level of total nitrogen in the soil. Of greater consequence to the nitrogen level is the removal of this element through leaching, soil erosion, or the harvest of plant and animal produce. Replacement of the nitrogen lost by these means is one of the main functions of adding fertilizers or other sources of supplemental nitrogen to the soil.

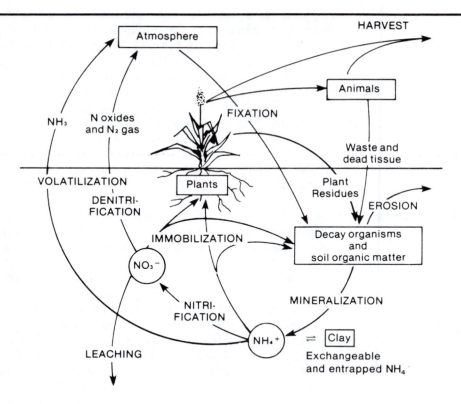

Figure 14.9.
The principal features of the nitrogen cycle.

Sulfur

Sulfur is much like nitrogen in a number of its basic chemical properties and behavioral characteristics. Both of these elements are essential components of plant and animal protein, and both accumulate in proteinaceous substances in soil organic matter. Each is released from organic combination by paralleling mineralization reactions, and both occur as stable, mobile anions, nitrate and sulfate, in the soil solution. One important difference between these two nutrients is that sulfur has no cationic form corresponding to the ammonium ion. Another difference relates to their respective sources in soils; nitrogen has been derived totally from the atmosphere through fixation, whereas the sole source of sulfur is minerals. Over long periods of time, therefore, the average content of nitrogen in soils has increased by accumulation from the atmosphere. At the same time, the average content of sulfur has decreased as a result of mineral weathering and loss through leaching.

Sources of soil sulfur

Most sulfur in soils has been supplied from a group of sulfide minerals distributed widely in rocks. The principal mineral is iron sulfide, FeS_2, commonly referred to as *pyrite*. Sulfur is released from sulfide minerals through oxidation by an autotrophic bacterium, *Thiobacillus*. The reaction supplies the *Thiobacillus* with energy and releases the sulfur as sulfate, SO_4^{2-}. It also produces two H^+ ions for each sulfur atom oxidized. Sulfur oxidation therefore has an acidifying effect on the weathering medium. The reaction is:

$$FeS_2 + 3O_2 + 2H_2O \rightarrow 4H^+ + 2SO_4^{2-} + Fe^{2+} + Energy \quad (14.5)$$

Sulfate occurs generally as a soluble, negatively charged anion. As such, it is repelled by negatively charged clays and organic matter in most soils. An exception is in weathered soils high in kaolinite and hydrous oxide clays that are capable of developing sites of positive charge under strongly acid conditions (see page 156). In these soils, some sulfate appears to be held in adsorbed form, particularly in subsurface horizons of high clay content. In other soils, sulfate-sulfur remains mobile and can be easily leached from the soil if it is not absorbed by plants. The high potential for leaching accounts for the disappearance of much of the sulfur from soils and subsequent transport to ocean water, where its concentration is about half that of the chloride ion. In extensively weathered and leached soils, the greater part of the total sulfur present is in the organic fraction.

An important source of sulfur in some soils is gypsum, $CaSO_2 \cdot 2H_2O$. Accumulation of this moderately soluble mineral is usually the result of evaporation of impounded water containing dissolved calcium and sulfate ions. Gypsum tends to remain in arid-region soils because of limited leaching, although it may be concentrated in the subsoil roughly at the depth of maximum water penetration from natural precipitation. Gypsum is readily dissolved and leached from soils of the humid regions.

Atmospheric sulfur dioxide, SO_2, is a continuing source of sulfur in soils. This gas comes largely from high-sulfur fuels, either coal or oil, which give off sulfur dioxide on burning in air:

$$S + O_2 \rightarrow SO_2 \quad (14.6)$$

When dissolved in water, sulfur dioxide forms sulfurous acid, H_2SO_3:

$$SO_2 + H_2O \rightarrow H_2SO_3 \quad (14.7)$$

When this reaction takes place in the atmosphere, the sulfurous acid is eventually carried to the ground as a causitive agent of acid rain. On entering the soil, the SO_3^{2-} in sulfurous acid is oxidized to sulfate, with sulfuric acid, H_2SO_4, being the final product.

Sulfur dioxide gas can also be absorbed by plants and is toxic in high concentrations. In some industrialized areas, excess sulfur dioxide produced by burning high-sulfur fuels or smelting sulfide ores has caused serious damage to both plants and soils (see fig. 14.10).

Storage and release of organic sulfur

Sulfur absorbed by plants is returned to the soil in residues and ultimately incorporated into the soil organic fraction. So long as the sulfur remains in this form, it is neither available to plants nor subject to loss by leaching. However, on mineralization, organic sulfur is released as soluble sulfate ions and can then be taken up by the plant or leached.

Figure 14.10.
A Tennessee landscape, once covered with lush vegetation, rendered almost completely barren by sulfur dioxide fumes from a nearby copper smelter. (U.S. Geological Survey photo by A. Keith.)

The amount of sulfur stored in soil organic matter is considerably less than the amount of nitrogen, the content of the latter being some 6 to 10 times that of sulfur. Plants require roughly 10 to 15 times more nitrogen than sulfur, however. As a result, sulfur released from decaying soil organic matter comes nearer to meeting plant needs than does the nitrogen released. In unfertilized soils, therefore, the nitrogen released from soil organic matter, rather than sulfur, is more likely to limit plant growth. Further, the supply of available sulfur can be supplemented from mineral sources. However, stimulation of plant growth with nitrogen fertilizer may so increase the need for sulfur that it becomes limiting to plant growth. Under some conditions, the addition of nitrogen may cause an increase in plant growth only if sulfur is added as well (see fig. 14.11).

Under some conditions of low sulfur supply, particularly in the Pacific Northwest, treatment of soils with nitrogen fertilizer not only fails to bring about a positive growth response but may even result in a reduction in yield. Such an effect on one crop, wheat, is shown by the lower curve in figure 14.12. According to the figure, yields are progressively decreased with increasing rates of applied fertilizer nitrogen when no sulfur is added. Apparently, the reduction in yield results from the immobilization of available soil sulfur by decay organisms when they are stimulated by added nitrogen. If sulfur is also applied, a significant response to fertilizer nitrogen results.

Nutrient relationships in soils and plants

Figure 14.11.
Response of corn to fertilization of a soil deficient in both nitrogen and sulfur. Benefits from added nitrogen are obtained only if the soil is also treated with sulfur. (Photo courtesy of Dr. W. E. Martin, University of California.)

Figure 14.12.
The response of wheat to nitrogen fertilization with and without added sulfur. Data are from eastern Washington. (From H. V. Jordan, and H. M. Reisenauer, in *Soil, 1957 Yearbook of Agriculture,* pp. 107–11. Washington, D.C.: Government Printing Office.)

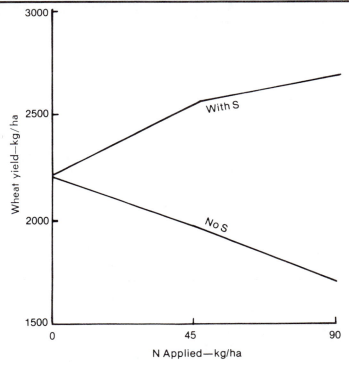

Sulfur and the plant

The relative availability of sulfur to plants usually depends on the amount present as free sulfate in the soil solution. Sulfate is the principal form for both absorption and translocation in the plant. When sulfur is in abundance, half of that in the plant may be as the sulfate ion.

Sulfur is a part of several plant components. It occurs in the amino acids methionine, cystine, and cysteine, in certain essential[3] oils, and in some vitamins. Sulfur also plays a part in chlorophyll formation and in the production of oils in seed crops such as soybeans and flax. If the supply of sulfur is limited, the synthesis of sulfur-bearing amino acids and protein is restricted, as is the formation of chlorophyll. Further, when sulfur is deficient, it is immobilized in protein in older tissue. Signs of sulfur deficiency therefore appear first in newly forming leaves, as shown for pears in figure 14.13. These symptoms include a general chlorosis of the leaves due to a lack of chlorophyll and a reduction in growth caused by limited protein synthesis.

Where a shortage of sulfur limits protein synthesis, nitrogen absorbed by the plant may accumulate as free ammonium or nitrate ions. The accumulation of nitrate seems to do no harm, but excesses of ammonium can be toxic. The problem is easily corrected by increasing the supply of available sulfur to the plant.

Sometimes, sulfur deficiencies that appear early in crop plants clear up later in the season. This happens where more available sulfur occurs in

[3]From the word essence. These oils provide the distinctive odor and flavor of onions, mustard seed, and members of the cabbage family.

Figure 14.14.
Barren coal-mine spoil from a high-sulfur coal deposit in Indiana County, Pennsylvania. The lack of vegetation, due to extreme acidity of the spoil, allows serious erosion and the delivery of acid leachate and sediment downstream. (U.S.D.A. Soil Conservation Service photo by A. Snyder.)

the subsoil than in the surface soil. The deficiency is overcome when the roots of the plants enter the deeper soil of higher sulfur availablity. The sulfur may be retained in the subsoil as gypsum-enriched layers in arid-region soils or as adsorbed sulfate in soils of warmer, wetter regions.

Soil acidification with sulfur

Sulfur, usually in elemental form (agricultural sulfur), is sometimes intentionally used to acidify soils. One reason is to improve growing conditions for plants such as azaleas and blueberries. These and similar plants favor a pH below 5. Above this pH, they often suffer from a shortage of iron, which is most soluble under strongly acid conditions. Soil acidification has also been used to control potato scab, a disease that disfigures the skin of the tubers and reduces their marketability. The scab is caused by an actinomycetes that will not grow well if the pH is below about 5.

Since soil acidification can be expensive, its use is limited mostly to small tracts of land, or to soil that is already relatively acid and with a low buffering capacity (limited resistance to pH change). The reaction involved, which is essentially identical to the oxidation of sulfur in pyrite, is shown in simplified form:

$$2S + 3O_2 + 2H_2O \rightarrow 2H_2SO_4 \qquad (14.8)$$

This reaction is responsible for the strong acidification of the spoil (waste) obtained from mining coal of high-sulfur content (see fig. 14.14). The intense acidity makes reclamation of some spoil fields very difficult.

Phosphorus

Phosphorus deficiencies are relatively widespread in soils. Two things account for this: the amount present in the minerals of soil parent materials is often low, and it has a natural inclination to form inorganic compounds of very low solubility in soils. In many soils, the supply of available phosphorus is sustained by release from the organic fraction, and in some, the only satisfactory way to maintain adequate available phosphorus is through the addition of fertilizers.

Forms of phosphorus in soils

The bulk of the phosphorus in soils comes from a group of calcium phosphate minerals known as the *apatites*. Principal among these is *fluorapatite*, which has the chemical formula of $Ca_{10}(PO_4)_6F_2$. The apatites are crystalline compounds that are relatively stable in neutral to basic environments, but they are more soluble, at least slowly, under acid conditions. Because of this, soils that have developed in dry regions and have a neutral to basic reaction retain much of their original phosphorus in apatite form. In the more acid soils of humid regions, on the other hand, a large part of the apatite may have been broken down under weathering influences. Where weathering has been intense and of long duration, little, if any, apatite may remain in the soil.

On dissolving, apatite minerals release phosphorus as the trivalent (PO_4^{3-}) phosphate ion. Once in solution, however, this ion has a strong tendency to combine with H^+ ions to form either the monohydrogen (HPO_4^{2-}) or the dihydrogen ($H_2PO_4^-$) phosphate ion. Which of these two ions is produced depends on soil pH. In basic soils, the HPO_4^{2-} ion dominates, and at a pH of 9, is essentially the only phosphate ion in the soil solution. With an increase in acidity and concentration of H^+ ions, the proportion of HPO_4^- in the soil solution decreases and the proportion of $H_2PO_4^-$ increases. At neutrality, soluble phosphorus occurs about equally in these two forms, but with a further increase in acidity, transformation of HPO_4^{2-} to $H_2PO_4^-$ continues and, at a pH of 5, $H_2PO_4^-$ becomes the only phosphate ion of significance in solution.

Except where it has been recently added to the soil in readily soluble fertilizers, phosphorus rarely occurs in the soil solution in more than trace amounts. One reason for this is the inherently low solubility of apatite minerals. Another is the ability of phosphorus to form other slightly soluble compounds over a wide range of soil conditions. In basic soils, soluble calcium is usually plentiful and keeps the level of phosphorus low by combining with HPO_4^{2-} to form slightly soluble dicalcium phosphate, $CaHPO_4$. In acid soils, where iron and aluminum are relatively soluble and phosphorus in solution occurs predominantly as the $H_2PO_4^-$ ion, precipitation is principally as iron and aluminum phosphates. Typical are such compounds as *strengite*, $Fe(OH)_2H_2PO_4$, and *variscite*, $Al(OH)_2H_2PO_4$, both of which are highly stable and, therefore, only very slightly soluble in acid media. The precipitation of soluble phosphorus can be undesirable where it causes a serious reduction in the availability of applied fertilizer phosphorus to plants.

The precipitation of soluble phosphorus by calcium in basic soils is much less serious than precipitation by iron and aluminum in acid soils. In basic

Figure 14.15.
The relationship between pH and the availability of soil phosphorus. Acid-stable iron and aluminum phosphates control phosphorus solubility and availability in the acid range. In the basic range, it is controlled by very slightly soluble calcium phosphate compounds.

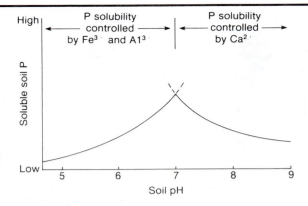

soils, the principal compound precipitated is dicalcium phosphate. This compound is stable in a basic environment, but it is soluble under acid conditions. Thus, on contact with roots, which create a localized acidic environment, dicalcium phosphate dissolves and releases phosphorus in available form. Such an interaction between roots and iron and aluminum phosphates does not take place, however. These compounds are acid-stable, so contact with roots does not increase their solubility. This problem with precipitation by iron and aluminum can be reduced by applying soluble phosphate fertilizers by a method called band placement, which is described on page 404.[4]

Although precipitation of phosphorus as dicalcium phosphate does not greatly reduce its immediate availability to plants, availability will decrease with time as the dicalcium phosphate gradually transforms to other, less soluble calcium phosphate compounds, including secondary apatite. High pH and abundant soluble calcium encourage this type of transformation.

Because phosporus forms very slightly soluble compounds at either high or low pH, its solubility and availability to plants tends to be greatest in the near-neutral range, as suggested in figure 14.15. Iron and aluminum lose their effectiveness in precipitating phosphorus with increasing pH because they are more inclined to precipitate as virtually insoluble oxides and hydroxides. Calcium is less effective in precipitating phosphorus as the pH drops because of the increasing tendency of phosphorus to occur in solution as the $H_2PO_4^-$ ion, which forms a highly soluble compound with calcium.

Phosphorus content of soils

Factors important to the total amount of phosphorus in a soil include: (1) the nature of the parent material, (2) the extent to which it has been altered by weathering and leaching, and (3) the amount of organic matter present. Parent materials formed from granites and other rocks high in SiO_2 are normally low in phosphorus. Basalts and similar dark-colored rocks often contain about twice as much phosphorus as do granites. However, many tropical

[4]The precipitation of soluble phosphorus, particularly that from fertilizers, is commonly referred to as *phosphorus fixation*. Use of this term is discouraged, since it can be misconstrued as having the same connotation as nitrogen fixation. Further, the term is often interpreted to mean a sharp reduction in the availability of applied fertilizer phosphorus under all conditions. As suggested in the text, this is not always the case.

Figure 14.16.
The phosphorus
content of the surface
30-cm of soils in the
continental United
States. (U.S.D.A.
map.)

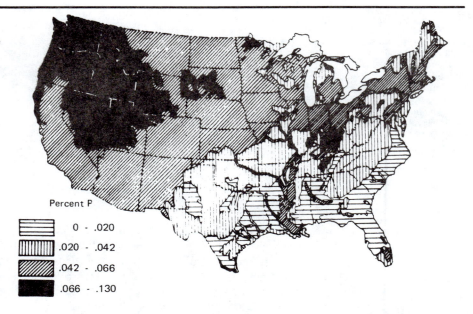

Percent P

0 - .020	
.020 - .042	
.042 - .066	
.066 - .130	

soils that have formed under conditions of intense weathering and leaching from basaltic parent materials of rather high phosphorus content contain comparatively little of this element, either in total quantity or in readily available form. Most of the phosphorus in these soils is in organic combination and is concentrated primarily in the surface horizon.

The general distribution of total phosphorus in surface soils of the continental United States is given in figure 14.16. Although the lowest phosphorus contents are associated with the Ultisols of the East and South, there are localized areas within these regions where the native phosphorus is much higher than indicated by the map. Further, extensive use of fertilizers has caused marked changes in the level of soil phosphorus in many instances. Indeed, some sandy soils supporting citrus groves in Florida, originally very low in phosphorus, have been heavily fertilized through the years and now contain as much as six times the phosphorus initially present. The water-soluble phosphorus in these soils appears to be nearly 100 times that normal to the soil solution of other highly productive soils.

Figure 14.16 indicates that there is a gradual increase in the average amount of soil phosphorus along a transect extending from the southeast to the northwestern part of the United States. Changes along this transect are associated in part with differences in parent materials and in part with differences in weathering intensities. Parent materials account for soils of high phosphorus content in Tennessee, Kentucky, and Ohio. Parent materials derived from the weathering of basalt and modified by volcanic ash are extensively distributed in the Pacific Northwest, which has much to do with the generally higher phosphorus content of soils in that section of the country. Recognize, however, that total phosphorus is referred to here, not the quantities available to growing plants.

Phosphorus is essential to plants in a number of ways, but two roles illustrate its truly indispensible nature. One is as a component of genetic tissue, where it occurs in deoxyribonucleic acid, DNA, and ribonucleic acid, RNA, both of these compounds consisting of long, threadlike molecules linked together by PO_4^{3-}. A second highly important role is in the storage and transfer of energy produced by the oxidation of sugar in plant cells. The energy is stored in the organo-phosphate compound, adenosine triphosphate, ATP. This compound is formed by combining PO_4^{3-} with adenosine diphosphate, ADP, the energy required for its formation being stored in the ATP molecule. On release of the energy for use in other biochemical reactions, the ATP reverts to ADP and free PO_4^{3-} ions.

Plant use of soil phosphorus

Phosphorus is taken up by plants mainly as the $H_2PO_4^-$ and, to a lesser extent, the HPO_4^{2-} ion. Some absorption of organically combined phsophorus may occur, but probably not in significant amounts. Phosphorus is absorbed continuously throughout the period of active plant development, but as plants mature, a substantial part of the absorbed phosphorus moves into the seed or fruit, the pattern being similar to that shown for nitrogen in figure 14.3. Thus, there is a gradual decline in the phosphorus concentration in the vegetative tissue during the growing season. As a general rule, the concentration of phosphorus in the vegetative plant parts ranges between 0.1 and 0.4 percent, on a dry-weight basis. A value as low as 0.1 percent, especially in the earlier part of the growing season, is commonly taken as a sign of deficiency.

Phosphorus deficiencies produce only a limited number of symptoms in plants. The principal effect is plant stunting, but this can be caused by other nutrients as well. A somewhat pale, bluish-green color of leaves has been noted in phosphorus-deficient alfalfa, and corn often develops a purplish coloration along leaf margins. The symptoms in corn may be more pronounced early in the season when root activity and the release of phosphorus from organic matter are slow because of low soil temperatures.

Because of its normally very low concentration in soluble form, phosphorus is not subject to much leaching loss except over long periods of time. Soluble phosphorus added to soils, even in relatively large amounts, simply will not move far before it is made immobile by precipitation. The low solubility of phosphorus also limits its movement to plant roots. As a consequence, the continual growth of new roots into previously unexplored parts of the soil is extremely important to maintaining an uninterrupted flow of this element into the plant.

Potassium, calcium, and magnesium

These three nutrients are discussed together because they are alike in several respects. All three are macronutrients, and they have a common source in primary aluminosilicate and ferromagnesian minerals. Each of these ions is retained in substantial quantities in exchangeable form, although the amount of exchangeable calcium and magnesium is usually several times greater than the amount of exchangeable potassium (see table 14.1). This

Table 14.1. The distribution of exchangeable calcium, magnesium, and potassium in two soils forming under different climatic and vegetational regimes. Concentrations of the elements are in milliequivalents per 100 g soil. (From *Soil Classification, A Comprehensive System,* 7th Approximation. Washington, D.C.: Government Printing Office, 1960).

Mollisol (Iowa)					Ultisol (Georgia)				
Depth (cm)	Horizon	Exchangeable ion			Depth (cm)	Horizon	Exchangeable ion		
		Ca	Mg	K			Ca	Mg	K
0–23	A	13.3	2.9	0.27	0–3	A	2.7	1.3	.1
23–33	AB	9.1	3.2	.15	3–13	E	.4	.5	.1
33–51	BA	11.7	5.4	.12	13–20	AB	.2	.2	.1
51–74	B1t	22.6	12.2	.38	20–30	BAt	.1	.3	T[a]
74–89	B2t	22.4	12.2	.45	30–36	B1t	.1	.4	.1
89–104	BCt	19.3	10.5	.38	36–58	B2tg	.1	.6	.1
104–120	C1	18.4	10.0	.38	58–81	Bctg	.1	.8	.1
120–135	C2	18.2	10.2	.30	81–112	IIC	.1	.6	.1

[a]Trace

Table 14.2 The range in percentage composition (dry-weight basis) of potassium, calcium, and magnesium in grasses and legumes grown under average fertility conditions. (Data from various sources).

Element	Grasses	Legumes
K	1.50–3.00	1.50–3.00
Ca	0.20–0.40	1.00–1.50
Mg	0.10–0.25	0.20–0.40

results primarily from the greater release of calcium and magnesium from weathering minerals as well as their stronger attraction to the negative charge of clays and organic matter. In contrast, the demand for and utilization of potassium by plants exceeds that of the other two nutrients (see table 14.2). Potassium is also deficient more often and added to soils more frequently in fertilizers. In terms of the amount added to soils in fertilizers, potassium is exceeded only by nitrogen.

The retention of potassium, calcium, and magnesium in exchangeable form is important because it reduces their potential loss by leaching. Leaching is not avoided, however, evident in the fact that of the total soluble inorganic materials in river water, 20 percent is calcium, 3.5 percent is magnesium, and 2 percent is potassium. Essentially all of these ions are derived from leachates moving out of soils and to rivers in groundwater.

Potassium

The range in content of total potassium in soils of the continental United States is shown in figure 14.17. As shown in this figure, well over half the soils have a total-potassium content of from 1.5 to 2.5 percent. Since the average potassium content of unweathered mineral matter is around 2.6 percent, it would appear that the majority of soils in the United States have suffered only a minimal loss of this element. This does not assure a high level of available potassium, however, for the greater part of the total quantity present is retained in minerals of low solubility. This element does not occur in organic combination either in plants or in soils.

Figure 14.17.
Average total-potassium content of the surface 30-cm of soil for different regions of the United States. Values are as a percentage of the weight of whole soil. (U.S.D.A. map.)

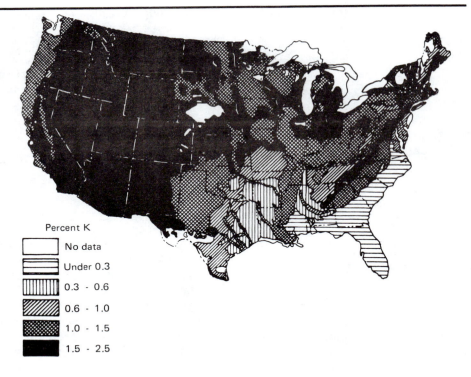

Percent K

	No data
	Under 0.3
	0.3 - 0.6
	0.6 - 1.0
	1.0 - 1.5
	1.5 - 2.5

Table 14.3. The release to plants of nonexchangeable potassium by 10 soils. All values are in kg potassium per 2 million kg of soil. (From E. H. Stewart and N. J. Volk, *Soil Sci.* 61:125–28, 1946).

Exchangeable plus soluble K		Decrease in exchangeable plus soluble K	K removed by crop	K obtained from nonexchangeable form
Initial	Final			
80	37	43	119	76

Important mineral sources of potassium in soils are the feldspars and the 2:1 layer minerals, mica and illite, the illite occurring as clay-sized particles. The potassium in both mica and illite is held in so-called nonexchangeable form; that is, between the stacked platelets making up particles of these nonexpanding minerals. Whereas potassium removal from micas is ordinarily a very slow process, removal from illite can be sufficiently rapid to be of significance to the ongoing potassium nutrition of plants. Complete removal of potassium from micas or illite alters these minerals to the 2:1 expanding clay, vermiculite.

The potential for the release of nonexchangeable potassium is demonstrated by the data in table 14.3, which are average values for 10 soils used in a greenhouse study. The total potassium taken up by plants grown on these soils in pots was in a ratio of 119 kg per 2 million kg of soil. However, during the growing period, only 43 of the total 119 kg came from readily available (exchangeable + soluble) forms of potassium. The remaining 76 kg, or about 60 percent of the total, was supplied from minerals, largely from sites holding nonexchangeable potassium. Since the release of the nonexchangeable form is gradual, it is referred to as *slowly available* potassium.

Nonexchangeable potassium benefits plant nutrition in soils high in micaceous or illitic minerals because of its contribution to the supply of potassium in readily available form. Nonexchangeable potassium aids in maintaining this supply because of its tendency to equilibrate with exchangeable and soluble potassium as depicted in the equation:

$$\text{Nonexchangeable K} \rightleftharpoons \text{Exchangeable K} \rightleftharpoons \text{Soluble K} \tag{14.9}$$

Due to the tendency for the three forms to maintain an equilibrium, a change in the concentration of one causes a subsequent change in the concentration of the other two. For example, the loss of exchangeable and soluble potassium by plant uptake or leaching causes a spontaneous release of nonexchangeable potassium, thus moderating the decrease in the others. Similarly, an increase in the soluble and exchangeable forms, as may follow soil fertilization, results in their partial transformation to the less readily available, nonexchangeable form. This transformation, referred to earlier as entrapment, reduces the effectiveness of added potassium fertilizer, but it increases the level of reserve potassium, which can be later released in available form. The entrapment of potassium, commonly called potassium fixation, is most likely to occur in soils containing illite or vermiculite clay.

Potassium
in plants

Potassium remains totally soluble and mobile in plants. Although its role in plant development is not known for certain, potassium is believed to be necessary for (1) the synthesis of protein, carbohydrates, and chlorophyll, (2) the translocation and storage of carbohydrates, and (3) the absorption of anions, such as nitrate and phosphate, from the soil. Potassium is particularly important to crops of high sugar or starch content. For example, both the yield and quality of plants such as sugar beets, sugar cane, and potatoes may be impaired where potassium deficiency occurs. Similarly, the lodging, or falling over, of wheat and other small grains is often associated with potassium deficiencies that limit the development of structural tissue required for strong stems. The principal components supplying strength to plant tissue are cellulose, a carbohydrate, and lignin.

The concentration of potassium in vegetative plant parts usually ranges between 1 and 3 percent of the dry-tissue weight, but it may exceed 5 percent in young tissue. As a plant ages, the concentration of potassium declines, reaching a minimum late in the season. As with other nutrients, maturation of the plant results in the translocation of potassium from vegetative parts into seed, fruit, or other storage tissue. For seed crops, such as wheat and corn, however, only a small part of the total potassium is translocated into the forming grain. In contrast, the concentration of postassium in sugar beet roots, potato tubers, and the fruit of tree, vine, and various vegetable crops tends often to far exceed that remaining in the vegetative parts of the plant.

Because of its high mobility, potassium tends to concentrate in the younger tissue of growing plants. As a consequence, shortages of this element cause visual deficiency symptoms to appear first in the older leaves. Modest deficiencies may cause little more than a dull green color in the leaves, but severe deficiencies cause leaf margins and tips to turn yellow and finally die. These symptoms are called *tip burn* in grain crops. In legumes, such as alfalfa and clover, the yellowing occurs around the margins of the leaves. An intense purplish-black color (*black leaf*) is characteristic of potassium deficiency in grapes and potatoes.

Potassium and sodium relationships in plants

Evidence exists that sodium is sometimes able to partially assume the role of potassium in plants. This has been demonstrated by correcting mild potassium deficiencies through the addition of sodium salts to soils containing limited amounts of available potassium and sodium. Severe deficiencies seem not to be corrected by sodium alone, however. Some reports suggest improvement in plant growth from sodium even where potassium is known to be adequate. For this reason, sodium cannot be entirely eliminated as an essential plant nutrient. In general, however, plants appear to thrive just as well without as with sodium in the rooting medium.

Calcium

The calcium content of soils varies greatly, ranging from less than 0.05 to more than 25 percent of the soil weight. The higher of these values normally applies to unleached, arid-region soils formed in highly calcareous (lime-bearing) parent materials. The total-calcium content of noncalcareous soils averages below one percent, or considerably less than the 3.6 percent estimated for average unweathered mineral matter. This low content attests to the extensive loss of calcium by weathering and leaching, especially in humid regions. It also reflects the low weathering resistance of the principal minerals supplying calcium: calcium feldspars and lime. Lime minerals are particularly susceptible to chemical decomposition in acid-soil environments.

The supply of calcium to plants depends on its release from minerals and its retention in exchangeable and soluble form. Except in strongly acid soils, calcium is usually relatively abundant in both exchangeable and soluble form. Because of its high average concentration in the soil solution, calcium is transferred to roots mainly by mass flow. In some soils of high pH and low lime solubility, more calcium may be transferred to roots by mass flow than can be absorbed by the plant. In such cases, the excess accumulates outside the root, often as a tubular deposit of lime surrounding the root.

The calcium requirement of plants is highly variable. Among crop plants, legumes generally have a greater need for calcium than do nonlegumes. Because of this, legumes are especially responsive to the correction of acid-soil conditions through the addition of calcium-bearing amendments, such as crushed limestone. In comparison, many trees, especially conifers, have a low calcium requirement, which explains in part the prevalence of coniferous forests on strongly acid Spodosols of low calcium-supplying power.

The exact physiological function of calcium in plants has not been determined. It is known to play a major role in the development of new tissue, however. Because of this, and because calcium is not very mobile in plants,

Figure 14.18.
Misshapen leaves of
tobacco grown on
calcium-deficient, acid
soil. (U.S.D.A.
Agricultural Research
Service photo.)

deficiencies show up first in new tissue, both in roots and in leaves and stems. Typical deficiency symptoms include poorly developed roots, small and misshapen young leaves, and failure of leaf and flower buds to form (see fig. 14.18). The effect of low calcium supplies on growth may explain the failure of the roots of sensitive plants to grow well in the lower horizons of strongly acid, leached soils. Since calcium is not involved in chlorophyll formation, calcium-deficient plants tend to retain a normal green color.

Magnesium

Similarities between calcium and magnesium in soils are many. Both occur in dolomite, which is a lime mineral of widespread distribution over the earth. The principal original source of magnesium is the ferromagnesian minerals,

Nutrient relationships in soils and plants

which, like calcium feldspars, are among the most easily weathered primary minerals. As exchangeable cations, calcium and magnesium are held with about the same tenacity by clay and organic particles, and together they make up the bulk of exchangeable ions in soils from about pH 5.5 to 8.5. On the average, the amount of exchangeable magnesium is from one-fifth to one-third of the amount of exchangeable calcium (see table 14.1).

Magnesium tends to be leached from soils less readily than calcium. This is due in part to its lower average content in exchangeable and soluble form. In addition, however, magnesium released during primary mineral weathering frequently becomes a part of relatively stable secondary silicate compounds, montmorillonite and chlorite clays, for example. Although the total content of calcium often exceeds that of magnesium in soils, this is not the general rule, particularly for soils of humid regions. The magnesium content varies from negligible amounts in highly leached, sandy soils to as much as several percent in calcareous soils.

In the plant, magnesium is a component of chlorophyll and is thought to aid in both the translocation of starch and the formation of fats and oils. It also appears to affect the absorption, movement, and utilization of phosphorus within plants.

Retarded growth is the first sign of magnesium deficiency. As the deficiency becomes more acute, the tissue between the veins of leaves tends to become chlorotic (*interveinal chlorosis*), as shown in figure 14.19. These symptoms appear first and are more pronounced in the older leaves, which suggests a fair degree of mobility for this element in the plant. In affected leaves, on the other hand, chlorosis is most pronounced at the outer tip. This, coupled with the tendency for interveinal tissue to show chlorosis first, suggests that mobility of magnesium in leaves is somewhat restricted.

The magnesium content of plants is usually less than that for calcium (see table 14.2). However, certain plant parts may actually contain more magnesium than calcium, a fact that relates to the higher mobility of magnesium. Potato tubers often accumulate more magnesium than calcium, and more magnesium is found in the grain than in the straw of crops such as wheat. The reverse is true of calcium.

The ratio of available calcium to magnesium influences their relative absorption by plants. Sometimes an imbalance occurs, with an excess of one inducing a deficiency of the other. Perhaps the best example of this occurs on so-called *serpentine soils* formed from metamorphic rocks high in the mineral serpentine, $Mg_6Si_4O_{10}(OH)_8$. These soils often contain such an excess of magnesium over both calcium and potassium that they may support little, if any, vegetation due to restricted uptake of the latter two nutrients (see fig. 14.20).

Because magnesium is not absorbed in large quantities by plants, it is not cycled, or returned to the surface soil in plant residues, to the degree that calcium is. As a consequence, the concentration of exchangeable magnesium is frequently lower in the surface than in subsurface soil horizons. This may also be true for calcium, but not so often as for magnesium.

Figure 14.19.
Severe interveinal chlorosis caused by magnesium deficiency in tobacco. (U.S.D.A. Agricultural Research Service photo.)

Special contributions of calcium and magnesium to soils

A high level of exchangeable calcium and magnesium is usually associated with a more nearly ideal, neutral to slightly basic pH in soils. It is therefore preferable to have these two elements as the dominant exchangeable ions, because, at a neutral pH, neither deficiencies of a number of the nutrients nor toxic effects caused by some ions are likely to be as pronounced as they are at extremes of pH. This condition also assures a more adequate supply of calcium and magnesium to plants. An undesirably low level of available calcium and magnesium should always be suspect in soils of abnormally high or low pH.

Exchangeable calcium and magnesium usually have a beneficial effect on the physical state of soils. This relates in part to the ability of these two

Figure 14.20.
A serpentine soil formed from high-magnesium metamorphic rock and devoid of vegetation, presumably because of excess magnesium. The adjacent soil supporting trees is from Swauk sandstone. View is in the Wenatchee National Forest, Washington. (Photo courtesy of A. R. Kruckeberg, University of Washington, Seattle.)

elements, when in exchangeable form, to stabilize soil aggregates. It also reflects their influence on the activity of decay organisms. It may be recalled that both the formation and stabilization of soil aggregates are encouraged by the active decay of organic residues in soils.

The micronutrients

Seven elements are classed as micronutrients because they are used in such small quantities by plants. Of the seven, four are cations. These are iron, manganese, copper, and zinc. The other three occur as anions; chlorine as chloride; boron as borate in boric acid (H_3BO_3), and molybdenum as the molybdate ion. Chemical symbols of the micronutrient ions are shown in table 14.4.

The primary sources of the micronutrients, along with the general range of total content in soils, are also listed in table 14.4. Except for iron and manganese, micronutrients are present in comparatively low total concentration in soils, for they make up very little of mineral matter in general. Iron and manganese are not only more abundant in common minerals, they tend to accumulate in soils by forming highly stable secondary compounds, usually oxides, that persist as other minerals are weathered away.

Table 14.4. Forms, initial sources, and some representative ranges in the concentration of micronutrients in soils.

Nutrient	Common soluble forms	Principal sources	Total content in soils (usual range)
Iron	Fe^{2+}, Fe^{3+}	Ferromagnesian minerals; sulfides	1–5%
Manganese	Mn^{2+}	As substituted ion in ferromagnesian minerals	100–4000 ppm
Copper	Cu^{2+}	Sulfides; substituted ion in common minerals	2–100 ppm
Zinc	Zn^{2+}	Sulfides; substituted ion in common minerals	10–30 ppm
Boron	H_3BO_3	Tourmaline; soluble borates	2–100 ppm
Molybdenum	MoO_4^{2-}	Sulfides; substituted ion in common minerals	0.2–5 ppm
Chlorine	Cl^-	Minor constituent of some silicates; substituted ion for fluorine in apatite; soluble chlorides	50–500 ppm[a]

[a]Higher concentrations of chloride may occur in soils troubled with accumulated salts.

Micronutrient cations in soils

Iron, manganese, copper, and zinc in soils are similar in behavior to the extent that the solubility and availability of each decreases with increasing pH. This effect is most pronounced for iron; it may be deficient for some plants at a pH as low as 5 even when its total content in the soil is high. Manganese availability is less affected by increasing pH, although it may be made deficient by raising the pH of acid soils above neutrality (see fig. 14.21). Such an effect on copper or zinc availability is relatively rare. However, deficiencies of these two elements are not uncommon in soils that are basic in reaction and also calcareous.

Reduced availability of micronutrient cations with rising pH is attributed to their formation of very slightly soluble hydroxide and oxide compounds as the hydroxyl-ion concentration of the soil solution is increased. This effect does not seriously impair the availability of micronutrients to all plants, however. Those most severely affected are so-called acid-loving species, as typified by blueberries and certain other bush, vine, and tree fruits, and a number of ornamental plants such as azaleas and rhododendrons. Even with these plants, iron is the nutrient most likely to become deficient as the pH is raised.

As an increase in pH reduces iron and manganese solubility in soils, a reduction in pH raises it, sometimes to the point of making these elements toxic to plants. Toxicity effects are much more common with manganese than with iron. However, acid conditions that produce manganese toxicities may also bring sufficient aluminum into solution to make it toxic to plants. One of the main reasons for raising the pH of strongly acid soils through the addition of crushed limestone is to correct or prevent toxicities due to manganese and aluminum.

The state of soil aeration is a cofactor with pH in controlling the availability of two of the micronutrient cations, iron and manganese. Aeration is a determinant of the oxidation (valence) state of iron and manganese, and the oxidation state, in turn, influences the solubility of these two elements. Good aeration, as characterized by a high level of free oxygen in soil air, promotes the oxidized ionic forms (Fe^{3+}, Mn^{4+}) of iron and magnesium; poor aeration encourages transformation to the reduced (Fe^{2+}, Mn^{2+}) ionic forms.[5] Since reduced iron and manganese are more soluble than the oxidized forms, reducing conditions associated with low oxygen supply enhance the solubility of these two elements. However, the effect on iron is much less than on manganese. Indeed, manganese toxicities can occur in soils of low pH even where aeration is good, but they are likely to be more severe if poor aeration also prevails.

Micronutrient cations are adsorbed to clay and organic particles through cation exchange. The quantity in exchangeable form is always small, however, for these cations invariably occur in very low concentration in the soil solution. One reason for the low concentration is the tendency of the cations to form compounds of extremely low solubility. Another is their ability to form *complexes*, or *chelation compounds*, (from *chela*, meaning claw) with negatively charged organic anions (see fig. 14.22). The bonding between the

---

[5]Ion reduction and oxidation result, respectively, from the loss or gain of negatively charged electrons by ions. On oxidation, the loss of electrons makes the valence more positive. Reduction has the reverse effect. Plants apparently absorb iron and manganese mainly in the reduced form.

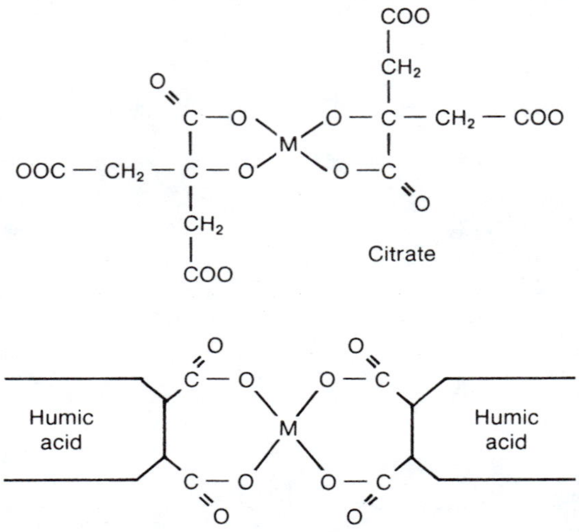

Figure 14.22.
The chelation of metal cations by negatively charged citrate and soluble humic acid anions.

micronutrient cations and the complexing anions is very strong, so that dissociation, or splitting apart, of the complex molecules into the separate ions is very limited. Yet, some micronutrient chelates are soluble and therefore able to migrate freely through the soil solution. As a consequence, micronutrient cations that might otherwise be inactivated through precipitation reactions or adsorption to soil particle surfaces are retained in a plant-available form. Apparently, roots are able to absorb nutrients from a wide range of soluble chelate compounds, both natural and synthetic.

Chelation can reduce toxicities due to excesses of micronutrient cations by reducing the concentration of the free cations in the soil solution. By the same token, excessive chelation may induce a deficiency of micronutrient cations, particularly of copper and zinc. Deficiencies of copper and zinc on organic soils is thought to be due to strong chelation by organic matter. Even certain plant residues or animal manures appear capable of supplying chelating agents that reduce the availability of some micronutrients. This probably explains why corn sometimes shows zinc deficiency when grown on land following sugar beets or where animal corrals or stored manure have been located.

Micronutrient anions in soils

Few similarities in behavior exist among the micronutrients chlorine, boron, and molybdenum. In the soil solution, boron occurs in uncharged boric acid molecules, and chlorine and molybednum as negatively charged anions. Thus, none of these elements is adsorbed by negatively charged soil particles. Except where chloride has accumulated in salt-affected soils, all three anionic micronutrients are usually present in only trace amount in the soil solution, which is consistent with their general scarcity in soil minerals. Neither chlorine nor boron form difficulty soluble compounds in soils, so both are rather easily lost by leaching. The chloride ion forms readily soluble compounds with all common cations in soils, and its concentration in the soil solution

Nutrient relationships in soils and plants

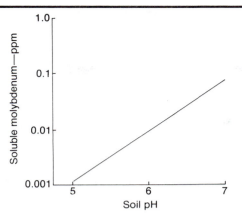

Figure 14.23. The general relationship between pH and the concentration of soluble molybdenum in the soil solution of four molybdenum-deficient soils in California. (From Reisenauer, H. M. *et. al., Soil Sci. Soc. Amer. Proc.* 26:23–27, 1962. By permission of the Soil Science Society of America.)

can be very high. The solubility of boron is assumed to be controlled by the solubility of slightly soluble boric acid, but the concentration of soluble boron in soils is virtually always far below that allowed by this compound. This element may be low in basic soils high in lime, or where the pH of an acid, marginally boron-deficient soil has been raised through liming. The effect here is attributed to a calcium-boron imbalance induced by the lime.

The behavior of molybdate (MoO_4^{2-}) is pH-dependent (see fig. 14.23), and in this regard, is similar to the behavior of phosphate. Like phosphorus, molybdenum is more readily available in soils of neutral or basic reaction than under acid conditions. The low availability of molybdenum in acid soils is attributed to the formation of iron and aluminum molybdate compounds of extremely low solubility. The solubility of these compounds can be increased by raising the pH of the soil.

Micronutrient relationships in plants

Micronutrients make their major contribution to plant growth as constituents of enzyme and hormone systems and as participants in biochemical oxidation-reduction reactions. Iron and manganese are especially important to the latter because of the ease with which they are in themselves oxidized and reduced.

Enzymes are specialized protein molecules whose function is to provide a site for specific biochemical reactions. For example, the enzyme *oxidase* serves as the site where oxygen and organic carbon are combined into carbon dioxide in the energy-supplying respiration process. Similarly, the enzyme *nitrogenase* provides a site for combining nitrogen with hydrogen to form ammonium in the fixation of atmospheric nitrogen. The enzymes are not destroyed in these reactions, and so may be used over and over again.

Hormones control the rate of biochemical reactions. For example, the hormone *auxin* controls the rate of cell expansion and, therefore, the rate of growth in plants. Hormones actually enter into biochemical reactions and are consumed as a result. Thus, the synthesis of hormones must go on continuously in plants.

Table 14.5. Representative concentrations in plant tissue (dry-weight basis) and some characteristic deficiency symptoms of the micronutrient elements. (Data from various sources).

Nutrient	Concentration			Deficiency symptoms
	Deficiency	Normal	Excess	
Fe, ppm	10–80	30–150	Generally nontoxic to plants	Interveinal chlorosis
Mn, ppm	5–20	15–100	Depends on Fe:Mn ratio; may be >1000	Interveinal chlorosis
Cu, ppm	3–5	5–15	>20; depends on Fe: Cu ratio	Gum boils on shoots and cracking of fruit rind in citrus; die-back of apple branches; low leaf turgor
Zn, ppm	5–15	10–50	200–500	Interveinal chlorosis; shortened internodes of stems; reduced size (little leaf) and rosetting of leaves
B, ppm	2–15	5–50	75–300	Interveinal chlorosis, often reddish; short internodes; leaf rosetting; abcission of flowers; cork-like tissue, heart rot in beets
Mo, ppm	0.1–0.3	1–100	Generally nontoxic to plants	N-deficiency symptoms in legumes; yellow spot of citrus leaves; whiptail of cauliflower
Cl,%	<0.2	0.2–1.0	0.5–2.5	Stubby roots; plant wilting; some chlorosis of new leaves

The status of micronutrients in plants is judged either from analyses run on plant tissue or from the appearance of the plant. Usually, plant abnormalities associated with micronutrients occur in the leaves, but in some instances, they affect stems, fruit, and even the roots. Deficiencies are the major problem, but some micronutrients are fairly often present in toxic concentrations. Toxicities are caused most often by manganese and boron, and less often by copper and zinc. Chloride toxicities may be relatively common in salt-affected soils, but usually only if the concentration in the plant is very high. For example, some plants native to dry-region soils may accumulate several percent of chloride by weight without showing any adverse effects. Chloride toxicities have been noted in tobacco and potatoes when heavily fertilized with potassium chloride fertilizer.

Representative concentrations of micronutrients in plants are listed in table 14.5. Three categories of concentration are shown: deficient, normal, and toxic. Each range is relatively broad, for it applies to many different plants displaying widely different requirements or tolerances for a given nutrient. For example, a concentration of 50 parts per million of boron might be handily tolerated by a plant of high boron requirement but not by one of low requirement.

Visually detectable micronutrient-deficiency symptoms often relate to the specific role of the micronutrient in the plant. For example, iron, manganese, zinc, and boron are all involved in chlorophyll synthesis, so a shortage

Figure 14.24.
Iron-deficiency symptoms in an ornamental *Spirea* grown on a soil at pH 6.2. Correction of this deficiency by localized soil acidification would be the most practical.

of any one of them can cause chlorosis, or yellowing, of leaf tissue. These elements are not very mobile in plants. Thus, when in short supply, the farther they must move after being taken up by roots, the more intense the chlorosis. This relationship is illustrated by iron-deficient leaves in figure 14.24, for the most severely chlorotic (light-colored) tissue is in new leaves at the far end, or growing tip, of the stem. Similarly, the most chlorotic tissue on the leaves is that farthest from the conductive tissue in the veins; that is, it occurs in the interveinal tissue.

A deficiency symptom identified with the specific role of zinc in plants is the limited expansion of leaves and the failure of stems or branches to elongate. Zinc has this effect because it is essential to the synthesis of the growth hormone, *auxin*. Where the shortage is severe, branches of some fruit trees may show virtually no elongation during the year, and normally widely spaced leaves are crowded together on the short section of new stem. This leaf pattern is referred to as *rosetting* (see fig. 14.25). Zinc deficiencies also result in interveinal chlorosis in many plants, but if severe, veins may also be chlorotic. As shown in figure 14.26, this results in the development of a broad band of chlorotic tissue running parallel to the veins in corn leaves.

Some micronutrients produce deficiency symptoms of a unique character. One of these, *whiptail* in cauliflower, is caused by a molybdenum deficiency (see fig. 14.27). A lack of this nutrient can also cause a nitrogen deficiency in legumes, for it is a part of nitrogenase, an enzyme involved in the fixation of nitrogen in root nodules of legumes.

Deficiency symptoms identifiable only with a shortage of boron include heart rot of beets, corklike tissue on the petioles of cauliflower, broccoli, and sugar beets, as well as in apples and stone fruits (see fig.14.28). The loss of leaves along the larger branches of apple and similar trees (*die-back*) is a symptom typical of copper deficiency. Distinctive characteristics such as

Figure 14.25.
Rosetting in peaches caused by zinc deficiency. (A) Light-colored
whorls of chlorotic leaves may be seen on shortened stem tips.
(B) Zinc-deficient peach stem and leaves at the left, and normal
stem and leaves on the right. (Photos courtesy of Wayne B.
Sherman, University of Florida, Gainesville.)

these greatly facilitate field diagnosis of micronutrient deficiency problems
and are sometimes the only practical way of determining that a micronutrient deficiency exists.

**Micronutrient
antagonism**

A number of nutrients, when present in excessive amounts in the soil solution, appear capable of reducing the absorption or utilization of some micronutrients by plants. These effects, referred to as nutrient antagonism, may
result in a deficiency of a micronutrient even though its availability in the
soil would otherwise be adequate for normal plant growth. Such interactions
become particularly important where they are caused by the overapplication of fertilizers or other chemical agents to the soil.

Many interactions affecting micronutrient availability have been reported. Some that have gained widespread attention include: (1) boron deficiencies resulting from the overapplication of lime to acid soils, (2) zinc
and iron deficiencies caused by heavy applications of phosphate fertilizers
(see fig. 14.29), and (3) iron deficiencies caused by the accumulation of ex-

Nutrient relationships in soils and plants

Figure 14.26.
Zinc deficiency in corn associated with leveled irrigated land in the Columbia Basin, Washington. (Photo courtesy of L. C. Boawn, U.S.D.A., Prosser, Washington.)

Figure 14.27.
Whiptail disease of cauliflower associated with molybdenum deficiency. (Photo courtesy Florida Potato Investigations Laboratory, and Climax Molybdenum Co.)

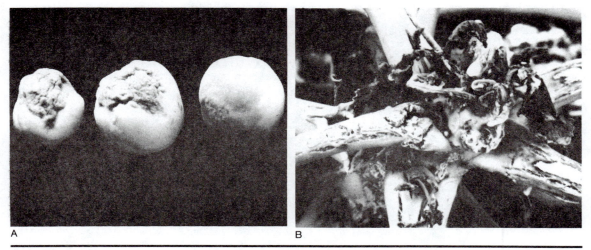

A   B

Figure 14.28.
Boron deficiency in the fruit of apricots (A) and in cauliflower
petioles (B). (Photos courtesy of N. R. Benson, Washington State
University, Wenatchee.)

Figure 14.29.
Severe stunting of
potatoes due to a zinc
shortage caused by
the application of
excess phosphate
fertilizer. Potatoes in
the background are
normal. (From Boawn,
L. C. and Leggett,
G. E., *Soil Sci.*
95:137–41. © 1963,
Williams & Wilkins,
Baltimore.

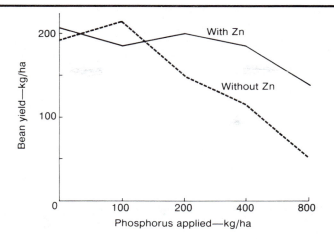

Figure 14.30.
Effect of zinc and phosphorus fertilization on the yield of pea beans grown on Kawkawlin loam. (From Judy, W. J. *et al.*, Mich. State Univ. Agri. Exp. Sta. Res. Rep. 33, 1965.)

cess copper or manganese in the soil. In most instances, this type of problem can be corrected by the addition of the deficient nutrient. For example, as shown in figure 14.30 for a trial in Michigan, a marked decreased in bean yields caused by excess phosphate fertilization was compensated for by the simultaneous addition of zinc fertilizer, provided the excess of phosphorus was not too large. However, it is usually better to avoid such problems through the judicious use of fertilizers and soil amendments.

Summary

Of the nutrients supplied to plants by the soil, nitrogen, phosphorus, and potassium are required in greatest amounts and are the most often deficient. Of these three, nitrogen is stored only in soil organic matter, where it has accumulated through fixation. Phosphorus is also stored in soil organic matter, but not potassium. Both phosphorus and potassium come originally from soil minerals.

Sulfur is also stored in soil organic matter. With nitrogen and phosphorus, it is released from organic combination by mineralization. Each of these elements occurs in protein or proteinlike compounds in plants, their residues, and in the organic fraction of soils.

The availability of plant nutrients is dependent on their total content in soils and their rate of release from unavailable forms. It also depends on the potential for leaching, which is greater for soluble anions than for cations retained by exchange adsorption. The low solubility of inorganic phosphorus, molybdenum, and the micronutrient cations tends to keep their availability to plants and loss by leaching low. Chelation of micronutrient cations sometimes increases and sometimes decreases their availability in the soil. The solubility and availability of micronutrient cations, especially iron and manganese, is also a function of soil pH and state of aeration.

Nutrient shortages are often apparent in plant deficiency symptoms, such as stunting and an off-color in leaves, commonly chlorosis. Other symptoms include poorly developed or misshapen leaves, cracking of stems or leaf petioles, and deformed fruit, sometimes with skin abnormalities. Deficiency symptoms more often than not reflect the role of the nutrient in the plant.

Review questions

1. Why can the organic matter content of soils be used to estimate the total content of nitrogen but not of other plant nutrients?
2. What is chlorosis, and deficiencies of what plant nutrients cause this symptom in plants?
3. Define nitrogen fixation, nitrification, and denitrification, and explain how each of these processes affects the nitrogen-fertility status of soils.
4. What soil conditions encourage the loss of inorganic nitrogen as ammonia gas?
5. Point out the major similarities in behavior of nitrogen and sulfur in soils and plants.
6. Why does oxidation of ammonium ions and elemental, organic, or sulfide-sulfur increase soil acidity?
7. Why does the content of nutrients in vegetative plant parts tend to decline as the plant matures?
8. Why are the phosphorus-bearing minerals in young soils or soils of arid regions likely to be different from those in leached, acid soils?
9. Why does the precipitation of soluble phosphorus in neutral to basic soils not reduce the availability of the phosphorus to plants as much as does precipitation in strongly acid soils?
10. Why does the potassium-supplying power of illitic soils differ from that of soils lacking illitic clay?
11. Compare potassium, calcium, and magnesium with respect to their content in exchangeable form in soils and their relative uptake by plants.
12. Why do calcium and magnesium benefit plant growth for nonnutritional as well as for nutritional reasons?
13. Why do micronutrients differ in their potential for leaching?
14. Explain how and why variation in soil pH and oxidation state affects the plant availability of iron and manganese.
15. Why does chelation not always have the same effect on the availability of micronutrients?

Selected references

Bartholomew, W. V., and Clark, F. E. eds. *Soil Nitrogen*. Madison, WI: American Society of Agronomy, 1965.
Brown, J. C. Iron Chlorosis in Plants. *Advances in Agron.* 13:329–69. New York: Academic Press, 1961.

Hemwall, J. D. The Fixation of Phosphorus by Soils. *Advances in Agron.* 8:95–112. New York: Academic Press, Inc., 1957.

Hodgson, J. F. Chemistry of the Micronutrient Elements in Soils. *Advances in Agron.* 15:119–59. New York: Academic Press, 1963.

Kilmer, V. J., Younts, S. E., and Brady, N. C. *The Role of Potassium in Agriculture.* Madison, WI.: American Society of Agronomy, 1968.

Sprague, H. B., ed. *Hunger Signs in Crops.* New York: David McKay Company, Inc., 1964.

Stefferud, A., ed. *Soil: 1957 Yearbook of Agriculture.* Washington, D.C.: Government Printing Office.

Tisdale, S. L., Nelson, W. L., and Beaton, J. D. *Soil Fertility and Fertilizers.* New York: The Macmillan Company, 1984.

15

Fertilizers and fertilizer use

Fertilizers are materials used to enhance the supply of available nutrients in the soil. Most commercially prepared fertilizers are inorganic products with a fixed composition and predictable behavioral characteristics. State laws require that each fertilizer be labeled to show the kinds and amounts of nutrients it contains. Without this information the user would be unable to estimate the quantity of material needed to meet the nutrient requirement of specific crop plants.

Organic materials, including manures, composts, and food-industry wastes, are also used to supplement the soil nutrient supply, but they are rarely marketed as fertilizer. The reason for this is that they are largely of local origin and are highly variable in composition and behavior. Organic supplements are normally added to build up the overall level of nutrients in the soil without regard to specific needs of crops. Often, these materials are applied primarily to improve the physical state, or tilth, of the soil. Because of the slow release of nutrients, especially nitrogen, the effect of organic materials is long-term as well as immediate.

General types of fertilizers

Commercially prepared fertilizers come in a wide variety of forms. When classified on the basis of chemical composition, they fall into two broad groups: *single materials* and *mixed fertilizers*. The term single material is applied to individual compounds, such as ammonia, ammonium nitrate, or potassium sulfate, or to products of somewhat more complex composition developed primarily to supply a single nutrient. Mixed fertilizers are combinations of single materials, usually of those that supply nitrogen, phosphorus, or potassium. Mixtures of single materials containing all three of these nutrients are called *complete* fertilizers.

Fertilizers vary markedly in their physical characteristics. They come as gases, solutions, water suspensions, and as solids, with the latter occurring in particle sizes ranging from fine powders to coarse granular or crystalline materials. Most finely divided products contain compounds of low solubility, so they are crushed to increase the rate at which they will dissolve in the soil. The coarser materials may consist of large crystals of fertilizer salts or of artificially prepared granules or pellets containing one or more single materials. The granular materials are usually preferred because they are dust-free and are relatively easy to apply with standard fertilizer equipment.

Nitrogen fertilizers

Most nitrogen fertilizers are readily soluble compounds containing either ammonium or nitrate ions. Both of these ions are chemically available forms but, since ammonium becomes adsorbed by the exchange complex, nitrate is by far the most mobile in the soil. It is easier for nitrate-nitrogen to become positionally available, therefore. However, in warm, moist, well aerated soils, ammonium converts rather quickly to nitrate, so even nitrogen added in this form does not remain immobile in the soil for long. In their general effect on plants, therefore, ammonium and nitrate fertilizers behave in much the same way, provided the ammonium can be rapidly nitrified. Rapid nitrification will not occur, for example, if the fertilizer is spread on the surface of a soil that remains dry for a long time.

Inorganic nitrogen fertilizers

The principal nitrogen fertilizers are listed in table 15.1. Also shown is their total-nitrogen content as a percent by weight. With but two exceptions, all of these compounds dissolve quickly once they are mixed with moist soil. The exceptions are sulfur-coated urea and ureaform, which are specialty products designed to release nitrogen gradually to long-season plants. Befitting their behavior when added to soils, sulfur-coated urea and ureaform are called *slow-release* nitrogen fertilizers. All of the nitrogen they contain does eventually become available to plants, however.

Most nitrogen fertilizers are derived from synthetic ammonia prepared by the Haber process, which involves combining nitrogen and hydrogen gases under high heat and pressure (see fig. 15.1). The reaction is:

$$N_2 + 3H_2 \rightarrow 2NH_3 \tag{15.1}$$

The major source of the hydrogen is methane, CH_4, the principal component of natural gas. The ammonia gas may be used directly as a fertilizer, or it may be transformed to other kinds of nitrogen fertilizers. Examples are ammonium nitrate, ammonium sulfate, and ammonium phosphate formed respectively by neutralizing nitric (HNO_3), sulfuric (H_2SO_4), and phosphoric (H_3PO_4) acids with ammonia. The ammonia can also be oxidized to nitric acid, which is then used in preparing the ammonium, calcium, and potassium nitrates. Of the latter three materials, only the first, ammonium nitrate, is used to any great extent in this country.

Another very important nitrogen fertilizer produced from ammonia is urea. This compound is prepared by combining ammonia and carbon dioxide at elevated temperature and pressure. The reaction is:

$$2NH_3 + CO_2 \rightarrow CO(NH_2)_2 + H_2O \tag{15.2}$$

Table 15.1. Some important nitrogen fertilizers. Nitrogen contents are percentages by weight.

Name	Chemical formula	Total nitrogen content
		%
Anhydrous ammonia	NH_3	82
Aqua ammonia	NH_4OH	25[a]
Urea	$CO(NH_2)_2$	45
Ammonium nitrate	NH_4NO_3	33
Ammonium sulfate	$(NH_4)_2SO_4$	21
Urea-Ammonium nitrate solutions	See above	32
Sulfur-coated urea	See above	35
Ureaform	Variable[b]	38[c]

[a]The maximum practical concentration of ammonia in water is 30 percent by weight, which provides a nitrogen content of 25 percent.
[b]Depends on formulation (ratio of formaldehyde to urea).
[c]Nitrogen content of the usual formulation.

Figure 15.1.
A reformer furnace, the large structure left of center, with stripper columns behind used to remove and clean (strip) hydrogen gas derived from hydrocarbon fuels. The hydrogen and nitrogen gases are then fed to compressors for conversion to ammonia. (Photo courtesy of the Chevron Chemical Corporation.)

Fertilizers and fertilizer use

Urea and ammonium nitrate are the two most widely used solid nitrogen fertilizers. Both have a comparatively high nitrogen content, which is important because it reduces transportation costs for these materials. Ammonium nitrate is sometimes preferred because it releases highly mobile nitrate as soon as it dissolves in the soil solution.

The use of slow-release nitrogen fertilizers is limited by their relatively high cost. The slow-release property of sulfur-coated urea is due to the sulfur coating, which is applied over pellets of urea in combination with a wax sealant. The coatings vary in thickness so that penetration of water takes place at different rates. Complete release of the urea usually requires several months in warm, moist soils.

Ureaform is a combination of urea and formaldehyde, and its solubility is a function of the ratio of the two materials in the mixture. The best use of the slow-release materials is with plants that have a long and continuing demand for nitrogen. Grasses in lawns and golf greens are examples.

Organic sources of nitrogen

Organic materials used to supply nitrogen to soils include green manures, animal manures, and other plant and animal wastes. Where a choice is possible, materials of the highest possible nitrogen content are preferred for agronomic use. Also important is the rate at which the nitrogen becomes available to plants, which is a function of the rates of mineralization and immobilization of the nitrogen once the material has been applied to the soil. Only if the rate of mineralization exceeds the rate of immobilization will the organic material cause an immediate build-up of available inorganic nitrogen in the soil solution.

It is possible to predict the relative rates of mineralization and immobilization of nitrogen for a given organic material from knowledge of its total-nitrogen content. As a general rule, these two processes proceed at about equal rates with materials containing from 1.5 to 2 percent nitrogen. Above this range, mineralization normally exceeds immobilization. Thus, when an organic material with more than this amount of total nitrogen is worked into the soil, there is usually a build-up in soluble nitrogen, as detected by the level of nitrate in the soil solution. Incorporation of materials containing less than 1.5 percent nitrogen ordinarily results in a depression in the nitrate level, however.

Several kinds of organic materials capable of supplying nitrogen to soils are listed in table 15.2. Also shown are the total-nitrogen contents, the percent of the total nitrogen that is soluble in water, and percentages of the total nitrogen transformed to nitrate after 20- and 40-day incubation periods in a laboratory experiment. The comparisons were made by applying the materials to the soil at a rate that would supply the same amount of total nitrogen.

Table 15.2. The total-nitrogen content and the rate of release of nitrate from several organic materials. (From E. J. Rubins, and F. E. Bear, *Soil Sci.* 54:411–23, ©1942, Williams & Wilkins Co., Baltimore, MD.)

Organic material	Total N	Water-soluble N	Conversion to NO_3^- in 20 days	in 40 days
	%	% of total N	% of total N	
Cottonseed meal	7.2	7	49	54
Alfalfa	2.8	48	24	32
Bone meal (raw)	4.2	0	7	10
Dried blood	13.8	2	60	66
Horse manure	1.5	20	−19	−16
Chicken manure	2.3	67	22	30
Tobacco stems	1.0	47	−14	5
Wheat straw	0.3	38	−16	−15

Data in table 15.2 show that, in general, the higher the nitrogen content of the organic source the greater the build-up of nitrate during the early stages of its decomposition. However, this relationship does not hold in all cases. For example, the data in the table show bone meal yielding less nitrate during a 40-day period than does alfalfa of lower total-nitrogen content. This may reflect either the presence of more resistant nitrogen compounds or their lower accessibility to decay organisms in the bone meal than in the alfalfa. Alfalfa and similar legumes are noted for their ability to rapidly release available nitrogen when they undergo decay in the soil.

Three of the materials in table 15.2—horse manure, tobacco stems, and wheat straw—actually caused a reduction in nitrate when first mixed with the soil. All of these materials are low in nitrogen (1.5 percent or less) and supply substantial energy carbon to stimulate the activity of and need for nitrogen by decay organisms. Thus, their initial effect is to reduce the level of available nitrogen by causing greater immobilization than mineralization. Upon depletion of the excess energy carbon, however, these materials would add to the supply of available nitrogen in the soil solution. The time required for this is often several months after incorporation into the soil.

Phosphorus fertilizers

Phosphorus fertilizers are prepared almost entirely from the mineral fluor-apatite, which is found in commercially exploitable sedimentary beds in several parts of the world. Since this mineral dissolves slowly in soils, its conversion to fertilizer involves steps to increase its solubility. These include (1) fine-grinding, (2) heating to a fused state, and then cooling and grinding, which yields an amorphous product that is more soluble that the original crystalline apatite, (3) treatment with strong acids to produce more soluble phosphate compounds, and (4) extraction of the phosphorus in elemental form by means of an electric furnace process and then combining the phosphorus into relatively pure compounds that dissolve readily in water. In this country, only the first, third, and fourth approaches are used extensively in phosphate-fertilizer manufacture.

Expressing the phosphorus content of fertilizers

Phosphorus fertilizers differ markedly in their solubility and availability to plants. They vary from water-soluble compounds, which dissolve quickly and totally in moist soil, to others, such as finely ground apatite, that dissolve slowly and release little phosphorus in plant-available form in a single year's time. For this reason some measure other than the total-phosphorus content must be used to estimate the availability of phosphorus in fertilizers.

In this country, the standard means of assessing the phosphorus-supplying power of fertilizers is a specially devised laboratory test in which a sample of the fertilizer is subjected to a two-step extraction, first with water and then with a neutral, 15–percent ammonium citrate solution. In effect, the water extraction detects the immediately available phosphorus in the fertilizer while the ammonium citrate extracts a fraction that is released more slowly in the soil. The total, or combined, amount of phosphorus removed in these two steps, when converted to a percent of the fertilizer weight, is termed *available phosphorus* and is used as the basis for expressing the phosphorus content of fertilizers sold in this country. In most cases the available-phosphorus content as determined in the laboratory is a reasonably accurate estimate of the actual amount of phosphorus a fertilizer will release in a chemically available form to the crop to which it is applied.

Because of a long-standing convention, the phophorus content of fertilizers is commonly expressed on the oxide basis, that is, as a percent P_2O_5 (phosphorus pentoxide), as opposed to the percent of elemental P. This means of expressing phosphorus contents is based on an early system used in rock and mineral analysis. For example, one may find the composition of the lime mineral calcite, $CaCO_3$, shown as 56 percent calcium oxide (CaO) and 44 percent carbon dioxide (CO_2), which is consistent with expressing the formula for this compound as $CaO \cdot CO_2$ rather than $CaCO_3$. On an elemental basis the composition of calcite would be 40 percent Ca, 12 percent C, and 48 percent O.

A similar approach can be used for expressing the composition of phosphate compounds. For example, the formula for tricalcium phosphate can be written either as $Ca_3(PO_4)_2$ or $3CaO \cdot P_2O_5$. The composition of tricalcium phosphate can therefore be expressed either as 56 percent CaO and 46 percent P_2O_5 (oxide basis), or as 40 percent Ca, 40 percent O, and 20 percent P (elemental basis).[1]

Whether the phosphorus content of fertilizers is expressed on the oxide or the elemental basis is of minor consequence, since either expression can be easily converted to the other. For example, as seen for tricalcium phosphate above, an elemental-P content of 20 percent is equivalent to a P_2O_5 content of 44 percent. Thus, the P_2O_5 content of a fertilizer is approximately 2.3 times that of elemental P, or:

$$\text{Percent P} \times 2.3 = \text{percent } P_2O_5 \qquad (15.3)$$

Conversely, conversion of P_2O_5 to elemental-P contents is by means of the equation:

$$\text{Percent } P_2O_5 \times .43 = \text{Percent P} \qquad (15.4)$$

[1]These percentages are approximate only, for they are computed using atomic weights rounded to the nearest whole number.

Table 15.3. Some important phosphorus fertilizers and their available-phosphorus (both elemental P and P_2O_5) and total-N contents as a percent of the fertilizer weight.

Fertilizer	Principal components	Available phosphorus as P	Available phosphorus as P_2O_5	Nitrogen
		%	%	%
Ordinary superphosphate	$Ca(H_2PO_4)_2$; $CaSO_4 \cdot 2H_2O$	8–9	18–21	
Concentrated superphosphate	$Ca(H_2PO_4)$	17–21	39–48	
Monoammonium phosphate	$NH_4H_2PO_4$	26	60	11[a]
Diammonium phosphate	$(NH_4)_2HPO_4$	21	48	18[a]
Ammonium polyphosphate	$(NH_4)_3HP_2O_7$	24–29	55–66	10–20[b]
Phosphoric acid	H_3PO_4	24	55	
Superphosphoric acid	$H_4P_2O_7$	34	78	
Rock phosphate	Fluorapatite	11–15[c]	25–35[c]	

[a]A popular material, known as 11–48 fertilizer (11% N, 48% P_2O_5), is prepared as a mixture of mono- and diammonium phosphates. Another, termed 16–20 fertilizer (16% N, 20% P_2O_5), consists of a mixture of monoammonium phosphate and ammonium sulfate.
[b]The amount of ammonium varies, yielding variable nitrogen and phosphorus contents.
[c]These are total- rather than available-phosphorus contents.

Important phosphorus fertilizers

Phosphate fertilizers that are currently most favored in the United States are listed in table 15.3. Also shown is their available-phosphorus content on both an elemental and oxide basis, and where applicable, their total-nitrogen content. Principal differences among these materials relate primarily to their method of manufacture.

One of the least valued phosphate fertilizers is finely ground apatite, referred to as *rock phosphate*. The principal limitation of this material is its low solubility. To compensate for this limitation, rock phosphate is usually applied to land at relatively high rates, and then mainly only to acid soils in which apatitie is most soluble. Many experiments have shown that rock phosphate is of negligible value when applied to neutral or basic soils.

The most extensively used phosphorus fertilizers are the superphosphates, which are prepared by treating finely ground apatite with acid, either sulfuric (H_2SO_4) or phosphoric (H_3PO_4). Treatment with sulfuric acid yields *ordinary superphosphate*, but if phosphoric acid is used, the product is *concentrated superphosphate*. In either case, acid added to apatite transforms trivalent phosphate (PO_4^{3-}) ions to dihydrogen phosphate ($H_2PO_4^-$) ions, which then crystallize with calcium, also from the apatite, as monocalcium phosphate, $Ca(H_2PO_4)_2$. This compound is readily soluble in water and accounts for the high availability of the phosphorus in either ordinary or concentrated superphosphate. However, because concentrated superphosphate is made with phosphoric rather than sulfuric acid, it contains roughly twice as much phosphorus (as monocalcium phophate) as does ordinary superphosphate. Indeed, half the weight of ordinary superphosphate comes from gypsum, $CaSO_4 \cdot 2H_2O$, which is formed by the reaction of sulfuric acid with calcium from apatite. For soils that also need added sulfur, ordinary superphosphate may be preferred over concentrated superphosphate.

The manufacture of superphosphate is carefully controlled to avoid an excess of sulfuric or phosphoric acid, since an excess of either yields a product of undesirable storage and handling properties. As a result, total conversion of the phosphate from apatite to monocalcium phosphate does not take place, but instead, a small amount of the phosphorus is transformed to less acidic dicalcium phosphate, $CaHPO_4$. Dicalcium phosphate is not water soluble, but it does dissolve in 15–percent ammonium citrate solution. The citrate test therefore shows the phosphorus in superphosphate to be essentially totally available to plants even though it is only from 75 to 85 percent water-soluble. This is consistent with the fact that the dicalcium phosphate dissolves on contact with acidic root environments. Similarly, when monocalcium phosphate in the superphosphates reacts with calcium and hydroxyl ions in basic soils, it precipitates largely as dicalcium phosphate. Again, however, since this compound dissolves on root contact, precipitation in basic soils has only a marginal effect on availability of the phosphorus from the superphosphates.

Phosphoric acid is prepared in two ways. One is by the extraction of elemental phosphorus from apatite in a high-temperature electric furnace. Following its extraction, the phosphorus is burned in air to form phosphorus pentoxide, P_2O_5, and then, on treatment with water, is converted to phosphoric acid. If a limited amount of water is used, a *polyphosphoric acid*, sometimes called *superphosphoric acid* and symbolized $H_4P_2O_7$, is produced. With more water, the normal *orthophosphoric* acid, H_3PO_4, results. Polyphosphoric acid has the higher phosphorus content because it contains less water, but it converts to orthophosphoric acid on reaction with soil water. Either ortho- or polyphosphoric acid may be applied directly to the soil as a source of phosphorus. Both types of acid can also be combined with ammonia to produce ammonium phosphate fertilizer salts (see table 15.3). All ammonium phosphate fertilizers are relatively pure salts that dissolve readily in water or in the soil solution.

The second method of phosphoric acid production, called the wet-process, is by treatment of apatite with excess sulfuric acid. This causes the calcium from the apatite to precipitate as gypsum. The released phosphorus combines with the hydrogen from sulfuric acid to form phosphoric acid. The acid remains liquid and can be drained off.

Properties important to phosphate fertilizer selection

The relative popularity of superphosphate fertilizers relates for the most part of their comparatively high solubility, a moderate-to-high phosphorus content, and a cost that is somewhat below that of other suitable phosphate sources. Ammonium phosphates and the phosphoric acids supply available phosphorus just as quickly as the superphosphates and have a somewhat higher phosphorus content, but they are more expensive. The added expense reflects more costly methods of manufacture, including the addition of nitrogen to the ammonium phosphates.

Whereas high solubility of phosphate fertilizers more nearly assures the rapid release of applied phosphorus in the soil, it also exposes this nutrient to precipitation reactions with calcium in neutral or basic soils or with iron and aluminum in strongly acid soils. Precipitation with iron and aluminum can seriously reduce the availability of the applied phosphorus to plants. However, this problem can be greatly reduced by using a method of fertilizer application that retards reaction between applied phosphorus and soluble iron and aluminum. This method, called band-placement, is discussed shortly.

Fertilizers supplying potassium and magnesium

Most potassium and magnesium fertilizer materials are obtained from underground salt deposits formed by the natural evaporation of sea water (see fig. 15.2). These deposits vary in composition, but many contain a preponderance of potassium salts with lesser amounts of magnesium chloride and sulfate and the common salt, sodium chloride. Various processes are used to separate these water-soluble minerals into relatively pure potassium and magnesium fertilizer compounds. Any sodium chloride present is removed and discarded as waste. Major fertilizer industries have been built around deposits of potassium or potassium-magnesium salts in Germany, the Canadian province of Saskatchewan, as well as New Mexico and Utah in the United States.

The principal compounds supplying potassium and magnesium in fertilizers are listed in table 15.4. The potassium salts are used far more extensively than are those of magnesium, and of the potassium salts, the chloride is used far more often than the sulfate. The compositional values shown in the table are for relatively pure, fertilizer-grade materials.

The values for potassium in table 15.4 are for quantities of this element soluble in water (water-soluble potassium) expressed as a percent by weight of the element (K) and of its oxide (K_2O). Industry tends to favor the oxide expression, which is commonly referred to as the *potash* content of fertilizers. Contents of potassium oxide are readily converted to contents of elemental K upon multiplication by a factor of 0.83, or:

$$\text{Percent K} = \text{Percent } K_2O \times 0.83 \qquad (15.5)$$

Conversely, elemental-K contents are converted to their equivalent of the oxide if multiplied by a factor of 1.2, or:

$$\text{Percent } K_2O = \text{Percent K} \times 1.2 \qquad (15.6)$$

The use of water-soluble values to express the potassium content of fertilizers is required by law in most states to assure that the potassium supplied is in readily available form. There is apparently no pressing need for similar laws for magnesium in fertilizers. Total contents of magnesium are usually indicated, as is the case in table 15.4.

All of the potassium and magnesium fertilizers listed in table 15.4 are salts that dissolve readily in the soil solution. Once dissolved, most of the potassium and magnesium becomes exchangeable, with only a relatively small proportion of the amount added remaining in soluble form. Major exceptions to this occur in soils of very low cation-exchange capacity, where most of the added potassium and magnesium remains in soluble and, therefore, leachable form.

Figure 15.2.
Mining potassium chloride with a continuous mining machine in a deep mine at Esterhazy, Sask., Canada. (Photo courtesy of the International Minerals and Chemical Corporation.)

Table 15.4. Common potassium and magnesium fertilizers. Nutrient contents are percentages by weight of water-soluble potassium and total magnesium, chlorine, and sulfur.

Material	Formula	Potassium content as K	Potassium content as K_2O	Magnesium content	Other nutrients
		%	%	%	%
Potassium chloride	KCl	50–52	60–62	—	Cl, 45
Potassium sulfate	K_2SO_4	42	50	—	S, 18
Potassium sulfate-Magnesium sulfate	$K_2SO_4 \cdot MgSO_4$	18	22	11	S, 22
Magnesium sulfate	$MgSO_4$	—	—	20	S, 13

Calcium, a nutrient that has many behavioral similarities to potassium and magnesium in soils, is rarely added specifically to correct a deficiency. Instead, it is usually supplied as a component of a liming material added to correct excess acidity in soils. Magnesium deficiencies are also frequently overcome in this way where the liming material consists of crushed dolomitic limestone. Dolomitic limestone consists of a mixture of calcium and magnesium carbonates.

Fertilizers supplying sulfur

Sulfur preparations for use in treating soils comprise a highly diverse group of materials. The more important of these are listed in table 15.5. Except for agricultural sulfur, all of the materials shown in the table are readily soluble and release sulfate-sulfur immediately on going into solution. Agricultural

Table 15.5. Some important sulfur-bearing fertilizer materials. Nutrient contents are percentages by weight of total sulfur and nitrogen, available P_2O_5, and water-soluble K_2O.

Material	Formula of S source	Nutrient content	
		S	Other
		%	%
Ammonium sulfate	$(NH_4)_2SO_4$	24	N, 21
Potassium sulfate	K_2SO_4	17	K_2O, 50
Ordinary superphosphate	$CaSO_4 \cdot 2H_2O$	9	P_2O_5, 18–21
Gypsum	$CaSO_4 \cdot 2H_2O$	18	—
Sulfur (agricultural)	S	100	—
Sulfur suspensions	S	Variable	—

sulfur takes more time, for it must first be transformed to sulfate by sulfur-oxidizing bacteria, *Thiobacillus*, present in all soils. In warm, moist soil, significant amounts of sulfate will form in a short period if agricultural sulfur is applied in a fine state of subdivision and mixed with the soil.

Agricultural sulfur is available in particle sizes ranging from fine powders to coarse, granular products. Finely ground sulfur is also supplied in water suspensions, or in solutions formed by dissolving it in ammonium hydroxide. Many types of sulfur-enhanced materials have been prepared by mixing fertilizer salts with molten sulfur, allowing the mixture to cool and solidify, and then grinding to a desired particle size. Since the sulfur oxidizes to sulfuric acid on reaction in the soil, its presence can increase the solubility of some materials. For example, rock phosphate can be made more soluble by this means. Sulfur is also used to acidify soils, either to improve soil conditions for acid-loving plants or to control acid-sensitive disease organisms.

Fertilizers supplying micronutrients

There are three principal kinds of micronutrient fertilizer materials: (1) readily soluble inorganic salts, (2) synthetic chelation compounds, and (3) a group of fritted materials (frits) prepared by fusing micronutrient salts with silicate mineral matter to form a glasslike product. The latter is finely ground for application to the soil.

The most widely used inorganic salts of the micronutrient cations are sulfates. Boron is usually applied as borax, a sodium borate, and molybdenum, as either ammonium or sodium molybdate. All of these compounds are readily soluble in water. For soil application, they are often mixed with other fertilizers supplying macronutrients, a step that simplifies application where very small quantities of micronutrient salts are needed.

Chelated micronutrient compounds contain only the catonic elements. They are prepared by combining the cations with chelating organic anions, called *ligands*. These products are readily soluble in the soil solution, and because they persist principally as undissociated molecules, they limit chemical precipitation of the cations. It may be recalled that precipitation of micronutrient cations often accounts for their low solubility and availability in the soil. In chelated form, micronutrients remain mobile and can move through the soil to plant roots. The chelated compounds are used primarily on high-value crops, principally fruit trees, with such use ordinarily being limited to chelates of iron or zinc. The relatively high cost of these materials limits their use on many other agriculture crops.

The use of fritted micronutrient fertilizers is not widespread. This is in part due to cost. It also reflects the rather low solubility of these products. However, low solubility can be of advantage with boron. If overapplied in a readily soluble salt such as borax, boron can be toxic to plants. Boron is also subject to leaching loss. The slow release of boron from fritted materials would eliminate both of these problems.

Mixed fertilizers

Mixed fertilizers are the most difficult to characterize and evaluate, for they are highly diverse with regard to composition and form. Methods of manufacture are also variable. One common procedure is to mix the materials in a slurry, which is then subjected to a combined drying-granulating treatment. This process has the advantage of producing a relatively homogeneous product. Some fertilizers are dried and then pulverized rather than being granulated.

Another means of preparing mixed fertilizers is the bulk-blending of materials in either dry or liquid form. Products prepared in this way are becoming increasingly popular because they can be compounded in small blending plants where the composition can be varied to fit the needs of a local clientele (see fig. 15.3). Bulk-handling also allows the materials to be prepared at a comparatively low cost. A substantial part of locally blended fertilizers is applied on a custom basis.

The fertilizer grade, analysis, and guarantee

As a protection against fraud, uniform state laws throughout this country require manufacturers of fertilizers to supply certain specific types of information about the materials they produce. Of particular importance in this regard is a statement of the nutrient content as expressed by the *grade*, or *guaranteed analysis*, of the fertilizer. For the so-called fertilizer elements, nitrogen, phosphorus, and potassium, this is indicated by a series of three numbers that express in order the weight-percentages of total N, available P_2O_5, and water-soluble K_2O. An example is 5–12–12, which is the grade or guaranteed analysis of a complete fertilizer containing 5 percent total N, 12 percent available P_2O_5, and 12 percent water-soluble K_2O.[2]

The convention used to indicate the nutrient content of mixed fertilizers is also applied to single materials. Thus, the grade of an ammonium nitrate fertilizer containing 33 percent total N would be 33–0–0, and that for

[2]A fertilizer grade of 5–12–12, which expresses phosphorus and potassium contents as oxides, would convert to 5–5–10, approximately, to express the contents of phosphorus and potassium on an elemental basis.

Figure 15.3.
Schematic drawing of
a bulk-blending plant.
Single fertilizer
materials are delivered
into the plant by
elevator A for storage
in bins, B. Fertilizer
mixtures are blended
in the rotating drum
mixer, C, and loaded
for delivery to the field
at D. (From the
*Western Fertilizer
Handbook,* 5th edition,
1975. Sacramento,
CA: California
Fertilizer Association.)

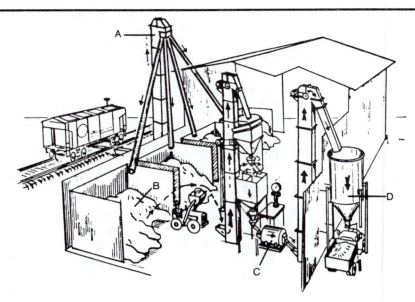

ordinary superphosphate containing 21 percent available P_2O_5 would be
0–21–0. Methods for expressing contents of nutrients other than nitrogen,
phosphorus, and potassium are not standardized, although some states specify
how their contents must be indicated. Usually it is the total quantity of the
element, which may or may not provide a good indication of how much nutrient is present in a form that is available to plants.

The grade of a fertilizer is determined by chemical analysis and is published by the manufacturer as a *guarantee* of the content of nutrients in his
product. Ordinarily, the actual analysis of a fertilizer is slightly higher than
the published guarantee to assure compliance with the law. For example, as
shown in table 15.6, the absolute contents of total N, available P_2O_5, and water-
soluble K_2O in a 10–10–10 fertilizer are somewhat above those needed to
satisfy the guarantee, or grade, of the fertilizer.

Table 15.6 lists several single materials that are commonly used together in mixed fertilizers. These may or may not be indicated by the manufacturer. Showing the component materials as well as the nutrient content
provides an *open-formula guarantee* for a fertilizer.

Over the years, the trend in fertilizer manufacture has been toward
products of high analysis, the advantage being a reduction in the cost of
storage, transport, and application. Until large-scale polyphosphate production became practical, high-analysis mixed fertilizers were limited to combinations of dry materials in which the combined content of total N, available
P_2O_5, and water-soluble K_2O was, at maximum, about 50 percent of the fertilizer weight. With the introduction of the polyphosphates, combined total
contents of 60 percent became possible. Examples of several high-analysis
mixed fertilizers are shown in table 15.7. All of these are dry-fertilizer mixes.

Table 15.6. An exemplary listing of ingredients and the content of nutrients in a 10–10–10 fertilizer. Nutrient contents are of total N, available P_2O_5, and water-soluble K_2O.

Name	Nutrient content	Fraction of total mixture	N	P_2O_5	K_2O
			Nutrient as a percent of fertilizer weight[b]		
Nitrogen solution	44.0% N	.115	5.05		
Ammonium sulfate	21.0% N	.255	5.35		
Concentrated superphosphate	46.7% P_2O_5	.075		3.5	
Ordinary superphosphate	20.0% P_2O_5	.336		7.0	
Potassium chloride	60.2% K_2O	.169			10.2
Conditioner[a]	—	.050			
Totals		1.000	10.40	10.5	10.2

[a]May consist of lime.
[b]The total contents purposely exceed the guarantee of 10–10–10.

Table 15.7. Percentage contents of total nitrogen, available phosphorus, and water-soluble potassium of representative, high-analysis fertilizers. The double listing shows phosphorus and potassium contents on both an oxide and an elemental basis.

$N-P_2O_5-K_2O$	N-P-K
6–24–24	6–10.6–20
10–20–20	10–8.8–16.6
15–15–15	15–6.6–12.5
12–24–24[a]	12–10.6–20
20–20–20[a]	20–8.8–16.6

[a]Mixtures prepared from ammonium polyphosphates.

The nutrient content of fertilizer solutions is below that of solid mixes. Concentrations as high as 10 percent each of total N, available P_2O_5, and water-soluble K_2O are practical and are in general use. Solutions containing as much as 10 percent total N and 35 percent available P_2O_5 can be prepared from ammonium polyphosphates, but such high concentrations cannot be maintained if potassium salts are added to the mixture.

Methods of fertilizer application

It is important that fertilizers be applied to soils in a manner that will minimize the chance of their being lost by leaching or volatilization, or of their being converted to an unavailable form through precipitation reactions. One of the most common methods of application is to *broadcast* the fertilizer uniformly over the soil surface prior to seedbed preparation (see fig. 15.4). The fertilizer may then be worked into the soil by plowing or discing (see fig. 15.5). Plowing is the preferred method of incorporation, because it places the fertilizer more deeply in the soil, thus assuring better contact with plant roots. Shallow placement, as by discing, may limit availability if the thin surface layer containing the fertilizer remains dry much of the time. Roots are unable to remove nutrients from dry soil.

Figure 15.4.
Applying liquid fertilizer by broadcasting. The large tires on the heavily laden vehicle minimize soil compaction. (Photo courtesy of the Ag-Chem Equipment Company.)

Figure 15.5.
The distribution of surface-applied (broadcast) fertilizer in the soil by plowing and discing. Discing distributes the fertilizer more uniformly but to a shallower depth.

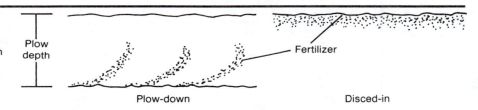

Fertilizers may also be broadcast over established stands of plants, a method called *top-dressing*. Since top-dressed fertilizer is not worked into the soil by cultivation, it must be washed in by rain or irrigation water. Top-dressing works most satisfactorily with nutrients that remain mobile in the soil, nitrate-nitrogen being the foremost example. Even nutrients of low mobility that penetrate the soil to a shallow depth only may be taken up in sufficient quantity if the soil remains continually moist at the surface. This fact makes top-dressing a reasonably satisfactory method of application for closely spaced plants, such as small grains and forages. These plants help maintain soil moisture by shading the soil. Range, forest, and pasturelands are normally fertilized by top-dressing, and where the area is large, the application may be by airplane or helicopter (see fig. 15.6).

One of the more widely used methods of fertilizing machine-planted crops is *band placement* at the time of seeding. In this method, the fertilizer is placed directly into the soil in a concentrated band near the seed (see figs. 15.7 and 15.8). Having the fertilizer near the seed permits early contact by emerging roots, which is imporptant to getting the young plants off to a good start. However, some separation between the seed and the fertilizer is normally wanted, otherwise the fertilizer, if readily soluble, may dehydrate and

Figure 15.6.
The aerial application of fertilizer to timberland. (Photo courtesy of the Weyerhaeuser Company.)

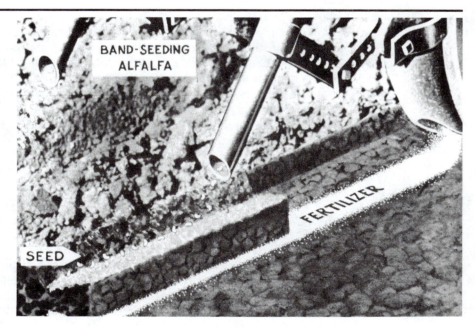

Figure 15.7.
Cutaway diagram illustrating the band placement of fertilizer beneath small-seeded legumes, such as alfalfa. The fertilizer is delivered into a furrow and then covered. A second shallower furrow is then formed to receive the seed. Finally, the seed furrow is filled and the soil compacted to assure good seed-soil contact. (Courtesy of the Potash-Phosphate Institute.)

BAND-SEEDING ALFALFA

FERTILIZER

SEED

Figure 15.8.
An 18-row planter with an attachment for fertilizer placement. The front row of boxes contains the fertilizer; the rear boxes, the seed. The side arms mark the soil to guide the driver during the succeeding trip across the field. (Photo courtesy of Deere and Company.)

kill the seed or young roots. For legumes or small grains, on narrow spacings, placement about 5 cm directly below the seed is usually satisfactory. For corn and other large-seeded crops planted in widely spaced rows, placement about 5 cm to the side and 5 cm below the seed is standard.

The amount of fertilizer applied to row crops at seeding time may be restricted to minimize possible salt damage to the seed, or it may be to guard against leaching of a soluble nutrient, such as nitrate-nitrogen, before roots have a chance to grow and absorb it. Later additions of fertilizer may therefore be necessary, and they may be applied by *side-dressing*, that is, in bands alongside the established plant rows. These applications are usually delayed until the root system is well developed and can quickly take up the added fertilizer. Applying fertilizer in a succession of treatments is called *split-application*, and it is a standard method of supplying nitrogen to lawns, golf greens, pastures, and hayland that have a continuing demand for this nutrient throughout the growing season.

Highly volatile anhydrous or aqua ammonia are usually applied to soils by *injection*, which requires special equipment such as that shown in figure 15.9. The ammonia solution or gas is carried downward in tubes and injected into the soil in slots formed by knifelike blades. The slots are immediately closed to prevent escape of the gas into the air. Once in the soil, the ammonia diffuses outward from the point of application, but the distance of movement is limited because of absorption of the gas by soil water. The depth at which the ammonia is placed must be great enough to prevent diffusion all the way to the soil surface. Typically, it is at about 15 cm, but it may be greater in coarse, sandy soils that place less restriction on the diffusion of ammonia.

Fertilizers and fertilizer use

Figure 15.9.
Equipment for injecting anhydrous ammonia into the soil. The
ammonia is delivered from the high-pressure tank through a flexible
hose to a metal tube attached at the rear of the injection knife (see
inset). (Photos courtesy of Dempster Industries.)

**Spray application
of fertilizers**

It is sometimes of advantage to apply nutrients by dissolving them in water
and spraying them directly on plants. Absorption of the nutrients can be
through leaf or stem tissue, or even through the bark of woody plants. To
avoid damage to tender leaf tissue, spray solutions must be relatively dilute,
which greatly limits the amount of nutrient that can be applied at any one
time. This makes the application of macronutrients by spraying generally
impractical, although some success has been had in using urea spray to overcome modest nitrogen deficiencies. Spray application is quite suitable for
micronutrients, however. Micronutrients can be applied successfully in sprays

Figure 15.10.
A tractor-drawn sprayer used to apply pesticides and fertilizer solutions to fruit trees. (Photo courtesy of the Goodfruit Grower, Yakima, Washington.)

added in dilute solutions that will not harm plant tissue but are still able to supply adequate nutrient to the plants. Sprays are especially beneficial when they can be used to quickly correct a deficiency that might respond to soil treatment only after a prolonged delay.

Iron is probably applied as a spray more often than other element, commonly as a 1-to-2 percent solution of ferrous sulfate ($FeSO_4$). Because of the strong tendency for iron to precipitate, overcoming a deficiency by direct application of iron salts to the soil is rarely successful. For soil application, greater benefits are usually seen where chelated iron is used, particularly on coarse, sandy soils. Greatest success is usually obtained by increasing the solubility of native iron through soil acidification, as with agricultural sulfur. This is a reasonably common practice where shrubs and other ornamentals are likely to become iron deficient because the soil pH is too high.

Spraying is especially useful in the application of micronutrients to fruit trees and vines (see fig. 15.10). These plants are particularly sensitive to shortages of some micronutrients, and often those that are difficult to maintain in available form when applied to the soil. The sprays can be applied at almost any time of the year to the fruit plants, and they may even be mixed with solutions used for insect or other pest control.

Applying fertilizer in irrigation water

Readily soluble fertilizers may be added to the soil in irrigation water. As suggested by figure 15.11, this method of application is relatively simple where commercially prepared fertilizer solutions or anhydrous ammonia are used.

Figure 15.11.
Application of anhydrous ammonia in irrigation water. The rate of ammonia delivery is controlled by a valve on the top of the tank. (Photo courtesy of the Chevron Chemical Corporation.)

Either the solutions or the gas may be metered into the water just prior to its delivery onto the field. Solutions of fertilizer salts may also be injected into lines feeding sprinkler systems. Sprinkler application of ammonia is not advised, however, for this gas is lost too easily by volatilization. The application of fertilizer in irrigation water is particularly useful during the growing season when plants are large and it is difficult to get onto the field with conventional fertilizer machinery.

Runoff should be avoided when fertilizer is being applied in irrigation water. This can be done by adding the fertilizer only during the earlier part of an irrigation; water applied later can then wash the fertilizer into the soil. Where volatile ammonia is added in the water, its concentration should be kept low, and the application should be made during the cooler part of the day in order to minimize volatilization loss.

Some important fertilizer reactions in soils

A number of common fertilizer materials behave in soils in such a way that if they are not applied correctly their availability to plants may be jeopardized. Some indication of this has already been given. For example, it has been suggested that aqua or anhydrous ammonia should be injected into the soil to minimize volatilization loss of ammonia gas, and that nitrogen-fertilizer loss may be reduced in areas if high natural precipitation by split application of the fertilizers. At this point, the behavior of urea, the acidifying effect of ammonium fertilizers, and the variable requirements for the placement of phosphorus fertilizers are considered.

Behavior of urea in soils

Urea, $CO(NH_2)_2$, is a unique fertilizer because it undergoes a succession of transformations that have a material effect on its behavior in the soil. In solution, urea remains in an undissociated, or molecular, form. Having no charge, it is not attracted to clay or organic matter and can therefore be leached from the soil under extreme conditions. Yet, such loss will not occur normally. The reason is that in the presence of a microbially produced enzyme, *urease*, urea in soils is very quickly hydrolyzed to relatively immobile ammonium carbonate:

$$CO(NH_2)_2 + 2H_2O \rightarrow (NH_4)_2CO_3 \qquad (15.7)$$

This transformation does not ensure against nitrogen loss, however, for ammonium carbonate is unstable and, under the right conditions, decomposes spontaneously to gaseous ammonia, carbon dioxide, and water:

$$(NH_4)_2CO_3 \rightarrow 2NH_3\uparrow + CO_2\uparrow + H_2O \qquad (15.8)$$

Nitrogen from urea spread on the surface of soils is therefore subject to loss by volatilization as ammonia. For this reason, it is strongly recommended that urea fertilizer be worked into the soil in some manner. Incorporation to a depth of only a few centimeters is normally adequate to avoid ammonia loss.

Reactions of phosphate fertilizers in soils

The principal reaction of phosphate fertilizers that affects their availability to plants is the precipitation of the phosphate ion in practically all soils. It may be recalled that phosphorus forms compounds of low solubility either with calcium in neutral to basic soils or with iron and aluminum in soils of relatively strong acidity, say below a pH of about 5.5. Precipitation by iron and aluminum greatly reduces the solubility of phosphorus, as well as its availability to plants. To limit precipitation by this means, band placement of phosphorus fertilizer is recommended for acid soils. Band placement concentrates the soluble phosphorus in a restricted soil volume, thus reducing contact with soluble iron and aluminum and thereby slowing the precipitation reaction.

Phosphorus fertilizers that benefit most from band placement are the more soluble materials, such as the superphosphates and the ammonium phosphates. Slowly soluble phosphates, especially rock phosphate, should not be band-placed. Instead, they should be mixed thoroughly with the soil to ensure maximum contact between the fertilizer particles and plant roots. The greater the quantity of the fertilizer and the finer its state of subdivision the greater the contact with roots.

Readily soluble phosphates ordinarily are not band-placed in soils where the phosphorus is precipitated by calcium. The initial precipitation by calcium is as dicalcium phosphate, $CaHPO_4$, a compound that dissolves readily on root contact. As with rock phosphate in acid soils, therefore, a more widespread distribution of the precipitated dicalcium phosphate will assure better contact with roots. Band placement will slow precipitation of the phosphorus by calcium, but it will also reduce the ability of roots to contact the phos-

Figure 15.12.
Response of field beans to broadcast and band-placed phosphate fertilizer on a calcareous, irrigated soil. Plots showing little growth received no fertilizer or were fertilized by band placement to the side and below the seed. In other plots, the fertilizer was mixed with the soil before planting. (Photo courtesy of A. I. Dow, Washington State University, Prosser.)

phorus. For this reason, band placement of soluble phosphates actually supplies less available phosphorus to plants than where the fertilizer is thoroughly mixed with the soil (see fig. 15.12).

Superphosphates, which contain monocalcium phosphate, $Ca(H_2PO_4)_2$, as the phosphorus source, behave uniquely in soils. When a granule of superphosphate fertilizer dissolves in moist soil, a large number of $H_2PO_4^-$ ions are released. Once in solution they dissociate to yield H^+ and HPO_4^{2-} ions:

$$H_2PO_4^- \rightleftharpoons H^- + HPO_4^2 \tag{15.9}$$

The latter of these two ions, in turn, recombines with calcium, also released from the dissolving monocalcium phosphate, to precipitate as slightly soluble dicalcium phosphate, $CaHPO_4$, at the granule site. The H^+ ions, on the other hand, combine with H_2PO_4 from the monocalcium phosphate to form readily soluble phosphoric acid, H_3PO_4. As shown by the following equation, for each molecule of dicalcium phosphate produced, one molecule of phosphoric acid is formed:

$$Ca(H_2PO_4)_2 \rightleftharpoons CaHPO_4 + H_3PO_4 \tag{15.10}$$

Whereas the dicalcium phosphate formed by this reaction is precipitated at the granule site, the phosphoric acid moves away from the granule in soil water. In basic soils, the acid will ultimately be neutralized and precipitated, most likely as dicalcium phosphate. In acid soils, however, it will probably precipitate as iron or aluminum phosphate. Because of the very low solubility of the latter compounds, only the dicalcium phosphate precipitated at the granule site will remain as a ready source of phosphorus to plants.

Table 15.8. The potential acidity of common fertilizers expressed in equivalents of pure calcium carbonate. (Adapted from *Western Fertilizer Handbook,* 5th Ed., 1975. Courtesy of the California Fertilizer Association, Sacramento, CA.)

Fertilizer	Nutrient content[a]			Equivalent acidity Kilograms of CaCO$_3$ per:	
	N	P$_2$O$_5$	K$_2$O	100 kg of fertilizer	100 kg of nitrogen
	%	%	%		
Ammonium nitrate	33	—	—	62	185
Monoammonium phosphate	11	48	—	58	528[b]
Diammonium phosphate	18	46	—	70	390
Ammonium sulfate	21	—	—	110	525
Anhydrous ammonia	82	—	—	147	180
Aqua ammonia	20	—	—	36	180
Urea	45	—	—	71	160
Ordinary superphosphate	—	18–20	—	Neutral	—
Concentrated superphosphate	—	45–46	—	Neutral	—
Phosphoric acid	—	52–54	—	110	—
Superphosphoric acid	—	76–83	—	160	—
Potassium chloride	—	—	60–62	Neutral	—
Potassium sulfate	—	—	50–53	Neutral	—

[a]Percentages of total N, available P$_2$O$_5$, and water-soluble K$_2$O.
[b]Half of the acidity comes from the phosphate ion, H$_2$PO$_4$.

Acidifying effects of fertilizers

Many fertilizers produce an acid reaction in soils. Because such an effect can have important long-term consequences on soil chemical properties, especially where acidity is already excessive, state laws often require fertilizers to be labeled to show their acid-producing potential. This potential is usually expressed in terms of the quantity of pure calcium carbonate that would be needed to neutralize the acid derived from a unit quantity of the fertilizer (see table 15.8).

Of the more widely used fertilizers, those supplying nitrogen in the ammonium form have the greatest capacity for producing acidity. The acidifying effect of ammonium fertilizers is due to the release of free H$^+$ ions as the ammonium undergoes nitrification. In simplified form, the reaction is:

$$NH_4^+ + 2O_2 \rightarrow 2H^+ + NO_3^- + H_2O \qquad (15.11)$$

The equation shows 2 H$^+$ ions produced for each ammonium ion nitrified. Where large amounts of ammonium fertilizers are added to soils of low buffering capacity, the H$^+$ produced can have a profound effect on soil pH even in a single growing season.

Of the various nitrogen fertilizers, ammonium sulfate, (NH$_4$)$_2$SO$_4$, has the greatest potential for increasing soil acidity. All of the nitrogen in this fertilizer is in the ammonium form. Anhydrous and aqua ammonia, along with urea after it has hydrolyzed to ammonium carbonate, also contain nitrogen solely in the ammonium form. However, these materials have less of an acidifying effect than ammonium sulfate because, being inherently basic, they neutralize part of the acid produced by nitrification. Ammonium nitrate

Figure 15.13.
The change in pH of a slightly basic loamy sand treated with three different nitrogen fertilizer salts and incubated in the laboratory. Nitrogen was added at a rate of 200 kg per 2 million kg of soil.

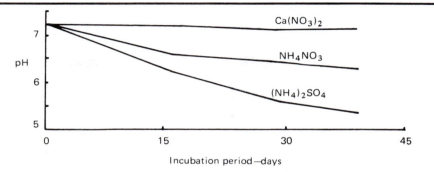

Figure 15.14.
The effect of soil acidification by ammonium fertilizers on iron availability to petunias. Maximum acidification by ammonium sulfate resulted in total elimination of iron-deficiency symptoms. (From Lucas, R. E. and Knezek, B. D., *Micronutrients in Agriculture*, 1972, pages 265–88 by permission of the American Society of Agronomy. Photo by James Boodley, Cornell University, Ithaca, NY.)

produces less acid than does ammonium sulfate because only half of the total nitrogen present is in nitrifiable ammonium form. Nitrogen fertilizers containing only nitrate have no acidifying effect on the soil.

Differences in the acidifying effect of three nitrogen fertilizers on a loamy sand of very low buffering capacity are shown in figure 15.13. The fertilizers were added at rates supplying the same total amount of nitrogen, after which the soil-fertilizer mixtures were incubated for a six-week period to allow nitrification to take place. As may be seen in the figure, ammonium sulfate caused the greatest depression in pH; ammonium nitrate was next. Calcium nitrate produced no significant change in pH. A striking example of the effect of different nitrogen sources on soil acidity, as it in turn affects iron availability, may be seen in figure 15.14.

Summary

Fertilizers comprise a diverse group of compounds prepared in solid, liquid, and even gaseous form. Principal among these are compounds of nitrogen, phosphorus, and potassium, a number of which contain substantial amounts of sulfur. Through the selection of appropriate compounds or mixtures, any nutrient may be supplied to the soil in fertilizer form.

The contents of nitrogen, phosphorus, and potassium in fertilizers are expressed by the grade, or guarantee, of the fertilizer. The standard expression is percent by weight of total nitrogen, available phosphorus, and water-soluble potassium. For other nutrients, as with nitrogen, the total amount present, expressed as a percent of the fertilizer weight, is indicated. With but two exceptions, the composition of fertilizers is computed using the elemental weights of the nutrients. The exceptions are phosphorus and potassium. Contents of these nutrients are traditionally computed on the basis of the weight of their oxides, P_2O_5 and K_2O.

The ability of a fertilizer to supply nutrients to plants is a function of its nutrient content, its solubility, and the method by which it is applied to the soil. Methods of application include broadcasting, band placement, spraying directly on plants, and delivery in irrigation water. The most appropriate method is one that is practical to use under existing conditions and that will assure maximum availability of the nutrient to plants once the fertilizer has been applied.

Review questions

1. Why do the ammonium and nitrate forms of nitrogen, though quite different chemically, have about the same effect on plant growth once applied to the soil in fertilizers?
2. Why is the manufacture of ammonia gas so important to the nitrogen-fertilizer industry?
3. What is likely to happen to the level of available nitrogen in a soil, at least initially, when the soil is treated with a large amount of organic material containing around 1 percent nitrogen by weight? Explain your answer.
4. Compare ordinary and concentrated superphosphates in terms of the available P_2O_5 contents and the form of phosphorus they release to the soil.
5. Why are the water-soluble and total contents of potash (K_2O) in a potassium chloride fertilizer likely to be the same?
6. Which of the materials in table 15.5 supply sulfur in available form immediately upon being added to moist soil?
7. What is the principal advantage of adding micronutrient cations to soils in chelate fertilizers?
8. Why would the guaranteed analysis of a 4-8-4 mixed fertilizer, in which the contents of phosphorus and potassium are expressed on the elemental basis, become 4-18.4-4.8 if the contents of phosphorus and potassium were expressed as oxides?

9. What is the value of knowing the open-formula guarantee of a mixed fertilizer?
10. What is the major advantage over broadcasting of band-placing water-soluble phosphate fertilizers in strongly acid soils?
11. The split-application of fertilizer is beneficial primarily for which single fertilizer nutrient?
12. Outline the changes in urea-nitrogen that leads ultimately to its conversion to the nitrate form in soils.
13. Why does applying the same quantity of nitrogen as ammonium sulfate fertilizer have a greater acidifying effect on soils than if it is applied in either ammonium nitrate or anhydrous ammonia?

Selected references

Bartholomew, W. V., and Clark, F. E., eds. *Soil Nitrogen*. Madison, WI: American Society of Agronomy, 1965.

McVikar, M. H., Martin, W. P., Miles, I. E., and Tucker, H. H., eds. *Agricultural Anhydrous Ammonia*. Madison, WI.: American Society of Agronomy, 1966.

Mortvedt, J. J., Giordano, P. M., and Lindsay, W. L., eds. *Micronutrients in Agriculture*. Madison, WI.: Soil Science Society of America, Inc., 1972.

Olson, R. A., Army, T. J., Hanway, J. J., and Kilmer, V. J. *Fertilizer Technology and Use*. Madison, WI.: American Society of Agronomy, 1971.

Stefferud, A., ed. *Soil, 1957 Yearbook of Agriculture*. Washington, D.C.: Government Printing Office.

Terman, G. L., Hoffman, W. M., and Wright, B. C. Crop Response to Fertilizers in Relation to Content of "Available" Phosphorus. *Advances in Agron.* 16:59–100. New York: Academic Press, Inc., 1964.

Tisdale, S. L., Nelson, W. L., and Beaton, J. D. *Soil Fertility and Fertilizers*. New York: The Macmillan Company, 1984.

16

Fertility management

Maintenance or improvement of the fertility status of soils is but one of the many steps important to sustaining maximum productivity in forest, range, and cropland. Unlike certain other productivity factors, however, fertility is subject to man's control. The correction of serious nutritional problems can be intensely gratifying, but more importantly, it is a fundamental requirement in most successful agricultural operations.

Maximum benefits are derived from a fertility program only where it is coupled with other treatments or practices that ensure the most suitable conditions for plant growth. These include the use of adapted, highly productive plant varieties, and where needed, the correction of problems due to salts or poor drainage, the control of diseases and insect pests, the conservation of moisture, and the minimization of soil loss by erosion. If limiting factors of this type go unheeded, the effectiveness of any fertility program can be reduced if not eliminated altogether.

A well-designed fertilizer program is based on three fundamental requirements: (1) the use of proper fertilizers, (2) the use of methods of application that ensure maximum utilization of added nutrients by plants, and (3) the application of fertilizer at rates most appropriate to the plant and to growing conditions. Factors that may modify the fertilizer program include: (1) the use of animal and green manures, (2) the cropping sequence, if rotations are employed, and (3) the control of pH, the latter being important to the availability of a number of nutrients.

Determining the proper fertilizer

A fertilizer may be considered suitable if it provides readily available nutrients that are in short supply in the soil. There is little point in applying nutrients when no benefit may be expected from them.

A number of means are available for determining which of the several essential elements may be limiting plant growth. Among them are: (1) the visual identification of deficiency symptoms, (2) soil analysis, especially the

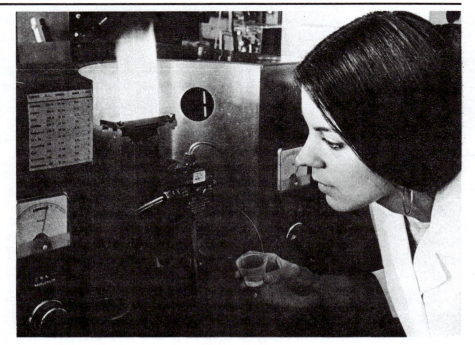

Figure 16.1.
Reading an atomic absorption spectrophotometer, an instrument widely used in the precise determination of nutrient elements in soil and plant-tissue analyses. (University of California, Davis, Cooperative Extension Service Laboratory photo. From the *Western Fertilizer Handbook,* 6th ed., 1980. Sacramento, CA: California Fertilizer Association.)

rapid soil test, (3) plant analysis, (4) field trials with fertilizers, and (5) greenhouse pot tests. Of these, the first three hold most promise in the practical diagnosis of nutritional problems. The greenhouse test is largely a research tool. The same may be said for the field trial, except where productivity is initially so restricted or deficiency symptoms so pronounced that response to a fertilizer can be judged correctly by simple visual means.

The use of deficiency symptoms in solving fertility problems is complex and may appear confusing to the layman. However, diagnosis of fertility problems within a given locale is usually limited to a relatively small number of nutrients. Thus, it is not difficult for an individual to learn to recognize the few symptoms normally encountered. The task is more complicated in areas of diversified farming where manifestations representing a single nutritional disorder may vary considerably from one plant type to another. Under these circumstances the advice of an expert may be required.

Most major agricultural areas in the United States have state or commercial laboratories capable of performing reliable nutrient tests on soil and plant samples (see fig. 16.1). Soil tests are more widely used than plant analyses, principally because they are simple, economical, and their reliability is more clearly defined. Furthermore, soil tests can be run almost any time of the year, whereas the use of plant analysis is limited to the growing season.

Nonetheless, the latter means of assessing the fertility status is useful with nutrients that are not amenable to soil testing; it is commonly applied to high-value crops where close control of nutrition is wanted during the growing season.

The interpretation of either soil test or plant analysis data is complex and requires the judgment of a person with broad experience. Soil testing as an aid to fertility control is strongly encouraged. However, since this diagnostic tool is not too well understood, some of its principal features are worthy of consideration.

The nature of soil tests

Soil tests serve two interrelated functions: (1) they aid in determining whether a particular nutrient is responsible for poor productivity, and (2) they provide a basis for judging the amount of fertilizer needed to correct a nutrient shortage. A practical soil test must be rapid, and it must estimate with reasonable accuracy the amount of a soil nutrient that will be made available to plants during a growing season.

The soil test procedure

Rapid soil tests involve the extraction of a small soil sample with water or a salt or acid solution and measurement of the amount of the nutrient elements released during the extraction. The quantity of nutrients measured is normally expressed in kilograms per hectare or parts per million parts of soil. Since, on the average, a hectare furrow slice of soil weighs approximately 2 million kilograms, one part of extracted element per million parts of sample is equivalent to 2 kilograms per hectare furrow slice.

Salt solutions are used in soil tests because they provide a cation, usually sodium, that displaces at least a portion of the nutrients present as exchangeable cations in the soil sample. Tests for phosphorus depend upon the solubility of phosphorus-bearing compounds in the sample and appear to work best if the extracting solution has a pH near the pH of the soil. Thus, acid-soil samples are extracted with acidic salt solutions, whereas the reverse is true for samples on the basic side. Through careful control of the pH, a single solution may be used to extract both phosphorus and the nutrients held in exchangeable form. Other types of extractions are used for nitrate-nitrogen, sulfate-sulfur, and the micronutrients.

Soil test correlations

The results of soil tests are meaningless unless they can be related to conditions in the field. The actual relationships are established through soil test *correlations*, or calibrations, based on fertilizer trials conducted in the field on soils of widely varying nutrient-supplying power (see fig. 16.2). For the correlations, the response to an added nutrient is related to the levels of extractable nutrient in soil samples taken from the various experimental sites.

Data from a well correlated soil test for potassium are shown in figure 16.3. Each point on the graph shows the relative yield of corn on unfertilized soil at a single experimental site plotted against the soil-test value (level of exchangeable potassium) for that site. Thus, the 22 points on the graph rep-

Figure 16.2.
Harvesting pasture plots used in soil-test correlation studies. Relating yields from unfertilized and fertilized test plots to soil test values for the unfertilized soil at numerous locations is the basis for soil test correlations.

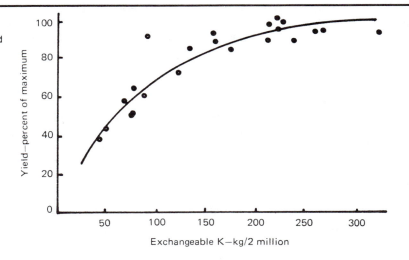

Figure 16.3.
Relationship between the yield of unfertilized corn, expressed as a percent of the yield obtained with potassium fertilizer, and the level of exchangeable potassium in the unfertilized soil. Data are for Illinois. (After R. H. Bray, *Soil Sci.* 58:305–24, 1944.)

resent 22 different field locations covering a wide range in soil-test potassium and corn yield. Since the plotted corn-yield and soil-test values vary in a close relationship, a high degree of reliability can be placed on the soil test; that is, the soil test serves as a relatively accurate means of predicting corn yield and need for potassium fertilizer in the climatic-soil region where the test correlation was carried out. Better correlation would result if all 22 plotted points fell on a smooth curve of the type shown in the figure, but such perfect agreement should never be expected. If the points were scattered indiscriminantly over the graph, then the soil test correlation would be poor and the test would not be suitable for predicting fertilizer need.

Most routine soil tests are run on surface soil samples taken to slightly beyond the depth of plowing. Thus, a measure of subsoil fertility is not obtained, although plants utilize nutrients from this part of the profile. No great problem is encountered if the distribution of available nutrients within the profile is consistent from soil to soil. Frequently, this is not the case, however. Soils often show similar soil test values based on surface soil sampling, yet they respond differently to fertilization because of marked differences in their ability to supply available nutrients from the subsoil.

There are soil tests that measure nutrient availability in more than just the surface, or plow layer, of soil. A prime example is a test for nitrate-nitrogen in soils used for small grain production under dryland (natural rainfall) conditions. For such a test, the soil is sampled early in the season to depths of up to 180 cm. The total nitrate-nitrogen present within this depth is determined on the assumption that it will be totally used during the growing season. Stored water is also measured and, along with anticipated rainfall, is used as the basis for predicting grain yield and the total need for nitrogen. The amount of nitrogen to be added as fertilizer is then computed as the difference between the total need and that present as nitrate plus an estimated quantity of N that will be released by the soil during the growing season.

The soil sample   The soil sample used for the laboratory test ordinarily consists of a composite of several small samples taken from various places within a field. If the test is to be of value, the composited sample should represent only one condition in the field. Thus, areas of different growth potential should be sampled separately (see fig. 16.4). Differences in such things as (1) slope, (2) texture, (3) degree of erosion, (4) past lime, fertilizer, and manure treatments, and (5) cropping history are logical bases for collecting separate composite samples. Small, unusual areas, as where hay or straw has been stacked, should be avoided or sampled separately. Recommendations based on soil tests for each composite sample will then be made after taking into account the type of crop to be grown. If the fertility pattern in a field is highly complex, one may be forced to sample only those areas of the poorest apparent nutritional status and treat the entire field on the basis of results obtained for these areas.

The small samples collected for compositing may consist of a thin slice of soil cut from the side of a freshly dug hole, or they may be obtained with a soil auger or sampling tube (see fig. 16.5). Where the field has been previously fertilized, sampling should extend slightly beyond the depth of plowing. If the fertilizer has been applied in bands, a relatively large number of samples should be taken for compositing, preferably after the residual fertilizer has been mixed with the soil by plowing.

Figure 16.4.
Diagram illustrating
factors important to
soil sampling for soil
tests. Each of the four
fields should be
sampled separately if
they have different
cropping and
fertilization histories.
Method of collecting a
composite sample is
shown for site 6.
(From Nebraska Ag.
Ext. Ser. Circ.
56–116.)

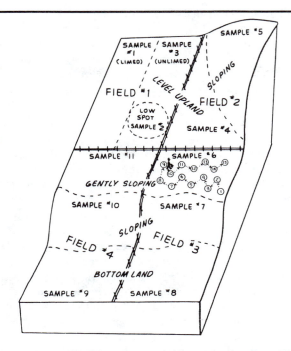

A   B

Figure 16.5.
Obtaining soil samples for compositing (A) with an open-sided
sampling tube, and (B) from a thin spade slice. Each composite
should contain 10 to 15 individual samples representing a given
level of productivity.

Soil tests have been used successfully with but a few of the several plant nutrients. Unquestionably, tests for phosphorus and the three cationic nutrients, potassium, calcium, and magnesium, are of greatest value. Neither nitrogen or sulfur lend themselves to soil testing in most instances, the reason being that the rate at which they are released from soil organic matter cannot always be accurately predicted. Even so, where cropping and climatic patterns are much the same from year to year, as in some dryland, wheat-growing regions, reasonable success has been achieved in predicting N-fertilizer needs from soil analyses.

One of the more serious limitations in soil testing has been a general lack of suitable tests for micronutrient availability. A test for boron has been widely used for years, but only recently have procedures for zinc, copper, manganese, and iron been introduced. Whether or not the proposed tests for the latter elements have widespread application must be determined by extensive correlation trials. No test has been devised that predicts the availability of molybdenum, and there seems to be little need to develop one for chlorine, for it is adequately supplied in the vast majority of soils.

In most well equipped laboratories the determination of pH is routine, since the diagnosis of nutritional problems is often aided by knowledge of soil reaction. An estimate of texture and organic matter content may also be made if knowledge of these characteristics is helpful in fertility diagnosis. Depending upon the area served by a laboratory, lime and gypsum requirements, salt content, and tests for toxic elements such as arsenic may be run on request.

Fertilizer recommendations for nutrients not amenable to soil testing may be based on response data obtained from fertilizer-rate experiments carried out under conditions typical for a given combination of crop, soil, and climate. Such recommendations are subject to many uncertainties because of soil variability. Where the same crop is grown repeatedly, nitrogen may be added at the same rate year after year, if this practice appears advisable. Micronutrients may be applied when plant deficiency symptoms or actual field trials show a need for them, or they may be applied occasionally as an insurance against possible deficiency. Application of micronutrients on this basis must be with great care, however, because excesses of these elements can so easily cause toxicities, either in the plants or in animals that use them for food. The need for micronutrients can sometimes be judged from plant analysis.

Frequently, a soil test shows the quantity of a nutrient to be far in excess of that needed by a crop. Under these circumstances the application of the nutrient should be discontinued, for numerous problems can be introduced by the accumulation of excess fertilizer materials in the soil. In some areas, soils used for citrus, tobacco, and potatoes and other vegetables have accumulated enough phosphorus from repeated fertilization to supply plant needs for a number of generations to come. Similarly, tests on heavily fertilized soils have shown nitrate-nitrogen in the plow layer to be up to ten times greater than the amount normally recommended for an annual application. If nothing else, such large concentrations of nitrogen can have serious effects on crop quality.

The soil test is a highly useful aid in the development of practical soil fertility programs. These tests are not infallible, however, and recommendations based on them should always be held in question. A constant search should be conducted for signs that either support or disprove the accuracy of the recommendations. Only in this way can confidence in the soil test be established.

The nature of plant tissue tests

Plant analysis as a measure of soil fertility depends on the relationship between the nutrient concentration in plant tissue and the nutrient-supplying power of soils. As with soil tests, chemical tests involving the plant must also be calibrated for use. However, the calibration is more complex than for soil tests. The reason it is more complex is that measured nutrient concentrations, which are the basis of the tests, vary with the stage of plant development; the concentration of a nutrient tends to decrease as the maturity of the plant increases.

The general relationship between nutrient content and plant age is shown in figure 16.6, where the four curves on the left plot changing levels of a nutrient in a plant as caused by four different rates of fertilization, R_0 through R_3. The effect of fertilization on yield is shown in the right-hand diagram. According to this diagram, an optimum or near optimum yield is produced by fertilization rate R_1. As a consequence, the nutrient-content curve R_1 shows the desired level of the nutrient in the plant throughout the growing season. Nutrient-content values plotting below curve R_1 would suggest a need for fertilization, whereas those occurring above curve R_1 would indicate a level exceeding the needs of the plant.

For the assessment of the fertility status through plant analysis, curves of the type in figure 16.6 must be developed for each plant and nutrient, usually through repeated field trials so that the normal range of variability in the content of a nutrient can be determined. Analyses may be for the total quantity of a nutrient or for some fraction of the total, such as that extracted from a sample by a salt or acid extracting solution. Extractions of this type remove only nutrients present in the plant sap, but they are usually quite sensitive since changes in the availability of nutrients in the soil reflect quickly in the content of soluble nutrients in the sap.

The part of the plant that is used for tissue tests varies depending on the plant and on the nutrient under consideration. In a great many instances, the blade or petiole of the most recently matured leaf is used (see fig. 16.7). Petiole analysis is common where the test is for salt- or acid-extractable nu-

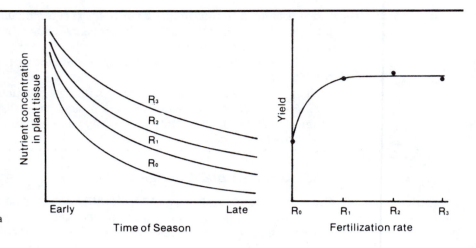

Figure 16.6.
Curves showing the trend in the concentration of a nutrient in plant tissue with time during the growing season as affected by four levels of available nutrient in the soil. As determined from the righthand diagram, fertilizer rate R_1 provides adequate, though not excess, nutrient for maximum yield and is therefore assumed to maintain a near optimum concentration of nutrient in the plant.

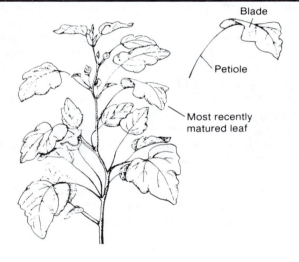

Figure 16.7.
Illustrating the most recently matured leaf of a plant and the distinction between leaf blades and petioles.

trients. This is logical because all nutrients that enter the leaf must pass first through the petiole. The leaf blade is more often used where the analysis is for the total quantity of a nutrient in plant tissue.

The greatest limitation to the use of plant analysis in fertility diagnosis and control is due to a lack of correlation data for many plants and for many growing conditions. Nonetheless, where good correlations exist, this technique has been successfully applied in the identification of both deficiencies and excesses of nutrients. Especially useful applications are with vine and tree fruits, for which good soil test correlations are difficult to obtain, and in the control of the nitrogen nutrition of long-season crops such as sugar beets, sugar cane, and pineapple. Continual monitoring of the nitrate concentration in these plants permits periodic adjustment of the nitrogen supply so that it is always adequate but not excessive. Such control is possible because these crops have a long growing season and, therefore, time to respond to a fertilizer application, and because the nitrogen can be supplied in a readily available form that is quickly absorbed and utilized by the plants.

Important factors in fertilizer selection

Assuming the kind and amount of nutrients to apply are known, a choice of the type of fertilizer must still be made. Logically, the one selected should provide the maximum return for the investment. The choice is not always an easy one, however, for a number of variables determine the relationship between fertilizer cost and value received. In general these resolve into two factors: (1) the cost per unit of nutrient applied, and (2) the ability of the fertilizer to provide nutrients that are readily available to plants. The latter factor relates to both the solubility characteristics of the fertilizer and its ultimate reaction with the soil. Sometimes a fertilizer is chosen because it has properties compatible with a particular method of application.

Fertilizer costs
How much one pays to fertilize a crop depends primarily on two things: (1) the cost of the fertilizer material, and (2) the expenses involved in transporting, handling, and applying the fertilizer in the field. The latter expenses are a function largely of the relative content of nutrients in the fertilizer, with those of highest analysis (nutrient content) generally being the least expensive to transport and apply. Often, the fertilizer is applied on a custom basis; that is, the dealer not only sells the fertilizer but also delivers and applies it in the field for a predetermined price.

In judging the overall cost of fertilization, an initial step is to compare prices to determine which material will supply the desired quantity of nutrients at the lowest possible cost. The basis for the comparison is the cost per unit weight of nutrient, the unit weight being a quantity such as the kilogram. This is shown by an example that compares the cost per kilogram of nitrogen in two fertilizers: ammonium nitrate and anhydrous ammonia.

The basic formula for computing the cost per unit weight (kilogram, pound) of nutrient in a given quantity of fertilizer is:

$$\text{Nutrient (cost/unit)} = \frac{\text{Fertilizer cost}}{\text{Fertilizer weight}} \times \frac{100}{\text{Nutrient content-\%}} \quad (16.1)$$

where the nutrient cost is in dollars per unit weight, and the fertilizer cost is in dollars for a specified number of unit weights (fertilizer weight), such as 1000 kg (1 metric ton) or 2000 lbs (1 British ton). Applying this equation to our example and considering first the cost per kilogram of nitrogen in ammonium nitrate containing 33 percent N and priced at $241 per metric ton:

$$\text{N (cost/unit)} = \frac{\$241}{1000} \times \frac{100}{33} = \frac{\$241}{330 \text{ kg}} = \$0.73/\text{kg} \quad (16.2)$$

For anhydrous ammonia at $435 per metric ton and containing 82 percent total N:

$$N \text{ (cost/unit)} = \frac{\$435}{1000} \times \frac{100}{82} = \frac{\$435}{820 \text{ kg}} = \$0.53/\text{kg} \qquad (16.3)$$

According to these computations, the anhydrous ammonia would be the least expensive.[1] If it could be applied as cheaply as the ammonium nitrate and was as satisfactory in other respects, the anhydrous ammonia would be the material to buy.

The computation of nutrient costs in fertilizers containing more than one nutrient is more complicated than where only a single nutrient is present. The simplest approach is to determine the relative cost of each nutrient in the mixture in terms of equivalent costs in single materials. For example, consider a 10–10–10 fertilizer containing 10 percent each, or 100 kg per metric ton, of N, P_2O_5, and K_2O. Assume that a fair market value for these nutrients is, respectively, $0.68, 0.55, and 0.44 per kilogram. On this basis, their net value in a 10–10–10 fertilizer would be computed as:

N	$100 \times 0.68 =$	$68
P_2O_5	$100 \times 0.55 =$	55
K_2O	$100 \times 0.44 =$	44
		$167

The basis for comparison would therefore be $167 per metric ton of 10–10–10 fertilizer. If the values assigned to the three nutrients were indeed minimal, the cost of the mixed fertilizer would probably exceed $167, since it is more expensive to manufacture mixed than single fertilizer materials. However, the extra charge may be more than compensated for by savings in transportation and application costs; it would be less expensive to apply all three nutrients at one time in a mixed fertilizer than to apply them separately in three different materials.

Where two nutrients are present in a fertilizer, a reasonably accurate estimate of the unit cost of one can be obtained by substituting a fair market value for the other in the cost of the fertilizer. This can be illustrated by estimating the unit cost of phosphorus in diammonium phosphate fertilizer with an analysis of 18–46–0 and priced at $350 per metric ton. A metric ton of this fertilizer will contain 180 kg of N and 460 kg of P_2O_5. For the computation, assume a fair value for N to be that in anhydrous ammonia as calculated in equation (16.3), or $0.53 per kilogram. At this cost, the total value of the nitrogen would be $180 \times 0.53 = \$95.40$, and when subtracted from the cost per metric ton of fertilizer would yield a total cost for phosphorus of $350.00 - 95.40 = \$254.60$. At $254.60 for 460 kg of P_2O_5, the unit cost would be $254.60/460 = \$0.55$ per kilogram of P_2O_5.

---

[1]At the prices per metric ton quoted above, the cost per British ton (2000 lbs) of ammonium nitrate would be $218; for anhydrous ammonia, it would be $394. Substituting these values in equation (16.1) would give costs per pound of nitrogen of $0.33 and $0.24, respectively, in ammonium nitrate and anhydrous ammonia.

Figure 16.8.
The relationship between costs of production and net financial return from increasing rates of fertilizer. Profitable yields are obtained at any level of productivity between the fertilization rates of X_1 and X_4. Maximum return is obtained at the X_2 rate. (From J. T. Pesek, and E. O. Heady, *Proc. Soil Sci. Soc. Amer.* 22:419–23, 1958.)

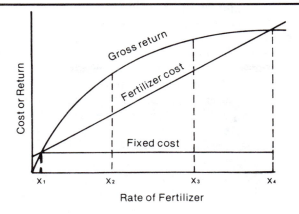

Fertilizer application rates

Customarily, economics control of the use of fertilizers. Many times fertilizers are not applied because the cost is thought to be in excess of the benefits gained. This assumption is probably incorrect in the majority of instances, for most agricultural soils are deficient in one or more nutrients that can be profitably applied in fertilizer.

Where fertilizer is used, the rate selected should be one that results in the maximum return from the investment. Normally, this rate provides something less than maximum yield, the reason being that as the maximum is approached the cost necessary to sustain continuously expanding yields increases more rapidly than the resulting financial return. As shown in figure 16.8, a linear increase in fertilizer costs is associated with increases in yield and financial return that are substantial at first but then tend to level off at higher rates of fertilizer application. The greatest gain is obtained when the difference between the total return and the cost of fertilizer is maximum. This corresponds to an application rate of X_2 in the diagram.

Fertilization that exceeds the optimum is of less serious economic consequence than where it is below the optimum. This fact can be demonstrated through use of figure 16.8. For example, if the rate were reduced from the optimum, X_2, to X_1 in the diagram, the net income would just equal total production costs. However, should the rate be increased above the optimum by the same amount, to X_3, the associated yield would still provide a return in excess of total production costs.

Time of fertilizer application

Sound fertility control often requires proper timing of the fertilizer application. The best time depends upon the potential for loss of nutrients through leaching or precipitation in unavailable form and on the feeding habit and value of the crop. Under most circumstances, fertilization at or just prior to planting is desirable, with split applications being used when necessary to protect against salt damage and leaching loss. The time of application of fertilizers that are broadcast and mixed with the soil is generally dependent upon plowing dates.

The fall application of fertilizers is sometimes recommended, principally as a means of spreading the work load throughout the year. This is a suitable procedure for potassium under most circumstances, and for phosphorus if its precipitation in unavailable form is not a problem. Ammonium-nitrogen may be applied in the fall after the soil temperature has dropped below the threshold for nitrification. If a soil remains warm and is subject to winter leaching, fall application of nitrogen as either ammonium or nitrate is risky, especially where a coarse texture allows rapid leaching.

It is possible to slow the leaching loss of nitrogen and thereby have greater latitude in the time of nitrogen fertilizer application through the use of slow-release fertilizers or *nitrification inhibitors*. The use of slow-release nitrogen fertilizers, as described on page 393, is limited because of their expense. Nitrification inhibitors are a practical alternative from an economic point of view.

Nitrification inhibitors slow the conversion of ammonium-nitrogen to the more mobile nitrate form. They are organic compounds that selectively block the action of *Nitrosomonas*, the bacterium that converts ammonium to the nitrite (NO_2^-) ions. The effect is temporary, however, for the inhibitors undergo gradual decay and eventually allow reinitiation of the nitrification process. Normally, the delay in nitrification is from a few weeks to a few months depending on the concentration of the inhibitor and on soil factors that determine the rate at which it decomposes. Nitrification inhibitors add to the expense of fertilization, but they may more than pay for themselves by preventing the loss of nitrogen through leaching. In addition, by slowing the rate of nitrification during the growing season, they may eliminate the need for multiple, or split, applications of nitrogen fertilizers.

Manures and plant residues in fertility programs

Previous discussions of the use of supplemental nutrients on soils have dealt almost entirely with commercial fertilizers. However, the improvement or maintenance of soil fertility in many instances depends also on the use of organic supplements. These materials may be derived from a number of sources but, for the most part, consist of manures and plant residues. The plant residues may be the materials left in a field following harvest, or they may be green manure or cover crops destined for incorporation into the soil.

The benefits derived from the use of organic materials on soils are many but relate principally to: (1) the contribution of nitrogen and other nutrients in readily available form, and (2) the improvement of the physical condition of mineral soils. Nutritional gains are substantial where plant residues and animal manures are applied to the soil in large quantities.

The nature and use of animal manures

The use of manures as a soil supplement has dwindled in recent decades with the increasing influence of mechanization on agriculture. Furthermore, a decrease in the number of farms, along with a higher degree of agricultural specialization, has caused the gradual disappearance of small milking operations that formerly provided manure for local consumption. Still,

Table 16.1. The moisture content, on a fresh-weight basis, and the nitrogen, phosphorus, and potassium contents, on both a fresh- and dry-weight basis, for various animal manures. All values are in percentages. (From Loehr, R. C. 1968. *Pollution Implications of Animal Wastes—a Forward-Oriented Review.* Ada, Okla.: Robert S. Kerr Water Research Center).

Animal	Moisture	N		P		K	
		Fresh	Dry	Fresh	Dry	Fresh	Dry
Dairy cattle	79	.56	2.66	.10	.48	.50	2.38
Fattening cattle	80	.70	3.50	.20	1.00	.45	2.25
Hogs	75	.50	2.00	.14	.56	.38	1.52
Horses	60	.69	1.72	.10	.25	.60	1.50
Sheep	65	1.40	4.00	.21	.60	1.00	2.86
Broiler chickens	25	1.70	2.26	.81	1.08	1.25	1.66
Hens	37	1.30	3.51	1.20	3.24	1.14	3.08

manure production remains high in feedlot operations and on the larger poultry and dairy farms. Where such establishments provide an inexpensive and convenient source of manure, its use can be encouraged as one means of sustaining soil productivity. Actual gains made from applied manure depend upon the need for nutrients and organic matter, the specific character of the manure, and the manner in which it is handled prior to incorporation into the soil.

The composition of manure

Beef and dairy cattle are the principal source of manure. Commonly, the manure is collected in pens or sheds where the animals remain a large part of the time. In some cases, the holding areas are lined with litter, usually straw or sawdust, which contributes to the comfort of the animals as bedding and also acts as an absorbant for liquid excrement. Manure accumulating under these conditions is diluted by the litter of relatively low nutrient content. Under feedlot conditions used for fattening cattle, manure usually accumulates without litter (see fig. 16.9). Although some soluble nutrients are lost by seepage of liquid wastes into the ground, the lack of litter results in manure of generally higher nutrient content (compare dairy with fattening cattle in table 16.1).

The composition of manure varies widely. As seen in table 16.1, these differences can be due to the type of animal, but important as well is the type of feed given the animal and the conditions under which the manure is collected and stored. The quantity of bedding materials contained in the manure also affects the composition, mainly by adding large amounts of substances such as cellulose, hemicellulose, and lignin without adding much in the way of nutrients such as nitrogen, phosphorus, and potassium.

Nutrient forms in manure

Animal manures contain all the nutrients plants need, but the ability of manure to supply these nutrients depends on the total amounts present, the forms in which they occur, and on the rate at which they are released to the soil. Most of the nutrients are released quite rapidly. Potassium, for instance, occurs in wholly soluble form in manure and is therefore immediately available for plant use. In contrast, phosphorus and a number of the micro-

A

B

nutrients occur in organic combination in manure, but ordinarily they are rapidly released through mineralization once the manure is applied to the soil.

Nitrogen may be considered as the most valuable nutrient in animal manure, and it occurs in both the liquid and solid portions. The nitrogen in cattle manure is divided about equally between the liquid and solid portions, with that in the liquid occurring almost totally as urea. In poultry manure, which is a mixture of urinary and fecal excreta, the nitrogen is present predominantly as uric acid. Like urea, uric acid is readily soluble, and under warm, most conditions, hydrolyzes rapidly to ammonium carbonate with the ultimate release of the nitrogen as ammonia gas. The nitrogen in the solid portion of manures is contained primarily in undigested protein derived from the feed or in the tissue of microbes that develop in great numbers in the digestive tract of animals. A substantial part of the feces consists of lignin, cellulose, and similar relatively indigestible materials contained in the feed.

Storage and handling of manure

The form in which nutrients occur in manure is important not only because it determines the ease with which the nutrients are released for plant use but also because it can affect the retention of the nutrients in manure that must be stored for a period prior to its application to the land. Nutrient loss during manure storage can result either from leaching of soluble, mobile elements, such as potassium or boron, or from the volatilization loss of ammonia gas. Fortunately, phosphorus and most of the micronutrients are retained in relatively immobile form in manure and are therefore not susceptible to loss either by leaching or volatilization.

The loss of the nitrogen present originally as urea in manure is the most critical aspect of nutrient conservation. The hydrolysis of urea and formation of volatile ammonia can hardly be avoided [see equations (15.7) and (15.8) on page 410], so improper storage of manure, as in a loosely stacked pile, can lead to the rapid evolution of this gas. Furthermore, good aeration encourages microbial activity that causes the pile to heat excessively and dry rapidly. Both heating and drying hasten the loss of ammonia.

From these considerations we may conclude that storage of manure in a compacted state is desirable. Compaction not only retards heating and drying but also promotes anaerobic decomposition and the production of various organic acids capable of neutralizing and converting ammonia to nonvolatile salts. Keeping the manure moist aids in maintaining anaerobic conditions within the pile. Stored manure should be protected from excess water that would cause nutrient loss by leaching.

In dairy operations, one rather widely used method of manure storage is in tanks, pits, or lagoons, where both solid and liquid wastes accumulate, along with water used to cleanse holding or milking areas (see fig. 16.10A). Due to the excess water anaerobic conditions can be maintained. This tends to minimize odors so long as the stored materials are not disturbed. The wastes are removed by pumping, sometimes through pipelines for direct disposal on the land (see fig. 16.10B). The recovery of nitrogen may be fair to good, depending on the conditions of storage and the length of the storage period.

A   B

Figure 16.10.
(A) A sewage lagoon constructed to collect wastes from a Georgia
dairy operation. Solids are pushed into the lagoon from the ramp at
the right; liquids are delivered by pipe. (B) A manure gun used for
spraying liquids from a sewage lagoon. (Photo (A) by J. R. Brown,
U.S.D.A. Soil Conservation Service; (B), Washington State University
Engineering Extension Service photo.)

Relative value of
manure as a soil
supplement

Greatest benefits are derived from manure when it is used in a continuing
program. Only in this way can the user become familiar with the capabilities
of the material at his disposal. Since the true composition of manure will not
likely be known, its value must be judged largely from the overall effect it
has on soil properties and plant growth.

   Not all of the nitrogen in animal manure becomes available during the
season it is applied to the land. Estimates of the quantity released the first
year range from 90 percent for poultry manure to as little as 25 percent from
farmyard manure high in straw. If, for a given application, the amount of
nitrogen released is estimated to be less than plants need, the balance can
be supplied in fertilizer.

   Because all the nitrogen is not released quickly, manure has a residual
effect on the nitrogen-fertility status of soils; that is, it continues to supply

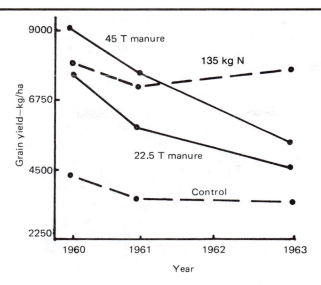

Figure 16.11.
The residual effect of manure in sustaining yields of irrigated grain sorghum at Garden City, Kansas. Manure applied only in 1960; N applied annually. Application rates are on a hectare basis. The manure contained 1.5 percent total nitrogen. (From G. M. Herron, and A. B. Erhart, *Proc. Soil Sci. Soc. Amer.* 29:278–81, 1965.)

available nitrogen for several years following its application. Successive additions of manure therefore tend to increase the soil nitrogen supply. Once an adequate level of nitrogen has been established, it can be maintained by relatively small additions of manure in subsequent years.

The residual effect of manure is demonstrated by the data in figure 16.11. These data are from an experiment in which manure was applied only once, in 1960, at 45 and 22.5 metric tons per hectare, and its effect on grain sorghum yield then measured for four successive seasons. Other treatments included a control receiving no supplemental nitrogen, and a fertilizer treatment supplying 135 kg N/ha/year. The residual effect of manure on the nitrogen supply is indicated by the higher yields of the manure treatments than of the control treatment during the latter years of the experiment. Since a different sorghum variety was grown in 1962, data for that year are not included.

A qualitative judgment of the rate at which nitrogen was released from manure can be made by comparing yields of the manure treatments with those due to treatments supplying 135 kg N/ha each year in fertilizer. Assuming yields reflect the amount of available nitrogen, then during the first year, the 22.5-ton manure treatment released less than 135 kg/ha of available nitrogen, for the yield of this treatment fell below that caused by the fertilizer. The manure contained 1.5 percent total N. Thus, the 22.5-ton treatment added a total of 340 kg N/ha, but, apparently, less than half of this amount of nitrogen was released in available form during the first year of the experiment.

The phosphorus and potassium in manure are judged to be relatively quickly released from manure, although the phosphorus, like nitrogen, will not be totally mineralized in a single year. The content of these two nutrients is comparatively low, however, so where they are very low in the soil and the manure is applied at the common rate of about 10 metric tons per hectare, supplementation of the manure with phosphorus or potassium fertilizers is often necessary. Since only a part of the phosphorus and potassium added to the soil in manure is absorbed by plants in any one year, repeated applications of manure to the land will cause the supply of these two elements to build up over time.

It is estimated that the nutrients in a metric ton of manure can be supplied in fertilizer costing from $10.00 to $12.00. The expense in spreading manure far exceeds the cost of applying the more concentrated fertilizer salts. Unless the purchase price of manure is appreciably less than the amount shown above, its use can be economically unsound. Numerous experiments have shown that the productivity of soils can be maintained just as well with commercial fertilizers as with manure. This is not to discount the desirability of the utilization of manure as a soil amendment where economically reasonable, or where such use serves an aesthetic function or solves a local waste-disposal problem. There are few practical methods of manure disposal that match application to the land in cost and convenience.

Experimental work has shown that the value of manure as a mulching material sometimes exceeds its value as a source of nutrients. An application of 20 metric tons of manure per hectare provides a ground cover that not only reduces the evaporative loss of water but also affords protection against rain-induced surface puddling and erosion. Maintenance of the soil in a granular, friable state is simplified by a surface cover of manure and is of particular value to root crops such as beets, potatoes, and peanuts. Except for possible loss of nitrogen by volatilization, the nutrients added in a manure mulch will ultimately find their way into the soil.

The use of manure on agricultural land can introduce some difficult soil management problems. When high in bedding material having a wide C/N ratio and, therefore, a high demand for available nitrogen as it undergoes decay, animal manures may depress plant growth unless supplemented with nitrogen fertilizer. They may also contain undigested seeds that could infest the land with weeds. Generally, these limitations are overlooked or improperly evaluated when manures are being considered as a soil amendment.

The nature and use of sewage

There is much current emphasis on the use of land for the disposal of sewage derived from industrial and municipal sewage treatment plants. Where such use is economically feasible, it can be encouraged as a practical means of supplying nutrients for plant growth. Further, disposal on the land results in the recovery of valuable plant nutrients that would otherwise be totally lost if disposal were by other common means. One point of caution is necessary, however. Sewage, especially the solid phase separated from the liquid during processing, often carries unwanted contaminants, most notably, heavy metals that can be toxic to both plants and animals. The use of sewage on agricultural land is therefore to be avoided unless one is assured that it contains harmless or manageable amounts of such contaminants.

A   B

Figure 16.12.
Treatment of municipal sewage. (A) Sludge removed from sewage
by screening. Application of sewage sludge to land is a convenient
means of disposal. (B) Aeration of sewage, aided by vigorous
stirring, reduces the content of readily oxidizable materials. (Photo
(A) courtesy of U.S.D.A. Agricultural Research Service.)

Two kinds of products are derived from typical sewage treatment. These
are a solid phase, termed *sludge*, and the liquid *effluent*. Much of the sludge
consists of solids separated from the raw sewage by a *primary treatment*
consisting of sedimentation and screening (see fig. 16.12A). The reactivity
of the remaining liquid and suspended solids is then reduced by a *secondary
treatment* that involves a period of biological oxidation (see fig. 16.12B).
Further sedimentation may follow, with the clarified supernatent liquid then
being chlorinated to kill pathogenic organisms and disposed of, most often
in natural waterways. The sludge is customarily buried or burned. Any one
of these methods of sewage disposal may be environmentally unsound at
times, which provides another reason for using land as the disposal medium.

Sewage contains all the nutrients needed by plants, but the one of
greatest value to agriculture is nitrogen. In addition, sewage can provide organic matter for the soil and water for plants, or even for the recharge of

Table 16.2. Typical ranges in the total nitrogen, phosphorus, and potassium contents of sewage sludge and effluent. (Data from various sources.)

Nutrient	Range in nutrient content of					
	Sludge[a]			Effluent[b]		
	%	kg/mt	lb/bt	ppm	kg/ha-cm	lb/ac-in
Nitrogen	2.0–8.0	20–80	40–160	10–30	1.0–3.0	2.3–6.8
Phosphorus	1.5–3.0	15–30	30–60	5–10	0.5–1.0	1.1–2.3
Potassium	0.2–0.8	2– 8	4–16	15–30	1.5–3.0	3.4–6.8

[a]Contents on a dry-weight basis in metric (mt) and British (bt) tons.
[b]Volumes are in hectare-centimeters or acre-inches; that is, a centimeter of depth over an area of 1 hectare, or an inch of depth over an area of 1 acre.

depleted groundwater supplies. Groundwater recharge occurs where sufficient effluent is applied to the land so that some filters through the soil and on to the level of the water table. Success with this procedure depends on the soil and plants to remove soluble organic and inorganic substances from the effluent that might otherwise pollute the ground water.

Composition of sewage

Plant nutrients present in greatest quantities in sewage are nitrogen, phosphorus, and potassium. Typical concentrations of these nutrients are listed in table 16.2. The rather wide variation in the concentrations shown results principally from differences in the composition of the raw sewage and in the type of treatment to which the sewage is subjected.

The contribution to soil fertility made by nutrients other than nitrogen, phosphorus and potassium in sewage is largely ignored. In sewage from industrialized areas, however, sufficient amounts of some of these nutrients may be present in amounts harmful to animal or human life. The same may be true for heavy metals and certain organic and inorganic compounds. The presence of these substances can often be anticipated if the source of the sewage is known, and if verified by chemical analysis, use of the sewage on land may have to be restricted. In fact, for fear of legal problems that might arise from foods containing toxic elements, some commercial food processors refuse to purchase produce grown on land treated with sewage from any source.

Nutrient forms in sewage

There are some important differences in the nature of the nutrients in sewage sludge and effluent. Nutrients in effluent are largely soluble, whereas the reverse is true for those in sewage sludge. There are exceptions, however. For example, effluents that are incompletely clarified during processing may contain nutrients in suspended solids. Conversely, some of the nutrients in sludge are soluble. Potassium and boron are examples, but their proportionate contents in sludge are comparatively small. Elements of this type tend to be drawn off in the effluent.

The nitrogen in sewage effluent is predominantly in the ammonium form. Substantial quantities of ammonium-nitrogen are also in the sludge, generally from one-third to one-half of the total nitrogen present. Once in the soil, the nitrogen in ammonium form tends normally to convert to nitrate, but if the sewage is not incorporated into the soil, it can be lost in part as volatile ammonia.

Although nutrients in sewage may be soluble to a degree, their movement in the soil tends to be limited, at least when they are first applied. The phosphorus in sewage is characteristically immobilized by adsorption or precipitation reactions, and the potassium by adsorption to exchange sites. Any nitrogen in ammonium form is also immobilized through exchange adsorption, but it becomes mobile once transformed to nitrate. Because of the strong tendency for ammonium-nitrogen to undergo nitrification, it is normal to assume that all the nitrogen added in ammonium form in sewage will be lost during the year of application, either through plant absorption or through leaching. Where leaching does not occur, however, soluble components may accumulate in the soil if sewage is applied to the land year after year.

Application of Sewage. Sewage sludges and effluents may be applied in various ways. Typically, the effluent is piped to the application site and then applied to the land by normal irrigation procedures. Because of the comparatively low concentrations of nutrients in effluent, it is usually impractical to meet the nutrient needs of plants without applying large excesses of effluent water to the soil. For example, where the nitrogen concentration of the effluent is in the low range, satisfying the nitrogen needs of a plant may require the application of twice as much water in effluent form as can be effectively used in evapotranspiration. At times, therefore, effluent application must be based on water use rather than nutrient use, and if the nutrients supplied are inadequate, the deficiency must be made up through fertilization. It is important to note, however, that nutrients such as phosphorus and potassium applied in effluent are never more than fractionally removed by plants in a single growing season; thus, these nutrients tend to accumulate in the soil from repeated application of effluent to the land.

Sewage sludge may be applied to the soil in either a dried or moist state. Drying is extremely expensive and likely to be avoided where disposal involves large amounts of sludge. Transportation costs of moist sludge are also a disadvantage. An alternative used extensively in Europe is to allow sewage sludge to accumulate, digest, and settle in lagoons and to pump the sediment, as a slurry, onto fields. As with other forms of sewage, rates of application of slurries are based on known or estimated concentrations of nutrients.

The use of sewage in soil fertility programs is, at times, both complex and expensive. Much of the expense must be borne by the municipality or agency producing the sewage, otherwise costs would far outweigh benefits to the land and crops. Because of the cost factor, the use of sewage on land can be justified more as a means of avoiding environmental pollution than as a practical alternative to the normal fertilization of plants. Further consideration of the use of land for the disposal of high volume wastes, such as sewage, is given in Chapter 19.

Figure 16.13.
Plowing down buckwheat as a green manure crop on potato land in Aroostook County, Maine. (U.S.D.A. Soil Conservation Service photo by J. C. Malley.)

Green manures, cover crops, and plant residues

Green manures and cover crops, both of which are ultimately returned to the soil (see fig. 16.13), have several overlapping functions. A distinction is made between them on the basis of the principal purpose for which each is grown. Cover crops are used primarily to protect the soil in the interval between normal cropping periods. Green manures are grown specifically for soil improvement. Like cover crops, green manure crops may be maintained for short periods only, as during the late fall and winter months. At other times an entire growing season may be sacrificed to green-manuring if the benefits seem to warrant it. Under such circumstances, however, a part of the green manure will probably be used as forage for animals.

Some of the major benefits derived from green manures and cover crops include: (1) the conservation of residual fertilizers that might otherwise be lost by leaching, (2) the conversion of slowly available soil nutrients to more readily available organic form, (3) the addition of organic matter to the soil, and (4) protection against erosion. Disadvantages are associated largely with the convenience factor. Even short-term cover crops grown during the winter sometimes interfere seriously with normal land preparation in the early spring, but this can often be overcome by proper planning.

Legumes are preferred as green manure crops because of their potential for adding fixed nitrogen to the soil. Small grains and grasses that are easily established and grow rapidly are used most frequently for winter cover crops. Even here, however, legumes play an important role, for they provide a highly satisfactory winter forage for grazing livestock.

More often than not, the value of a green manure crop is judged in terms of the quantity of nitrogen it supplies to the soil. Major differences among the legumes in nitrogen-supplying power relate primarily to rate of growth, the nitrogen content of the tissue, and the total plant residue produced. Nitrogen contents of some plants commonly used as green manure or cover crops are shown in table 16.3. These values are on a fresh-weight

Table 16.3. The nitrogen content, as a percent of fresh weight, of several crops used frequently as green manures.

Crop	N Content
	%
Alfalfa, vetch, Austrian winter peas	.75–1.00
Clovers	.70– .80
Cowpeas	.60– .75
Lespedeza	.50– .60
Small grains and grasses	.30– .40

basis and apply to succulent plant tissue that is subject to rapid decomposition in the soil. The values may be converted to a dry-weight basis if they are multiplied by a factor of 4, because, on the average, fresh plant materials of this type contain approximately 25 percent solid matter. The nitrogen content of the small grains and grasses can vary somewhat from the values shown in the table under extreme conditions of nitrogen availability.

Green manure and cover crops are of greatest benefit to the available-nitrogen status of soils if they are incorporated while still in a young, succulent state. This reflects the narrow C/N ratio of younger tissue, which permits the more rapid release of available nitrogen than is possible with wide-ratio tissue of high oxidizable carbon content. Since green manure and cover crops are often turned under just prior to seeding another crop, rapid release of nitrogen in available form is desirable.

Pasture and sod crops for green manuring

Where soils are highly erosive or subject to other deterioration, crop rotations are often used as an aid in sustaining productivity. In these systems cash crops are grown only periodically, and the land is utilized for pasture or hay production in intervening periods. The objective of such practices is to rehabilitate the soil, but at the same time, some return is obtained from the land in the harvested forages. Maximum benefits are derived if a sound fertility program is followed throughout the rotation.

The development and maintenance of hay or pastureland for a period of several years often fits more conveniently into a farming program than does short-term green-manuring. Benefits to the soil may also be appreciably greater. Soil granulation can often be markedly increased under dense, well-fertilized grass sod. Also, the nitrogen status of the soil may be greatly improved if legumes constitute a major part of the vegetation. Alfalfa, for instance, when grown for several years as a hay crop, may leave residues in the soil capable of releasing several hundred kilograms per hectare of nitrogen in available form over a period of time. Any contribution of this type must be taken into consideration in the evaluation of rotations.

Crop residues as soil amendments

Sound farming operations on cultivated land include the eventual incorporation of unharvested residues into the soil. Failure to follow this procedure hastens the decline of organic matter and the level of available nutrients in the soil. If erosion is a problem, the residues may be left as a surface cover between normal cropping periods (see fig. 16.14).

Figure 16.14.
Burning stubble not
only destroys a source
of nutrients and
organic matter but
also leaves the soil
unprotected against
erosion. (U.S.D.A. Soil
Conservation Service
photo by R.
Branstead.)

The amount of unharvested plant residues is a function of the type of plant grown and total crop production. Thus, sustaining a desired level of organic matter in the soil is simplified if productivity is maintained by proper fertilization. This probably explains in part why it is possible to continuously support as high a level of productivity with commercial fertilizers as it is with organic supplements such as manure.

Soils, fertilizers, and the nutritional quality of plants

Questions are often raised about the nutritional quality of plants used as food by animals and humans. Two extreme viewpoints are involved depending on whether one is concerned with shortages of nutrients or with excessive concentrations of elements that may be toxic. Shortages can ordinarily be corrected by fertilization of the soil. Elements that occur in excess and may accumulate in harmful quantities in plants are more difficult to cope with, unless their availability in the soil can be altered by precipitation reactions or by leaching.

Plant-related nutrient shortages in animals

All elements essential to plant growth are also needed by animals; thus, the fertilization of crops to increase yields will sometimes take care of animal needs as well. That this is not always the case is indicated by figure 16.15, which shows the phosphorus content of alfalfa and of oat grain and straw produced on the same fertility plots. The data on the content of phosphorus in the plant materials are of interest in that, to sustain normal growth, the food of many animals should contain about 0.3 percent phosphorus, a level indicated by the broken line in the graph. Only the oat grain and the heavily

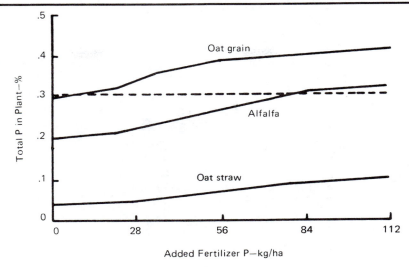

Figure 16.15.
The effect of
fertilization with
phosphorus on the
phosphorus content of
alfalfa and oat grain
and straw. (From
*U.S.D.A. Information
Bulletin* 299, 1965.)

fertilized alfalfa provide this concentration of phosphorus. Since the straw
of small grains is used extensively for animal feed in some parts of the world,
the potential for a phosphorus deficiency is clearly seen. Where such a deficiency is recognized, it can be corrected by supplementing the feed with
inexpensive inorganic phosphates.

The supply of nitrogen and sulfur in foods is extremely important to
animal nutrition; both elements are required in amino acid form for the synthesis of protein by all animals except ruminants.[2] Where diets are largely
vegetarian, the nitrogen and sulfur nutrition of the plant, as it affects amino
acid production, can greatly influence the value of the food in sustaining the
normal development of nonruminating animals and humans.

Plants also absorb and retain some elements essential to animals but
not to themselves. Principal among these are cobalt (Co), selenium (Se), and
iodine (I). Cobalt is necessary as a constituent of vitamin B_{12}, and iodine is
important for the prevention of goiter. The role of selenium in human and
animal nutrition has not been clearly defined.

Plants are good sources of dietary iodine in some areas, but not where
the supply of this element in the soil is lacking. However, problems of iodine
shortage can be largely eliminated through the use of iodized salt. Hence,
little attention has been paid to the treatment of soils as a means of increasing the iodine content of food plants.

Cobalt is required not only by higher animals but also by microorganisms including *Rhizobium*, the N-fixing bacteria associated with legume
plants. Since the need for cobalt by *Rhizobium* is small, it may be satisfied

[2]Cattle and sheep are ruminants. They contain a compound stomach, the first section of
which is the rumen, an organ used for the temporary storage of undigested food.
Microorganisms that abound in the rumen contribute to the digestive process. They are also
capable of synthesizing essential amino acids from inorganic nitrogen and sulfur. Urea is the
principal form in which nitrogen is added as a supplement to the diet of these animals.

even where so little cobalt is absorbed by the legume that it is deficient for sheep and cattle. Cobalt can be supplied to these animals in inorganic form as a dietary supplement or as an additive to salt. Where grown on soils of very low cobalt content, legumes can be treated with cobalt fertilizer, which not only benefits the *Rhizobium* but also the animals using the legume as food.

There are several areas in the United States where cobalt deficiencies have been noted in animals, and others where the low cobalt content of legumes indicates a potential deficiency. These areas are delineated in figure 16.16, which shows them to be largely concentrated in the eastern part of the United States. One potentially cobalt-deficient area is also located in Utah.

The third element often inadequately provided by plants, selenium, is required in very small amounts for the normal development of farm animals, and an excess can be very toxic. Humans appear not to be so sensitive to extremes in selenium concentration, but since their tolerance has not been established, the treatment of soils or livestock feed with selenium to overcome a dietary deficiency has been made illegal. Therefore, correction of the problem is by direct injection of selenium into the animal. As indicated in figure 16.16, known cases of selenium deficiency in animals have been recorded in many states, with the highest incidence being in the northeastern and northwestern parts of the country.

Plant-related nutrient excesses in animals

Two elements, selenium and molybdenum, are absorbed by forage plants in quantities that sometimes impair animal health. As shown in figure 16.16, selenium toxicities in animals have been recognized in all of the seventeen Western States with the exception of Washington. They occur most often on soils of dry regions that have formed from parent materials of marine origin. Selenium is readily leached from soils in areas of high rainfall, which reflects its chemical similarity to the element sulfur.

Selenium toxicities, which have several manifestations in animal and fowl (see fig. 16.17), are limited primarily to rangeland supporting certain plant species, often weeds, capable of accumulating relatively large quantities of this element. Grasses, which provide a preferred type of forage, do not have this capacity and, in general, are less tolerant of drouth than the selenium accumulators. Accordingly, the toxicity problem is often more acute in dry years, when reduced grass vegetation forces grazing animals to feed on the accumulator plants.

The best way to avoid selenium toxicities is through controlled grazing. Some of the more seriously affected areas have been purchased by the Federal Government and have been retired from all agricultural use.

Molybdenum toxicity in animals is observed most often in soils that are naturally neutral to basic in reaction or in limed acid soils. According to figure 16.16, the problem has been identified in three western states: Oregon, California, and Nevada; and in Florida. The solution to the problem rests with either lowering the pH of the soil with acidifying amendments, or reducing the level of available molybdenum by cropping the land repeatedly to plants capable of absorbing rather large amounts of this element.

Figure 16.16.
Locations of known or potential nutrient shortages or excesses, as relates to the nutritional quality of plants used for animal food. The legend is: large dots, plants of excessive selenium content; X's, occurrence of selenium deficiency in livestock; dark shading, historic areas of cobalt deficiency in animals; horizontal hatching, areas where legumes are moderately low in cobalt, suggesting a potential deficiency in animals; and vertical hatching, areas where molybdenum toxicities have been noted. (Modified from *U.S.D.A. Information Bulletin* 299, 1965.)

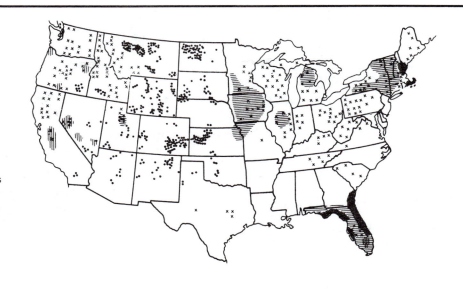

Figure 16.17.
Selenium-toxicity symptoms in animals: (A) cracked hoof of a cow, and (B) a malformed chick embryo. (From Olson, O. E. In *Effect of Poisonous Plants on Livestock*. New York: Academic Press, Inc. 1978.)

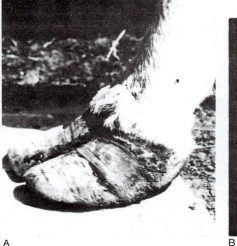 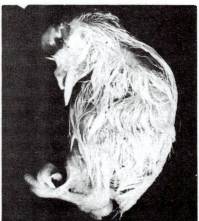

A   B

Summary

Fertility management is the control of plant yield and quality through adjustments in the nutrient status of the soil. It involves determining specific nutrient needs and the selection and proper application of fertilizers, amendments, and other supplements to overcome nutrient deficiencies, imbalances, and in some instances, toxicities. Good fertility management has its greatest impact where all factors of plant growth are optimal.

Nutrient needs of plants are judged quantitatively by soil and plant-tissue testing. Aids to the evaluation include observance of signs of nutrient deficiencies or excesses in plants and their response to applied fertilizers. The choice of fertilizer is then based on specific nutrient needs and cost. Cost is most suitably evaluated on the basis of price per unit weight of nutrient. Time and method of application are important to obtaining maximum benefits from fertilizers.

Manures, plant residues, and a variety of organic materials can be applied to improve the fertility status of the soil, often while concurrently enhancing soil tilth. Use of amendments, particularly those that reduce excess acidity or basicity, often correct certain nutrient-deficiency, and even some plant-toxicity, problems.

An important consideration of fertility management is the quality of produce used for food by man and animals. Quality is manifest in the nutritive value of the food, including its content of essential elements and other components, such as vitamins, proteins, carbohydrates, and fiber. Some elements present in plants used as food are essential to man and animals but not to the plants.

Review questions

1. Why is it necessary to correlate soil and plant-tissue tests?
2. Why should variation in productivity of a field be taken into account when obtaining samples for soil testing?
3. Show that the cost per pound of nitrogen in urea fertilizer costing $540 per ton (2000 lb) and containing 45 percent nitrogen is $.60/lb.
4. On the basis of figure 16.8, explain why the financial loss would be greater where fertilizer is underapplied by a given amount, relative to the optimum, than where it is overapplied by the same amount.
5. How do nitrification inhibitors added with fall-applied ammonium fertilizer help protect against the winter leaching loss of nitrogen?
6. Show why one ton (2000 lbs) of fresh dairy cattle manure listed in table 16.1 would supply 11.2 lbs of total N.
7. Why should the repeated application of manure, sewage, or other organic wastes cause a gradual build-up in the nitrogen content of the soil?
8. Why are legumes preferred as green manures?
9. Of what importance is the uptake of cobalt, iodine, and selenium by plants?

Selected references

Allaway, W. H. *The Effect of Soils and Fertilizers on Human and Animal Health.* U.S.D.A. Agriculture Information Bulletin 378. Washington, D.C.: Government Printing Office, 1975.

Chapman, H. D. ed. *Diagnostic Criteria for Crops and Soils.* Berkeley: University of California, Agricultural Division, 1966.

Kitchen, H. B. ed. *Diagnostic Techniques for Soils and Crops.* Washington, D.C.: American Potash Institute, 1948.

Mortvedt, J. J.; Giordano, P. M.; and Lindsay, W. L., eds. *Micronutrients in Agriculture.* Madison, Wis.: Soil Science Society of America, Inc., 1972.

Munson, R. D., and Doll, J. P. The Economics of Fertilizer Use in Crop Production. *Advances in Agron.* 11:133–69. New York: Academic Press, Inc., 1959.

Nelson, W. L., and Stanford, G. Changing Concepts of Plant Nutrient Behavior and Fertilizer Use. *Advances in Agron.* 10:68–141. New York: Academic Press, Inc., 1958.

Olson, R. A., ed.-in-chief. *Fertilizer Technology and Use.* Madison, Wis.: Soil Science Society of America, Inc., 1971.

Sprague, H. G. ed. *Hunger Signs in Crops.* New York: David McKay Company, 1964.

Stefferud, A. ed. In *Soil, Yearbook of Agriculture for 1957,* pp. 172–276. Washington, D.C.: Government Printing Office.

Tisdale, S. L., Nelson, W. L., and Beaton, J. D. *Soil Fertility and Fertilizers.* New York: The Macmillan Company, 1984.

Walsh, L. M., and Beaton, J. D. eds. *Soil Testing and Plant Analysis.* Madison, Wis.: American Society of Agronomy, 1973.

17

The use of lime on soils

Strongly acid soils are troubled by a variety of problems that interfere with the normal growth of many crop plants. By and large, the harmful effects are the result of unsatisfactory nutritional conditions. Sometimes, however, they relate to acid-induced toxicities associated with excess soluble aluminum or manganese in the soil, or to a poor physical or biological state resulting from insufficient calcium and magnesium. Soil acidity may, therefore, hinder the growth of plants either directly or indirectly. In any event, correction of adverse conditions attendant to a low pH consists of treatment of the soil with lime[1] to neutralize the excess acidity (see fig. 17.1).

Excessively acid soils are limited primarily to humid regions where soil development involves extensive weathering and leaching. In soils formed under these conditions, a lack of easily weathered minerals capable of supplying exchangeable bases, especially calcium and magnesium, is one of the main reasons why acidity becomes a permanent soil feature. Where easily weathered minerals occur in relative abundance, on the other hand, excess soil acidity is relatively rare, mainly because the acidity is largely neutralized as it reacts with weathering minerals.

The reaction of lime in soils

A wide range of substances can be used to neutralize soil acidity. Principal among them are (1) crushed limestone, which supplies calcium carbonate, $CaCO_3$, or calcium-magnesium carbonate, $CaMg(CO_3)_2$, as the neutralizing agent, (2) burnt lime, which is calcium oxide, CaO, and (3) slaked lime, or calcium hydroxide, $Ca(OH)_2$. Each of these compounds neutralize soil ac-

---

[1]Technically, lime is CaO. However, any amendment that supplies calcium or magnesium and is used to neutralize soil acidity may be referred to as lime.

idity by supplying hydroxyl (OH$^-$) ions for reaction with exchangeable hydrogen and aluminum. The reaction, using calcium hydroxide as the acid-neutralizing agent, is:

$$\boxed{\text{Clay}}\begin{matrix}\text{H}\\\text{Al}\end{matrix} + 2\text{Ca(OH)}_2 \rightarrow \boxed{\text{Clay}}\begin{matrix}\text{Ca}\\\text{Ca}\end{matrix} + \text{H}_2\text{O} + \text{Al(OH)}_3\downarrow \quad (17.1)$$

In the reaction, Ca^{2+} from the calcium hydroxide replaces the H$^+$ and Al^{3+} from exchange sites, and the latter ions are inactivated by combining with the OH$^-$ ions to form undissociated water and insoluble aluminum hydroxide.

Calcium hydroxide supplies OH$^-$ ions directly. The same may be said for calcium oxide, for it combines with water to form calcium hydroxide:

$$\text{CaO} + \text{H}_2\text{O} \rightarrow \text{Ca(OH)}_2 \quad\quad\quad (17.2)$$

Calcium and magnesium carbonates supply OH$^-$ ions through hydrolysis, or reaction with water, as shown for calcium carbonate:

$$\text{CaCO}_3 + \text{H}_2\text{O} \rightarrow \text{Ca}^{2+} + 2\text{OH}^- + \text{CO}_2\uparrow \quad\quad (17.3)$$

This equation illustrates hydrolysis under strongly acid conditions, which results in the complete decomposition of the calcium carbonate. The reaction releases one Ca^{2+} ion, two OH$^-$ ions, and a molecule of carbon dioxide for each molecule of calcium carbonate decomposed. Calcium hydroxide and calcium oxide also provide two OH$^-$ ions per molecule in neutralization reactions (see equations 17.1 and 17.2). Thus, whether the reactive compound in a liming material is calcium oxide, calcium hydroxide, or calcium (or magnesium) carbonate, all have the same capacity for neutralizing acidity when compared molecule for molecule.

Benefits of liming to plant growth

One of the foremost problems with plant growth on soils of very low pH is a toxic effect caused by excess soluble aluminum or manganese. Aluminum toxicities are generally more prevalent, since the concentration of this element needed to inhibit plant growth is substantially less than that needed for a manganese toxicity. By and large, raising the pH of a soil to 6.0 or above is sufficient to negate the undesirable effects of soluble aluminum and manganese. It does so by converting these elements to relatively insoluble oxides or hydroxides.

In addition to reducing toxicities, liming of strongly acid soils can be beneficial to plants by improving the nutritional status of the soil. One effect is to increase the solubility of phosphorus held in combined form as iron or aluminum phosphates; such compounds are relatively insoluble so long as the soil is strongly acid. On liming, however, the iron and aluminum in the phosphates are transformed to insoluble hydroxides, and the phosphorus is released as a soluble ion. This is illustrated by the reaction between variscite, an aluminum phosphate, and OH^- ions derived from lime:

$$Al(OH)_2H_2PO_4 + OH^- \rightarrow H_2PO_4^- + Al(OH)_3 \qquad (17.4)$$
$$\text{Variscite} \qquad\qquad \text{Soluble} \quad \text{Insoluble}$$

Similarly, liming tends to increase the availability of molybdenum in strongly acid soils, apparently by reacting with insoluble iron and aluminum molybdate compounds. Molybdenum deficiencies seem to occur only in soils of relatively low pH.

Liming can also increase the availability of nitrogen in soils. Under strongly acid conditions, nitrogen availability may be reduced by inhibition of both nitrogen fixation and the transformation of ammonium to nitrate through nitrification. Nitrogen fixation is limited simply because of the sensitivity of some nitrogen-fixing organisms to strongly acid conditions. In rare instances, it is due to a shortage of molybdenum. Root-nodule organisms on legumes are particularly sensitive to molybdenum deficiencies.

The transformation of ammonium to nitrate in soils is reduced at a pH of around 5.5, and it may stop altogether if the pH falls to 4.5 or below. Under conditions of low pH, therefore, the ammonium ion appears relatively stable, and if the bulk of inorganic nitrogen occurs in this form, its availability to plants can be greatly restricted. The reason is that ammonium ions, occurring primarily in exchangeable form, are not very mobile and must be sought out by roots to become positionally available. At best, roots may contact no more than about 15 percent of immobile nutrients during a single growing season.

Although a principal purpose of liming is to increase the level of exchangeable calcium, or calcium and magnesium, in acid soils, it also tends to raise the level of potassium held in exchangeable form. One reason for this is the increase in the cation-exchange capacity that results as liming raises the pH and activates more of the pH-dependent charge in the soil. With an increase in the cation-exchange capacity, more exchange sites become available for potassium adsorption and retention. Equally as important, however, is the displacement of exchangeable aluminum by calcium and magnesium, since potassium can displace calcium and magnesium more easily than aluminum from cation-exchange sites. Potassium added to soils that have been limed is therefore more likely to be taken up as an exchangeable ion. It is well known that the liming of acid soils results in a substantial reduction in the loss of soil potassium by leaching.

Nature and evaluation of liming materials

Most liming materials are derived from limestone. There are some exceptions, but they are usually of local importance only. Examples include blast-furnace slag, a glasslike, acid-soluble calcium silicate material, and shells of sea animals (oysters, clams), which are essentially pure calcium carbonate. Either of these substances is effective in neutralizing acid if crushed or ground to a fine particle size and then mixed with acid soil.

Limestone applied directly to the soil is the most widely used amendment for correcting acidity, but it too must be reduced to a finely divided state before it will react satisfactorily in the soil. Limestone is also the basis for the manufacture of calcium oxide and hydroxide materials. Calcium oxide is prepared by heating limestone to a high temperature to drive off carbon dioxide:

$$CaCO_3 \xrightarrow{\text{heat}} CaO + CO_2 \uparrow \qquad\qquad (17.5)$$

Calcium hydroxide is prepared by reacting calcium oxide with water. The reaction in this transformation is shown in equation (17.2).

The suitability of liming materials for correcting soil acidity is a function of two properties: (1) chemical composition, and (2) particle size. For materials derived from limestone, variation in composition depends on such things as the ratio of calcium to magnesium present, and whether these elements occur in the material as the oxide, hydroxide, or carbonate form. Composition is also dependent on the amount of nonreactive impurities present. These impurities are usually either sand or clay, which are common components of limestone. The chemical composition of a liming material determines its total ability to neutralize acidity in soils. This ability, when expressed on the basis of a unit weight of the material, is referred to as the *neutralizing value*.

Particle size determines the *reaction rate* of a liming material, with the rate increasing as the size of particles decreases. In actual use, the rate at which a liming material neutralizes acidity also depends on how thoroughly it is mixed with the soil. Mixing increases the total surface area of particles placed in direct contact with acid components in the soil. The reaction rate of liming materials is important because the usual intent is to reduce soil acidity (increase pH) as quickly as possible. At best, it may take a liming material up to six months to react completely with soil acidity, even if it has been finely ground and thoroughly mixed with the soil.

To assure the dependability of marketed liming materials, many states require that their sale be accompanied by a guarantee. Although guarantees differ among states, all generally must show the total capacity of the liming material to neutralize acid as well as some indication of its particle size analysis, or fineness of grind. At times, a statement on the calcium and magnesium contents of the material is required.

The neutralizing value of liming materials

The neutralizing value expresses the total capacity of a liming material to neutralize acid. It is measured by treating a sample of a material with excess acid and then determining the total amount of acid consumed once the reaction is complete. Although the neutralizing value could be expressed by the amount of acid consumed, it is conventionally indicated on a relative basis. The most common means is as the *calcium carbonate equivalent*, or the capacity to neutralize acidity expressed as a percent of this capacity in an equal quantity (weight) of pure calcium carbonate. The neutralizing value (NV) is computed by means of the equation:

$$NV(\%) = \frac{\text{Acid neutralized by liming material}}{\text{Acid neutralized by pure CaCO}_3} \times 100 \qquad (17.6)$$

where the weights of the liming material and calcium carbonate are the same. To illustrate, assume that equal weights of a liming material and of calcium carbonate consume, respectively, 0.075 and 0.10 g of H^+ in a neutralization reaction. Substituting these values in equation (17.6):

$$NV\ (\%) = \frac{0.075}{0.100} \times 100 = 0.75 \times 100 = 75\% \qquad (17.7)$$

Accordingly, the material in question would be only 75 percent as effective as pure calcium carbonate in neutralizing acidity. For crushed limestone, a neutralizing value below 100 percent would suggest the presence of non-reactive components, such as sand or clay, in the material.

The neutralizing value of some liming materials is above 100 percent. Such materials normally contain magnesium carbonate, calcium hydroxide, or calcium oxide as the active neutralizing agent. These compounds have a neutralizing value above 100 percent because they supply as many OH^- ions per molecule but have molecular weights that are less than the molecular weight of calcium carbonate (see table 17.1). For example, a molecule of magnesium carbonate weighs only 84 percent as much as a molecule of calcium carbonate, but a molecule of each neutralizes the same amount of acid. This relationship results in a neutralizing value of 119 percent for magnesium carbonate, as can be computed through use of the formula:

$$\text{NV of compound }(\%) = \frac{\text{Molecular weight of CaCO}_3}{\text{Molecular weight of compound}} \times 100 \qquad (17.8)$$

Applying this equation to the computation of the neutralizing value of magnesium carbonate:

$$\text{NV of MgCO}_3\ (\%) = \frac{100}{84} \times 100 = 1.19 \times 100 = 119\% \qquad (17.9)$$

which means that it takes 119 kg of $CaCO_3$ to do the same job as 100 kg of $MgCO_3$. Neutralizing values for $Ca(OH)_2$ and CaO in table 17.1 are also determined in this way.

Table 17.1. Molecular weights and neutralizing values of the principal active compounds in common liming materials.

Compound	Formula	Molecular weight[a]	Neutralizing value
			%
Calcium carbonate	$CaCO_3$	100	100
Magnesium carbonate	$MgCO_3$	84	119
Calcium hydroxide	$Ca(OH)_2$	74	135
Calcium oxide	CaO	56	179

[a]Computed as the sum of the weights of the ions making up the molecule. The ionic weights used: Ca^{2+}, 40; Mg^{2+}, 24; CO_3^{2-}, 60; OH^-, 17; and O^{2-}, 16.

The neutralizing value expresses only the total capacity of a liming material to neutralize soil acidity, but it gives no indication of how quickly the material will react with the acid. This property is largely a function of the degree of fineness of the material. Calcium oxide or hydroxide materials normally occur in a finely divided, highly reactive state, so there is little concern over their rate of reaction in soils. The degree of fineness of crushed limestone is highly variable, on the other hand. Thus, particle size and its relation to reactivity become important considerations in judging the suitability of crushed limestone for use on acid soils. Composition is also important in that the reaction rate is more rapid for calcitic than for dolomitic materials. These factors as determinates of the reactivity of crushed limestone materials are discussed in the sections that follow.

Particle-size relationships of crushed limestone

The relationship between reaction rate and particle size is sufficiently consistent so that it is possible to judge the potential reactivity of a crushed limestone from particle-size information. The basis for the judgment is the proportions of different size-groups of particles determined by passing a sample of the material through a series of screens, each having a specified mesh,[2] or hole size (see fig. 17.2). Screens of largest hole size are placed on top, with those of smallest hole size on the bottom. Typical sizes are 4-, 8-, 20-, 40-, and 60-mesh. As the material passes through the screens, it is sorted into groups having a range in particle size. For example, material collected on an 8-mesh screen after having passed through a 4-mesh screen will be 4-8 mesh material; that passing through an 8-mesh screen but accumulating on a 20-mesh screen immediately below would be 8-20 mesh material. Usually, the smallest openings used in the standard evaluation of crushed limestone is 60-mesh; the size of particles passing this screen is 60-mesh or finer.

[2]The mesh of a screen designates the number of openings per inch; for example, a 100–mesh screen has 100 openings per inch. Because the screen wire takes up part of the space, the holes in a 100–mesh screen are slightly less than 0.01 inch in size. Mesh sizes have been standardized according to the following listing, wherein the sizes are in millimeters:

Mesh	4	8	20	40	60	100	300
Opening (mm)	4.76	2.38	0.84	0.40	0.25	0.15	0.05

The opening of a 300–mesh screen corresponds to the lower size limit of sand, according to the U.S.D.A. system of particle-size classification.

Figure 17.2.
Screens used for separating particulate matter, such as crushed limestone, into different size fractions. (Photo courtesy of the Tyler Company.)

The general relationship between particle size and the rate at which crushed limestone neutralizes soil acidity is illustrated in figure 17.3. The figure shows the effect of adding equal quantities of three size fractions of a liming material on soil pH over a three-year period. As seen in the figure, 100-mesh material caused a rapid correction of acidity, but the coarsest group (20-30 mesh) reacted both slowly and incompletely with the soil. The benefit from use of lime of fine particle size is clearly demonstrated by these data. As a general rule, totally satisfactory results are obtained from limestone that is ground to completely pass a 60-mesh screen.

Data of the type presented in figure 17.3 provide a basis for evaluating lime reactivities from knowledge of the particle-size distribution for a given liming material. Methods of evaluation vary, but they tend to be similar in their basic approach. The method used in Iowa serves as an example.[3]

For the Iowa rating, a sample of lime is placed on a nest of three sieves: 4-, 8-, and 60-mesh. After an appropriate period of shaking, the amount of lime that passes through each screen is determined and expressed as a percentage of the total sample. The percentage of each size fraction is then multiplied by a factor representing its relative potential reaction rate. A factor

[3]For details of this procedure see R. D. Voss *et al. A New Approach to Liming Acid Soils.* Iowa Extension Pamphlet 315. Iowa State University, Ames, 1965.

The use of lime on soils

Figure 17.3.
Relationship between size of limestone particles and rate of reactivity in acid soil, as measured by change in soil pH. The quantity of lime added was the same in each case. (From *Plant Food Review* 2(3):19, 1956. By permission of The Fertilizer Institute, Washington, D.C.)

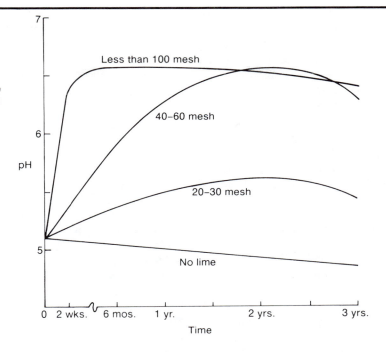

of 0.1 is used for all material that passes the 4-mesh screen, 0.3 for all material passing the 8-mesh screen, and 0.6 for that passing the 60-mesh screen. The three products obtained by the multiplications are added together and then used as an index of the reaction rate of the entire liming material. In effect, this index, referred to here as the *reactivity index*, estimates the fraction of the total neutralizing capacity of a liming material that will react quickly with soil acid. It is expressed as a percent of the total neutralizing capacity, or neutralizing value, of the material.

Application of this procedure is as follows. If 100 percent of a sample of crushed limestone passes a 4-mesh screen, and 90 and 55 percent pass 8- and 60-mesh screens, respectively, the reactivity index of the material would be computed as:

For material passing the 4-mesh screen	$100 \times 0.1 = 10$
For material passing the 8-mesh screen	$90 \times 0.3 = 27$
For material passing the 60-mesh screen	$55 \times 0.6 = \underline{33}$
	Reactivity index $= 70$

Based on the reactivity index, this material would be only 70 percent as effective as an equal weight of finely ground calcium carbonate in causing quick neutralization of soil acid. This same limestone, if ground to completely pass a 60-mesh screen, would show a reactivity index of 100, however. In this case, all of the material would pass the 4- and 8-mesh screens as well as the 60-mesh screen. Thus, each of the factors, 0.1, 0.3, and 0.6, would be multiplied by 100, and the sum of the products would be 100.

The quality factor is a rating that shows the overall effectiveness of a liming material in neutralizing soil acidity relative to this capacity in finely ground, pure $CaCO_3$. It is a computed value determined by multiplying the neutralizing value of a material by its reactivity index (both expressed as decimal fractions), and then by 100 to convert the final product to a percentage. For example, if a ground limestone has a neutralizing value of 90 and a reactivity index of 70, its quality would be calculated as:

$$0.90 \times 0.70 \times 100 = 63$$

In other words, this material would be 63 percent as effective as pure, finely ground $CaCO_3$ in correcting an acid-soil condition.

Lime recommendations are based on finely ground, pure $CaCO_3$ as the amendment. The recommended rate therefore applies only to a material having a quality rating of 100. If a material with a rating other than 100 is to be used, its rate of application must be adjusted to compensate for the difference in quality. Low quality materials must be applied at a higher rate while superior quality materials may be applied at a lower rate than is recommended. For example, a material having a quality rating of 75 should be applied at 133 percent of the recommended rate $(100/75 = 1.33)$; a material with a quality rating of 105 would be applied at 95 percent of the recommended rate $(100/105 = 0.95)$.

Methods of rating the quality of liming materials are not uniform from state to state, and in some, no standardized procedure has been established. In the latter instances, control may be exercised by specifying the extent to which the material must be ground to be acceptable. For example, some state laws stipulate that 80 percent of a material must pass an 8-mesh screen, the presumption being that this degree of grinding yields sufficient fine particles to provide a quality product. In other instances acceptability is based on the proportion of fine particles the material contains. Such specifications vary so widely that little can be gained by discussing them in greater detail.

Chemical composition as a factor of lime quality

The quality of a liming material cannot be fully assessed from measurements of neutralizing value and particle-size distribution alone. Chemical composition is also important. In this regard, the relative contents of calcium and magnesium are of greatest concern. The ratio of calcium to magnesium depends on whether the liming material has been derived from calcitic or dolomitic limestone.

Knowledge of the ratio of calcium and magnesium in a liming material is important for two reasons. One is that calcitic limestone will ordinarily add very little magnesium to the soil, and a shortage of magnesium may be one of the problems liming is intended to correct. The other is that dolomitic limestone reacts more slowly with soil acidity than does calcitic limestone. To be equally effective in neutralizing acidity, therefore, dolomitic limestone may have to be applied either at a higher rate or in a finer state of subdivision than is required for crushed calcitic limestone.

Figure 17.4 illustrates the general relationship between reactivity of crushed dolomitic and calcitic limestones and fineness of grind. If both materials are finely ground so that 80 percent or more passes a 60-mesh screen,

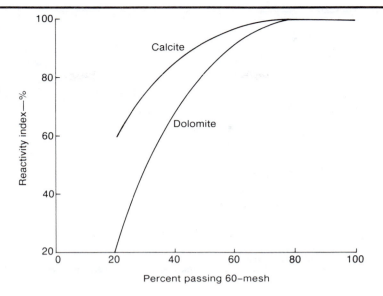

Figure 17.4.
Relationship between fineness of grind and relative reactivity rate for calcitic and dolomitic limestones. (Adapted from Barber, S. A. In *Soil Acidity and Liming*, pp. 125–60, 1967. By permission of The American Society of Agronomy.)

either material will show a maximum, or 100-percent, reactivity. With decreased fineness of grind, the rate of reaction with soil acidity drops for both, but more for dolomitic than for calcitic limestone. For example, if the materials are ground so that only 20 percent passes a 60-mesh screen, calcitic limestone is 60 percent as reactive as when finely ground, but dolomitic limestone is only 20 percent as reactive. If these coarsely ground materials had to be used to correct acidity, then the calcitic material at 60 percent reactivity should be applied at 1.67 times the rate recommended for finely ground calcium carbonate (100/60 = 1.67), but the dolomitic material, at 20 percent reactivity, should be applied at 5 times the recommended rate (100/20 = 5.0).

The basis for lime recommendations

Usually, lime is added to soils in an amount judged adequate to establish a pH optimum for plant growth. This quantity is known as the *lime requirement*. Commonly, the adjustment is to pH 6.5, but the type of plant involved has some bearing on whether or not this pH is appropriate. For instance, data from Iowa, which are plotted in figure 17.5, show that corn, oats, and alfalfa attain maximum yields at about pH 7.0.[4] On the other hand, red clover seems to do best at about pH 6.6. Although the differences in yield for the first three crops caused by only a half-unit increase in pH above 6.5 seem small, they are often sufficient to justify liming to a pH of 7.0. Because of the limited response to liming by red clover, however, an investment in lime for this crop might be questionable at times.

[4]These pH values are measured on a suspension of one part soil to two parts water and tend to be slightly higher than those obtained on a saturated soil paste.

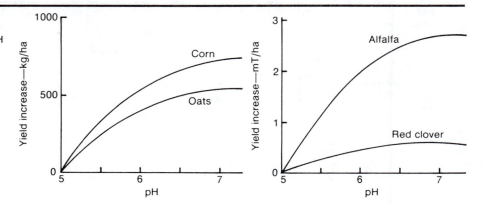

Figure 17.5.
The response of four crops to increasing pH above 5.0 induced by soil liming. (From Ross, R. D. *et al.*, Iowa Extension Pamphlet 315, 1965.)

Table 17.2. Ranges in pH suitable for optimum growth of a number of field and horticultural crop plants. (Modified from V. Ignatieff, *Efficient Use of Fertilizers*, London: Leonard Hill, Ltd., 1952).

Plant	Optimum pH Range	Plant	Optimum pH Range
Cereals		**Grasses**	
Barley	6.5–8.0	Bermuda grass	6.0–7.0
Corn	5.5–7.5	Blue grass, Kentucky	5.5–7.5
Oats	5.0–7.5	Fesque, meadow	4.5–7.0
Rice	5.0–6.5	Kafir corn	6.0–7.5
Rye	5.0–7.0	Millet	5.0–6.5
Wheat	5.5–7.5	Milo, dwarf yellow	5.5–7.5
		Orchard grass	6.0–7.0
Legumes		Sudan grass	5.0–6.5
Alfalfa	6.5–8.0	Timothy	5.5–8.0
Beans, field	6.0–7.5		
Beans, Soy	6.0–7.0	**Miscellaneous**	
Clover, Ladino	5.6–7.0	Azaleas	4.5–5.5
Clover, red	6.0–7.5	Beets, red	6.0–7.5
Clover, sweet	6.5–7.5	Beets, sugar	6.5–8.0
Lespedeza	4.5–6.5	Blueberries	4.5–5.5
Lupine	5.0–7.0	Cotton	5.0–6.0
Peas	6.0–7.5	Potatoes	4.8–6.5
Vetch	5.2–7.0	Sugar cane	6.0–8.0
		Strawberries	5.0–6.0
		Turnips	5.5–6.8

Most plants adapt well to a fairly wide range in pH, even though they tend to perform most satisfactorily at a specific pH. A wide variety of plants can be grown equally well whether the soil is moderately acid or slightly basic in reaction. Still, some definitely require strongly acid soils, whereas others may do very poorly under these conditions.

The desired range in pH for a number of important crop plants is shown in table 17.2. The use of some of these crops with certain soils may necessitate modification of the pH if maximum economic return from the land is expected. Where adjustment of the soil pH is not practical, plants that adapt well to the existing conditions should be grown.

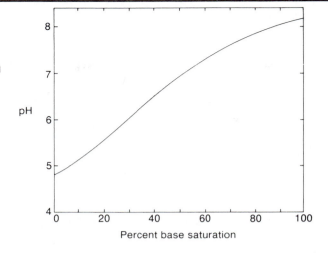

Figure 17.6.
Curve showing the relationship between pH and percent base saturation for a representative mineral soil.

Liming, pH, and percent base saturation in soils

Raising the pH of a soil increases the percent base saturation, that is, the quantity of exchangeable bases expressed relative to the cation-exchange capacity, say as measured at pH 7. The general relationship between pH and percent base saturation for one soil is shown in figure 17.6. It is established by treating an acid soil at zero base saturation with increasing quantities of calcium hydroxide and then measuring the pH after allowing time for the neutralization reaction to take place. A base saturation of 100 percent is produced in the sample receiving just enough base to neutralize all of the original exchangeable acid.

With a curve of the type shown in figure 17.6 and knowledge of the cation-exchange capacity of the soil, it is possible to compute the amount of calcium, and lime, needed to change the pH by a specified amount. This can be shown by an example using the soil in figure 17.6. As a basis for the computation we will assume a cation-exchange capacity of 10 me/100 g for the soil, that the proposed change in pH is from 4.8 to 6.5, and that the amount of soil in the field to be treated is a hectare furrow slice weighing 2,000,000 kg.

As can be judged from figure 17.6, raising the pH of this soil from 4.8 to 6.5 is accompanied by a rise in the percent base saturation from 0 to 40, or 40 percent. Since the cation-exchange capacity of the soil is 10 me/100 g, a 40-percent increase in base saturation will require 4 me of calcium per 100 g of soil. Multiplying the milliequivalents of calcium by its milliequivalent weight, or 0.02 g (see table 7.1, page 143), gives grams of calcium required per 100 g of soil. Multiplying this value by 10 then gives grams of calcium per kilogram of soil. Multiplying again by 2,000,000 gives grams of calcium per 2,000,000 kg of soil. Thus:

$$4 \text{ me} \times 0.02 \text{ g} \times 10 \times 2{,}000{,}000 = 1{,}600{,}000 \text{ g calcium}$$

The quantity, 1,600,000 g, translates to 1,600 kg or 1.6 metric tons of calcium per 2,000,000 kg of soil.

The next step is to determine the amount of $CaCO_3$ that must be added to supply 1,600 kg of calcium. Since $CaCO_3$ contains 40 percent calcium, it must be added at a rate that is 2.5 times the rate of needed calcium $(100/40 = 2.5)$. Thus:

1,600 kg calcium $\times$ 2.5 = 4,000 kg $CaCO_3$

which is equivalent to 4 metric tons of $CaCO_3$. The same result would be obtained using the data for calcium in table 7.7 (page 156), which shows that to change the calcium content of a soil by 1 me/100 g requires the addition of 1,000 kg (1 metric ton) of $CaCO_3$ per 2,000,000 kg of soil. To change the quantity of calcium by 4 me/100 g would require 4 metric tons of $CaCO_3$ per 2,000,000 kg of soil.

Measurement of lime requirements

Most laboratories equipped for examining the nutrient status of soils can determine lime requirements. Various procedures have been developed for this. One consists of establishing the direct relationship between pH and the amount of base (calcium) added in pure $CaCO_3$. Separate curves are developed for groups of soils similar in kind of clay, organic matter content, and cation-exchange capacity. Examples of such curves are shown for three soils of widely varying characteristics in figure 17.7. Two of the curves are for mineral soils, one of low and one of moderate cation-exchange capacity, and the third is for an organic soil of high cation-exchange capacity.

Figure 17.7 shows that the lime requirement of a soil depends on the change in pH wanted. For example, raising the pH of soil A from its initial value of about 4.8 to 7 would require roughly 5 metric tons of lime per hectare. If the pH were raised to only 6.5 rather than 7, then only about 4 metric tons would be needed. Likewise, the requirement would drop to about 2 metric tons of lime per hectare if the initial pH were 5.5 and was to be raised to 6.5.

A second widely used method for determining lime requirements is based on the ability of acid soil samples to lower the pH of a standard basic buffer solution, that is, a solution that neutralizes considerable acid with only a modest drop in pH. The relationship between the drop in pH of the buffer and the lime requirement is first established by mixing the buffer with a series of soil samples of known lime requirements covering a range from very high to very low. Once this relationship has been established, the buffer solution can be used to measure the lime requirement of any acid soil sample.

The estimation of lime needs from pH data

In the absence of a better basis for judging lime requirements, a rough estimate can be made based on knowledge of pH, texture, and the general nature of the soil as relates to type of clay and the amount of organic matter present. Lime applications may then conform to the suggested rates in table 17.3. Obviously, this arbitrary approach is subject to considerable error and should be tested in the field on a small scale before a large investment is made in a liming material.

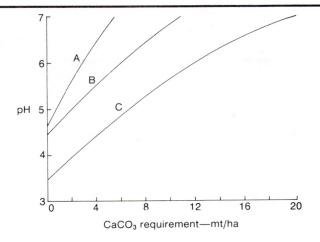

Figure 17.7.
Typical relationships between pH and lime requirement for soils of widely varying clay mineralogy and organic matter content: A and B, mineral soils of low and moderate cation-exchange capacity; C, organic soil of high cation-exchange capacity.

Table 17.3. Approximate metric tons of finely ground limestone[a] needed per hectare to raise the pH of a 17-cm layer of soil by the amount indicated. (Modified from V. Ignatieff, *Efficient Use of Fertilizers.* London: Leonard Hill, Ltd., 1952).

Climatic region and soil textural class	Limestone requirements		
	From pH 3.5 to pH 4.5	From pH 4.5 to pH 5.5	From pH 5.5 to pH 6.5
Soils of warm temperate and tropical regions[b]			
Sand and loamy sand	0.7	0.7	0.9
Sandy loam	—	1.1	1.3
Loam	—	1.0	2.0
Silt loam	—	2.7	2.9
Clay loam	—	3.1	4.0
Muck[c]	5.2	6.7	7.6
Soils of cool-temperate and temperate regions[d]			
Sand and loamy sand	0.9	1.1	1.3
Sandy loam	—	1.6	2.7
Loam	—	2.5	3.4
Silt loam	—	3.1	4.0
Clay loam	—	3.8	4.7
Muck[c]	5.8	7.6	8.7

[a]Fifty percent of the limestone should pass a 100-mesh screen. With coarser materials, applications need to be greater.
[b]Ultisols and Oxisols, primarily.
[c]The suggestions for muck soil are for those essentially free of sand or clay. For those containing much sand or clay, the amounts may be reduced to values midway between those given for muck and the corresponding textural class of mineral soil. If the mineral soils listed are unusually low in organic matter, the recommendations should be reduced about 25 percent; if unusually high, increase about 25 percent, or even more.
[d]Spodosols and acid Alfisols and Inceptisols, primarily.

The use of lime on soils   459

Table 17.3 is presented largely because it demonstrates some of the soil features that are highly important in determining lime requirements. Except for the initial pH, all relate in one way or another to the exchange capacity of the soil and therefore to the ability of the soil to retain exchangeable hydrogen and aluminum. This capacity decreases with increasing coarseness of texture and with decreasing organic matter content. For comparable textures, soils of the warmer regions have the lowest lime requirements because of the predominance of clays, kaolinite and sesquioxides principally, that have characteristically low cation-exchange capacities.

The application of lime

Quality, as judged by composition and fineness of grind, is not the sole determinant of the value of a liming material; equally important is the manner in which it is applied, especially how well it is mixed with the soil. Inadequate mixing limits contact between added lime and the acid soil components and thereby slows the rate at which acidity is reduced.

Incomplete mixing results from (1) inadequate stirring of the soil-lime mixture, and (2) failure to work the soil to a suitable depth. Because lime is usually applied in large amounts, it is normally spread on the land by broadcasting, often by commercial applicators (see fig. 17.8). Mixing is accomplished by cultivation, the method used largely determining the degree of mixing achieved. Rototilling, which results in vigorous stirring of the soil, is perhaps the most satisfactory, though not necessarily a commonly available method. Discing or a combination of discing and plowing are the procedures generally followed.

Discing alone is least satisfactory because it affects only shallow depths of soil (see fig. 15.5, page 404). The concentration of the material worked in by discing is highest near the surface and decreases sharply with depth. Plowing, if used alone, inverts the soil and places much of the lime at the bottom of the inverted layer. Plowing alone generally proves more satisfactory than discing alone, however, for it locates the lime in a soil zone that remains moist for longer periods during the growing season. Moisture aids in the continued reactivity between lime and soil and maintains actively growing roots in the zone where a reduction in acidity has been achieved.

A combination of discing and plowing is probably the best readily available means of incorporating lime into the soil. A standard recommendation is to apply half the lime prior to plowing, the remainder to be applied and incorporated by discing. Regardless of the method employed, it is seldom practical to achieve a fully homogeneous distribution of lime within the plow layer.

Where a limited amount of lime is used for reasons of economy, it is sometimes placed in bands near the seed at planting time. This type of application results in the highly localized correction of acidity and is most successful if roots of the seedling soon penetrate into deeper substrata where pH relationships are more favorable. If plant growth is restricted by a high

Figure 17.8.
Applying pulverized
lime with a broadcast
spreader. (Photo
courtesy of the Ag-
Chem Equipment
Company.)

concentration of toxic elements such as Mn and Al, however, band placement of small amounts of lime is usually not as effective as correcting acidity
throughout the plow depth of soil.

In other instances where the amount of added lime is less than sufficient to produce the desired change in pH throughout the plow layer, the
depth of mixing may be reduced. Inducing an optimum pH in a shallow depth
of soil sometimes gives better results than the partial correction of acidity
throughout a greater depth.

In correcting soil acidity, consideration should be given to the time of
application as well as to the rate. In rotations, treatments should precede
crops that are most responsive to lime. Because of the relatively slow reactivity between most liming materials and the soil, applications should be well
in advance of seeding; from six months to a year is often advised.

Minimal benefits may be expected from lime if it is applied as a top-
dressing but not mixed with the soil. However, such a procedure must be
followed on permanent pasture or hayland or where no-till farming is practiced. Under these conditions, incorporation is mainly by water that either
slowly dissolves the lime or washes the finer particles into soil pore spaces.
The affected soil zone will be quite shallow, and noticeable effects may not
appear for a number of years. Greatest penetration will occur with soils of
high infiltration rates and where the lime is applied in as fine a state of subdivision as possible.

Fate of lime added to acid soils

The rate at which a liming material dissolves when thoroughly mixed with an acid soil may be rapid at first, but it declines with the disappearance of the finer particles and as the pH rises due to the neutralization of the acidity. The reaction therefore tends toward an equilibrium state where the continued, though slow, release of calcium (or calcium and magnesium) from the residual lime particles just about keeps pace with the loss of soluble and exchangeable calcium through plant absorption and leaching. The nature of the equilibrium is depicted by the equation:

$$CaCO_3 \rightleftharpoons Ca^{2+} \rightleftharpoons Ca \boxed{Clay} \qquad (17.9)$$

Solid   Soluble   Exchangeable

Because of this relationship, the coarser particles in crushed limestone slow reacidification. With the gradual disappearance of the larger particles, however, exchangeable hydrogen and aluminum will again accumulate, causing the soil to slowly return to its original acid state. Thus, liming has only a temporary effect and must be repeated on occasion. The best means for determining the frequency of lime application is the soil test. Such tests should be repeated every three to five years depending on soil and climatic conditions.

Practical considerations in applying lime

Acidity should not be looked upon as an unquestionably undesirable characteristic in soils. Indeed, the adaptability of many plant species, particularly those that are native to an area, is determined in part by soil reaction. Some of these plants thrive only on soils that are strongly acid. Even many domesticated plants are acid-loving and do not grow well except under conditions of comparatively low pH.

Preliminary judgment as to the desirability of treating land with lime can be made from information such as that shown in table 17.2, which lists the apparent optimum pH range for a wide variety of plants. However, the highly general nature of the listing makes its application under specific conditions questionable, because tolerance of acidity sometimes varies more within a species than from one species to another. For example, small grains that grow well and produce high yields on neutral or near neutral soils may be decidedly inferior to acid-tolerant varieties when grown on acid soils (see fig. 17.9). Further, pH-dependent factors that affect plant growth are not everywhere the same. As a case in point, liming corn in the central United States often seems to be economically unsound, but figure 17.1 is clear evidence that corn responds dramatically to lime on some soils.

Because of variability in response to liming, little is gained by discussing the specific lime requirement of a wide range of plant types. However, some common agricultural plants do show reasonable consistency so far as their growth in relation to soil reaction is concerned. Alfalfa, for example, is almost universally responsive to lime when grown on acid soils. As

Figure 17.9.
Differential tolerance of (left to right) Thatcher, Monon, and Atlas 66 wheat to acid Bladen clay containing potentially toxic levels of aluminum. (Photo courtesy of C. D. Foy, U.S.D.A. Agricultural Research Service, Beltsville, Maryland.)

a result, this plant is widely treated with lime. Treatment of Corn Belt soils with lime has been principally for alfalfa, or other legume forages, grown in rotation with corn. A prime purpose of the legume has been to supply nitrogen to the soil. With a sharp reduction in the cost of nitrogen fertilizers following World War II, the use of legumes as a nitrogen source has decreased, as has the use of lime on land normally devoted to corn. However, with the increasing cost of energy and raw materials for nitrogen fertilizer manufacture, this trend may be reversed in the future and, therefore, the use of legumes in rotation with corn increased.

Many factors determine whether a treatment with lime is worthwhile. In most instances the controlling factor is economics. Costs of the treatment must be balanced against the estimated potential for increased earnings from the treatment. Treatment costs depend on soil characteristics, such as pH and buffering capacity, and on the crop to the extent that it determines the change in pH required. Evaluation of the benefit of lime necessitates an estimation of the anticipated increase in yield expressed in terms of the net value of the crop.

It should be recognized that a soil sufficiently weathered and leached so as to need lime will very likely be lacking in nutrients other than those supplied in lime. Applying lime will not correct such problems, although it may enhance the availability of native phosphorus and molybdenum by reducing soil acidity. Added lime may even induce a boron deficiency, probably because of a boron-calcium interaction, as well as reduce the availability of other pH-sensitive micronutrients such as iron and manganese. Low productivity associated with an acid-soil condition may not be overcome by liming, therefore, if this treatment leaves the soil deficient in other ways. Success in soil liming can be more nearly assured only if all factors of plant growth are as close to optimum as possible.

Although the benefits of liming acid soils are usually judged from immediate effects, more valid judgments normally depend on observations made over periods of several years. To be sure, changes in soil caused by added lime are not permanent, for the natural tendency of soils in humid climates is toward an acid condition. Nevertheless, once a desired pH is obtained by liming, additional treatment will not likely be needed for many years, except under very unusual circumstances. The benefit of a single application of lime can be truly determined only if its cumulative effects are taken into account.

Summary

Soils are limed to reduce excess acidity. Excess acidity in soils leads to possible toxic effects on plants from elements such as aluminum and manganese, it limits the chemical availability of some nutrients, and it interferes with a number of important biological processes. Most liming is with ground, or crushed, limestone, or with products manufactured from it. These materials supply hydroxyl ions that neutralize and inactivate exchangeable acids, namely, hydrogen and aluminum.

Liming materials are added in prescribed amounts to raise the pH of soils to a desired level. The quantity added is a function of need, as judged from pH data or soil tests, and the quality of the liming material. The quality of a liming material depends on its total ability to neutralize soil acidity, and its rate of reaction once added to the soil. In general, finely ground materials react more rapidly than do those that have a coarse particle size. Rapid reaction also requires that the material be thoroughly mixed with the soil.

Plants differ in soil pH requirement; some do best under strongly acid conditions, whereas others prefer a near-neutral reaction. Treatments designed to correct soil acidity must therefore take into account the plant to be grown.

Review questions

1. Show why a liming material such as calcium carbonate can inactivate exchangeable hydrogen and aluminum in soils.
2. How may liming increase the availability of nitrogen, phosphorus, and potassium in strongly acid soils?
3. Why does pure magnesium carbonate, $MgCO_3$, have a higher neutralizing value than does pure calcium carbonate, $CaCO_3$?
4. Show that limestone ground to where 80 percent passes a 60-mesh screen and all passes through 4- and 8-mesh screens has a reactivity index of 88.
5. To be equally reactive, why does dolomitic limestone need to be more finely ground than does calcitic limestone?

6. Why does variation in clay content, type of clay, and in organic matter content change the amount of lime needed to adjust the soil pH by a given amount, say from 4.5 to 5.5?
7. What considerations are important to the application of a liming material to the soil?
8. Why is it usually necessary to make repeated applications of lime to naturally acid soils?

Selected references

Coleman, N. T.; Kamprath, E. J.; and Weed, S. B. Liming. *Advances in Agron.* 10:475–522. New York: Academic Press, Inc., 1958.

Lawton, K. and Kurtz, L. T. Soil Reaction and Liming. In *Soil, 1957 Yearbook of Agriculture*, pp. 184–93. Washington, D.C.: Government Printing Office.

Pearson, R. W. and Adams, F. eds. *Soil Acidity and Liming*. Madison, Wis.: American Society of Agronomy, 1967.

Tisdale, S. L., Nelson, W. L., and Beaton, J. D. *Soil Fertility and Fertilizers*. New York: The Macmillan Company, 1984.

18

The salt problem in soils

The accumulation of soluble salts in soils results in one of the most serious problems associated with agriculture of the arid and semiarid regions. Harmful effects of salts include poor seed germination and plant growth resulting from one or both of two conditions: (1) a limited availability of water caused by ions dissolved in the soil solution, and (2) an unsuitable physical or nutritional state caused by a high level of exchangeable sodium. The first of these conditions can be induced by any salt so long as it is readily soluble. However, a high level of exchangeable sodium results only where the accumulated soluble compounds are principally sodium salts. Since soluble salts are more easily removed, the problems they create are normally considered to be less serious than those caused by excess sodium in exchangeable form.

The origin of salts in soils

In the usual pattern, salts accumulate when they are carried into a soil by water that is selectively lost by evaporation and transpiration. The origin of the salts is variable, but a principal source is the ocean. It contributes soluble materials to coastal soils by tidal action or through spray carried inland by the wind. It also provides salt indirectly where soils have formed in parent materials of marine origin. Although leaching tends to remove these salts in humid regions, they are more likely to remain a part of the parent material, or of the soil forming from it, where the climate is dry.

The build-up of salts ordinarily occurs in soils troubled with poor drainage, a condition that automatically enhances water loss by evaporation. These soils may be located in depressional areas or other low-lying positions that collect drainage water through seepage or surface flow. They may also be alluvial soils subject to seepage or flooding from a neighboring waterway. In many places a salted condition has resulted from the application of salt-laden irrigation water without suitable leaching, or, where the water has been applied in excess, by a rise in the level of the ground water, which then serves as a source of salts (see fig. 18.1).

Variability among the factors affecting salt accumulation normally results in a nonuniform distribution of salts over the landscape. Thus, as apparent in the vegetation pattern of figure 18.2, it is common for salt-affected

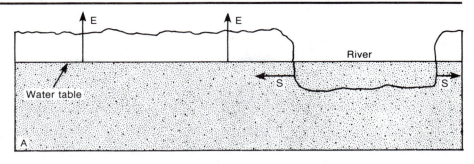

Figure 18.1.
Two conditions promoting salt accumulation in soils from evaporation (E) of water rising from a water table. In diagram A, the water table is maintained by seepage (S) from a river. In diagram B, the water table is raised by the slow lateral drainage (D) of irrigation water (I) applied in excess to the soil.

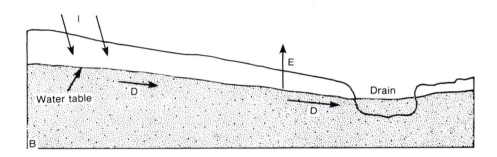

Figure 18.2.
A salt-affected field showing the irregularity of the vegetation pattern. (Photo courtesy of Dr. C. D. Moodie, Washington State University.)

Figure 18.3.
The flood plain of the Tomichi River in central Colorado showing the scattered distribution of standing water, as may occur following a heavy rain or a flood. Because of leaching by the standing water, soil in the lower-lying positions tends not to accumulate as much salt as will accumulate in neighboring, somewhat higher-lying positions on the flood plain. (From Rainey, M. B. and Hess, A. D. *Irrigation of Agriculture Lands*, 1967, pages 1070–81 by permission of The American Society of Agronomy.)

soils to be interspersed with others of relatively low salt content. Inequalities in salt distribution are most often associated with an uneven topography, which causes variation in the entry of surface water into the soil, or with changes in texture over a continuous landscape, which affects the pattern of water flow and salt redistribution within the soil (see fig. 18.3).

Some chemical aspects of salt accumulation

Salts common to arid-region soils vary in both kind and amount. In the vast majority of cases they consist of combinations of only three cations, sodium (Na^+), calcium (Ca^{2+}), and magnesium (Mg^{2+}), with two anions, chloride (Cl^-), and sulfate (SO_4^{2-}). With the exception of gypsum, $CaSO_4 \cdot 2H_2O$, combinations of these ions are readily soluble in water (see table 18.1); they therefore move readily through the soil, sometimes to form surficial salt crusts, as seen in figure 18.4.

Carbonate (CO_3^{2-}) and bicarbonate (HCO_3^-) ions may also occur in soil salts. The bicarbonate ion is seldom present in large quantities, for it tends to convert spontaneously to carbonate, with water and carbon dioxide being released by the reaction:

$$2HCO_3^- \rightarrow CO_3^{2-} + H_2O + CO_2 \uparrow \qquad (18.1)$$

This transformation continues so long as the gaseous carbon dioxide can escape from the system. The carbonate formed by the above reaction may accumulate as a soluble ion provided soluble calcium and magnesium are lacking. If these two cations are present in quantity, they combine with the

Figure 18.4.
Soil with a heavy incrustation of salt due to the rise of salt-laden water from a shallow water table over a period of many years.

Table 18.1. The solubility of several compounds found frequently in salt-affected soils. (Data from standard solubility tables)

Compound		Solubility in water[a]
Name	Formula	
Sodium chloride	$NaCl$	35.7
Sodium sulfate	Na_2SO_4	19.4
Sodium carbonate	Na_2CO_3	7.1
Sodium bicarbonate	$NaHCO_3$	6.9
Magnesium chloride	$MgCl_2$	52.8
Magnesium sulfate	$MgSO_4$	26.9
Magnesium carbonate	$MgCO_3$	0.01
Calcium chloride	$CaCl_2$	59.5
Calcium sulfate[b]	$CaSO_4 \cdot 2H_2O$	0.22
Calcium carbonate	$CaCO_3$	0.001

[a]Solubility as a percent by weight, or grams of compound per 100 g of water.
[b]Calcium sulfate as the dihydrate (gypsum).

Table 18.2. The effect variation in the ratio of sodium to calcium + magnesium in the soil solution on the exchangeable-sodium percentage. Values are averages that apply to a wide range of arid-region soils.

Ion concentrations (me/l)			Exchangeable-sodium percentage
Total[a]	Na$^+$	Ca^{2+} + Mg^{2+}	
15	7.5	7.5	4.2
15	10.0	5.0	7.4
15	12.0	3.0	11.7
15	14.0	1.0	22.0

[a]Sum of soluble sodium + calcium + magnesium.

carbonate and cause its precipitation as slightly soluble calcium and magnesium carbonates. If there is very little soluble calcium and magnesium in the accumulating salts, the concentration of soluble carbonate can be substantial, for it will occur in solution mainly with sodium, and as shown in table 18.1, sodium carbonate is a fairly soluble compound.

The principal problem arising from accumulating soluble salts in soils is the competition they offer seeds and plant roots for available water. This effect is due to the strong force of attraction of charged ions for polar water molecules, which leads to hydration of the ions and a reduction in the osmotic potential of soil water (see page 168). Ions in excess concentration cannot be absorbed by plants and seeds, nor can the water bound tightly to the ions. Sometimes, sodium and chloride ions in accumulated salts may be toxic to plants, but this is usually less harmful than the reduction in water availability.

Another, and often serious, problem accompanying salt accumulation in soils is the build-up of exchangeable sodium at the expense of more desirable exchangeable calcium and magnesium. Sodium is not particularly competitive with the other two ions for exchange sites unless its concentration in solution exceeds their's substantially. This fact is illustrated by the data in table 18.2, where it is seen that, at a constant total concentration of 15 me/liter,[1] variation in the sodium/calcium + magnesium ratio from 1/1 to 14/1 causes the exchangeable-sodium percentage (ESP), identified as the percent of exchange sites in the soil filled by sodium ions, to increase from about 4 to 22. It is noteworthy, however, that even when the concentration of soluble sodium is 14 times greater than that of calcium + magnesium, the sodium still occupies less than one-fourth of the exchange sites in the soil. Nonetheless, this amount of exchangeable sodium can have a decidedly adverse effect on soil properties important to plant growth. An ESP of 4, as shown for the first example in table 18.2, would not likely have an adverse effect on soil properties and plant growth.

[1]This means that in a liter of soil solution the combined concentrations of sodium + calcium + magnesium is 15 me; that is, together they supply a charge equal to that provided by 0.015 g of H$^+$ ion. It would take 4 kg of soil at a 25 percent gravimetric water content to provide a liter of soil solution.

Sodium may be the dominant ion in the soil solution because of the nature of the soluble salts added to the soil or because the calcium and magnesium in these salts are eventually precipitated by combining with ions such as carbonate or sulfate to form compounds of relatively low solubility; lime or gypsum, for example. Precipitation is encouraged by the selective removal of water through plant absorption and evaporation, which increases the concentrations of calcium, magnesium, carbonate, and sulfate to where they will precipitate as lime or gypsum.

Characterization of salt-affected soils

Salt-affected soils are characterized and classified on the basis of two chemical properties: (1) the content of soluble salts, and (2) the percentage of all exchange sites filled by sodium ions. On the basis of these properties, soils troubled with salts are assigned to one of three classes: *saline*, which denotes excess soluble salts; *sodic*, a term indicating that an excess of exchangeable sodium is the problem;[2] and *saline-sodic*, which recognizes the presence of both conditions.

Salinity assessment

The level of salinity in soils is judged indirectly by the *electrical conductivity* of *saturation-paste extracts*, which are obtained by mixing soil with enough water to yield a thin paste and then filtering. This method of salinity evaluation depends on the ability of ions in solution to carry an electrical current between two neighboring electrodes. The conductivity is measured in a small, glass conductivity cell containing electrodes spaced at a fixed distance from each other. Interpretation of the results is based on the known relationship between conductivity values and plant growth, including seed germination, over a wide range of salinity conditions.

The electrical conductivity of a solution is a function of two variables: (1) the *conductance*, which depends directly on the concentration of ions in solution, and (2) the *distance* of current flow between electrodes. Electrical conductivity of a solution varies directly with conductance but inversely with the distance of flow. Thus, conductivity is expressed by the conductance divided by the distance, with the standard unit of conductivity being *mhos* of conductance per centimeter of distance (mhos/cm). This is a much larger unit than is needed for soil extracts, however. For this reason, a smaller unit, millimhos per centimeter (mmho/cm) is used.[3] A millimho is one-thousandth of a mho. A still smaller unit, micromhos per centimeter (μmhos/cm) is used with irrigation and other waters of very low ion content. A micromho is one-millionth of a mho.

[2]In the literature, the term *alkali* may be used to designate excess exchangeable sodium in soils. Because this term has been misconstrued as representing merely an alkaline (basic) soil reaction, sodic has been adopted to indicate the presence of excess exchangeable sodium.

[3]In SI units (see Appendix C) conductivity is expressed in siemens (S) per meter. One millimho per centimeter is equal to 0.1 siemen per meter (0.1 S/m), or 1.0 decisiemen per meter (1.0 dS/m).

In this country, a soil is classed as saline if the conductivity of its saturation-paste extract exceeds 2 mmhos/cm. The kind of salts is not important, since all that contribute to electrical conductivity are essentially equally competitive with plants and seeds for soil water. A conductivity of 2 mmhos/cm has been chosen to separate saline from nonsaline soils because it represents a level of salts capable of reducing growth in a number of sensitive agricultural plants. Other plants, particularly those native to salt-affected soils, can tolerate salt levels many times higher than this.

Assessment of sodic-soil conditions

In concept, a sodic-soil condition results whenever exchangeable sodium is in sufficient quantity to have an adverse effect on plant growth. As with salinity, however, plants differ in their sensitivity to this condition; thus, the level of exchangeable sodium used to define a sodic-soil condition has been selected arbitrarily. Further, an indirect method is used to detect this condition; it depends on an analysis of the soil solution rather than on a measured quantity of exchangeable sodium. This approach is possible because of the close relationship between the quantities of sodium, calcium, and magnesium in the soil solution and the proportionate quantities of these elements present as exchangeable ions.

The evaluation of sodic-soil conditions is based on the *sodium-adsorption ratio* (SAR) of saturation-paste extracts. The SAR is a ratio of soluble sodium to soluble calcium + magnesium in the saturation-paste extract as expressed by the formula:

$$SAR = \frac{[Na^+]}{\sqrt{\frac{[Ca^{2+} + Mg^{2+}]}{2}}} \qquad (18.2)$$

where the symbols $[Na^+]$ and $[Ca^{2+} + Mg^{2+}]$ denote the respective concentrations of sodium and of calcium + magnesium in milliequivalents per liter of saturation-paste extract. According to the formula, the SAR, which parallels the ESP, varies directly with the concentration of soluble sodium but inversely with that of soluble calcium + magnesium. In other words, an increase in sodium in the soil solution causes an increase in the SAR and the ESP, whereas an increase in calcium or magnesium causes a reduction in the SAR and the ESP.

By definition, a sodic soil has a saturation-paste extract with an SAR of 15 or above.[4] At this limiting SAR value, soils have, on the average, about 17 percent exchangeable sodium, as can be judged from the diagonal line in figure 18.5. This figure shows the average relationship between SAR and ESP values. As stated in the caption, figure 18.5 is a nomogram that can be used to determine the SAR for both saturation-paste extracts and irrigation waters from measured concentrations of sodium and calcium + magnesium.

[4]For many years a sodic soil was defined as one having an exchangeable-Na percentage of 15. Use of SAR values for the definition was adopted, because these values closely approximate the percent exchangeable sodium but are far less complicated to determine in the laboratory.

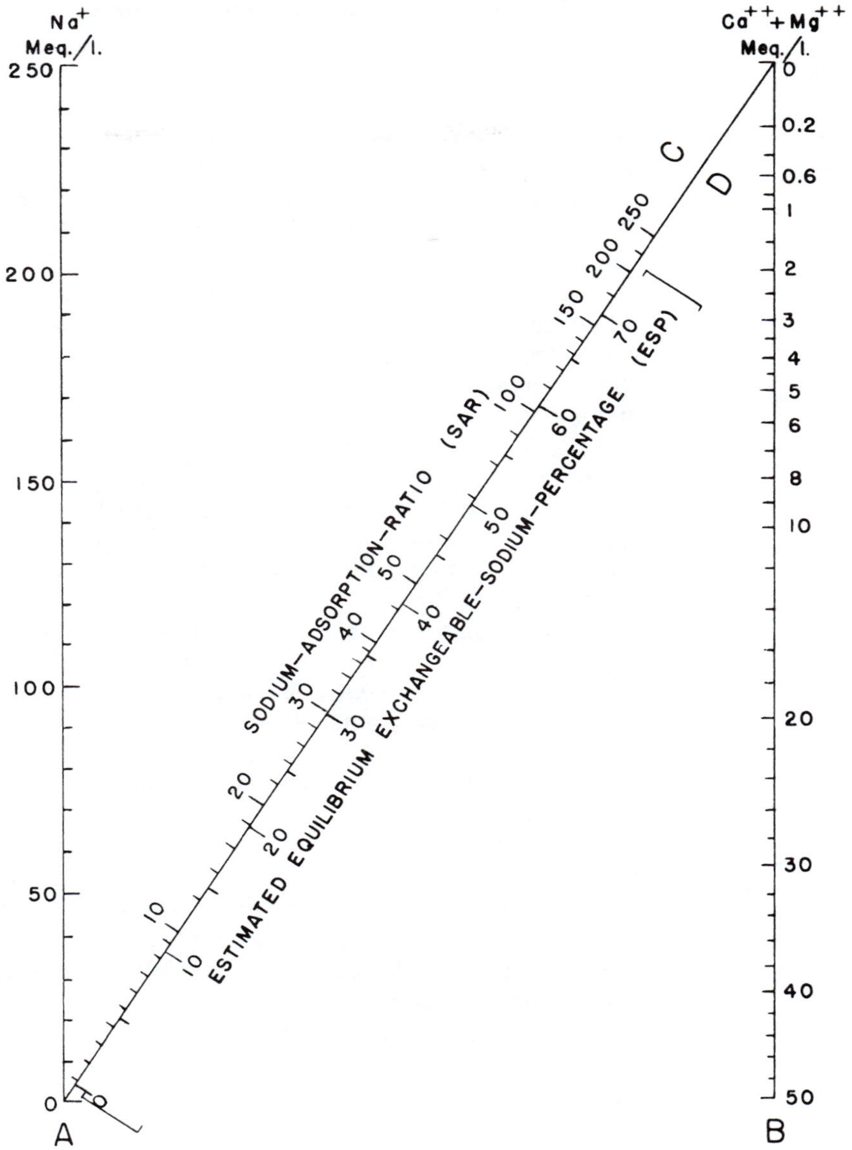

Figure 18.5.
Nomogram for determining SAR values from water analyses. To use, lay a straightedge between measured ion concentrations shown on vertical lines A and B. The SAR is shown where the straightedge intersects the diagonal line. The diagonal line also shows the relationship of SAR values to the exchangeable-sodium percentage. (From U.S.D.A. Agriculture Handbook 60, 1954.)

Plant growth relationships in salt-affected soils

The way in which plants respond to salted-soil conditions is variable. Sensitivity may be due to the low availability of water, to toxic effects caused by specific ions, or to adverse physical or nutritional conditions often associated with sodic soils. Plants may respond differently at different stages of development. Further, some may tolerate salinity better than a high level of exchangeable sodium. Because of the diverse effects of salt-affected soils on plant growth, only a few of the more general relationships can be considered here.

Salinity and plant growth

The principal harm to plants caused by soluble salts is a reduction in the availability of water. Ion-hydration forces add to matric forces in resisting the uptake of water by plants. This effect is not necessarily serious where irrigation water is available and can be added to the land at frequent intervals.

In general, salts of different ionic composition have about the same influence on water availability provided they occur at comparable osmotic concentrations. At times, however, a plant may be intolerant of a single ion species in the accumulating salts and suffers a reduction in growth because of a toxicity factor rather than a reduction in water uptake. Each of the elements common to soil salts may be toxic to some plants, but chlorine and sodium are among the more frequent offenders. Occasionally, a toxicity is caused by boron, which can reduce growth when in comparatively low concentration in the soil solution. Boron toxicities are primarily of local importance, since this element is not an important constituent of soil salts under most circumstances.

Most plants are more sensitive to salinity during germination than at any other time. Success in establishing and maintaining a stand of crop plants therefore strongly depends on conditions that prevail at planting time. The occurrence of barren patches in fields containing salt-affected soils is often the result of a localized salinity effect on germination.

Variation in the sensitivity of germinating seeds to salinity is demonstrated in figure 18.6, where the percentage germination of four plant species is plotted against conductivity as measured by standard procedures. As seen in the figure, beans and sugar beets are more sensitive to salts at germination time than are alfalfa and barley. It is of particular interest that germination of alfalfa and barley seeds is not significantly affected where the conductivity of the saturation-paste extract is 4 mmhos/cm, the point of division between saline and nonsaline conditions. This fact emphasizes the arbitrary nature of the definition of saline soils and indicates the need for taking into account the type of plant when considering the importance of a salinity factor on growth.

Exchangeable sodium as a plant growth factor

Excess exchangeable sodium is harmful to plants principally because it induces undesirable physical and chemical conditions in soils. One effect is the breakdown of aggregates and a reduction in pore size, effects that lower the permeability of the soil to air and water. This is because exchangeable sodium simply does not bind soil particles together sufficiently to provide stable aggregates. Often, just wetting a sodic soil causes the particles to separate, or disperse, so that the soil can be easily puddled. Once puddled, the soil may form a firm surface crust on drying (see fig. 18.7). Such crusts, when formed after seeds are planted, may prevent emergence of seedlings and cause barren patches in a field.

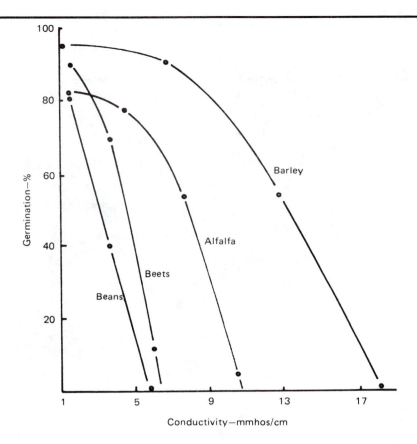

Figure 18.6.
Percentage germination of seeds of four crop plants in relation to the salt content of the soil, as indicated by the conductivity of the saturation-paste extract. Germination of less than 100 percent at the lowest level of salinity is caused by unviable seeds. (After A. D. Ayres, and H. E. Hayward, *Proc. Soil Sci. Soc. Amer.* 13:224–26, 1949.)

Figure 18.7.
A saline-sodic soil that has formed a very hard surface crust following leaching and drying.

The salt problem in soils   475

Figure 18.8.
The surface incrustation of salts is just as great at the bottom as in the upper, whitish portion of this photo, but it is black due to accumulated solubilized organic matter.

Soils high in exchangeable sodium commonly have a high pH, provided they are low in neutral salts. Some nonsaline sodic soils, for example, may have a pH of 10 or above. Whereas plants seem not to suffer materially under strongly basic conditions if adequately supplied with nutrients, they may suffer if these conditions decrease the availability of nutrients. For example, the availability of iron and manganese is often low in strongly basic soils because of the tendency for these two elements to precipitate as essentially insoluble hydroxides or oxides.

The high pH of sodic soils causes soil organic matter to dissolve. If the dissolved organic matter is carried upward by the capillary rise of water, it may be deposited as a dark incrustation on the surface of the soil (see fig. 18.8). When present, a dark-colored surface film is usually indicative of a sodic-soil condition. However, many sodic soils lack this partiuclar feature, so it is not a universal property.

The dispersive effect of exchangeable sodium on clay is due to the highly hydrated nature of this ion. Because of its thick water shell, exchangeable sodium is neither adsorbed tightly to clay surfaces, nor does it draw clay particles together in tightly knit, aggregated (flocculated) form. This effect is reversed in sodic soils if they contain a high content of soluble salts, however. When in solution, these salts compete with sodium for water and thereby effectively decrease the thickness of its water shell. As a consequence, saline-sodic soils usually occur in a more highly aggregated state than do nonsaline-sodic soils. In addition, if the salts are neutral, they suppress the hydrolysis of exchangeable sodium and basic salts and thereby prevent the soil from having an excessively high pH. Under some circumstances the pH of saline-sodic soils is no higher than 8.5. In spite of these apparent benefits from salts, they do not help plant growth on sodic soils, for they still compete with the plants for available water. Leaching neutral salts from saline-sodic soils normally causes a rise in pH and a reduction in soil aggregation.

Reclamation of salt-affected soils

The reclamation of salt-affected soils consists of the removal of soluble salts or exchangeable sodium, or both, to the extent necessary to return these soils to a normal, productive state. The discussions that follow deal with some of the requirements fundamental to these objectives. Each problem area must be examined individually and the reclamation program developed to cope with the conditions peculiar to the location in question. Reclamation is not always practical, but it is almost always expensive; thus, it should not be undertaken unless there is some assurance of success.

The removal of soluble salts

Soluble salts are rather easily removed from soils, provided they can be leached to a depth that will prevent their return to the rooting zone. Generally, several conditions must be met to accomplish satisfactory leaching. First, irrigation water must be available, and if reclamation is to be rapid, there should be some means of impounding the water on the surface to provide extensive infiltration into the soil. Ideally, the land should be flat so that it can be diked and flooded (see fig. 18.9). The period of flooding will then depend on the amount of leaching required. Whereas the leaching requirement cannot be stated precisely, a reasonable generalization is 1 cm of water for each centimeter of soil depth to be reclaimed.

A second requirement for successful leaching is adequate drainage. This means two things: textural and structural relations should be adequate for the rapid downward flow of water through the soil, and a water table should not be so near the soil surface that it can later regenerate a salted condition through the capillary rise of a salt-laden water. A satisfactory depth to a water table is 120 to 150 cm at minimum, with 180 cm or more being preferred. If a water table occurs at a shallow depth, provisions for improved drainage should be made before leaching is undertaken (see chapter 9).

Salt removal from saline-sodic soils presents a greater challenge than where salinity alone is the problem. Although sodic soils may be flocculated and permeable so long as salts are present, once the salt concentration is decreased to a low level, dispersion of clay may follow and so reduce permeability that further leaching becomes all but impossible. Leaching of a saline-sodic soil should not be attempted without making provisions to remove exchangeable sodium at the same time.

Removal of exchangeable sodium

The displacement of exchangeable sodium from sodic soils is normally accomplished by adding an amendment that will supply soluble calcium. Gypsum is the amendment used most widely for this purpose. Once in so-

Figure 18.9.
Soil diked and flooded
for the removal of
salts by leaching.
(Photo by Bob LeMert,
U.S. Salinity
Laboratory, Riverside,
California.)

lution, the calcium from gypsum undergoes exchange with sodium, which on displacement from exchange sites, remains as a soluble ion and can be removed from the soil by leaching. The exchange reaction is:

$$\boxed{\text{Soil}} \begin{matrix} \text{Na} \\ \text{Na} \end{matrix} + CaSO_4 \cdot 2H_2O \rightleftharpoons \boxed{\text{Soil}} Ca + 2Na^+ + SO_4^{2-} + 2H_2O$$

$$(18.3)$$

Since gypsum is not very soluble, it should be applied in a finely divided state and mixed thoroughly with the soil to assure quick solution and reaction with exchangeable sodium.

Agricultural sulfur or sulfuric acid may also be used for sodic-soil reclamation. Since sulfur converts to sulfuric acid upon its biological oxidation in the soil (see page 365), either of these amendments supplies hydrogen ions that are capable of displacing exchangeable sodium in sodic soils. However, most sodic soils are calcareous; thus they contain lime that will react with sulfuric acid to yield gypsum:

$$H_2SO_4 + CaCO_3 \rightarrow CaSO_4 \cdot 2H_2O + CO_2 \uparrow \qquad (18.4)$$

In soils that are calcareous, therefore, calcium will be supplied for the displacement of exchangeable sodium regardless of whether the added amendment is gypsum, sulfur, or sulfuric acid.

Soluble sodium carbonate, which is a common component of many saline-sodic soils, is capable of precipitating soluble calcium as calcium carbonate and therefore increases the amount of amendment required for reclamation. Soluble calcium lost by this means is not available for the dis-

The salt problem in soils

placement of exchangeable sodium. If a soil containing soluble sodium carbonate were highly permeable, a preliminary leaching treatment could be used to reduce its concentration in the soil solution. Otherwise, to assure complete reclamation, sufficient amendment must be added to react with both soluble sodium carbonate and exchangeable sodium.

One of the greatest challenges in reclamation is with medium- to fine-textured soils in which exchangeable sodium has resulted in low permeability in the subsoil as well as in the surface soil. The problem is due primarily to the difficulty of improving subsoil permeability, for unless this is done, the leaching of salts will be either extremely slow or impractical. Reclamation programs should be initiated only where internal drainage characteristics of the profile are known to be satisfactory.

Estimating amendment needs

The amount of amendment to apply for reclamation of a sodic soils can be judged by a direct determination of the quantity of exchangeable sodium and soluble sodium carbonate present, or indirectly by a rapid *gypsum requirement test*. For the first, the quantity of exchangeable sodium and soluble sodium carbonate is measured, and the total amount of sodium in these two forms is then expressed as a single quantity, usually in milliequivalents per 100 g of soil. For each milliequivalent of sodium to be removed 1 me of calcium, as from gypsum, must be supplied. According to table 7.7, on page 156, 1 me of calcium per 100 g of soil is equal to 400 kg/2,000,000 kg of soil. Also shown in table 7.7 is that 400 kg of calcium is contained in 1720 kg, or 1.72 metric tons, of gypsum.[5] Thus, the gypsum requirement of a soil can be computed as:

$$\text{Gypsum requirement} = 1.72 \times \text{NaX} \qquad (18.5)$$

where the gypsum requirement is in metric tons/2,000,000 kg of soil, and the amount of sodium in exchangeable form and as sodium carbonate, symbolized by NaX, is in milliequivalents/100 g of soil. In keeping with this, for every milliequivalent of sodium/100 g of soil, gypsum must be added at a rate of 1.72 metric tons/2,000,000 kg of soil.

A principal advantage of the gypsum requirement test is its simplicity. It consists of mixing a soil sample with a saturated gypsum solution and then measuring the reduction in calcium caused by either its adsorption to the exchange complex in place of sodium or its precipitation as calcium carbonate on reaction with soluble sodium carbonate. The gypsum requirement can then be computed by substituting the decrease in calcium concentration for NaX in equation (18.4). For example, assume that a soil sample weighing 10 g decreases the calcium concentration in a saturated gypsum solution by 0.8 me. This would be equivalent to 8 me/100 g of soil. Substituting 8 me for NaX in equation (18.4) yields a gypsum requirement of 8 × 1.72, or 13.76 metric tons/2,000,000 kg of soil.

---

[5]Gypsum has a molecular weight of 172 g, of which 40 g is due to the weight of calcium. Accordingly, the amount of gypsum containing 40 g of calcium is 172 g, and the amount containing 400 kg of calcium is 1720 kg. Based on this relationship, calcium makes up approximately 23.5 percent of the weight of gypsum.

Although the gypsum requirement test measures the specific need for gypsum in sodic-soil reclamation, it can be used with equal facility to predict the need for elemental sulfur or sulfuric acid when they are used for this purpose. Either one atom of sulfur, which oxidizes to a molecule of sulfuric acid, or a molecule of the acid applied directly to the soil, have the same ability for displacing exchangeable sodium as does one molecule of gypsum. Thus, as can be determined by dividing the atomic weight of sulfur (32) by the molecular weight of gypsum (172), it takes only 19 percent as much elemental sulfur as gypsum to provide the same sodium-displacing power. Dividing the molecular weight of sulfuric acid (98) by the molecular weight of gypsum indicates that approximately 57 percent as much of this compound as gypsum is needed to effect the same displacement of exchangeable sodium. In other words, the gypsum requirement $\times$ 0.19 estimates the elemental sulfur requirement for sodic-soil reclamation, and the gypsum requirement $\times$ 0.57 estimates the requirement for sulfuric acid.

Managing saline and sodic soils

If the reclamation of salt-affected soil cannot be undertaken, certain practices may be followed that provide a better soil environment for plant growth. Some of the procedures involved may gradually improve soil conditions, or at least prevent them from worsening.

Probably the most vital aspect of the management of saline soils is to avoid excessive concentration of salts in the root zone of plants. For this, frequent irrigations are recommended. When possible, excess water should be applied to flush salts more deeply into the soil or to counter the tendency for salts added in the water to accumulate in the soil.

Some control over the distribution of salts in soils is possible with careful irrigation management. When water is applied, it dissolves salts as it moves into and through the soil. Where sprinklers are used, salt movement is in a downward direction. Where furrow irrigation is used, movement is both downward and lateral, with the result that the soil immediately below and to the sides of the furrow may be flushed relatively free of salts. Advantage is sometimes taken of this by planting seeds very close to an irrigation furrow where the salt concentration is low (see fig. 18.10). After the plants are well-established, the furrow can be relocated midway between the planted rows. The tendency for salts to accumulate at a point midway between irrigation furrows is demonstated in figure 18.11.

An additional aid in the utilization of saline soils is to grow tolerant plant species. The salt tolerance of a number of common agricultural plants is shown in table 18.3.

There are very few management tactics available for assuring satisfactory plant growth on sodic soils. Tolerant crops may be grown, preferably perennials that need not be re-established each year. These may include some, though not all, of the species also tolerant of salts. Care should be exercised in planting, the objective being to prepare a good seed bed and then

Figure 18.10.
Method of locating seeds in irrigated soil to minimize salt injury. Movement of irrigation water concentrates salts beyond the seed, as shown by shading in the diagram.

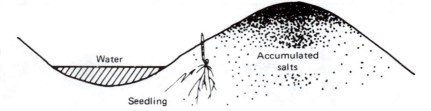

Table 18.3. The relative tolerance of some crop plants to salts in the soil. For a more complete listing refer to *U.S.D.A. Handbook* 60, p. 67, 1954.

Fruit Crops	Field Crops	Forages	Vegetables
		Plants of high salt tolerance	
Date Palm	Barley	Alkali sacaton	Garden beets
	Sugar beets	Saltgrass	Kale
	Rape	Nuttall alkali grass	Asparagus
	Cotton	Bermuda grass	Spinach
		Canada wildrye	
		Western wheatgrass	
		Tall wheatgrass	
		Birdsfoot trefoil	
		Plants of medium salt tolerance	
Pomegranate	Rye	Sweetclover	Tomato
Fig	Wheat	Perennial ryegrass	Broccoli
Olive	Oats	Strawberry clover	Cabbage
Grape	Rice	Sudan grass	Cauliflower
Cantaloupe	Sorghum	Dallis grass	Lettuce
	Corn	Alfalfa	Sweet corn
	Flax	Tall fescue	Potato
	Sunflower	Orchard grass	Carrot
			Onion
			Peas
			Squash
		Plants of low salt tolerance	
Pear	Field beans	White Dutch clover	Radish
Apple		Meadow foxtail	Celery
Citrus		Alsike clover	Green beans
Plum		Red clover	
Almond		Ladino clover	
Apricot			
Peach			

Figure 18.11.
The movement of
water from irrigation
furrows has
concentrated salts at
ridge centers causing
serious damage to this
carrot crop in the
Coachella Valley,
California. (Bureau of
Reclamation, U.S.
Department of the
Interior photo by E. E.
Hertzog.)

to plant and secure seedling emergence before the physical state of the soil changes. The application of irrigation water, for instance, may cause the soil to seal over and thereby prevent germination.

In due time the level of exchangeable sodium can be reduced and the physical state gradually improved if a sodic soil containing free lime is subjected to long periods of leaching without the addition of an amendment. Reclamation is effected by the small quantity of calcium derived from the lime. Sometimes the application of plant residues or manure is suggested as a means of improving the physical state of sodic soils.

The use of saline irrigation water can improve a sodic-soil condition, provided soluble calcium, magnesium, and sodium occur in proper balance and some leaching is possible to prevent the salts from accumulating in the soil. Other than the few treatments mentioned previously, however, little short of complete reclamation can be expected to greatly change the productivity of seriously affected sodic soils.

Irrigation water quality

In the majority of cases, the quality of irrigation water is judged from the kinds and concentration of soluble ions it contains. These ions are predominantly a mixture of sodium, calcium, and magnesium with chloride, sulfate, bicarbonate, and occasionally, small amounts of carbonate. Unless periodically leached, these ions accumulate in the soil as the applied water is selectively removed by evapotranspiration. Accumulating chlorides and sulfates add mainly to soil salinity, but bicarbonate and carbonate aid in the build-up of exchangeable sodium. This is because the bicarbonate ion, on soil drying, converts to carbonate [see equation (18.1)], and with any carbonate already present in the water, causes calcium and magnesium to precipitate as relatively insoluble carbonates. The precipitation of calcium and magnesium reduces their concentration relative to that of sodium and thus enhances the adsorption of sodium in place of calcium and magnesium.

Sometimes, the concentration of bicarbonate and carbonate in irrigation water exceeds the concentration of calcium and magnesium. Where this is the case, and where conversion of the bicarbonate to carbonate is complete, it is possible for virtually all of the calcium and magnesium in the water to be precipitated as carbonates in the soil. Removal of calcium and magnesium leaves sodium as essentially the only important cation supplied in the irrigation water to accumulate in the soil solution. Any carbonate not consumed in precipitating calcium and magnesium is therefore associated with sodium, the combination being referred to as *residual sodium carbonate*. Whether an irrigation water contains residual sodium carbonate can be computed from analyses for bicarbonate + carbonate and calcium + magnesium through use of the equation:

$$\text{Residual sodium carbonate} = [HCO_3^- + CO_3^{2-}] - [Ca^{2+} + Mg^{2+}] \quad (18.6)$$

where $[HCO_3^- + CO_3^{2-}]$ is the concentration of bicarbonate + carbonate in the irrigation water, and $[Ca^{2+} + Mg^{2+}]$, the concentration of calcium + magnesium, both quantities expressed in milliequivalents/liter of water.

Residual sodium carbonate in irrigation water has the potential for precipitating exchangeable calcium and magnesium and therefore permitting a build-up of exchangeable sodium in the soil:

$$\boxed{\text{Soil}} \, Ca + Na_2CO_3 \; \longrightarrow \; \boxed{\text{Soil}} \begin{matrix} Na \\ Na \end{matrix} + CaCO_3 \downarrow \quad (18.7)$$

In theory, all sodium from the residual sodium carbonate in irrigation water can become exchangeable. To minimize this hazard, pretreatment of irrigation water with gypsum or other source of soluble calcium is often recommended. The hazard can also be reduced by increasing the frequency of irrigation, as discussed under leaching requirements on page 486.

A classification of irrigation water

Numerous classifications to characterize the salt status of irrigation water have been proposed. The one used in this country has been developed by the U.S. Salinity Laboratory, located in Riverside, California, and is outlined in the chart of figure 18.12. This system takes two variables into account: (1) the *sodium hazard*, as expressed by the SAR of the irrigation water, and (2) the *salinity hazard*, as indicated by the conductivity of the water in micromhos per centimeter. Classes of salinity hazard are defined by fixed ranges of conductivity, but the sodium hazard varies depending on the conductivity. For example, if the conductivity were 500 μmhos/cm, an SAR of 6 would result in a medium sodium hazard. If the conductivity were 1000 μmhos/cm, on the other hand, an SAR of 6 would result in a high sodium hazard.

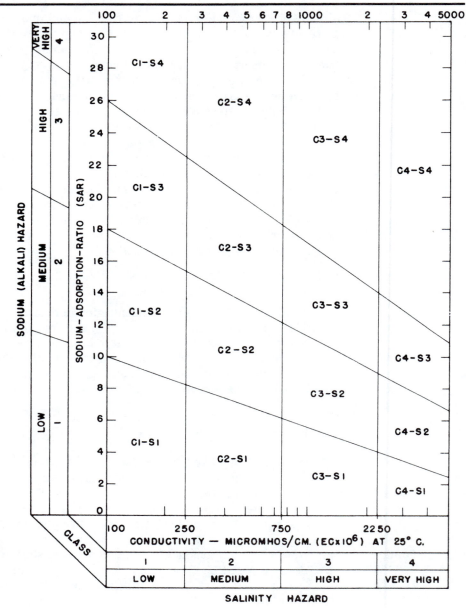

Figure 18.12.
The classification of irrigation water according to the system of the U.S. Salinity Laboratory. (From U.S.D.A. Agriculture Handbook 60, 1954.)

The quality classification for 14 river waters from the south-central and western parts of the United States is shown in table 18.4. This classification is based on the chart in figure 18.12. Also shown in the table are electrical conductivities, SAR and residual-sodium carbonate values, and analytical data for various soluble ions. Seventeen separate analyses are listed, since samples for three of the rivers were obtained from two different locations.

The salt problem in soils

Table 18.4. Chemical composition of some river waters used for irrigation in the western United States (Selected from *U.S.D.A. Agriculture Handbook* 60, 1954).

River	Location	EC[a]	Sum of cations (me/l)	SAR	Quality class[b]	Milliequivalents per liter								
						Ca	Mg	Na	K	CO_3	HCO_3	SO_4	Cl	Residual Na_2CO_3
Missouri	Williston, ND	838	9.48	2.0	C3–S1	3.49	2.38	3.48	0.13	0	3.54	5.39	0.34	0
Platte	Aurora, NE	800	7.98	2.2	C3–S1	2.96	1.67	3.35	—	.20	2.85	4.56	.76	0
Arkansas	LaJunta, CO	1210	14.38	1.5	C3–S1	7.18	3.49	3.47	.24	Tr	3.95	9.80	.62	0
Do	Ralston, OK	1670	14.52	4.5	C3–S2	4.34	2.14	—8.04—		0	2.79	4.39	7.28	0
Rio Grande	Otowi Bridge, NM	340	3.39	0.7	C2–S1	1.86	.70	.83	—	Tr	1.77	1.50	.14	0
Do	El Paso, TX	1160	11.54	3.6	C3–S1	4.16	1.42	5.96	—	.05	3.59	5.00	3.10	0
Pecos	Carlsbad, NM	3210	38.00	3.2	C4–S1	17.27	9.21	—11.52—		0	3.18	23.11	11.99	0
Gila	Florence, AZ	1720	16.85	6.7	C3–S2	3.59	1.99	11.27	—	.20	3.68	3.26	9.95	0
Colorado	Yuma, AZ	1060	10.96	2.2	C3–S1	4.79	2.11	4.06	—	Tr	2.64	6.39	2.05	0
Sevier	Central, UT	580	5.47	1.1	C2–S1	2.50	1.23	1.57	.17	.10	4.10	1.12	.74	.47
Do	Delta, UT	2400	25.81	6.8	C4–S2	3.14	6.90	15.31	.46	.33	4.76	8.44	12.52	1.56
Humboldt	Rye Patch, NV	1173	11.55	5.9	C3–S2	1.75	1.89	7.91	—	0	5.20	2.17	4.46	.12
Sacramento	Tisdale, CA	162	1.73	0.6	C1–S1	.66	.57	.45	.05	0	1.35	.14	.20	.27
Kern	Bakersfield, CA	234	2.36	1.3	C1–S1	1.00	.24	1.06	.06	0	1.51	.49	.40	0
Columbia	Wenatchee, WA	151	1.48	0.2	C1–S1	.90	.39	.19	—	0	1.26	.21	.07	0
Snake	Minidoka, ID	410	4.54	0.6	C1–S1	2.15	1.29	.84	.26	.34	2.59	.91	.74	0
Rogue	Medford, OR	108	1.15	0.5	C1–S1	.54	.26	.33	.02	0	.85	Tr	.25	.05

[a]Electrical conductivity in micromhos per centimeter.
[b]Based on figure 18.12.

The salt problem in soils   485

As may be judged from table 18.4, the main problem arising from use of these river waters is potential soil salinity. The sodium hazard tends to be low, with only four of the samples falling above the low (S1) range. In comparison, eight have a high (C3) and two a very high (C4) salinity-hazard rating. Four of the rivers are of excellent quality, being low in both sodium and salinity hazard.

The classification of water quality according to the system in figure 18.12 does not take residual sodium carbonate into account. This is in part the result of the highly variable effect residual sodium carbonate can have on soil calcium and magnesium after irrigation water has been applied to the land.

In table 18.4, the salt contents (conductivities) for the Arkansas, Rio Grande, and Sevier Rivers increase from the first to the second listing, which is in a downstream direction. It is characteristic for the composition of river water to change from one location to another, and in these examples, it probably results from the influx of salt-laden drainage water from irrigated land paralleling the rivers. However, rivers that pass from dry to more humid regions may show a decline in salt content as the salts carried from the drier region are diluted by the inflow of low-salt drainage from the wetter areas.

The data in table 18.4 typify the range in chemical composition of river waters used for irrigation in the western United States. A similar table for well waters would show a much wider range in variation. Some wells produce very high quality water, but others are unsuitable by virtue of extreme concentrations of salts, sodium, and at times, other contaminants. Since well waters can differ markedly in composition, even over relatively short distances, it is usually advised that they be tested for quality before being applied to the land.

Water quality and leaching requirements

Because of the potential for salt influx from irrigation water, leaching of irrigated soils is often essential to their continued use. Without leaching, salts accumulate in proportion to the amount of water applied, and all bicarbonate and carbonate added can be effective in the precipitation of calcium and magnesium. With adequate leaching, however, water of rather high salt content can be used indefinitely without causing serious deterioration of the soil or harm to crop plants grown on it.

Leaching requirements depend on the total salt content of irrigation water and its potential to precipitate calcium and magnesium carbonates when applied to the land. Where salts are rather low and there is little tendency for calcium and magnesium to precipitate, standard irrigation practices, which usually result in limited leaching each time water is applied, provide ample protection against the development of a serious salt problem. However, where either the salt content or the potential for calcium and magnesium precipitation is high, special care in irrigation is usually necessary to prevent salt accumulation or a build-up of exchangeable sodium. In general, the usual precaution is an increase in the frequency with which irrigation water is applied, with the result that only a fraction of the stored available water is used between irrigations. Because of the increased frequency, each addition of water is made before the salt concentration has attained damaging proportions or before the soil solution has become saturated with respect to calcium and magnesium carbonates. The quantity of water added should be sufficient to flush residual salts beyond the depth of rooting. Ultimately, they will be lost from the soil in drainage.

Generally speaking, the main problem with low-quality irrigation water is the development of salinity in soils, and prevention of the problem is the addition of excess irrigation water to promote leaching. Estimates of the leaching requirement, or the amount of excess water needed to prevent the build-up of soluble salts, may be made using the electrical conductivity values for the irrigation water (EC_{iw}) and for the soil solution when at its maximum allowable conductivity (EC_{ss}). The computation is based on the equation:

$$\text{Leaching requirement} = \frac{EC_{iw}}{EC_{ss}} \qquad (18.8)$$

where the leaching requirement is the proportion of applied irrigation water to be lost in drainage. To illustrate use of this equation, assume that an irrigation water with an electrical conductivity of 1000 μmhos/cm (1.0 mmhos/cm) is applied to land where the maximum allowable electrical conductivity of the soil solution is 4 mmhos/cm. Substituting these values in equation (18.8), the leaching requirement for the irrigation water would be computed as:

$$\text{Leaching requirement} = \frac{EC_{iw}}{EC_{ss}} = \frac{1.0}{4.0} = \frac{1}{4} = 25 \text{ percent} \qquad (18.9)$$

According to this computation, each irrigation should be made after approximately 75 percent of the water from the previous irrigation has been selectively removed from the soil by evaporation and transpiration. This will reduce the water remaining in the soil to 25 percent, or to one-fourth, of its original volume, and will theoretically have caused a four-fold increase (from EC_{iw} of 1.0 mmhos/cm to EC_{ss} of 4.0 mmhos/cm) in the concentration of dissolved salts. As fresh irrigation water is applied, the residual soil water with its increased load of soluble salts would be moved downward as it is replaced by fresh irrigation water of lower salt content.

Under circumstances where the quality of irrigation water is limited because of residual sodium carbonate, the leaching requirement must be based on the potential of the water to precipitate exchangeable calcium and magnesium. From analytical data, such as that shown in table 18.4, a judgment is made of what proportion of applied irrigation water can be used by combined evaporation and transpiration before all soluble calcium and magnesium are precipitated as carbonates. At this point, residual sodium carbonate will start precipitating exchangeable calcium and magnesium, and the soil solution with the residual sodium carbonate should be displaced with fresh irrigation water. An alternative is to add sufficient gypsum to the irrigation water to remove all residual sodium carbonate before the water is applied to the land.

As with most of the problems of salt-affected soils, the development of an irrigation-management program to prevent salt accumulation must be geared to local conditions. Involved in decisions regarding the program are such things as the composition of the water and its availability for leaching, the tolerance of crops to salts and exchangeable sodium, and soil characteristics, including internal drainage, that influence the potential for successful leaching. Satisfactory drainage is particularly important, for without it, subjection of arid-region land to irrigation will likely result in its eventual abandonment because of accumulated salts.

Summary

Damaging quantities of soluble salts accumulate principally under conditions of inadequate drainage in soils of the drier regions. Excess salts, which produce saline soils, compete with plants for available water and sometimes contain ions that are toxic to plants. Where the salts are predominantly compounds of sodium, they also promote the accumulation of exchangeable sodium. Excess exchangeable sodium produces sodic soils. They ordinarily have poor tilth because of a lack of structure. Many sodic soils also have a high pH, which may limit the availability of some plant nutrients.

Correction of salted-soil conditions usually involves improved internal drainage. Salinity can then more effectively be reduced by leaching. If a sodic-soil condition exists, leaching may be preceded by soil treatment with an amendment supplying soluble calcium. Added calcium displaces exchangeable sodium, thus permitting its removal by leaching.

Some alleviation of a salinity problem is possible where, through the controlled application of irrigation water, soluble salts are kept at a minimum in the vicinity of seeds and sensitive seedlings. No such effect is possible with sodium held in exchangeable form by soil particles.

Salts and exchangeable sodium can accumulate in soils from irrigation water. This problem is minimized by frequent irrigations that are sufficiently large to move salts below the root zone. Treatment of high-sodium irrigation water with a calcium-supplying amendment helps prevent the buildup of exchangeable sodium.

Review questions

1. Why has it been necessary to assign arbitrary values for definitions of saline- and sodic-soil conditions?
2. Why is it practical to evaluate the level of salinity in soils by measuring the electrical conductivity of saturation-paste extracts?
3. What is the advantage of using sodium-adsorption ratios rather than measured exchangeable-sodium percentages in characterizing sodic-soil conditions?
4. Explain why improved drainage is advised as an initial step in the reclamation of salt-affected soils.

5. Why can the controlled application of irrigation water reduce the damaging effect of soluble salts but not of excess exchangeable sodium on germinating seeds and seedlings?
6. What is meant by residual sodium carbonate, and why is it undesirable in irrigation water?
7. What is the leaching requirement of soils, and how does it relate to the evapotranspirational use of irrigation water applied to the soil?

Selected references

Bower, C. A. and Fireman, M. Saline and Alkali Soils. In *Soils, 1957 Yearbook of Agriculture*, pp. 282–90. Washington, D.C.: Government Printing Office.

Hagan, R. M.; Haise, H. R.; and Edminster, T. W., eds. *Irrigation of Agricultural Lands*. Madison, Wis.: American Society of Agronomy, 1967.

Kelley, W. P. *Alkali Soils*. New York: Reinhold Publishing Corporation, 1948.

Salinity Laboratory Staff, U.S. Diagnosis and Improvement of Saline and Alkali Soils. *U.S.D.A. Handbook 60*, Washington, D.C.: Government Printing Office, 1954.

19

Soils and the quality
of the environment

Environmental quality is or should be of concern to every segment of society. The reason is clear. The purity of the environment, that is, of air, water, and land, is a fundamental determinant of human existence. That the level of environmental purity is not all that it should be is currently being heard from all quarters, and rightly so. Many of our rivers, streams, and lakes have become choked with sewage, industrial refuse, and sediment; the usefulness of much land has been impaired by chemical pollutants; and the air in many localities is at times hazardous to both plant and animal life. Although these conditions cannot be wholly dissociated from the natural state of things, they have become unbearable in many instances because of careless use of our land, water, and air resources, particularly from the standpoint of waste disposal. Revulsion against a polluted environment is very frequently an aesthetic reaction. On the other hand, the recent great wave of protest against the state of the environment results largely from the realization that our surroundings have been degraded to the point that they now pose a serious threat to our very existence.

Much pollution is caused by man's careless use of the soil. Through his failure to control erosion, the soil has become a major pollutant of air and water. In turn, man has seriously polluted many soils by the intentional or unintentional addition of a wide range of chemical substances, some of which are lethal to plants. Further, he has not effectively utilized the soil as a medium for the disposal of wastes, particularly those that are organic in nature. Because of interrelationships such as these, the soil, as a major component of the environment, has an inescapable role in the control of environmental quality. Whether or not this role is to be beneficial rather than harmful depends to a large extent on how well we manage our soil resources in the future.

The pollution of soils

Soil deterioration can result from the accumulation of pollutants having various origins. Soil pollutants include common agricultural chemicals used to enhance soil fertility or to control plant pests, salts or toxic ions supplied in irrigation water, and toxic dusts or radioactive fallout coming from the atmosphere. Of these the last, radioactive fallout, is of least immediate concern and should remain so barring further widespread nuclear testing in the atmosphere. Localized harm that has come from fallout is due to the radioactive isotopes cesium-137, strontium-90, and iodine-131. These elements can be absorbed from the soil by plants and are injurious to man and other animals that use these plants for food. Although iodine-131, with a half-life of 8 days, disintegrates and loses its harmful effect in a few months, strontium-90 and cesium-137, with half-lives of 28 and 30 years, respectively, could affect many generations of man. The half-life is the period required for a radioactive isotope to lose half its radioactivity through natural decay and conversion to a nonradioactive form.

The effects of excessive fertilizer elements on soils have been discussed in earlier chapters; they are mentioned here chiefly as a matter of emphasis. It may be recalled that excesses of some nutrients can cause a reduction in either the quality or yield of plants, or in both. The logic of investing in fertilizer that may reduce the net financial return from land while impairing the quality of harvested produce is obviously open to question.

Pesticides become soil pollutants if they accumulate in amounts that interfere with plant growth. Usually, an excess of pesticide occurs in soil for two reasons: (1) over-application to the target crop, or (2) the repeated application of a material that persists for long periods in the soil. An excess of a pesticide may reduce plant growth or quality, or the pesticide may be absorbed in such quantities that it makes the plant or its edible parts unacceptable for human consumption.

Most pesticides are organic compounds that can be destroyed by organic-decay processes. Persistent pesticides tend not to decay rapidly, however. This may reflect the chemical nature of the pesticide, or it may result from environmental conditions, such as low temperature or drouth, which inhibit microbial activity in soils. Persistence may be a desirable feature for some pesticides used for insect and disease control, for it prolongs the protection provided. On the other hand, persistence would not be a desirable trait for herbicides, since it could result in damage to sensitive crop plants grown long after the herbicide has been applied. For this reason most herbicides are organics that decompose rather quickly in the soil. However, there are inorganic compounds used for weed control, notably borates and chlorates, that are not altered by biological means; instead, they persist until leached from the soil. Since these compounds are customarily added in amounts sufficient to kill all plants, their eventual removal by leaching is necessary.

The fact that soils can be damaged by inferior irrigation water has been discussed earlier. The harmful effects may result from the build-up of salts, of excess exchangeable sodium, or of biologically toxic ions. Sometimes the contaminants in irrigation water are derived from municipal sewage, sewage-plant effluent, or industrial wastes discarded in rivers. Even good quality

Table 19.1. Recommended maximum concentrations, in parts per million, of toxic elements in good quality drinking and irrigation water. (From data compiled by H.B. Peterson. In *Municipal Sewage Effluent for Irrigation,* pp. 31–44, Ruston, Louisiana: Louisiana Tech Alumni Foundation, 1968).

Element	Drinking water	Irrigation Water	
		Coarse Soils	Fine Soils
Arsenic	.01–.05	1.0	10.0
Beryllium	—	.5	1.0
Boron	20.0	.75	2.0
Cadmium	.01	.005	.05
Chromium	.05	5.0	20.0
Cobalt	—	.2	10.0
Copper	1.0	.2	5.0
Lead	.05	5.0	20.0
Lithium	5.0	5.0	5.0
Molybdenum	—	.005	.05
Nickel	—	.5	2.0
Vanadium	—	10.0	10.0
Zinc	5.0	5.0	10.0

water, when used in excess, may cause the rise of salty ground water to the point at which it may damage the soil. Failure to establish adequate drainage is one of the prime causes of pollution of irrigated land with salts.

Plant-toxic elements carried in water are usually dissolved metals, and the concentration of some of them need not be very high to injure sensitive plants. In general, water with a dissolved metal content considered safe for human consumption is usually also safe for use on soils. However, as shown in table 19.1, this relationship does not hold for boron. The concentration of boron allowed in domestic water may exceed the recommended limit for irrigation water by ten times or more. On this basis, the water used by some municipalities in the southwestern United States would be judged unsafe for use on agricultural land, and it could seriously harm lawns, gardens, and potted plants if the added boron were not leached from the soil periodically. Waste water from these municipalities would be even less tolerable where it has been further contaminated by such things as household and industrial cleansing agents of high boron content.

The last type of soil pollutant to be considered is air-borne substances that cause the build-up of undesirable elements or compounds in soils. Examples are smokes and dusts originating with industry, and lead emitted in automotive exhaust. Lead is known to have accumulated in large quantities in soils adjacent to well traveled highways. The potential for lead poisoning exists where livestock is fed lead-enriched forage.

Most air-borne dusts affect agriculture adversely by settling on plants. Dust coatings restrict the use of available sunlight in photosynthesis and slow the interchange of gases between plant leaves and the surrounding atmosphere. Occasionally, industrial dusts such as those from metal-processing plants may build up to harmful levels in the soil. Also, in some localities gases occurring in industrial fumes or those derived from automobile exhaust have a direct toxic effect on plants without polluting the soil. More is said about these gases in a later section.

Soils and the quality of the environment

Soil erosion and pollution

Although soils have been and continue to be seriously harmed by pollution, through erosion and redeposition on lower-lying land or in ditches or other drainageways, they are in themselves a major source of pollutants (see fig. 19.1). Further, windblown dust originating with the soil has adversely affected man by inflicting or intensifying eye and respiratory ailments and by obstructing vision. The annual cost for the repair of machinery damaged by dust is inestimable, as is the expense incurred in the removal of this unsightly menace. Most people are aware of the soil as a source of pollutants, but few recognize that it is a source subject to much better control than has been provided in the past. The state of affairs will probably change as our understanding of the total pollution problem increases.

It is estimated that over 3.5 billion metric tons of sediment wash into the streams and rivers of the United States each year. This amount is equivalent to the plow depth of soil on about 1.5 million hectares of land, or about 1 percent of the cultivated land in this country. Approximately one-fourth of the sediment is moved on to the sea, which means that the remaining three-fourths is deposited along the way on flood plains or in lakes or reservoirs (see fig. 19.2).

As much as half of the sediment delivered into the streams and rivers of this country is judged to come from cultivated agricultural land. The remainder is contributed by erosion of untilled forest and rangeland, or of urban, highway, and industrial construction sites (see fig. 19.3). Erosion connected with construction activities makes up but a small part of the total, although it can have severe local effects, especially in areas of high population density. Because the source of such a problem can usually be identified, those responsible for it can more often than not be held accountable. Paradoxically, no one in particular is blamed for the massive pollution caused by the erosion of agricultural land, although virtually everyone suffers from it to some extent.

Since soil erosion in one way or another adversely affects society as a whole, it is justifiably within the rights of society to insist on the continued improvement of erosion-control measures and on more extensive application of these measures to the nation's agricultural lands. The benefits derived would include not only a better environment but also a reduction in the rate at which our vital soil resources continue to deteriorate.

Soil-derived chemicals as pollutants

A problem that has been drawn increasingly to our attention in recent years is the pollution of ground and surface waters with eroded or dissolved chemical substances from soils. These include plant nutrients, natural salts, pesticides, and biologically active wastes, the latter being primarily derived from poultry and livestock. In many cases the pollutants are a health hazard. Sometimes, however, their effect in reducing water quality is judged wholly from an aesthetic point of view.

A

B

Figure 19.1.
(A) A low-lying field, in Walla Walla Co., Washington, completely covered by sediment eroded from surrounding slopes, (B) a roadside ditch filled with windblown sand, and (C) the removal of water-laid sediment from a roadside ditch, which takes more than its fair share of roadway maintenance funds. (U.S.D.A. Soil Conservation Service photos by Earl Baker, B. McGuire, and E. E. Rowland.)

C

Soils and the quality of the environment

Figure 19.2.
The former Lake Ballinger, Texas, which was used as a city water supply for 22 years until it became completely clogged with silt eroded from the surrounding watershed. (U.S.D.A. Soil Conservation Service photo.)

Plant nutrients as water pollutants

Two of the ways in which plant nutrients impair water quality are (1) by *eutrophication*, which results from the build-up of nutrients in water until they support an undesirable level of plant and animal life, and (2) by the accumulation of hazardous concentrations of nitrate-nitrogen in drinking water. Eutrophication is limited mostly to surface water, but excessive concentrations of nitrate-nitrogen occur primarily in ground-water supplies.

One of the principal indicators of eutrophication is the growth of plants in water. Among the more important of these plants are microscopic algae. When present in proper ecological balance, algae form a vital link in the food chain of aquatic organisms and, at the same time, make a major contribution to the regeneration of atmospheric oxygen from carbon dioxide. Still, when present in excess, algae can have an undesirable effect on water quality; they produce green scums and may impart an offensive odor or flavor to the water. Where other factors favorable for growth exist, an increase in algae to undesirable levels is usually associated with an increase in dissolved inorganic nutrients above the normal concentration in unpolluted surface waters (see fig 19.4).

Where eutrophication has caused an abundant growth of algae, the death and decay of these organisms can seriously reduce the level of dissolved oxygen in water. There are many instances on record where the reduction in oxygen by this means has caused extensive kills of fish and other aquatic life (see fig. 19.5). This effect is seen most often in ponded water, where aeration is naturally poor at best.

Figure 19.3.
Two construction sites severely damaged by erosion. (A) An area leveled by grading, and (B) sloping land smoothed and shaped for landscaping but then left unprotected against winter precipitation. (U.S.D.A. Soil Conservation photos by Herb Gaines and E. E. Rowland.)

A

B

Figure 19.4.
Aquatic vegetation, principally algae, in the Snake River, Idaho. Nutrients responsible for eutrophication of the water have come from the combined inputs of industrial, municipal, and agricultural wastes. (Photograph courtesy of the U.S. Environmental Protection Agency.)

Figure 19.5.
A fish kill caused by stream pollution with food-processing wastes of excessive biological oxygen demand for decay. (Photograph courtesy of the U.S. Federal Environmental Protection Agency.)

Eutrophication almost invariably involves a build-up in the concentration of soluble phosphorus in water. The reason for this is that most surface waters, including those derived by drainage of fertilized land, have a phosphorus content below that required for the abundant growth of algae.[1] The soil therefore escapes most of the blame for excessive phosphorus in eutrophic water. A more important source of soluble phosphorus seems to be sewage or industrial waste discharged into streams and rivers. Much of the phosphorus in municipal sewage is derived from detergents and other cleansing or water-softening agents.

Nitrogen is a second nutrient important to the growth of algae. Unlike the phosphorus content, the nitrogen content of drainage water from most land areas is adequate for an abundant growth of algae, and little can be done to change this fact. Where the drainage is from virgin land, the nitrogen is provided mostly by the mineralization of soil organic matter. In this instance the loss of nitrogen from the soil tends to counterbalance gains that result from the fixation of nitrogen in the soil or its addition to the soil in precipitation. The greatest losses of soil nitrogen in drainage water may be expected from cultivated agricultural lands, especially from those treated with nitrogen fertilizers.

As noted above, accumulating nitrate-nitrogen can impair the usefulness of water for drinking purposes. According to the U.S. Public Health Service, water containing more than 10 ppm nitrogen in the nitrate form should be considered unsafe for human consumption. Most surface waters contain far less than this amount, but occasionally ground water tapped by wells contains considerably more. Concentrations of nitrate-nitrogen in excess of 1000 ppm have been reported for some well waters.

Soils that have been heavily manured or treated with large quantities of nitrogen fertilizer provide one possible source of nitrate-nitrogen in ground water. For example, it is known that water draining beyond the root zone of heavily fertilized land often has sufficient soluble nitrate-nitrogen to exceed the limit of safety set by the U.S. Public Health Service. However, whether this will seriously lower the quality of ground water beneath cannot be judged from the composition of the soil solution as it moves downward beyond the root zone. Two reasons account for this. First, the nitrate-nitrogen content of the percolating water can be changed drastically by denitrification, although this is unlikely in well aerated soils that lack organic materials of high biochemical oxygen demand (BOD). Second, once a percolate reaches the water table, it may be diluted by ground water. Because of the complex nature of the reactions invovled, it is possible to determine the extent to which soluble soil nitrogen contaminates ground water only by a thorough study of conditions as they exist in the field.

The potential for movement of nitrate-nitrogen from agricultural land into ground water varies widely depending on such things as farming prac-

---

[1]Most natural surface waters contain no more than a trace of phosphorus, which means that the phosphorus content is too low to be accurately measured. Minimum concentrations of phosphorus in water capable of supporting a large population of algae usually fall within the range of 0.01 to 0.05 ppm.

Figure 19.6.
The effect of rate of fertilization with nitrogen on crop yield and on the quantity of unused, residual nitrogen in the soil following harvest. Where leaching occurs, residual fertilizer nitrogen is a potential water pollutant. The rate of fertilization denoted by X in the diagram approximates a level of nitrogen in the soil that should result in maximum economic gain from the fertilizer. It should also result in a relatively small residue of unused nitrogen in the soil.

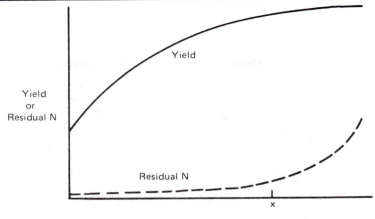

Yield
or
Residual N

Rate of Fertilization with N

tices, soil conditions, and the input of water, either by rainfall or irrigation. A first requirement for contamination is that some leaching must occur. In dry regions leaching is likely only under irrigated conditions. In humid regions, any soil is a possible source of leachable nitrate, with the potential increasing as the average level of soil fertility increases.

Because of the rising concern over pollution of ground water with nitrate-nitrogen, care in the use of nitrogen fertilizers will undoubtedly become a more critical part of soil management in the future. The care will consist largely of avoiding excess fertilizer applications, that is, the application of more fertilizer nitrogen than can be effectively absorbed by plants. It happens that the most appropriate rate of nitrogen usually is one that provides a maximum economic return from the fertilizer, although it will not necessarily result in maximum yields. This fact is illustrated in figure 19.6, which shows the relationship between yield, as controlled by fertilization with nitrogen, and the amount of residual, or unused, inorganic nitrogen remaining in the soil at the end of the growing season. Except where there is no deep percolation of water through a soil, inorganic nitrogen left in the soil at harvest time is subject to loss by leaching.

Natural salts as water pollutants

All streams and rivers contain at least small quantities of dissolved salts. The quantity of salt occurring naturally depends on several factors, but a principal one is the nature of the land mass from which drainage water is received. In general, rivers in humid regions are lowest in salts because they drain land that has been largely depleted of readily soluble components. Conversely, rivers in arid regions contain more salts on the average, the reason being that they drain lands that have undergone very little leaching. These lands therefore serve as a continuing source of salts (see fig. 19.7).

Figure 19.7.
The Green River as it
passes through Split
Mountain Gorge,
Dinosaur National
Monument, in arid
northwestern
Colorado, where it
receives salt-bearing
drainage from deep,
largely unleached
sedimentary rock
formations. (U.S.
Department of the
Interior, National Park
Service photo.)

Because of the limited precipitation in dry regions, the total quantity
of salts delivered into rivers by natural drainage is often low. Further, these
salts may be so diluted by the river that they cause no more than a modest
increase in its average salt content at any one site of input. However, this
situation may be drastically changed where water is diverted from the river
to irrigate neighboring land and is then returned in part as drainage from
the irrigated land. The overall effect of this process is to increase the salt
content of the river. Where the water is to be reused downstream, the increase in salts must be viewed as a potential pollution problem.

The effect that the use of water for irrigation can have on the composition of that water is illustrated in table 19.2. These data show the average
amount of water delivered per hectare of land in an irrigation project in
Washington, along with the amount of the water recovered in drainage. Shown
also are changes that occur in the total quantity of five different ions between the time the water was applied to the land and the time it was recovered as drainage. As may be judged from the input-outflow data, only half
of the water applied was recovered as drainage, but it contained about twice
as much calcium, magnesium, and potassium, 5 times as much nitrate, and
almost 13 times as much sulfate as was in the water initially. The net gain
in nitrogen is equivalent to about 13 kilograms per hectare.

There are several ways in which the pollution of streams and rivers by
soil salts can be minimized. One is by means of increased water-use efficiency; that is, a reduction in the amount of water applied to irrigated land

Table 19.2. The average input and outflow of irrigation water and certain soluble ions for the Sunnyside (Washington) Irrigation District for the year beginning March 1, 1943. (After C. S. Scofield, Mimeo. Report, U.S. Bur. of Plant Ind. to U.S. Bur. of Reclam., 1944).

Measured Item		Input in Irrigation Water	Outflow in Drainage Water
Water	cm	146	70
Ca^{2+}	kg/ha	151	338
Mg^{2+}	kg/ha	54	131
K^+	kg/ha	22	45
SO_4^{2-}	kg/ha	49	616
NO_3^-	kg/ha	15	74

in excess of plant needs. Another is to limit the application of excess water for salt removal to those periods when water reuse downstream is at its lowest ebb or when stream flow is maximum. A more drastic alternative is to dispose of salt-laden drainage water on wasteland rather than to return it to a river. Ultimate disposal of the diverted water would be by evaporation.

Pesticides as pollutants

So long as pesticides are used at proper rates and remain at the site of application they can effectively control weeds, insects, and plant diseases without critical side effects. Problems arise where improper methods of application, wrong pesticide formulations, or excessive amounts of pesticides are used, or where the applied material moves out of the target area to cause damage elsewhere. Movement of pesticides can be as a vapor, like that derived from volatile forms of 2,4–D, as a leachate, or as a suspended component of eroding soil. Since most persistent pesticides are neither volatile nor water-soluble, the third means of movement, erosion, is the one of greatest significance.

Insecticides make up most of the persistent materials that are widely used for pest control. Of these the best known example is DDT, one of a series of similar chlorinated hydrocarbon compounds. When present in excess, DDT can destroy not only insects but some plant and animal life as well. Because of its persistence and potential for harming plants and animals, DDT, along with other insecticides, has been banned from use in this country. It is probable that legal control is the only practical way in which environmental protection against pesticides can be achieved.

Animal wastes as pollutants

In the United States the disposal of animal wastes has become a major problem in recent years. It is estimated that domestic animals in this country produce about 1 billion tons of fecal wastes annually. Their liquid wastes amount to nearly 400 million tons. These, in combination with used bedding, carcasses, and wastes from the meat-packing industry yield a total waste from domestic animals equivalent to that of a human population of nearly 2 billion. Our inability to satisfactorily dispose of human wastes is indicative of the problem confronting agriculture and allied industries in the disposal of animal wastes.

In numerous local situations, the problem of animal-waste disposal results from the highly concentrated nature of many modern cattle and poultry operations. Production units containing 10,000 cattle or 100,000 birds are no

Figure 19.8.
An aerial view of a feedlot containing thousands of beef cattle. Such large operations require careful management if pollution of air or water by manural wastes is to be avoided. (Photograph courtesy of the U.S. Environmental Protection Agency.)

longer unusual (see fig. 19.8). A unit of 10,000 cattle often produces nearly 300 tons of manure daily, which obviously results in a mammoth disposal task. Too often, therefore, manure is allowed to accumulate in place, whereupon it becomes a site for fly-breeding and a source of offensive odors and dusts. Drainage from accumulating manure may also contaminate ground water, particularly where large quantities of nitrate-nitrogen are produced. Yet, soil compaction, high moisture, and an abundance of readily oxidizable carbon compounds often maintain anaerobic conditions in the soil beneath accumulating manure, with the result that denitrification and the loss of nitrogen in gaseous form may sometimes almost entirely eliminate nitrate as a potential ground-water pollutant.

Although ground-water pollution by manure is caused only by soluble components such as nitrate-nitrogen, pollution of surface waters is by both solids and liquids derived from the manure. The pollutants may be carried by normal surface runoff that feeds the natural drainage network of an area. Sometimes, however, pens or feeding sites are located immediately adjacent to a stream or river (see fig. 19.9), which greatly intensifies the potential for

Figure 19.9.
A riverside feedlot where animals have direct access to the water. Under such conditions, the animals pollute the water with intestinal organisms, some of them pathogenic, as well as with oxygen-demanding substances and soluble nutrients that aid eutrophication. Stricter regulations are gradually reducing this type of water pollution. (Photograph courtesy of the U.S. Environmental Protection Agency.)

water pollution. Not only do these sites yield runoff that is likely to contain a high content of organic and inorganic pollutants, their location makes it possible for animals to excrete wastes directly into the water.

Aside from aesthetic considerations, water pollution with animal manure is undesirable for several reasons. One is that manure is a source of infectious disease, particularly for other animals. Another is that the decay of these wastes places a large demand on the supply of oxygen in water. Manure also contributes nutrients necessary for eutrophication. As is true for any waste, contamination with manure can greatly increase the cost of purification of water needed for other uses.

Soil as an aid to waste disposal

Because of its widespread occurrence and capacity for retaining vast quantities of inorganic and organic substances, the soil provides perhaps the best facility we have for the disposal of many different kinds of waste. Where the wastes are nontoxic to plants and do not contain harmful substances that can be absorbed by food plants, their disposal in the soil may be a useful adjunct to the proper management of soil fertility and tilth. Although the capacity of air and water to accommodate wastes is frequently exceeded, we have hardly begun to utilize this capacity in soils.

The above facts notwithstanding, there is a reluctance to use the soil for waste disposal where the cost, either in money or inconvenience, outweighs the benefits gained. Yet, these costs must be borne if the trend toward ever greater environmental pollution is to be reversed. Further, since in the final analysis, the pollutants are largely by-products of efforts to supply the needs or demands of society, the financial responsibility for disposal must ultimately fall on society.

Soils used for plant production cannot be used for the disposal of all wastes. For example, many industrial wastes are toxic to plants and may destroy all vegetation if allowed to accumulate in excessive amounts. Wastes that are compatible with the soil are limited primarily to organic and liquid materials such as animal manure, sewage sludge and effluent, and some discarded products of the food-processing industry.

Animal manure is the single, most abundant waste that can be conveniently disposed of on land, and the bulk of manure produced in the United States is finally discarded in this way. However, where large animal production units are located at some distance from disposal sites, the cost due to transportation is often excessive. Further, alternate methods of disposal, such as burning, burying, or composting, have not been successful. As a consequence, manure management is often carried out improperly, with the result that the wastes become significant sources of environmental pollutants. Well enforced laws are often necessary to ensure the proper processing of all manure. However, enforcement must be accompanied by the development of improved methods and equipment for handling the wastes if their rapid and proper disposal is to be economically feasible. Otherwise, the price consumers pay for products of the animal industry will have to increase to compensate for increased production costs.

Problems of manure disposal on land

The use of manures to enhance soil fertility and tilth is an age-old practice. The quantities of manure used have normally been tied rather closely to plant requirements for nutrients, typical applications being from 20 to 25 metric tons per hectare per year. At this rate of application, manure presents no particular problem; decomposition proceeds smoothly and the nutrients released to the soil can easily be absorbed by plants.

Where a choice has been made to utilize land for manure disposal it is usually with the realization that the rate of application will be large, sometimes from 100 to 250 tons per hectare per year. At such high rates manure can induce a number of serious soil and plant problems. Its decomposition can create anaerobic conditions in the soil, and the release of mobile, mineralized nitrogen can be excessive. For example, 100 tons of fresh manure could contain as much as 600 to 800 kg of nitrogen. Even if only half of the added nitrogen were mineralized, more would be released into the soil solution than could be taken up by plants under most conditions. The excess nitrogen would remain as a potential water pollutant.

Another danger that arises from the use of large amounts of manure on land is salt accumulation (see fig. 19.10). Various inorganic ions are excreted by animals. Cattle manure is especially high in sodium chloride, one

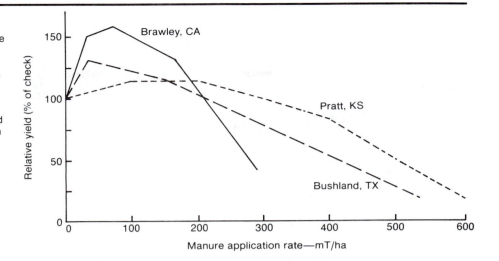

Figure 19.10.
The effect of high rates of manure on the yield of grain sorghum, at Brawley, CA and Bushland, TX, and on corn silage, at Pratt, KS. The decrease in yield at high rates is attributed to excess salts. (From Stewart, B. A. and Meek, B. D. in *Soils for Management of Organic Wastes and Waste Waters*, 1977, pages 219–32 by permission of The American Society of Agronomy.)

reason being that salt is fed to these animals as a dietary supplement. Too much sodium as an exchangeable ion affects the soil physical state adversely, this in spite of the normal beneficial effect of manure on soil structure formation and stabilization. The salts that accumulate affect growth by competing with seeds and plants for water.

Where manures are applied to the land, it is usually important to know the contents of phosphorus and potassium as well as of nitrogen. Representative contents of these three elements in a variety of animal manures have been listed earlier in table 16.1. The values in table 16.1 are percentages and are useful in estimating the total amount of nitrogen, phosphorus, and potassium that will be applied in a given quantity of manure. In some local dairy, cattle-feeding, or poultry operations, however, it is often simpler to base the rate of manure application on the quantities of nutrients excreted by the animals on a daily basis. Examples of this type of information are shown in table 19.3. With such data, the amount of land required to accomodate a herd or flock of known size can be computed. If the manure is stored prior to application, changes in composition must be determined and taken into account.

Although the data on total dry solids and biochemical oxygen demand,[2] as shown in table 19.3, are not used to determine the rate of manure application to land, they are indexes of the ability of the manure to deplete the supply of oxygen in the soil. Oxygen depletion can be of advantage to manure disposal on land. The reason is that it may cause denitrification and the volatilization loss of nitrogen as oxides and nitrogen gas. To the extent that this occurs, the rate of manure disposal on land can be increased without in-

---

[2]Conventionally, the biochemical oxygen demand is determined as the reduction in the oxygen content of the water-organic waste mixture during a 5-day period of incubation at 20°C. The 5-day biochemical oxygen demand, when multiplied by an experimentally determined factor, yields the total, or *ultimate oxygen demand*, of the waste.

Table 19.3. Average characteristics important to the disposal of animal wastes. (From R. C. Loehr, Animal Wastes—A National Problem. *Jour. San. Eng. Div. ASCE* 95:189–221, 1969).

Species	Kilograms per individual per day			
	Total dry solids	BOD[a]	Total N	Total P
Chickens	0.03	0.007	0.0014	0.0007
Swine	0.4	0.14	0.023	0.006
Dairy cattle	4.5	0.45	0.18	0.023
Beef cattle	4.5	0.45	0.14	0.023

[a]Biochemical oxygen demand, in kilograms, for a 5–day incubation of waste in an amount equivalent to that produced per day per individual.

creasing the potential for ground water pollution by nitrate. The surface application and drying of manure can have a similar effect in that, on drying, a substantial part of the nitrogen present in NH_4^+ form can be lost as volatile ammonia gas. However, these benefits may have other undesirable consequences. For example, nitrous oxide, though usually a minor product of denitrification, is blamed for reducing ozone (O_3) in the atmosphere; this atmospheric component offers protection against ultraviolet light in solar radiation. Another potential drawback is the volatilization of large amounts of ammonia, which, if absorbed by local bodies of water, may add to their eutrophication. The seriousness of these effects is extremely difficult to evaluate.

The use of soil for disposal of sewage effluent

Considerable experimental work has shown that the disposal of sewage effluent on land can be highly satisfactory. Sewage, either solid or liquid, contains all the nutrients plants require, and, in this respect, is much like animal manure. A particularly important advantage in applying sewage to land is that it avoids disposal in natural drainageways. A problem associated with disposal in drainageways is the eutrophic effect sewage may have on the water because of its generally rather high content of soluble nitrogen and phosphorus.

In general, there are more obstacles to the disposal of sewage effluent than of sewage sludge on agricultural land. The greatest problem with effluent disposal on land arises from the fact that the supply is continuous. Application of effluent in off-season, such as during the winter, is often inconvenient, and since the use of water in evapotranspiration is limited during the winter months, most of the water applied will percolate through the soil. The use of forestland, where available, has been found a practical means of overcoming this type of problem under most circumstances (see fig. 19.11).

The greatest restriction on effluent disposal occurs with cropped land where even during the growing season there are times when plants neither need nor can tolerate extra water. For reasons such as these, the successful use of sewage effluent on land usually reuqires facilities in which the effluent can be stored, sometimes for as long as 2 to 3 months. Ponds or lakes may be used for storage, but unless they are relatively leak-proof, they too can result in ground water contamination. The potential consumptive use of

Figure 19.11.
Forestland has a high capacity as a disposal site for sewage effluent and may often be used at all seasons during the year. Effluent application in this example is by sprinkler. (Photo courtesy of L. T. Kardos, University of Pennsylvania, College Park.)

water from effluent may be as low as 75 to 100 cm per year in cooler, humid regions, but it may be near 200 cm in hot, dry areas where the growing season and the need for water lasts throughout the greater part of the year.

Criteria for manure and sewage disposal on land

By and large, the disposal of manure and sewage wastes on land is tied to the ability of the soil or plants to remove or inactivate components that might otherwise cause environmental contamination. In the great majority of cases, nitrogen is the component that is most frequently of concern, and this is due to the high potential for the nitrogen, once in nitrate form, to move into surface or ground-water supplies. Less often the potential contaminant is an element such as boron or a heavy metal that may be toxic to either plants or animals. Even elements such as phosphorus and potassium may create plant-growth problems if allowed to accumulate to too high a level in soils. Such accumulations occur mainly from the long and repeated application of wastes to the same land.

Table 19.4. Annual utilization of nitrogen, phosphorus, and potassium, in kilograms per hectare, by selected plants at moderate to high levels of productivity. (Adapted from Knezek, B. D. and Miller, R. H., eds., *Ohio Agri. Res. Bull. 1090*, 1976.)

Plant	Yield[a]	N	P	K
Corn	9500 kg	200	40	200
Corn silage	72 tons	225	40	225
Soybeans	4000 kg	375	30	134
Grain sorghum	9000 kg	280	45	185
Wheat	5400 kg	210	25	150
Oats	3600 kg	170	25	140
Barley	5400 kg	170	25	140
Alfalfa	8 tons	500	40	450
Bermuda grass[b]	25 tons	560	60	440
Orchard grass	13 tons	335	50	350
Brome grass	11 tons	185	30	235
Tall fesque	8 tons	150	30	170
Blue grass	7 tons	225	25	165
Hardwood forest	—	95	9	30

[a]Yield per hectare in kilograms or metric tons.
[b]From Wilkinson, S. R. and Langdale G. W. In *Forage Fertilization* pp. 119–169. Madison, Wis.: American Society of Agronomy, 1974.

An indication of the ability of plants to remove nitrogen, phosphorus, and potassium from soils is given in table 19.4. In general, greatest absorption is by perennial forages, with alfalfa and Bermuda grass being outstanding examples. The very large uptake of nutrients shown for this pair of forages applies principally to warm climates where the growing season is long. Under these conditions, relatively large amounts of waste can normally be applied to the land without seriously threatening ground-water supplies. For example, the removal by alfalfa of 500 kg of nitrogen per hectare in a single growing season should permit an application of about 50 metric tons per hectare of a solid waste containing 1 percent nitrogen, even if all the nitrogen were mineralized during the year of application. Since only a part of the nitrogen in solid wastes is usually mineralized the first year, an application of more than 50 tons per hectare would be possible. However, at such high rates, other factors may become limiting. For example, heavy applications of poultry manure can cause plant damage because of the release of excessive amounts of ammonia gas, and a high level of sodium chloride in cattle manure may limit its rate of application. A similar effect might arise from heavy metals in industrial sewage sludge.

Large applications of sewage effluent can also be made where plants are able to remove substantial amounts of nitrogen from the soil. It would take over 600 cm of an effluent containing 20 ppm nitrogen to exceed a potential nitrogen uptake of 500 kg per hectare per year. In the case of effluent, essentially all of the nitrogen applied would appear in available form during the year of application. Table 16.2, on page 436, lists the nitrogen content of effluents.

Where nitrogen-bearing wastes are applied year after year to the same soil, a build-up in the nitrogen level requires a gradual reduction in the amount of waste applied each year. The build-up results from the failure of all nitrogen in solid wastes to be mineralized in a single year's time. As indicated earlier, the rate at which nitrogen in organic wastes mineralizes during the year of application varies from as little as 25 percent to as much as 90 percent of the total nitrogen applied. Materials with a slow mineralization rate are likely to contribute significantly to the level of available nitrogen over a period of years, however.

Where a waste contains a toxic component that can accumulate in the soil, the rate of application may have to be restricted. Such limitations are frequently encountered where heavy metals are present in sewage sludge produced in industrialized areas. The metals most often involved are lead, zinc, copper, nickel, and cadmium. Amounts of these elements that may be applied to land without causing a serious threat to the health of either plants or animals are conventionally expressed relative to the cation-exchange capacity of the soil; the higher the cation-exchange capacity the greater the level of tolerance. According to one rather widely accepted standard, for each milliequivalent of cation-exchange capacity, no more than the following amounts, in kilograms per hectare, of the above five elements should be allowed to accumulate: lead, 100; zinc, 50; copper 25; nickel, 10; and cadmium, 1. Realize, however, that some feel the use of land for sewage disposal of any sort should be avoided, particularly where the land is or eventually could be used to produce plants for human or animal consumption.

Other uses of soil in waste disposal Food processors comprise one industry that can make substantial use of the soil for waste disposal. The canning industry especially has a tremendous disposal problem because of the massive quantities of waste it produces. This problem is compounded where the processing of food involves extensive chemical treatment. For example, peeling operations, such as of potatoes, depend largely on the removal of skins with the aid of a lye (sodium hydroxide) solution. With lye as a component, these wastes require great care in their disposal. The reason is that sodium hydroxide, when present in high concentration, can cause rapid deterioration of land or water used as a disposal medium. The problem could be reduced if potassium hydroxide were used in place of sodium hydroxide and the wastes applied to land as a source of potassium for plants.

Another way in which the soil is being used for waste disposal is in sanitary land fills, that is, for the burial of solid wastes (see fig. 19.12). The main advantage of burial is aesthetic. In some places the wastes also provide a comparatively cheap source of fill material for the creation of artificial land. One problem is that burial must be limited to accessible land where little more than dumping and covering is required for disposal. Unfortunately, many of our densely populated centers are simply running out of this kind of land. Another problem with burial is the potential of water contamination by drainage from the land fill.

Other pollutants important to agriculture

One last topic to be considered is the ever-increasing influence of gaseous air pollutants on agricultural and forest crops. Although these agents have little if any direct relationship to the quality of the soil environment, they

Figure 19.12.
Burying waste in a sanitary land fill in Monroe Co., Wisconsin. The trash is spread in a deep trench, covered with soil each day, and compacted. The area will finally be capped with a deep layer of soil, shaped for drainage, so as to reduce leaching, and planted to grass, legumes or trees. (U.S.D.A. Soil Conservation Service photo by E. W. Cole.)

are indirect determinants of the usefulness of soil as a medium for plant growth. Two air pollutants that have long been known to damage plants when present in excess are fluoride and sulfur dioxide fumes (see fig. 14.10 on page 362). Excessive concentrations of these substances in air are usually traced to industry.

Smog-laden air usually contains several gases that damage plants. The more important are ozone (O_3) and peroxyacetyl nitrate, both of which are produced by the interaction of nitrogen oxides, air, and hydrocarbon vapors in sunlight. Plants sensitive to these gases include a wide range of fruits and vegetables. Ponderosa pine at elevations as high as 1.5 km have been damaged by smog from the nearby Los Angeles Basin in California (see fig. 19.13). Some of the phytotoxic gases in smog are also a hazard to human health.

From the foregoing, it should be apparent that land has a vital role in the control of environmental quality. How well the available resources can be utilized in the task will undoubtedly prove to be a severe test of man's ingenuity and patience. A problem developing from generations of neglect cannot be solved overnight. To place blame for the problem solely on any one segment of the population, or to make inordinate demands on a specific segment to bring about its solution, is wholly unrealistic. Pollution is the

Soils and the quality of the environment

Figure 19.13.
The east end of Los Angeles Basin, California, on a smoggy day.

product of society in general, even though contributions to it may vary widely among different individuals and under different conditions. This fact suggests that all society must share the burden of environmental preservation and reclamation if either is to be fully realized.

Summary

Environmental quality depends on the purity of soil, air, and water resources. Each of these environmental components can serve as a pollutant, or a source of pollutants, for the other two. In the great majority of cases, the serious pollution of the environment can be traced to man's activities.

Soil is a source of air and water pollution. It may be the polluting agent, or it may supply chemical substances in gaseous or soluble form that degrade the purity of air and water. Conversely, air and water can pollute the soil with noxious vapors and fumes or soluble salts and toxic elements. In each instance, avoidance of such problems usually requires control of the pollutants at their source.

The soil provides an important means of improving environmental quality where it is used for waste disposal. Relatively large amounts of organic materials and municipal and industrial wastewater can be disposed of safely on land used for agricultural crops, timber production, or grazing. The principal limitation of waste disposal on land is with materials containing chemicals that are toxic to plants or to man and animals that use the plants for food.

Review questions

1. In general, what determines the persistance of organic pesticides in soils?
2. What is the most likely harm to the soil that will result from using it as a site for sewage disposal?
3. Define eutrophication, and summarize its principal causes.
4. What property of nitrate-nitrogen accounts for the ease with which it can become a pollutant of groundwater?
5. Why does irrigation of land increase the potential for increased pollution of river waters with soluble salts?
6. What are the principal advantages and disadvantages of using agricultural land for the disposal of animal wastes and municipal sewage sludge and effluent?
7. Under what conditions is the denitrification process an aid in avoiding groundwater pollution?
8. What are the problems associated with using soils as sites for sanitary land-fills?

Selected references

Brady, N. C., ed. *Agriculture and the Quality of Our Environment*. AAAS Pub. 85. Washington, D.C.: American Association for the Advancement of Science, 1967.

Elliott, L. F. and Stevenson, F. J., eds. 1977. *Soils for Management of Organic Wastes and Waste Water*. Madison, Wis.: American Society of Agronomy.

Loehr, R. C. *Agricultural Waste Management*. New York: Academic Press, 1974.

————, ed. *Land as a Waste Management Alternative*. Ann Arbor, Mich.: Ann Arbor Science Publishers, Inc., 1977.

Sopper, W. E. and Kardos, L. T. eds. *Recycling Treated Municipal Wastewater and Sludge Through Forest and Cropland*. University Park, Pa.: The Pennsylvania State University Press, 1973.

Staff, Utah State University Foundation. *Characteristics and Pollution Problems of Irrigation Return Flow*. Ada, Okla.: Robert S. Kerr Water Research Center, 1969.

Subcommittee on Environmental Improvement. *Cleaning Our Environment—The Chemical Basis for Action*. Washington, D.C.: American Chemical Society, 1969.

Wadleigh, C. H. *Wastes in Relation to Agriculture and Forestry*. U.S.D.A. Misc. Pub. 1065. Washington, D.C.: Government Printing Office, 1968.

Wilson, C. W. and Beckett, F. E., eds. *Municipal Sewage Effluent for Irrigation*. Ruston, La.: Louisiana Tech Alumni Foundation, 1968.

20

Engineering properties of soils

Among the various properties of soils, some are of special importance to the planning, development, and maintenance of certain engineering works. These are referred to as engineering properties of soils. They determine the suitability of soils as sites for buildings, highways, and other structures, or for canals, ponds, or other water-impoundment areas. They also determine the suitability of soil for use as a construction material, such as the fill in highway embankments or in earthen dams or levees. Finally, engineering properties are basic to the design of irrigation and drainage systems, cultivation equipment, and erosion-control terraces and diversions. Whereas engineering properties are largely physical in nature, they may also reflect the chemical, mineralogical, or biological character of the soil.

Among the soil properties important to engineering works, three stand out above all others. These are *compressibility, shear strength*, and *permeability*. Compressibility determines the extent to which a soil will undergo a reduction in volume and thereby allow settlement of a load applied to it. Shear strength is a determinant of the maximum load-carrying capacity of a soil and of energy requirements for soil cultivation and excavation. Permeability is a characteristic crucial to the design of irrigation and drainage systems, or to the suitability of soil for the retention of water, as in canals or behind dams. Although, in most engineering applications, one of these properties may be more important than either of the other two, there are relatively few such applications where any one of them can be totally ignored. Because of their great importance, these properties are discussed in some detail.

Soil compressibility

Most soils undergo some compression when subjected to an applied load. This effect is due mainly to a reduction in average pore size, and in porosity, as soil particles are reshuffled or squeezed in to a more closely packed arrangement. However, compression may also involve the bending of flexible soil particles, such as fibrous organic fragments, micaceous mineral grains, or diatomaceous earth. If these materials are present, they may cause a soil

B

Figure 20.1.
(A) A hand-operated soil compactor driven by a gasoline engine, and
(B) a sheepsfoot roller used to compact an earthen dam. (Photo (B)
courtesy of U.S.D.A. Soil Conservation Service.)

to display *elastic compressibility*, that is, the tendency to expand once a compressive force is removed. Elastic soils are relatively unstable, so they are normally avoided for most engineering uses.

Excessive compressibility, whether elastic or nonelastic, is an undesirable characteristic in soils used for engineering purposes. This is because it may allow undue settlement and damage to buildings or other man-made structures. However, such problems can often be minimized by first artificially compacting the soil, provided it is nonelastic. Various means are used to compact soils, among them, vibration, tamping, and rolling (see fig. 20.1). In general, the finer the texture the greater the effort required to bring about satisfactory compaction.

The degree to which medium- and fine-textured soils can be compacted depends on their water content. Maximum compaction is achieved within a fairly narrow range in water content, and less than maximum compaction is obtained at either a higher or lower water content. The moisture level permitting maximum compaction is termed the *optimum water content*. As

Engineering properties of soils

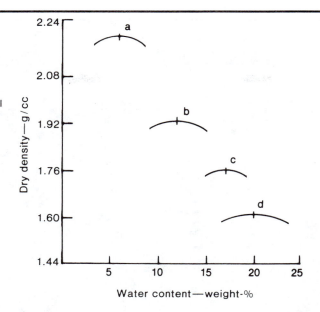

Figure 20.2. Moisture-density curves for (a) well graded sand with some clay, (b) nonplastic clay, (c) silt, and (d) plastic clay. The short vertical lines indicate the optimum water content for each soil.

shown in figure 20.2, the optimum water content varies from one soil material to another and is judged in terms of the maximum density that can be produced under a standard compaction effort.

Shear strength of soils

Soil shear is the movement, or slippage, of one part of a soil mass relative to another. Shear occurs along a plane (*shear plane*), commonly at the boundary between two layers where cohesion between neighboring soil particles is less than within the layers.

Slippage of soil on slopes is probably the most frequent manifestation of soil shear. Typical are landslides, or the collapse of buildings, retaining walls, or other structures built on sloping surfaces as shown in figure 20.3. Slippage under these circumstances is most likely to happen where a soil overlying bedrock or other impermeable subsurface layer becomes saturated with water. The excess water reduces soil cohesion and strength and allows gravity to pull the soil downslope.

Even soil on a flat surface may undergo shear if subjected to an excessive load. An example is the oozing of soil underfoot when one steps on muddy ground. Another is the failure of soil to support the weight of heavy structures, as illustrated in figure 20.4. As indicated in the figure, shear is preceded by compaction of the soil immediately beneath the load. As the load increases, forces transmitted through the compacted soil act in a lateral as well as in a downward direction, much like a wedge driven into the soil. Slippage occurs when the lateral force exceeds the shear strength of the soil. When slippage occurs, it can be both extensive and abrupt and may cause severe damage to man-made structures.

Figure 20.3.
Failure of a residence foundation caused by soil slippage on a steep slope. Such slippage usually occurs when a soil overlaying bedrock or an impermeable subsurface layer becomes saturated with water. (U.S.D.A. Soil Conservation Service photo.)

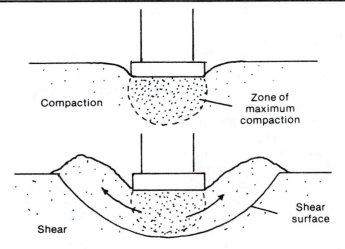

Figure 20.4.
Settlement of a compressive load, such as a building foundation, due to soil compaction and shear.

Compaction

Zone of maximum compaction

Shear surface

Shear

Texture and shear strength of soils

Shear strength in soils is a function of two forces: (1) those due to internal friction as particles slide past one another, and (2) those associated with the cohesive bonding of soil particles to each other. Internal friction contributes to soil strength principally in coarse-grained materials, such as gravels and coarse sands, whereas cohesive forces are more important in fine materials.

Maximum internal friction is associated with soils in which the individual soil grains occur in an interlocking pattern as illustrated in figure 20.5. In this example, shear occurs as the upper (unshaded) particles are moved

Engineering properties of soils

Figure 20.5.
Diagram of a soil shear box. The lower part of the box is fixed, but the upper part can be moved by a shearing force, as applied at A. Letter B denotes a load, which is a determinant of the force required to cause shear. The general path followed by the unshaded particles during shear is indicated by the heavy arrow.

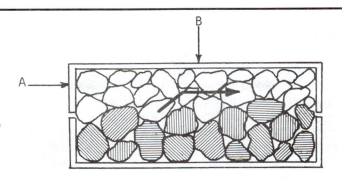

Figure 20.6.
Diagrams illustrating gradation in coarse materials. Finer particles, by filling the spaces between those of larger size, add strength to well graded materials.

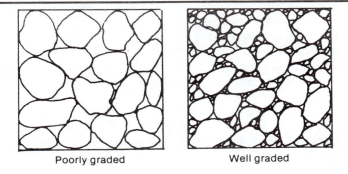

Poorly graded   Well graded

to the right relative to the lower (shaded) particles. However, before the upper particles can move freely to the right, they must first move upward against the compressive force B. Friction is therefore developed as the moving particles slide upward over the face of the stationary particles, and it tends to be greatest where both the compressive force and the interlocked particles are large.

As might be imagined, particle-interlocking in fine material such as clays is of minor importance. Shear strength in these fine materials therefore depends on cohesion. Shear strength attributable to cohesion can be very high in dense, compacted, fine soils. However, it may be negligibly small in soils that occur in a wet, plastic state. Evidence of this is seen in the inability of muddy roads or foot paths to support much weight.

In terms of their ability to support heavy loads, the strength of coarse soil materials is also a function of the proportions of different sized particles present, a property referred to as *grade* by the engineers. Coarse materials display greatest strength when they consist of mixtures of different sized particles and, for this reason, are said to be *well graded* (see fig. 20.6). Materials made up mostly of single-sized particles are termed *poorly graded*. They are less stable than well graded soils, particularly if they are made up of rounded particles, such as waterworn gravel.

Reasons why gradation influences strength are considered in figure 20.6. As shown at the left in this figure, particles of a poorly graded material contact each other at relatively few points. Not much total friction is developed at these contact points, so the particles can be moved about rather easily under loading. If the material is well graded, as shown at the right in the figure, smaller particles tend to fill pore spaces and increase strength by increasing the number of contact points. Furthermore, if some of the fine particles are clay, cohesional binding will also strengthen the material. This is particularly important to soil materials used as surfacing on unpaved roads. The clay content should be small, however, otherwise the roadway may become rutted during wet periods. The content of the fines (silt and clay) should not be greater than 10 to 15 percent of the total weight of the material.

Soil permeability as an engineering property

The ease with which soils transmit water, especially when saturated, is highly important to many engineering applications. In most instances, it is desirable for the soil to be readily permeable, but sometimes it is not. For example, soils of low permeability are preferred wherever seepage is to be avoided. Thus, at least part of the material used in earthen dams should be of low permeability, as should the soil where sewage lagoons, reservoirs, and canals are located.

Restricted permeability is particularly undesirable where soils associated with engineering works must be drained. Drainage of these soils normally requires elaborate and expensive drainage systems. Even where these systems are well designed and properly installed, some slowly permeable soils may still be excessively wet on occasion, and therefore, unsuited for certain uses. Often, the reason for this is that slowly permeable soils tend to be fine-textured and display unsatisfactory physical characteristics when wet.

A unique relationship between soil strength and permeability occurs where the right combination of soil, water pressure, and saturated flow result in the soil at seepage sites having negligibly low strength. A well known example is *quicksand*. Quicksand occurs where readily permeable sand, usually fine sand, is subjected to substantial water flow in an upward direction and out of the soil at the seepage site. The upward flow has a buoyant effect on the sand particles often making them essentially weightless. Under these conditions, the sand grains cannot form an interlocked, weight-supporting system. Thus, the ability of the quicksand to support a load may be very little greater than that provided by water alone. Conditions typical of quicksand are shown in figure 20.7.

If soil particles lack cohesion and are comparatively small, they can be carried away by seepage water. This results in the formation of shallow pools sometimes seen at springs. It is also responsible for the process known as *soil piping*, which is the development of tubelike channels through earthen dams or in the soil along gullies or similar erosion channels. Piping starts with surface erosion at seepage sites near the base of saturated slopes and then progresses toward the source of seepage water. If a channel is formed all the way back to the water source, highly erosive water flow can take

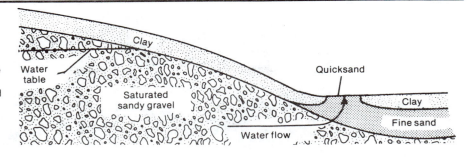

Figure 20.7.
Conditions typical of quicksand. Water standing at the elevation of the water table, as shown on the left, results in a submergence potential that causes the comparatively rapid flow of water at the seepage site where quicksand is located.

place through it, which may result in the failure of earthen dams or in a dramatic increase in erosion along gullies. Destructive water flow can also take place through cracks in the soil, such as those formed when poorly compacted earthen dams settle. Internal erosion associated with piping or cracking is most common in noncoherent soils high in silt or very fine sand.

Soil permeability is measured as the saturated flow through undisturbed soil cores; that is, soil cores taken carefully in the field so as to preserve their natural physical state. The permeability rate is expressed as the volume of water flow per unit of time. This type of information is published for soils in standard soil survey reports. The defined permeability class of a soil may also be shown, each such class corresponding to a range in permeability as follows:

Permeability class	Range in Permeability	
	in/hr	cm/hr
Very slow	<0.06	<0.15
Slow	0.06– 0.2	0.15– 0.5
Moderately slow	0.2 – 0.6	0.5 – 1.5
Moderate	0.6 – 2.0	1.5 – 5.0
Moderately rapid	2.0 – 6.0	5.0 –15.0
Rapid	6.0 –20.0	15.0 –50.0
Very rapid	>20.0	>50.0

Other soil properties important to engineering uses

Two additional properties of soils are mentioned because of their importance to several kinds of engineering uses. These are the *shrink-swell potential* and *corrosivity*. The shrink-swell potential is a physical characteristic of soils, whereas corrosivity is a chemically related property.

Shrink-swell potential

Some soils high in clay, particularly in expanding clay, undergo extensive and repeated shrinkage and swelling with cyclic wetting and drying. This reaction to changing water content is undesirable because of its effect on man-made structures. One example is the gradual loss of alignment in fence posts and telephone or electrical power poles set in Vertisols. Another is the

Figure 20.8.
Cracking and separation of the wall of a house due to a shift in the foundation caused by the shrinking and expansion of clay soil. (U.S.D.A. Soil Conservation Service photo.)

distortion or destruction of walls, buildings, or other structures located on these soils (see fig. 20.8). However, problems may arise without repeated shrinkage and swelling. For example, shrinkage cracks formed on the bottom of pond or reservoirs that are intermittently dry may result in serious water loss when these storage sites are recharged with water.

takes place when a dry soil clod is wetted to a specified moisture content. Soils that expand by less than 3 percent of their dry soil volume are rated as having a *low* shrink-swell potential. The shrink-swell potential is *moderate* if the volume increase is between 3 and 6 percent, and *high* if greater than 6 percent. Soils of high shrink-swell potential should be avoided for engineering applications if possible, although they may be used if appropriate safeguards against damage to structures are provided.

Corrosivity

This soil property relates to the deterioration of uncoated steel or cement pipe buried in the soil. Deterioration is caused largely by chemical solution, but for steel, it may also be due to rusting. Either steel or cement tends to deteriorate quite rapidly in strongly acid soils or in soils high in salts. Sulfate salts are especially harmful to cement.

Iron pipes react uniquely with salts in soil by forming an electrical cell that induces the flow of current between the pipe and the soil. In the associated reaction, metallic iron is converted to soluble ions. This effect is greatest where the pipe passes through soils of differing chemical characteristics. Under these conditions, the flow of current, which is through the pipe in one direction and through the soil in the other, can cause the rather rapid deterioration of the pipe. Corrosion is prevented mainly by coating pipes with a waterproof material.

Figure 20.9.
A common problem with building in low-lying positions, such as flood plains, is the rise of ground water during the wetter season of the year. Information on soils published in standard soil surveys usually identifies this type of problem. (U.S.D.A. Soil Conservation Service photo.)

Drainage condition and engineering use of soils

Engineering works of many types are hindered by poor or inadequate soil drainage. Waterlogged soils are difficult and expensive to excavate, and if fine-textured, they tend to be very low in strength. Soils troubled with a high water table cannot be satifactorily used for the construction of buildings with basements (see fig. 20.9), nor are they suitable as sites for septic tanks used for small-scale sewage disposal (see page 531). Septic tanks placed in waterlogged soil fail to function because of their inability to drain freely into saturated soil. They may also cause contamination of ground water if located over a shallow water table.

Another problem in the use of waterlogged soils for engineering works is *frost heave*. Frost heave results from the gradual accumulation of ice layers or lenses within the soil profile. As shown in figure 20.10, ice layers form and enlarge as water is carried upward by capillarity to the frost line in the profile. Water moves upward under these conditions, because freezing has the effect of soil-drying and thereby lowers the water potential at the frost line below that of the underlying saturated soil. As water reaches the frost line, it freezes and causes the forming ice layer to thicken. If freezing temperatures persist for long periods, the soil surface may be pushed upward until it stands several cm above normal. Structures built above the frost line can be pushed upward with the soil and can be seriously damaged as a result.

Soils high in silt are the most susceptible to frost heave. This phenomenon is less common in fine-textured soil of slow saturated conductivity and in sands or gravels, which show little ability to lift water by capillary transfer.

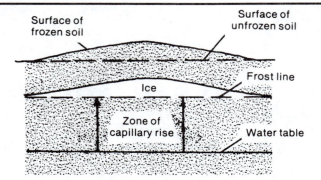

Figure 20.10.
Typical conditions
causing frost heave
due to the formation
of a thick ice layer or
lens in the soil.

The engineering classification of soils

Planning the engineering use of soils requires an understanding of soil properties and behavior. Often, this understanding is gained by practical experience in the field. Perhaps equally as often, however, work is with soil of
unknown qualities. Success depends on how the soil will react when put to a
specified use, and this may not be known until the use is actually imposed.
Obviously, if there were no way to predict soil behavior under these circumstances, there would be a good chance that many engineering projects involving soil would fail.

To limit failures of this type, engineers have established systems of soil
testing and classification to guide them in soil use. Although several such
systems have been devised, only two are considered here. One is the *Unified
system;* it is used by engineers from several specialized engineering fields
including agricultural and civil engineers. The other, referred to as the
AASHTO (pronounced ash-toe) *system,* has been adopted by the American
Association of State Highway and Transportation Officials for soil materials
used in highway construction. Of the two systems, the Unified system is the
most widely used and is therefore given major attention here.

The classification of soils into either the Unified or AASHTO system
depends mainly on two properties: *particle-size distribution* and *plasticity.*
On the basis of these properties, soils are placed into classes representing
fairly narrow ranges of suitability for one type of engineering use or another.
The procedures used for classification are described below.

**Engineering
particle-size
classification**

Engineers and soil scientists use different systems of particle-size classification. These differences are illustrated in figure 20.11. As shown by the
figure, the system used by soil scientists in this country (USDA system) has
10 particle-size fractions, but the Unified and AASHTO systems have only 7
and 8, respectively. Further, whereas two of the systems recognize the four
general particle-size classes of gravel, sand, silt, and clay, the third system
(Unified) combines silt and clay into a single class termed *fines.* Regardless
of these differences, all three systems consider the division of coarse and
fine particles to be at the lower size limit of sand, and they place this limit
at the same or almost the same point.

Figure 20.11.
Size fractions of the
U.S. Department of
Agriculture, AASHTO,
and Unified systems
of particle-size
classification.

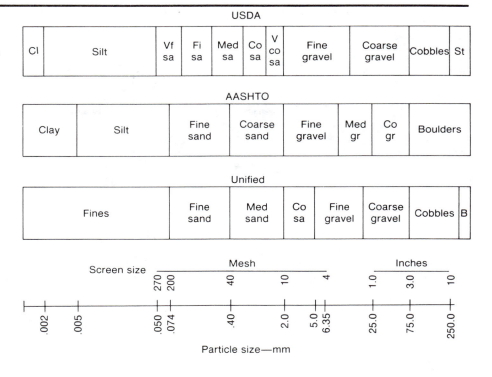

Often, engineering applications require only the most general understanding of particle-size distribution, perhaps nothing more than the proportions of coarse and fine fractions in a soil material. This can be suitably determined at times by visual means or by hand-working the material. It is more accurately determined by measuring the quantity of a sample that passes a 200–mesh (0.074 mm) screen, which, as indicated in figure 20.11, separates silt and clay from sand and gravel. In a more complete characterization, however, a sample is passed through a series of screens that not only separates the fine and coarse fractions but also divides the coarser material into the various sand and gravel subfractions. Information obtained by this type of screening is then combined with plasticity data to determine the engineering class of the soil material.

Plastic properties used in engineering soil classification

Two measurements are made to characterize plasticity for the engineering classification of soils: (1) the *liquid limit* and (2) the *plastic limit*. Each is a water content expressed as a percent of the oven-dry weight of the soil. These properties are determined only if there are sufficient fines in a soil to materially affect its physical behavior.

The liquid and plastic limits correspond to the upper and lower limits of the plastic range, that is, the range in moisture content over which the soil can be molded without showing either crumbliness or stickiness. Crumbliness in a soil is a sign that it is too dry to be plastic. If the water content

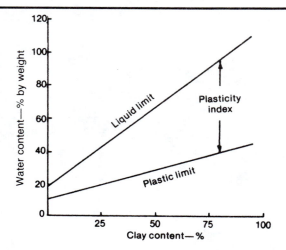

Figure 20.12.
The general relationship between clay content and the liquid and plastic limits of silt-clay mixtures (soil fines) containing expanding clay. Lower values would be obtained for mixtures containing nonexpanding clay.

of such a soil is increased, a point is eventually reached where it transforms from a crumbly to a plastic state. The water content at this point is referred to as the plastic limit. As more water is added, the soil becomes gradually softer, ultimately losing its plasticity by transforming to a semifluid, sticky state. The water content at this point marks the upper, or liquid, limit of the plastic range. The mathematical difference between the plastic and liquid limits provides the *plasticity index* of a soil. Of these three values, only two, the plasticity index and the liquid limit are used directly in the engineering classification of soils.

The principal source of plastic properties in soils is clay. As shown in figure 20.12, increasing clay causes all three indices of plasticity to increase, with this effect being greater for expanding than for nonexpanding clays. Since organic matter retains a large amount of water per unit of weight, but is relatively nonplastic, its presence in large amounts results in high plastic and liquid limits but comparatively small plasticity indexes.

Engineering classification systems

Although texture and plasticity are used to classify soils in both the Unified and AASHTO systems the way these properties are used is described only for the Unified system.

Unified system

The main features of the Unified system are outlined in table 20.1. It contains 15 classes, 8 for coarse-grained materials, 6 for fine-grained materials, and 1 for highly organic materials. The classes are identified by two-letter symbols from which some inference of class properties can be made, as shown in the footnote of the table. Brief descriptive statements noting the general character of each class are also given in the table. The descriptions identify the dominant size fractions of the class, and for the fine-grained materials, the degree of plasticity.

The initial separation of soils into coarse- and fine-grained classes of the Unified system is based on sieving soil samples using a 200–mesh screen with openings 0.074 mm in diameter. Soils containing a predominance of par-

Table 20.1. Soil classes in the Unified system of soil classification.

Major divisions		Class symbol[a]	Typical compositional and consistence characteristics
Coarse-grained soils (<50% passes 200–mesh screen)	Gravels or gravelly soils	GW	Well graded gravel or gravel-sand mixtures
		GP	Poorly graded gravel or gravel-sand mixtures
		GM	Silty gravels; gravel-sand-silt mixtures
		GC	Clayey gravels; gravel-sand-clay mixtures
	Sands or sandy soils	SW	Well-graded sands or gravelly sands
		SP	Poorly graded sands or gravelly sands
		SM	Silty sands; sand-silt mixtures
		SC	Clayey sands; sand-clay mixtures
Fine-grained soils (>50% passes 200–mesh screen)	Liquid limit of fines below 50	ML	Silts and very fine sands; silty or clayey fine sands; clayey silts of low plasticity
		CL	Clays of low to medium plasticity; gravelly, sandy, or silty clays
		OL	Organic silts or silty clays of low plasticity
	Liquid limit of fines above 50	MH	Silts; micaceous or diatomaceous fine sandy or silty soils; elastic silts
		CH	Clays of high plasticity
		OH	Organic clays of medium to high plasticity
		Pt	Peat or other highly organic soils

[a]These symbols indicate composition, gradation, or liquid limits according to the following scheme:

Compositional symbols	Gradation symbols	Liquid limit symbols
G —gravel	W —well graded	H —high (>50)
S —sand	P —poorly graded	L —low (<50)
M —silt		
C —clay		
O —organic (humus)		
Pt —peat or muck		

ticles larger than 200–mesh are classed as coarse-grained. The coarse-grained soils, in turn, are classified as sands or gravels depending on which of these two size fractions is most prevalent. Further subdivisions within coarse-grained materials are made according to whether they contain more than or less than 12 percent fines.

Gravels and sands that contain less than 12 percent fines are placed into well-graded classes of GW or SW if they consist of a range of particle sizes. If either consists principally of single-sized particles, on the other hand, they are classed as poorly-graded gravel or sand, which are symbolized GP or SP. Coarse materials containing more than 12 percent fines are classified to show the influence of the fine fraction. Differences are based on the plastic properties of the fines. If the fines display low plasticity, they are interpreted

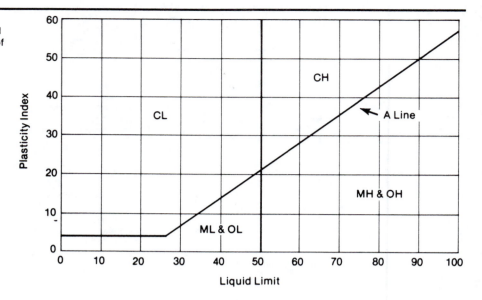

Figure 20.13.
A plasticity chart used for the classification of fine soil materials according to the Unified system of engineering soil classification.

as silty, and the assigned coarse-grained classes become either GM or SM.[1] Coarse-grained soils containing more than 12 percent highly plastic fines are placed in classes GC or SC.

Fine-grained soils are those containing more than 50 percent fines. These soils are placed into subclasses on the basis of the plasticity index and liquid limit of the fines through use of the chart in figure 20.13. The chart shows plasticity index on the vertical axis and liquid limit on the horizontal axis. The diagonal, identified as the A line, separates clayey soils (classes CL and CH) from silty and organic soils (classes ML and OL, or MH and OH). The location of the A line is arbitrary, but it is based on broad experience relating soil plasticity to other behavioral characteristics. The last letters, L and H, in the class symbols of fine-grained soils indicate whether the soil has a liquid limit above or below 50.

Engineering classes OL and OH recognize mineral soils of relatively low plasticity due to the presence of substantial organic matter. These soils fall in the same range of plasticity as silty soils designated M. Having similar plasticity properties, soils in these two groups are distinguished from each other on the basis of other properties such as color and odor. Soils of the OH and OL classes are usually regarded as inferior for engineering purposes.

At times the engineering classification of soil is indicated by a compound symbol such as CL-ML or SM-ML. These are used where the soil falls near the boundary between two classes. For example, the symbol CL-ML would be used for a soil with a liquid limit below 50 that plots very near the A line in the plasticity chart of figure 20.13. Similarly, the symbol SM-ML would apply to a soil containing nearly equal quantities of coarse and fine particles, with the fine fraction plotting in the ML box of the plasticity chart.

[1]The symbol M comes from the Swedish word mo, meaning silt.

Engineering properties of soils

AASHTO system Only a brief description of this system is given. In it there are seven general classes identified by symbols ranging from A-1 to A-7. Class A-1 includes coarse gravels and sands largely free of fines. At the other extreme, class A-7 contains plastic clays for the most part. The classes between A-1 and A-7 represent gradations between the two extremes.

Classes in the AASHTO system express the suitability of soil materials for use as the base or subgrade under highways. In general, the most suitable class for this purpose is A-1, and the least suitable, A-7.[2] The principal limitations of clayey A-7 materials are their low load-carrying capacity when wet and their high shrink-swell potential.

The seven general classes of the AASHTO system are very broad, and provisions are therefore made for subdivisions within them. These subdivisions recognize differences in the proportions of the major size fractions present, the degree of gradation of coarse-grained materials, and the total quantity, liquid limit, and plasticity index of the fines present. Twenty-one such subdivisions have been established, each identified by a *group index number*. Group index numbers range from 0 to 20, and they are shown in parentheses after the more general class symbol, as for example, A-6(13). Typically, high group index numbers indicate either a high content of fines or the presence of fines with high liquid-limit and plasticity-index values, or both. For the most part, group index numbers identify subdivisions in classes A-4 through A-7. A group index of zero is characteristic of gravels or sands containing little if any fines.

Engineering classes and soil use

The use of soils in engineering applications depends on two things: (1) the character of the soil as implied by the engineering soil class, and (2) the conditions under which the soil exists in the field. Both of these factors must be considered in the final determination of proper soil use. In general, the more limiting the conditions in the field the greater the care needed to determine whether a selected material is right for the use in question.

Although engineering soil classes are established on the basis of particle-size distribution and plasticity properties, these characteristics relate to, and can therefore be used as an index of, other major properties such as permeability, compressibility, and shear strength. Thus, some judgment about these latter properties can be made from knowledge of the engineering class of a soil, the general relationship being that shown for the Unified system in table 20.2. Also shown is the relationship between the Unified classes and potential frost action (frost heaving), along with relative ratings of each class for four engineering uses: earthen dams, canals, building foundations, and roadways, both as fill material under hard-surfaced roads, or surfacing for unpaved roads.

---

[2]The AASHTO system is more complex and less straightforward than this simple description suggests. However, to outline it in detail would require an unjustifiably long discussion.

Table 20.2. Relative ratings for soil classes in the Unified system of engineering soil classification.[a] (Adapted from data of the Bureau of Reclamation, U.S. Department of Interior.)

Class symbol	Drainage characteristics	Compressibility and expansion	Shear strength when saturated	Potential frost action	Relative suitability for various uses[b]								
					Rolled earthen dams			Canals	Foundations		Roadways		
					Uniform fill	Core	Shell		Seepage present	No seepage	Fill		Surfacing
											No frost heave	Frost heave possible	
GW	Ex	Neg	H	V SI	—	—	1	—	—	1	1	1	3
GP	Ex	Neg	M-H	V SI	—	—	2	—	—	3	3	3	—
GM	F-VP	SI-V SI	M-H	SI-M	2	4	—	4	1	4	4	9	5
GC	P-VP	SI	M	SI-M	1	1	—	1	2	6	5	5	1
SW	Ex	Neg	H	V SI	—	—	3[d]	—	—	2	2	2	4
SP	Ex	Neg	M-H	V SI	—	—	4[d]	—	—	5	6	4	—
SM	F-P	M-V SI	M-H	SI-H	4	5	—	5	3	7	8	10	6
SC	P-VP	SI-M	M	SI-H	3	2	—	2	4	8	7	6	2
ML	F-P	SI-M	ML	M-VH	6	6	—	6	6	9	10	11	—
CL	VP	M	ML	M-H	5	3	—	3	5	10	9	7	7
OL	P	M-H	L	M-H	8	8	—	7	7	11	11	12	—
MH	F-P	H	ML	M-VH	9	9	—	—	8	12	12	13	—
CH	VP	H	L	M	7	7	—	8	9	13	13	8	—
OH	VP	H	L	M	10	10	—	—	10	14	14	14	—
Pt	F-P	VH	L	SI	—	—	—	—	—	—	—	—	—

[a]Ratings in the left-hand columns are by means of the following abbreviations: Ex = excellent, Neg = negligible, H = high, L = low, F = fair, P = poor, SI = slight, M = moderate or moderately, and V = very.

[b]For the suitability ratings, number 1 means most suitable; a dash means unsuitable.

[c]An earthen dam may contain an impermeable core covered by a permeable, nonerodible shell.

[d]Not suitable unless gravelly.

In its entirety, table 20.2 permits a rather detailed comparison of the soil classes in the Unified system. In addition, it allows preliminary judgments to be made regarding certain engineering uses of soil. The tentative nature of these judgments should be stressed, however. For any project of significant magnitude and cost, the final judgment of soil use in engineering applications should be made by an experienced engineer.

Engineering interpretations of soil survey information

Much of the information collected in standard soil surveys is about properties important to engineering uses of soils. This information is presented in two forms: (1) as measured or estimated engineering properties, and (2) as interpretations of these properties to show the suitability or limitations of soils for a variety of uses.[3] Since most persons using soil surveys are not trained to make their own interpretations of engineering properties, they must rely on the interpretations supplied in the survey reports.

Because of the many different engineering uses made of soils, many different types of interpretations are required. Some of them deal with soil use in its undisturbed state, as for a dwelling site. Others are for soil excavated from one location, known as a *borrow site*, for use in another. As the uses differ so do the kinds of properties that must be considered in making the interpretations.

Several examples of engineering interpretations are shown in table 20.3 as they appear in soil survey reports. As may be judged from column 1 of the table, interpretations are on a soil-to-soil basis, phases in this instance, and they may characterize a specific soil property (column 2), or indicate the relative suitability of a soil for certain uses (columns 3–6), or they may do no more than identify features of a soil important to a particular use (columns 7–9).

The interpretations in Table 20.3 are only representative of the various kinds published in current soil survey reports. A more complete listing follows. It is organized under five general types of soil use:

Sanitary facilities	*Community development*	*As source of*	*Water management*	*Recreation*
Septic tank filter fields	Shallow excavations	Roadfill	Pond reservoir areas	Picnic areas
Sewage lagoons	Dwelling sites	Sand	Embankments, dikes, and levees	Camp areas
Sanitary landfills	Local roads and streets	Gravel	Land drainage	Golf fairways
		Topsoil	Terraces and diversions	Playgrounds
				Paths and trails

[3]Here, interpretation means the translation of knowledge about soil properties into a judgment of soil performance under a specified use or treatment. Interpretive information for engineering soil uses is available for most soils in the United States and has been included as a standard part of soil surveys published since aobut 1960.

Table 20.3. Example interpretations of engineering soil properties.[a] (Adapted from McGee, Dale A., *Soil Survey of Clark County, Washington*, 1972 Superintendent of Documents, U.S. Government Printing Office, Washington, D.C.).

Soil phase name and map symbol (1)	Shrink-swell potential (2)	Limitation for use as		Suitability for			Soil features affecting engineering practices — Farm ponds	
		Septic tank filter field (3)	Site for residences (4)	Topsoil (5)	Road fill (6)	Highway location (7)	Reservoir area (8)	Embankment (9)
Bear Prairie silt loam, 0–8% slope, (BpB)	Low	Slight where slope is 0–5%; moderate where slope is 5–8%	Slight	Good	Poor: low shear strength; high susceptibility to frost action; A-6 material	High susceptibility to frost action; slope range is 0–8%	Moderate permeability; slope range is 0–8%	Low shear strength; semipervious to impervious when compacted; medium compressibility
Cove silty clay loam, 0–3% slope, (CvA)	High	Severe: very slow permeability and ponding	Severe: high shrink-swell potential	Unsuitable: high clay content	Poor: low shear strength; A-7 material	High shrink-swell potential; ponded during winter months except where drained; slope range is 0–3%	Very slow permeability; slope range is 0–3%	Low shear strength; impervious when compacted; high compressibility
Larchmount very stony silt loam, 30–75% slope, (LcG)	Low	Severe: slope	Severe: slope	Unsuitable: cobbly or very stony	Fair: low shear strength; A-4 material	Slope range is 30–75%	Moderate permeability; slope range is 30–75%	Low shear strength; semipervious when compacted; slight compressibility
Semiahmoo muck, (Sr)	High shrink, low swell	Severe: seasonal high water table	Severe: organic soil; very poorly drained	Poor: organic soil	Unsuitable: organic soil	Low density; low strength; water table at or near surface in winter if not drained; slope range is 0–1%	Moderate permeability; organic soil; slope range is 0–1%	Organic soil

[a]Interpretations are derived from three separate tables, which results in some repetition, such as the statements on slope range. This repetition has been allowed to retain the original character of the interpretations.

Interpretations for engineering uses of soils, as appear in soil survey reports, are based largely on a set of guides published by the United States Department of Agriculture.[4] Where practical, these guides permit the direct translation of soil data into specific ratings, such as those shown in columns 3 through 6 of table 20.3. Guides of this kind can be developed for soil uses that depend primarily on soil properties and conditions. However, some engineering soil uses are undertaken without much regard as to whether the soils involved have limitations for such use. One example is in the choice of highway location, which is based initially on terrain features and the relative location of points to be joined by the highway. Once the location of the highway has been established, information on soil limitations is then wanted so that plans to compensate for or overcome them can be made. Because of this, engineering interpretations for highway location, as in column 7 of table 20.3, list properties important to the use but make no attempt to rate the soil relative to this use.

Examples of interpretation guides that do permit the rating of soils for specific engineering uses are given in the following sections. Four different soil uses are considered: sites for septic tank filter fields, dwellings, sewage lagoons, and as roadfill material. These uses have been selected in part because of their importance to a wide range of individuals and in part because, together, they recognize most of the soil properties and conditions important to engineering soil use.

Soil use for septic tank filter (absorption) fields

Septic tanks are used for the disposal of sewage where connections to regular sewage disposal systems are not available. These tanks, along with one or more lines of tile pipe connected to them, are buried in the soil, normally adjacent to a residence or other source of sewage. As suggested in figure 20.14, sewage is delivered first to the septic tank, where solids settle out and soluble organic materials undergo rapid decomposition. Liquid effluent that drains from the tank is distributed in the surrounding soil through the tile pipes, which are loosely joined to each other. The soil receiving the effluent is referred to as the *absorption* or *filter field*. A principal function of the soil is to filter out solids, including micoorganisms, as the effluent moves toward the ground water or an open drainageway.

Areas suitable for septic tank installation have a number of basic requirements. They must contain soil that is permeable (see fig. 20.14), but not so permeable as to deliver unfiltered effluent to the groundwater or a nearby stream. They must not be troubled with a high water table nor be subject to flooding.[5] Either flood water percolating through the soil or groundwater rising into the filter field may result in groundwater pollution. Further, saturation of the soil in the filter field prevents absorption of the effluent and thereby interferes with the functioning of the system. In addition to the above

---

[4]*See Guide for Interpreting Engineering Uses of Soils*, November, 1971, an unnumbered publication of the Soil Conservation Service, USDA, Washington, D.C.

[5]For the characterization of soil drainage classes and flooding incidence, see Appendix B.

Figure 20.14.
A diagram of a septic tank installation. The percolation test holes are used according to a standard procedure for judging the permeability of the soil as it occurs in the field. (From A. A. Klingebiel, U.S.D.A. Agriculture Information Handbook 320, 1967.)

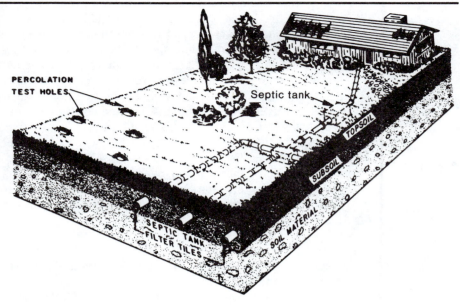

Table 20.4. Soil limitation ratings for septic tank filter fields. (From *Guide for Interpreting Engineering Uses of Soils,* 1971, Soil Conservation Service, U.S.D.A., Washington, D.C.).

Item affecting use	Degree of soil limitation		
	Slight	Moderate	Severe
Permeability class[a]	Rapid, moderately rapid, and upper end of moderate	Lower end of moderate	Moderately slow and slow
Permeability in inches (centimeters) per hour	More than 1 in (2.5 cm)	0.6–1 in (1.5–2.5 cm)	Less than 0.6 in (1.5 cm)
Depth to water table	More than 72 in (180 cm)	48–72 in (120–180 cm)	Less than 48 in (120 cm)
Flooding incidence	None	Rare	Occasional or frequent
Slope class	0–8%	8–15%	More than 15%
Depth to bedrock	More than 72 in (180 cm)	48–72 in (120–180 cm)	Less than 48 in (120 cm)
Stoniness class[b]	0 and 1	2	3, 4 and 5
Rockiness class[b]	0	1	2,3,4, and 5

[a] Definitions of permeability classes are given on page 519.
[b] Determined from soil textural class or miscellaneous land type names (see Appendix table A2).

Engineering properties of soils

Table 20.5. Interpretation of soil survey information on three soils for use as septic tank filter fields.

Soil Feature	Cashmere sandy loam				Konner silty clay loam	
	3–8% slope		8–15% slope		Nearly level	
	Data	Limitation	Data	Limitation	Data	Limitation
Permeability class	Moderately rapid	Slight	Moderately rapid	Slight	Slow to moderate	Moderate to severe
Permeability (in/hr)	2.0–6.3	Slight	2.0–6.3	Slight	.05–0.8	Moderate to severe
Depth to water table	>72 in	Slight	>72 in	Slight	18–30 in	Severe
Flooding incidence	None	Slight	None	Slight	Occasional	Severe
Slope class	3–8%	Slight	8–15%	Moderate	2%	Slight
Depth to bedrock	>72 in	Slight	>72 in	Slight	>72 in	Slight
Stoniness class	0	Slight	0	Slight	0	Slight
Rockiness class	0	Slight	0	Slight	0	Slight

requirements, the best sites for septic tanks have no more than a moderate slope and consist of deep soils having only a limited degree of stoniness or rockiness.[6]

The interpretation guide for soil use as a septic tank filter field is shown in table 20.4. The items listed first in the table (permeability class, permeability rate, depth to a water table, and incidence of flooding) are properties or soil conditions important to the proper functioning of the septic tank. The remaining items are properties that largely determine the ease of septic tank and filter field installation and development.

Three levels of limitation to soil use as a septic tank filter field are indicated in table 20.4, namely: *slight, moderate* and *severe*. A slight limitation means that none of the properties of a soil places a restriction on this use. However, if a soil has only one property that will interfere with septic tank installation or functioning, it must be placed in one of the other two limitation classes. Class selection in this instance depends on how that property fits the guidelines in table 20.4.

Use of the guide in table 20.4, as well as others that follow this discussion, is illustrated by the interpretations of data for three soil phases, as shown in table 20.5. Two soil series are considered, the Cashmere sandy loam and the Konner silty clay loam. The Cashmere soil occurs on nearly level to moderately sloping, sandy, well drained terraces. Data pertinent to septic tank filter fields are given for two slope phases, 3–8 and 8–15 percent slopes, of this soil. The Konner soil occurs only on nearly level land of the flood plains. Its parent material is stratified alluvium. Because of its position on flood plains, the Konner is subject to occasional flooding and has a high water table.

Interpretations based on the guide in table 20.4 show that the 3–8 percent slope phase of the Cashmere soil has no more than a slight limitation for use as a septic tank filter field. The other, or 8–15 percent, slope phase is assigned a moderate limitation rating, however, the difference being due

[6]For the characterization of stoniness and rockiness in soils, see Appendix A.

Figure 20.15.
Installing a septic tank filter field in Pewamo soil, in Macomb Co., Michigan. This soil has a severe limitation due to slow permeability and a high water table. Even with over 300 meters of tile line, this system will have problems at certain times of the year. (U.S.D.A. Soil Conservation Service photo by R. L. Larson.)

to the steeper slope. The principal reason for this is that as the slope increases greater attention to the design and installation of the tile filter system is necessary, which leads to greater cost.

In comparison to the Cashmere soil, the Konner is decidedly inferior as a site for a septic tank filter field. This soil has a permeability that is too low, a water table that is too high, and it is subject to occasional flooding. Slow permeability could probably be overcome by increasing the number of tile lines in the filter field (see fig. 20.15). This would, of course, increase the expense. It would also require that sufficient land be available for tile line installation, which could impose an insurmountable problem on a small urban lot. Problems due to a high water table and flooding could be even more serious than that due to slow permeability. Often such problems can be dealt with only through the development of area-wide flood-control and drainage measures.

Soil use for dwelling sites

The suitability of soils for dwellings and similar structures depends on properties important to both construction and maintenance. A common dilemma in the choice of sites for dwellings arises from the fact that sites that are highly desirable for aesthetic reasons often have serious construction and maintenance limitations. For example, steep hillsides may provide a home with a view, but they may also afford considerable difficulty in the building phase and in maintaining the home once in place. Maintenance may be an extremely critical factor where, due to steepness of slope, soil slippage occurs. The principal alternatives are to avoid such sites or to include construction features that minimize the chance of slippage. A severe limitation, such as a steep slope, does not eliminate a site for use in dwelling construction, but the use may be either at great expense or great risk.

The more important features of soils important to their use for dwellings are listed in table 20.6. Within this listing, five features—soil drainage class, slope, depth to bedrock, stoniness, and rockiness—are of major con-

Table 20.6. Soil limitation ratings for dwellings.[a] (From *Guide for Interpreting Engineering Uses of Soils*, 1971, Soil Conservation Service, U.S.D.A., Washington, D.C.).

Item affecting use	Degree of soil limitation		
	Slight	Moderate	Severe
Soil drainage class[b]	*With basements:* Excessively drained, somewhat excessively drained, well drained *Without basements:* Excessively drained, somewhat excessively drained, well drained, moderately well drained	*With basements:* Moderately well drained *Without basements:* Somewhat poorly drained	*With basements:* Somewhat poorly drained, poorly drained, very poorly drained *Without basements:* Poorly drained, very poorly drained
Dept to seasonal water table (Seasonal means for 1 month or more)	*With basements:* More than 60 in (150 cm) *Without basements:* More than 30 in (75 cm)	*With basements:* More than 30 in (75 cm) *Without basements:* More than 20 in (50 cm)	*With basements:* Less than 30 in (75 cm) *Without basements:* Less than 20 in (50 cm)
Flooding[b]	None	Not applicable	Rare, occasional, or frequent
Slope[c]	0–8%	8–15%	More than 15%
Shrink-swell potential	Low	Moderate	High
Unified soil group	GW, GP, SW, SP, GM, GC, SM, SC, CL with plasticity index below 15	ML, CL with index of 15 or more	CH, MH, OL, OH, Pt
Potential frost action[d]	Low	Moderate	High
Stoniness class[e]	0 and 1	2	3,4, and 5
Rockiness class[e]	0	1	2,3,4, and 5
Depth to bedrock[f]	*With basements:* More than 60 in (150 cm) *Without basements:* More than 40 in (100 cm)	*With basements:* 40–60 in (100–150 cm) *Without basements:* 20–40 in (50–100 cm)	*With basements:* Less than 40 in (100 cm) *Without basements:* Less than 20 in (50 cm)

[a]If slope limits are reduced by 50% (i.e., from 0–8 to 0–4), this table can be used for evaluating soil limitations for shopping centers and for small industrial buildings with foundation requirements not exceeding those of ordinary three-story dwellings. Some soils given limitation ratings of moderate or severe may be good sites aesthetically, but in use, they will require more preparation and maintenance.
[b]For soil drainage class and flooding incidence, see Appendix B.
[c]Reduce slope limits by 50% for those soils susceptible to hillside slippage.
[d]Use this item only where frost penetrates to the assumed depth of foundation footings and where soil is moist during freezing weather.
[e]Determined from textural class or miscellaneous land type names (see Appendix table A2).
[f]If bedrock is soft enough so that it can be dug out with light power equipment, such as backhoes, reduce ratings of moderate and severe by one step.

cern to construction, particularly to soil excavation. Properties or conditions important to maintenance are those affecting soil strength, settlement, and seepage. These include drainage, flooding, shrink-swell potential, potential frost action, as well as those characteristics that determine the engineering class of a soil. For example, figure 20.16 shows a dwelling with a serious settlement problem due to subsidence of an organic soil (Unified class Pt). Regardless of other characteristics, organic soils are always assigned a severe limitation as a site for dwellings because of their instability and tendency to subside.

It should be noted that many of the properties of a site important to dwellings are also important to the installation and maintenance of septic tanks, sewer and water lines, and walks and driveways. Septic tanks, along with sewer and water lines, as well as other buried utilities, are installed most readily where soil is easily excavated. Maintenance problems for these installations occur if there is soil slippage or settlement, or with pipes, if they can be easily corroded by the soil.

Normally, all the soil variation that occurs over the landscape is not shown on soil maps developed primarily for agricultural land use. Because of this, maps produced by the U.S. Soil Survey should not be used as the sole basis for selecting a dwelling site. Final site selection should be made only after tapping all other reliable sources of information. Helpful in this respect are county agents, employees of the Soil Conservation Service, contractors of extensive local experience, and in developed areas, local home owners.

Where assurances about the suitability of a soil as a dwelling site are not forthcoming from other sources, it may be necessary to carry out an on-site inspection of the land. By this means, topographic and surface soil conditions can be rather easily evaluated. However, subsurface conditions that might hinder construction work can be determined only by examining the soil profile to an appropriate depth. Questions to be answered by an on-site study are suggested by the items listed in table 20.6.

Engineering properties of soils

Table 20.7. Guide for determining suitability of soil for use as roadfill. (From *Guide for Interpreting Engineering Uses of Soils*, 1971, Soil Conservation Service, U.S.D.A., Washington, D.C.).

Item affecting use	Degree of suitability		
	Good	Fair	Poor
Properties affecting serviceability under use			
Unified soil class	GW, GP, SW, SP, GC[a], GM[a], SM[a], SC[a]	ML, CL with plasticity index below 15	Pt, OL, OH, MH CH, CL with plasticity index above 15
AASHTO group	0–4	5–8	Above 8
Shrink-swell potential	Low	Moderate	High
Susceptibility to frost action	Low	Moderate	High
Properties affecting handling at borrow site			
Slope	0–15%	15–25%	Above 25%
Stoniness class[b]	0, 1, and 2	3	4 and 5
Rockiness class[b]	0 and 1	2	3,4, and 5
Soil drainage class[c]	Excessively drained to moderately well drained	Somewhat poorly drained	Poorly drained and very poorly drained

[a]Downgrade the suitability rating to fair if the material contains more than 30% fines.
[b]Determined from textural class or miscellaneous land type names (see Appendix table A2).
[c]See Appendix B.

The use of soil for roadfill

Roadfill is the material used in the construction of embankments on which highways are commonly located. These embankments provide greater uniformity of grade and, in some instances, elevate the roadway to a more desirable level above a water table. The soil used in the embankments may come from cuts through neighboring hills, or it may come from more distant borrow sites.

The choice of a roadfill material depends mainly on the properties outlined in table 20.7. Two different sets of properties are identified. Those shown first in the table are properties important to the serviceability of the soil once it is in place in the embankment. They include engineering soil class, shrink-swell potential, and susceptibility to frost action. The second set of properties relates primarily to the accessibility and handling of the material at the borrow site. These are slope, stoniness, rockiness, and drainage condition.

It is sometimes necessary to modify the soil ratings specified in table 20.7. For example, the Unified class ML is shown to have a fair rating in the table, but some soil materials of this class may be subject to severe frost action under the right set of conditions. Where this is known to be true, a poor rather than a fair rating would be assigned to the material. On the other hand, no change in rating would be necessary if a soil of this nature were used where soil freezing did not occur.

Suitability ratings for soil used as roadfill usually apply to fill beneath walkways, patios, driveways, and parking areas. However, failure of these structures often depends as much on how the fill is handled as on its inherent suitability. A major problem in such use results from improper compaction, or the incorporation of trash, including construction wastes, in the fill, which interferes with compaction.

Soil use for sewage lagoons

Sewage lagoons are used extensively for the disposal of human and animal wastes. They consist of one or more basins in which the solid and liquid wastes are held and allowed to decompose, largely under anaerobic conditions. Excess water delivered into the lagoon may be lost be evaporation, or it may be disposed of on the land, as by irrigation. Some water loss through seepage can be expected, but it should be minimal to prevent the pollution of local ground water.

Lagoons are often constructed through a combination of excavation and diking. The soil at the site will preferably be one of low permeability in both its natural state, which is retained beneath the lagoon, and in its reworked, compacted state in the dike outlining the lagoon. High permeability may not only allow ground water or stream pollution, it may also increase the difficulty of maintaining a depth of water necessary for the proper functioning of the lagoon. Suitable water depths range between about 1 and 2 meters.

Soil limitation ratings for sewage lagoons are given in table 20.8. These ratings refer principally to site characteristics important to the construction phase and ultimate functioning of the lagoon and not to the suitability of the soil material for the dikes. The soil used in the dike should have low permeability and a high shear strength when compacted and saturated with water. It should not display elastic compressibility, nor should it be susceptible to piping. A preliminary judgment of soil suitability for dike construction can be made from table 20.2 on page 528.

Of the soil properties shown in table 20.8, permeability ratings, depth to bedrock, organic matter content, and the Unified soil classes relate mostly to seepage loss through the lagoon floor. Depth over bedrock is included because, if the depth is not adequate, serious seepage losses through cracks or fissures in the bedrock are likely to occur. An adequate depth over a water table is required to prevent ground-water pollution. Flooding also poses a pollution problem, especially if flood waters overtop the lagoon or cause serious erosion of the dikes. If only nonerosive, shallow flood waters are expected, flooding can often be overlooked in the selection of a lagoon site.

The information in table 20.8 indicates that the content of gravel, cobbles, or stones should not be too high at the site of lagoon construction.[7] The reason for this is that large fragments interfere with soil excavation, and perhaps even more importantly, with compaction of the lagoon floor to reduce its permeability.

[7]For the classification of soils with coarse fragments such as gravel, cobbles, or stones, see Appendix A.

Table 20.8. Soil limitation ratings for sewage lagoons. (From *Guide for Interpreting Engineering Uses of Soils*, 1971, Soil Conservation Service, U.S.D.A., Washington, D.C.).

Item affecting use	Degree of soil limitation		
	Slight	Moderate	Severe
Depth to water table (seasonal or year-round)	More than 60 in (150 cm)	40–60 in[a] (100–150 cm)	Less than 40 in[a] (100 cm)
Permeability in inches (centimeters) per hour[b]	Less than 0.6 in (1.5 cm)	0.6–2.0 in (1.5–5.0 cm)	More than 2.0 in (5.0 cm)
Depth to bedrock	More than 60 in (150 cm)	40–60 in (100–150 cm)	Less than 40 in (100 cm)
Slope	Less than 2%	2–7%	More than 7%
Coarse fragments less than 25 cm (10 inches) in diameter: percent by volume	Less than 20%	20–50%	More than 50%
Percent of surface area covered by coarse fragments more than 250 cm (10 inches) in diameter	Less than 3%	3–15%	More than 15%
Organic matter content	Less than 2%	2–15%	More than 15%
Flooding	None	Not applicable	Soils subject to flooding[c]
Unified soil class	GC, SC, CL and CH	GM, SM, ML and MH	GP, GW, SP, SW, OL, OH and Pt

[a]If the floor of the lagoon is nearly impermeable and at least 2 ft. (60 cm) thick, disregard depth to the water table.
[b]For the measurement of permeability see page 519.
[c]Disregard flooding if flood water is not likely to enter or damage the lagoon.

Interpretive maps for engineering use of soils

Soil survey information important to the engineering use of soils has its greatest application, generally speaking, in land use planning as carried out by local, state, or regional planning officials. Because of the complexity of the information, however, it is usually preferable to have it presented on interpretive maps from which the degree of limitations for specific engineering uses can be quickly judged. Four such maps are shown in figure 20.17. They are for soil use as sites for dwellings with basements, septic tank filter fields, and sewage lagoons, and as a source of roadfill material. The area shown on the map is the same as in figures 12.19 through 12.21, which are for soil use for agricultural and recreational purposes. It may be recalled that soils in the upper part of the map are in upland positions, some with moderate to steep slopes, while the others are essentially flat bottomland soils subject to excessive wetness when in a natural, undrained condition. Two of the lowland soils, Ss and Sp, are high in organic matter, with Sp being classed as a peat.

Only a brief glance at the maps in figure 20.17 indicates that the soils in this area are of limited value for various engineering uses. The basic problems in most cases are steepness of slope for the upland soils and excessive

Figure 20.17.
Interpretive maps for soil use as sites for dwellings with basements, septic tank filter fields, and sewage lagoons, and as a source of roadfill.

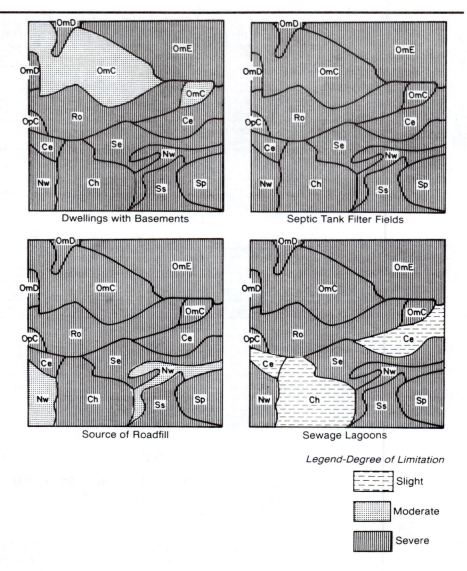

Dwellings with Basements

Septic Tank Filter Fields

Source of Roadfill

Sewage Lagoons

Legend-Degree of Limitation

Slight

Moderate

Severe

wettness on the lowlands. Further, some of the soils in the bottomlands have severe limitations because of texturally related shrink-swell or frost-action properties, or because of their high organic matter content. For the area, only one use has a slight limitation, and this is with phases Ce and Ch for use as sites for sewage lagoons. Otherwise, the limitations are either moderate or severe, with the latter being considerably the more dominant.

Although soil survey information is of considerable value for planning various engineering works, its use will normally be limited to highly generalized judgments or as a starting point for planning of specific projects. For the most part, soil survey information collected for agricultural soil use is not sufficiently precise for most engineering works. This is due primarily to the variability of the soil that may occur in a given mapping unit without the variation being identified. It is also the result of the broad variation in slope that may occur within a given mapping unit, such as a soil phase, which may be of little significance to agricultural soil use but of great importance to engineering use. For reasons such as these, then, final planning for major engineering projects usually involves considerable on-site examination of the soil as it occurs in the field.

Summary

Engineering properties are important to soil use for construction purposes, either as a site for structures or as a building material. Most engineering properties relate to the sizes of mineral grains making up the soil. Mineral-grain size determines such behavioral characteristics as strength, workability, and permeability to water. Strength is important to the ability of the soil to support heavy loads, as well as to its workability, as in excavation or cultivation. Permeability determines the suitability of the soil for water-retention structures, as earthen dams or man-made levees, or the ease of drainage of wetlands selected as sites for buildings or other structures.

Engineers classify soils into groups of like behavioral characteristics under use. The classsification is based on particle-size distribution of the soil and the plasticity properties of finer fractions. A separate group of organic soils is also recognized. These soils are usually unsuitable for engineering uses. Several somewhat different systems of engineering classification have been established, but the one most widely used is the Unified system. Both civil and agricultural engineers use the Unified system.

Information collected in the standard U.S.D.A. soil surveys allows a number of judgments to be made about the suitabilty of soils for engineering uses. Textural and plasticity data provide a basis for estimating the engineering class of the soil. Also of value to the engineer is soil depth over bedrock or a water table, its position on the landscape, slope, and permeability. A property not determinable from soil survey information is compressibility.

Review questions

1. Why do the compressibility properties of highly organic soils make them unsuitable for a wide range of engineering uses?
2. Compare the contributions of coarse soil grains and clays to the shear strength of soils.
3. Distinguish between poorly and well graded sands and gravels, and show why, of the two, the well graded groups have the greatest strength.
4. Why does soil high in expanding clay make a poor material for many engineering uses.
5. Explain the process referred to as frost heave.
6. What is the plasticity index of soils, and why does it increase with increasing clay content?

7. Why will a soil containing 90 percent sand particles of varying size and 10 percent clay be placed in the Unified soil class of SW, whereas the class would be SC if the sand made up 85 percent of the sample and clay the rest?

8. Using the plasticity chart in table 20.3, show that a soil containing more than 50 percent fines with a plasticity index of 40 and a liquid limit of 60 would be placed in the Unified class CH.

9. According to table 20.2, Unified soil class GC is better than class GW as surfacing material for an unpaved road. Why is this?

10. From table 20.4 judge which of the following conditions or soil properties places the greatest limitation on a soil for use as a septic tank filter field: (a) rapid permeability class, (b) soil slope of 10 percent, or (c) depth to a water table of 40 inches.

11. According to table 20.6, what is the minimum depth to a water table that should be allowed for a slight limitation where a soil is to be used as a site for a dwelling with a basement? Where the use is for a dwelling without a basement?

12. What is the general relationship between the permeability of a soil and its suitability for use as a site for a sewage lagoon? Explain why this is so.

Selected references

Hough, B. K. *Basic Soils Engineering*. New York: The Ronald Press Company, 1969.

Portland Cement Association. *PCA Soil Primer*. Chicago: Portland Cement Association, 1973.

Terzaghi, K. and Peck, R. B. *Soil Mechanics in Engineering Practice*. New York: John Wiley and Sons, 1967.

The Asphalt Institute, *Soils Manual for Design of Asphalt Pavement Structures*. College Park, Maryland: The Asphalt Institute, 1963.

Appendix A

Classification of soils with coarse fragments

Coarse fragments are mineral particles larger than sand. As shown in table A1, there are various types of coarse fragments depending on their size, shape, and in some cases, mineral make-up. In soil classification and mapping, a different approach is used to measure and express the content of large coarse fragments such as stones (see table A2) than is used for smaller coarse fragments such as gravel or cobbles.

Classification of soils containing small coarse fragments

The content of the smaller coarse fragments (smaller than stones) in soils is expressed as a percent by volume. If the fragments in this size range exceed 20 percent of the soil volume, the textural class name of the soil is modified through use of the adjectives in table A1. For contents between 20 and 50 percent, the terms in the table are used alone in modifying the names, as for example, *gravelly* sandy loam or *slaty* clay (see fig. A1). For contents above 50 but less than 90 percent, the above names would be further modified to *very* gravelly sandy loam or *very* slaty clay. If the content exceeds 90 percent, the material is no longer considered to be soil, but rather one of several *miscellaneous land types*. Examples are *gravelly riverwash*, which corresponds to gravel bars, and *gravelly colluvial land* consisting of coarse erosional debris located at the base of eroded slopes.

Classification of stony and rocky soils

Stony and rocky conditions are somewhat similar, the main difference being that stoniness signifies the presence of large, loose fragments, whereas rockiness implies the presence of bedrock, either exposed at the surface or lying at a shallow depth beneath the surface. The degree of stoniness and rockiness is expressed by the percent of surface area taken up by stones or rocks. As with coarse fragments of smaller size, the content of rocks and stones is then indicated through use of modified textural class names or appropriate miscellaneous land type names (see fig. A2). The relationship between these names and the rock or stone content is outlined in table A2.

Figure A1.
Profile of the Barneston gravelly coarse sandy loam in which the dominant soil separate is coarse sand. The gravel content of this soil is between 20 and 50 percent, as may be judged from the modified textural class name. Some cobble-sized fragments are also present, but the content is insufficient to warrant recognition. (U.S.D.A. Soil Conservation Service photo by Al Zulauf.)

Table A1. Terms used to indicate the presence of loose, coarse fragments in soils. (Adapted from *Soil Survey Manual*, U.S.D.A. Agriculture Handbook 18, 1951, Superintendent of Documents, Government Printing Office, Washington, D.C.).

Nature of fragment		Dimensions of fragment and descriptive term		
Shape	Material	Diameter		
		<7.5cm	*7.5–25 cm*	*>25 cm*
Rounded	Various	Gravelly	Cobbly	Stony[a]
Angular	Chert[b]	Cherty	Coarse cherty	Stony
	Other than chert	Angular gravelly	Angular cobbly	Stony
		Length of longest dimension		
		<15 cm	*15–38 cm*	*>38 cm*
Flat or platelike	Limestone, sandstone, or schist[c]	Channery	Flaggy	Stony
	Slate	Slaty	Flaggy	Stony
	Shale	Shaly	Flaggy	Stony

[a]Bouldery is sometimes used in place of stony for fragments larger than 60 cm in diameter.
[b]Chert is a massive rock of precipitated silica (SiO_2) that fractures into irregularly shaped fragments.
[c]Schist is a coarse-textured, foliated metamorphic rock, usually high in mica minerals. It has a platelike cleavage.

Figure A2.
Profile of the Hezeltine stony silt loam. As may be determined from Table A2, stones cover from 0.01 to 0.1 percent of the surface of this soil and result in its being assigned to stoniness class 1. An example of a soil belonging to stoniness class 2 is shown in figure 4.1, page 61. (U.S.D.A. Soil Conservation Service photo by C. C. Dillon.)

Table A2. General relationships between mapping unit names and the degree of rockiness and stoniness. (Adapted from Soil Survey Manual, U.S.D.A. Agriculture Handbook 18, 1951. Superintendent of Documents, Government Printing Office, Washington, D.C.).

Mapping unit names	Approximate surface area covered by stones	Stoniness class	Mapping unit names	Approximate surface area covered by exposed rock[c]	Rockiness class
	%			%	
Unmodified textural name. Example: Gloucester loam	0 to 0.01	0	Unmodified textural name. Example: Hagerstown loam	0 to 2	0
Stony phase. Example: Gloucester stony loam[a]	0.01 to 0.1	1	Rocky phase. Example: Hagerstown rocky loam[c]	2 to 10	1
Very stony phase. Example: Gloucester very stony loam	0.1 to 3	2	Very rocky phase. Example: Hagerstown very rocky loam	10 to 25	2
Stony land	3 to 90[b]	3 & 4	Rock land	25 to 90[d]	3 & 4
Rubble land	90 to 100	5	Rock outcrop	90 to 100	5

[a]This is a simplified means of indicating a phase name.
[b]May be split into two ranges: 3–15%, identified as class 3 stoniness and as an extremely stony phase, and 15–90%, identified as class 4 stoniness and by the miscellaneous land type name of stony land.
[c]Includes areas too shallow over rock for useful plant growth.
[d]May be split into two ranges: 25–50%, identified as class 3 rockiness and as an extremely rocky phase; and 50–90%, identified as class 4 rockiness and by the miscellaneous land type name of rock land.

The degree of rockiness or stoniness may also be indicated by means of rockiness or stoniness classes. Although these classes are defined in terms of specific ranges in the content of rocks or stones, they have been established to show the extent to which stony or rocky conditions limit the use of agricultural machinery in the production of crop plants. There are six such classes, and their potential agricultural use, as relates to limitations placed on tillage and harvest equipment, is as follows:

Rockiness or stoniness class	Potential use
0	Cultivated crops without equipment limitations
1	Cultivated crops, but with some limitations to the use of tillage equipment
2	Hay crops requiring very limited or infrequent tillage, but frequent machine harvest
3	Improved pastures (fertilization, weed control)
4	Unimproved pastures
5	Unsuited for economic plant production

As may be noted in table A2, a rather low level of stoniness or rockiness results in mapped areas being named as miscellaneous land types rather than as soils. For example, where stones cover only 3 percent of the land surface, an area may be mapped as *stony land*. However, up to 25 percent of the surface may be covered by rock outcrops before soil names are supplanted by the miscellaneous land type name of *rock land*.

Because of the extreme variability associated with stoniness and rockiness in soils, accurate mapping of these conditions is often impractical. Frequently, small isolated stony or rocky areas are ignored in mapping, in which case they occur as inclusions in other mapping units. At other times, the specific location of these sites is indicated by special mapping symbols. Where a moderate to high proportion of the landscape is covered by rocky or stony areas that are too small to be shown individually on a soil map, mapping may require the combination of these areas with other soils or miscellaneous land types into complexes. Examples of this latter type of mapping unit are Dragoon very rocky complex and Cheney extremely rocky complex. In these two examples, the terms very rocky and extremely rocky indicate differences in the proportions of the complexes that are rock outcrops, and the names Dragoon and Cheney are for the soil series intermixed with the rocky areas to form the indicated complexes.

Appendix B

Soil drainage classes and flooding incidence

Most all soils uses in the field are at least indirectly influenced by soil water relationships as characterized by the soil drainage condition. The use of soils in low lying positions may also be influenced by flooding. Because of this, the drainage condition and potential for flooding are evaluated during soil survey, and the findings are included in soil survey reports. In the reports, the drainage condition is indicated by assigning a soil to a defined drainage class. Normally, little is said about flooding other than to indicate whether or not it occurs. Flooding is limited primarily to soils located on flood plains.

Soil drainage classes

Soil drainage classes are defined on the basis of the duration or potential duration of excessive wetness. The duration of excessive wetness depends on the balance between the rate at which water is taken in by a soil and the rate at which it is lost by internal drainage, plant use, and evaporation. Thus, the drainage condition depends in part on such things as rainfall, irrigation, and seepage. It is also a function of surface slope, which determines water loss by runoff, of profile characteristics that determine infiltration and percolation rates, and of topographic position that determines the rate and direction of internal water flow within the soil. Regardless of the specific cause of soil wetness, the longer its duration the poorer the drainage condition of the soil.

As opposed to soils that are inadequately drained, some soils lose water so rapidly by surface runoff or internal drainage that they are considered to be excessively drained. These soils tend either to occur on steep slopes where runoff is rapid or they may have a coarse texture that promotes rapid percolation of water through the profile. Excessively drained soils lie well above a water table. A principal problem with excessively drained soils is the difficulty in maintaining an adequate supply of available water for plant use.

Drainage classes of soils are as follows:

Very poorly drained—A water table at or near the surface most of the time.

Poorly drained—A water table at or near the surface during a significant part of the year, but even when the water table is at its lowest level, the soil above remains wet most of the time.

Imperfectly (somewhat poorly) drained—The soil remains wet for significant periods but not all the time. Periods of wetness may or may not be associated with a high water table.

Moderately well-drained—The profile remains wet for a small but significant part of the time. Wetness is not associated with a water table.

Well-drained—Water is removed readily but not rapidly. Well-drained soils tend to have a medium texture with a capacity for retaining optimum amounts of water for plant growth following a soaking rain or irrigation.

Somewhat excessively drained—Water is removed rapidly either by surface or internal flow, alone or in combination. Unless the supply is replenished frequently, water stored in these soils may be too limited to support optimum plant growth.

Excessively drained—Water loss is very rapid. Failure to retain water places severe limitations on plant growth unless the soil is watered frequently.

Flooding

Flooding strongly influences soil use in flood plain and delta positions. Frequent flooding may require that land be left in a wild state, although it may sometimes be subjected to limited, occasional agricultural use, such as grazing. With a decrease in the incidence of flooding, more intense agricultural land use is possible. However, the development of urban or industrial settlements on land subject to any flooding is not wise. Such developments may be severely damaged or destroyed by flood waters, or they may so obstruct and divert water flow at floodtime as to greatly increase the overall damage caused by floods.

Descriptions of soils, as presented in soil survey reports, include statements on the potential for flooding where this potential exists. Often, these statements are brief and in themselves may not fully indicate the severity or frequency of flooding. Thus, final judgments of potential flood problems often have to be made from historical records or other sources of information, including on-site inspection of the land under consideration.

Appendix C

The international system of units

The International System of Units (SI units) was adopted in 1960 by the Eleventh General Conference of Weights and Measures and provides a worldwide standard for expressing units of measure in publications. Use of this system is currently being recommended and will eventually be required in papers submitted to journals published by the American Society of Agronomy, including the Soil Science Society of America Journal. Reference is made to this system at various places in the preceding text material, but primarily only as an introduction to it. As this system becomes more firmly entrenched, total conversion to it will undoubtedly become necessary.

Two sets of units have been established in the new system. One set, termed *base units* (see table C1), is for fundamental units: distance, mass, time, temperature, electrical current, and quantities of chemical substances or of light. The others are called *derived units*, and they are named for the basic units from which they are derived, as shown in table C2, or they are given special names, as listed in table C3. The right-hand columns in tables C2 and C3 give mathematical definitions of derived units in terms of the base units.

Many scientific measurements result in very large whole numbers or in decimals that are very long and unwieldy. These numbers may be simplified through use of a prefix to the symbol for a unit of measure, as, for example, the use of kilo- with grams in kilogram. A list of these prefixes and their symbols are given in table C4. In practice, a prefix will normally be selected so that the numerical value of a unit falls between 0.1 and 1000.

Table C1. Base SI units.

Quantity	Unit	Symbol
Length	meter	m
Mass	kilogram	kg
Time	second	s
Electric current	ampere	A
Thermodynamic temperature	kelvin	K
Amount of substance	mole	mol
Luminous intensity	candela	cd

(Adapted from Agronomy News, March–April 1982, pp. 10–14 by permission of the American Society of Agronomy, Madison, Wisc.)

Table C2. Examples of SI derived units expressed in terms of base units.

Quantity	SI unit	
	Name	Symbol
Area	square meter	m^2
Volume	cubic meter	m^3
Velocity, speed	meter per second	m/s
Acceleration	meter per second squared	m/s^2
Density	kilogram per cubic meter	kg/m^3
Concentration	mole per cubic meter	mol/m^3
Luminance	candela per square meter	cd/m^2

(Adapted from Agronomy News, March–April, 1982, pages 10–14 by permission of the American Society of Agronomy, Madison, Wisc.)

Table C3. Examples of SI derived units expressed by special names.

Quantity	SI unit			
	Name	Symbol	Expression in terms of other units	Expression in terms of SI base units
Frequency	hertz	Hz		1/S
Force	newton	N		$m \cdot kg/s^2$
Pressure, stress	pascal	Pa	N/m^2	$kg/m \cdot s^2$
Energy, work, quantity of heat	joule	J	$N \cdot m$	$m^2 \cdot kg/s^2$
Power, radiant flux	watt	W	J/s	$m^2 \cdot kg/s^3$
Electrical potential	volt	V	W/A	$m^2 \cdot kg/s^3 \cdot A$
Electrical resistance	ohm	Ω	V/A	$m^2 \cdot kg/s^3 \cdot A^2$
Electrical conductance	siemans	S	A/V	$s^3 \cdot A^2/m^2 \cdot kg$
Celsius temperature	degree Celsius	°C		°K
Surface tension	newton per meter	N/m		kg/s^2
Heat capacity	joule per kelvin	J/°K		$m^2 \cdot kg/s^2 \cdot °K$

(Adapted from Agronomy News, March–April, 1982, pages 10–14 by permission of the American Society of Agronomy, Madison, Wisc.)

Table C4. SI Prefixes.

Multiplication factor		Prefix	Symbol
1 000 000 000 000 000 000	$= 10^{18}$	exa	E
1 000 000 000 000 000	$= 10^{15}$	peta	P
1 000 000 000 000	$= 10^{12}$	tera	T
1 000 000 000	$= 10^{9}$	giga	G
1 000 000	$= 10^{6}$	mega	M
1 000	$= 10^{3}$	kilo	k
100	$= 10^{2}$	hecto	h
10	$= 10^{1}$	deka	da
0.1	$= 10^{-1}$	deci	d
0.01	$= 10^{-2}$	centi	c
0.001	$= 10^{-3}$	milli	m
0.000 001	$= 10^{-6}$	micro	μ
0.000 000 001	$= 10^{-9}$	nano	n
0.000 000 000 001	$= 10^{-12}$	pico	p
0.000 000 000 000 001	$= 10^{-15}$	femto	t
0.000 000 000 000 000 001	$= 10^{-18}$	atto	a

(Adapted from Agronomy News, March–Arpil, 1982, pages 10–14 by permission of the American Society of Agronomy, Madison, Wisc.)

Appendix D

The Canadian system of soil classification

The discussion of this system of soil classification is in two parts: (1) definition and symbolization of soil horizons, and (2) descriptions of the major classes in the genetic system. The classification of horizons is dealt with first, since the nature of soil horizons in a profile determines the class of a soil in the genetic system.

Canadian system of soil horizon classification

Soil horizons in the Canadian system of soil classification are identified by letter symbols (capital or capital plus lower-case). Each symbol designates a specific kind of property or combination of properties. One set of symbols is for mineral horizons (<17 percent organic carbon by weight) and another for organic horizons (>17 percent organic carbon by weight), with certain of these commonly being used together in describing a profile. The term layer rather than horizon is sometimes used. This applies to bedrock, to soil zones saturated with groundwater, or to mineral layers within or beneath organic soil profiles.

Mineral horizons and layers

Five different kinds of mineral horizons or layers are recognized. These are the A, B, and C horizons, which occur in succession down through the profile, and the R (rock) and W (water) layers. General characteristics of these horizons are as follows:

A — A horizon at or near the surface of the mineral soil showing maximum effects of organic matter accumulation, eluviation, and leaching.

B — A mineral horizon characterized (1) as a zone of enrichment by illuvial clay, organic matter, or sesquioxides, (2) by the development of soil structure, or (3) by a change in color due to hydrolysis, reduction, or oxidation.

C   A horizon generally unaffected by processes responsible for A and B horizon formation. May show effects of gleying or of lime or salt accumulation.

R   Consolidated bedrock.

W   Water layer in waterlogged mineral or organic soils (called hydric in the latter).

Lower case letters are used as suffixes to the capital-letter symbols to denote specific kinds of properties, conditions, or components of the master horizons listed above, or of their subdivisions. Abbreviated definitions of these symbols follow:

b   Buried horizon.

c   Irreversible cementation.

ca   Horizon at least 10 cm thick caused by lime accumulation. The amount of lime should exceed that of the parent material by 5 percent or more depending on the lime content of the parent material.

cc   Cemented concretions, typically in pellet form.

e   Eluvial horizon. Used only with A, as in Ae.

f   Enriched with illuvial iron and aluminum, as determined by a chemical extraction procedure in the laboratory; level of extracted metals must exceed a specified limit.

g   Gleying, as evident in grayish to bluish colors, mottling, or both, produced by chemical reducing conditions in mineral soils.

h   A surface (Ah) horizon enriched by organic matter as a result of accumulation in place or biological mixing, or a subsurface (Bh) horizon enriched by humus through eluviation-illuviation.

j   Modifier used to denote limited change due to processes that would otherwise produce properties specified by the symbols e, f, g, n, and t. For example, a Bfj horizon contains extractable, illuviated iron and aluminum, but too little to produce a Bf horizon.

k   Carbonate in an amount that causes effervescence with dilute hydrochloric acid, but too little for a ca horizon. May be an accumulation caused by translocation from above or residual lime from a calcareous parent material.

m   Used with B to denote too little change to produce a Bt or a Podzolic B horizon (see table D1).

n   Accumulation of exchangeable sodium (ratio of exchangeable calcium to exchangeable sodium should be 10 or less).

p   An A or O horizon, as in Ap or Op, disturbed by cultivation, logging, or habitation.

s   Presence of soluble salts, including gypsum, visible in crystals.

sa   Horizon at least 10 cm thick with sufficient soluble salts to yield a saturation-paste extract with a conductivity of 4 mmhos/cm (4 dS/m) or more.

t   Used with B, as in Bt, to denote a horizon enriched in illuvial, silicate clay. The clay content of a Bt must exceed that in an overlying Ae horizon by from 3 to 20 percent depending on the clay content of the Ae horizon.

u   Horizon disturbed by mixing other than cryoturbation (see y).

x   Fragipan layer.

y   A horizon affected by cryoturbation (frost action).

z   Frozen layer.

Organic horizons   Organic horizons occur in organic soils and as surface or buried horizons in mineral soils. Two types exist: those forming in swampy or boggy environments from mosses, sedges, and similar plants (O horizons), and those accumulating as surface litter from woody plants (L, F, and H horizons). There are three types of O horizons depending on the state of decomposition of the organic materials.

Of   Horizon of fibrous (fibric) materials that are readily identifiable as to botanical origin.

Om   Horizon of mesic material at an intermediate stage of decomposition.

Oh   Horizon consisting of humic material in an advanced state of decomposition.

L   Largely undecomposed leaf, needle, or woody tissue.

F   Partially decomposed leaf, needle, or woody tissue, with some of the original material difficult to recognize.

H   Well decomposed leaf, needle, or woody material in which the original materials cannot be recognized.

Categories of the canadian soil classification system

The system of soil classification used in Canada is a genetic, or taxonomic, system based on profile properties that relate to the conditions of soil formation. In it, soil individuals (polypedons) are organized into classes at five levels of generalization, the combined classes at each level being referred to as a category of classification. Listed in an order of decreasing generalization, the categories are:

Order

Great group

Subgroup

Family

Series

As suggested by the previous listing, orders are the most generalized classes. Soil individuals making up an order need be alike in only one, or at most, very few properties. Below the orders, classes become progressively more rigidly defined (defined by a greater number of profile properties), with definitions of the series being the most specific. Since soil series are combined into classes of the higher categories, they are viewed as the basic units of the classification system.[1]

Due to space limitations, only the orders and great groups of the Canadian system can be discussed, and then only in the briefest of terms.[2] Classes within these two categories, and in subgroups and families as well, are defined by specific kinds or sets of *diagnostic properties*. Usually, though not always, these are combinations of properties that characterize unique types of *diagnostic horizons*. Diagnostic horizons for soil orders in the Canadian classification system are listed in table D1. Most of these horizons can be assigned standard alphabetical symbols as described above. However, as noted in the table, some must meet more specific requirements than is indicated by the alphabetical symbol alone.

Table D1. Named diagnostic horizons of soil orders in the Canadian Soil classification system. (Adapted from Canada Department of Agriculture Publication 1646).

Horizon	Description
Chernozemic A	An Ah horizon at least 10 cm thick and meeting specific color requirements relating to organic matter content. Organic carbon content between 1 and 17 percent; percent base saturation is 80 or greater, with calcium as the dominant exchangeable cation; horizon should be neither massive nor hard when dry.
Solonetzic B	A Bn or Bnt horizon with columnar or prismatic structure that breaks in a blocky structure. Peds very hard when dry.
Illuvial B[a]	A Bt horizon at least 5 cm thick.
Podzolic B	Bh, Bhf, or Bf horizons, alone or in combination, exceeding 10 cm in total thickness; texture is coarser than clay.
Organic horizon	An O, L, F, or H horizon.

[a]From Robinson, J. A. (Ed.), Laboratory Manual for Introduction to Soil Science. Edmonton: University of Alberta, 1980.

Variation in properties of the diagnostic horizons, or in other properties of the profile, provide the basis for subdividing orders into great groups of soils, as is indicated in the discussions of the orders, which follow. The first two orders considered are of immature soils that have been only slightly to moderately changed by soil-forming processes. This is followed by discussions of the mature-soil orders, starting with those of drier climates. The last three orders considered occur under unusual conditions that impart unique properties of great significance to soil behavior and use. They include soils with severe natural drainage problems, soils with permafrost, and organic soils.

[1]For a discussion of soil series, see page 287.

[2]For greater detail see Canada Department of Agriculture Publication 1646 and the two-volume Soils of Canada cited on page 570.

Table D2. Names and definitive properties of the nine orders in the Canadian system of soil classification. Numbers in parenthesis are estimated percentages of land area in Canada in which the named soil is the dominant order (see Bentley, C. F., in Selected References).

Order	Principal Properties
Regosol (1.3%)	Soils of limited development. Lack solonetzic, illuvial, or podzolic B horizons, evidence of strong gleying, or permafrost within 1 m of the surface, or 2 m if cryoturbated. May have L, F, and H or O horizons, or an Ah horizon if less than 10 cm thick.
Brunisol (8.8%)	Soils of intermediate development. Typically, have brownish Bm horizons. May have L, F, and H horizons, Ah, Ae, and weakly developed B horizons, such as Btj, Bfj, or Bf horizons less than 10 cm thick. May not have permafrost within 1 m of the soil surface, or 2 m if cryoturbated, or evidence of strong gleying.
Chernozem (5.1%)	Soils of the cool, subarid to subhumid grasslands with a chernozemic A horizon, but lacking solonetzic or podzolic B horizons, permafrost within 2 m of the surface, or evidence of strong gleying. Mean annual soil temperature must be 0°C or above.
Solonetz (0.8%)	Soils of drier regions with a solonetzic B horizon but lacking evidence of strong gleying.
Luvisol (10.3%)	Soils of deciduous or mixed deciduous-coniferous forests with an Ae and an illuvial B (Bt) horizon. They may not have a solonetzic B horizon, permafrost within 1 m of the surface, or 2 m if cryoturbated, or evidence of strong gleying. May have a chernozemic A horizon if an Ae horizon is also present.
Podzol (22.6%)	Soils of coniferous forests formed in coarse parent materials and containing a podzolic B horizon. Evidence of strong gleying is permitted.
Gleysol (1.9%)	Soils, typically on flat lowlands or in depressional areas, showing signs of strong gleying. May have properties diagnostic of other orders, except for a podzolic B horizon, or permafrost within 1 m of the soil surface, or 2 m if cryoturbated.
Cryosol (45.0%)	Soils of very cold regions with permafrost within 1 m of the soil surface, or 2 m if cryoturbated.
Organic soils (4.2%)	Soils consisting of organic (O) horizons exceeding 40 cm in thickness if well decomposed, or 60 cm if incompletely decomposed, but lacking permafrost within 1 m of the soil surface.

Major properties of the nine orders in the Canadian system of soil classification are summarized in table D2. For cross-reference, soil classes in the U.S. system of soil taxonomy (see Chapter 12) that correspond approximately to the Canadian soil orders are listed in table D3. In addition, tables D4 and D5 show soil moisture and soil temperature regimes important to soil classification based on the Canadian system.

The general distribution of soil orders in Canada is shown in figure D1. Basic subdivisions, as indicated by solid lines on the map, are of the major physiographic regions listed in the map legend. Soils tend to differ among these regions because each has its own somewhat specific set of topographic, geologic, climatic, and drainage conditions. Soil variation is most closely cor-

Table D3. Taxonomic correlation between orders in the Canadian system of soil classification and soil classes in the U.S. system of soil taxonomy. (From Canada Department of Agriculture Publication 1646).

Canadian	United States
Regosol	Entisols
Brunisol	Inceptisols, some Psamments
Chernozem	Borolls
Solonetz	Natric great groups of Mollisols and Alfisols
Luvisol	Boralfs and Udalfs
Podzol	Spodosol, some Inceptisols
Gleysol	Various Aquic suborders
Cryosol	Pergelic subgroups
Organic soils	Histosols

Table D4. Soil moisture classes in the Canadian system of soil classification. (From Canada Department of Agriculture Publication 1646).

Moisture regime	Definition
Aqueous regime	Free water standing continuously on the soil surface.
Aquic Regime	
Peraquic	Soil saturated for very long periods.
Aquic	Soil saturated for moderately long periods.
Subaquic	Soil saturated for short periods.
Moist unsaturated regime	
Perhumid	No significant water deficits in the growing season.
Humid	Very slight deficits in the growing season.
Subhumid	Significant deficits in the growing season.
Semiarid	Moderately severe deficits in the growing season.
Subarid	Severe deficits in the growing season.
Arid	Very severe deficits in the growing season.

Table D5. Brief definitions of soil temperature classes in the Canadian system of soil classification. (From Canada Department of Agriculture Publication 1649).

Class	Mean annual soil temperature, °C	Degree days > 5° C[a]
Extremely cold	<-7	< 15
Very cold	-7 - 2	< 550
Cold	2 - 8	550–1250
Cool	5 - 8	1250–1700
Mild	8 - 15	1700–2800

[a]Degree day $> 5° C = \dfrac{T_{max} - T_{min}}{2} - 5$, where T_{max} and T_{min} are the daily maximum and minimum temperatures. This value multiplied by the number of days per year having a temperature above 5° C gives the values in the third column above.

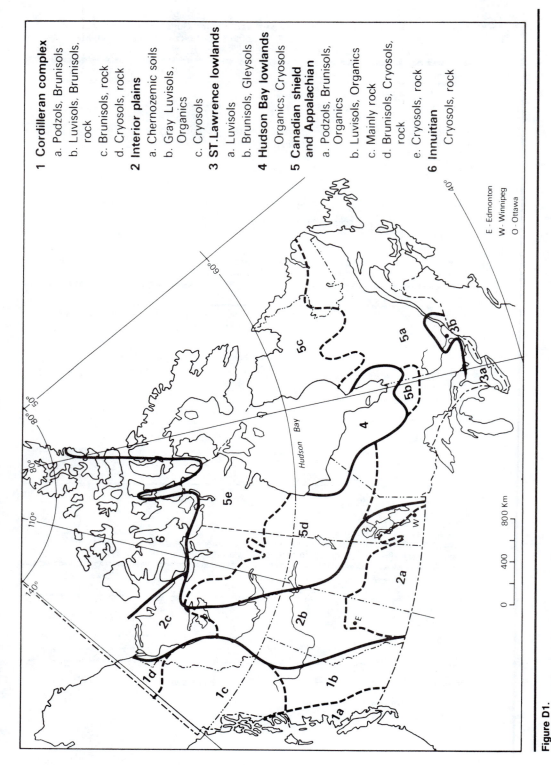

1 **Cordilleran complex**
  a. Podzols, Brunisols
  b. Luvisols, Brunisols, rock
  c. Brunisols, rock
  d. Cryosols, rock
2 **Interior plains**
  a. Chernozemic soils
  b. Gray Luvisols, Organics
  c. Cryosols
3 **ST.Lawrence lowlands**
  a. Luvisols
  b. Brunisols, Gleysols
4 **Hudson Bay lowlands**
  Organics, Cryosols
5 **Canadian shield and Appalachian**
  a. Podzols, Brunisols, Organics
  b. Luvisols, Organics
  c. Mainly rock
  d. Brunisols, Cryosols, rock
  e. Cryosols, rock
6 **Innuitian**
  Cryosols, rock

E - Edmonton
W - Winnipeg
O - Ottawa

0   400   800 Km

Figure D1.
Principal occurrences of soil orders in relation to the major physiographic regions of Canada. (Map source: W. W. Pettapiece, Land Resource Research Institute, Agriculture Canada, Ottawa.)

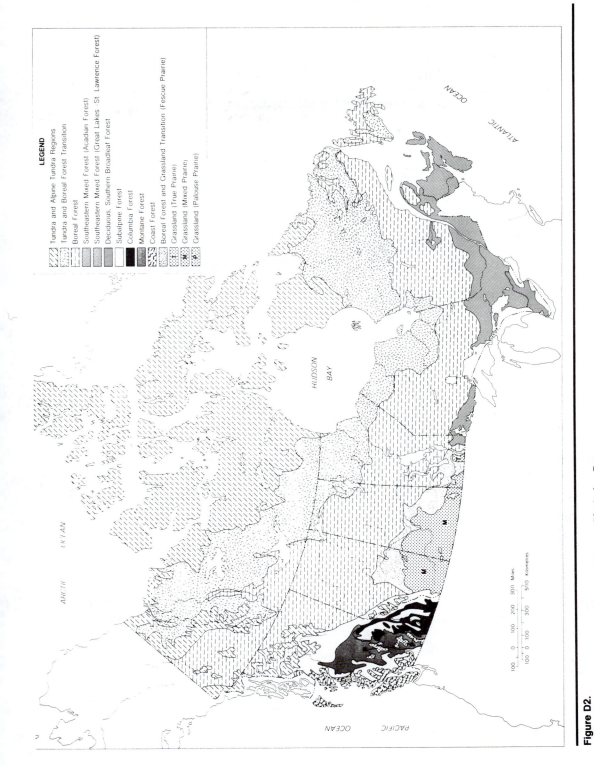

Figure D2.
Major vegetational zones of Canada. (Map modified after Rowe,
Canadian Forestry Service. From *Soils of Canada*. Ottawa: Supply
and Services Canada, 1977.)

LEGEND

Tundra and Alpine Tundra Regions
Tundra and Boreal Forest Transition
Boreal Forest
Southeastern Mixed Forest (Acadian Forest)
Southeastern Mixed Forest (Great Lakes - St Lawrence Forest)
Deciduous, Southern Broadleaf Forest
Subalpine Forest
Columbia Forest
Montane Forest
Coast Forest
Boreal Forest and Grassland Transition (Fescue Prairie)
Grassland (True Prairie)
Grassland (Mixed Prairie)
Grassland (Palouse Prairie)

related with the climatic-vegetational factors of soil formation, however. This type of variation can be judged in a general way from the vegetation map shown in figure D2.

Regosolic Order

Regosolic (Gr. *rhegos*, blanket)[3] soils show little, if any, effect of normal developmental processes (see fig. D3). This may be due to (1) a lack of time for horizon formation, (2) climatic or parent material conditions that do not promote rapid change in soil formation, or (3) erosion that removes the soil essentially as rapidly as it forms. Typical regosols are thin residual soils, soils on young parent materials, such as alluvium, sandy beaches, or shifting sand dunes, soils on steep, erosive slopes, or soils in cold, dry regions where a lack of both heat and water prevents much change in parent material over time. Regosols are widespread and occur under all climatic conditions. An exception is where the presence of permafrost results in the placement of a soil in the Cryosol order.

Regosols can have an Ah but not a B horizon. If an Ah horizon is absent, then only a C horizon is recognized. There are two great groups of Regosols, and they are distinguished from each other depending on whether an Ah horizon (not Chernozemic) is present:

Regosol	An Ah horizon is lacking, or if present, is less than 5 cm thick. The profile commonly is totally C horizon.
Humic Regosol	Has an Ah horizon > 5 cm thick. The horizon sequence is Ah-C.

Some Regosols contain buried horizons caused by the deposition of new parent material over a pre-existing soil.

Brunisolic order

Brunisols (Fr. *brun*, brown) are soils exhibiting an intermediate degree of development. Most Brunisols contain L, F, or H horizons, alone or in combination, and Ah horizons, which can form quickly. All contain weakly developed B horizons (Bm, Btj, Bfj), consistent with their intermediate stage of maturity. The central concept of this order is a soil formed under forest conditions with a brownish-colored Bm horizon resulting from faint iron staining of other mineral grains. Some may contain Ae horizons where rapid weathering, leaching, and eluviation remove finer particles from the surface soil. Illuviated iron and aluminum may accumulate in the subsoil to form Bfj or Bf horizons. However, the Bf must be less than 10 cm thick, otherwise it would be classed as a Podzolic B, which is diagnostic of the Podzolic order. Some Brunisols also show evidence of gleying, but less than that required of the Gleysolic order.

[3]The origin or significance of soil order and great group names in the Canadian classification system is indicated parenthetically throughout these discussions. For example, the root of Regosol suggests a thin soil over bedrock. Some of the names have their origin in the Russian system of soil classification. Where this is the case, basic properties associated with the names are the same in the Canadian and Russian systems.

Figure D3.
A weakly developed Regosol forming in deep windblown sand in Southeastern British Columbia.

As may be judged from figure D1, Brunisols occur under a wide range of climatic conditions, although they tend to be concentrated primarily in the wetter regions. The principal associates of Brunisols are Podzols, Luvisols, and to a lesser extent, Cryosols, and Gleysols.

Four great groups of Brunisols are recognized, and they are distinguished on the basis of the presence or absence of an Ah horizon and on the degree of base saturation, as judged from pH:

Melanic Brunisol (Gr., *melas*, black) Thick Ah horizon with a high base status (pH > 5.5).

Eutric Brunisol (Gr., *eu trophein*, well nourished) No or thin Ah horizon; high base status (pH > 5.5).

Sombric Brunisol (Fr., *sombre*, dark) Thick Ah horizon; low base status (pH < 5.5).

Dystric Brunisol (Gr., *dys trophe*, difficult nourishment). No or thin Ah; low base status (pH < 5.5).

An Ah horizon in Brunisols is usually produced by biological mixing of surface litter with the mineral soil. This horizon, along with L, F, or H horizons, provides a source of organically stored nutrients. The presence of both an Ah horizon and a high base status more nearly assures a soil with a reasonably good level of fertility.

Figure D4.
A Solodized Solonetz, from Saskatchewan, with the Ah horizon partially removed to expose a whitish Ae horizon that overlies a Bnt (solonetzic B) horizon. The latter horizon has a strongly developed columnar structure. (From *The Canadian System of Soil Classification,* Canada Department of Agriculture Publication 1646, 1977. Photo by John Day. Reproduced by permission of the Minister of Supply and Services Canada.)

Solonetzic order   Soils of the Solonetzic (Russ., *solonetz*) order are distinguished from other orders by the presence of a solonetzic B horizon high in exchangeable sodium, or sodium and magnesium (see fig. D4). These soils occur generally in association with Chernozems, but in locations where restricted drainage and limited leaching allow salts to accumulate. The salts may come from the parent materal, or they may build up from drainage water that flows into low-lying areas and is then lost by evaporation. Leaching of Solonetzic soils is usually sufficient to remove salts from the A and B but not from the C horizon. It also results in an A horizon with a neutral to slightly acid pH.

Illuvial clay normally accumulates with exchangeable sodium in the diagnostic solonetzic B horizon. Whereas the presence of exchangeable sodium enhances the translocation of clay in the profile, it also reduces the stability of peds in the B horizon when the soil is wet. Sufficient organic matter accumulates in the surface soil to produce an Ah horizon. Eluviation-illuviation of the organic matter may also take place, which results in the formation of the Ahe or Ae horizons and the deposition of illuviated humus as thin coatings on ped surfaces in the B horizon.

There are three great groups of Solonetzic soils, and they differ in the degree of profile development caused by leaching and eluviation-illuviation:

Solonetz	Very little leaching. Distinct Ah and Bn or Bnt horizons. If present, an Ae must be less than 2 cm thick.
Solodized Solonetz	Leaching and eluviation sufficient to produce an Ae horizon 2 cm thick or more between the Ah and Bn or Bnt horizons.
Solod	Advanced stage of leaching. Salt and sodium removal causes degradation of the upper part of the B and its conversion to transitional AB and BA horizons between the Ae and B horizons. The Bn or Bnt horizons may have converted to a Bt due to the loss of sodium.

In their natural state, Solonetzic soils usually support only salt-tolerant grasses and forbes. The main limitation to crop plants is the B horizon, which greatly hinders both root and water penetration. Exposure of the B horizon through loss of topsoil can result in a soil of extremely poor physical state and high erodibility.

Chernozemic order

Chernozems (Russ., *chernozem*) are soils typical of steppe to prairie grassland regions; they are the principal soils of the great wheat-growing region in southern Alberta, Saskatchewan, and Manitoba (see fig. D1). Some may form in association with shrubs and forbes, or even where trees are present, as in the forest-grassland transition. However, a moisture deficiency in the summer of most years greatly limits the potential for tree growth. In general, the climate for Chernozemic soil development is subarid to subhumid.

The predominant process of Chernozemic soil formation is the accumulation of organic matter from grass tops and roots, which produces a diagnostic chernozemic Ah horizon that is usually well granulated and friable (see fig. D5). Silicate mineral weathering is limited, as is the downward translocation of clay. Lime, if present in the parent material, is easily moved out of the surface horizon, commonly to accumulate as Bk, Cca, or Ck subsurface horizons. In the wettest Chernozems, however, carbonates are removed completely from the profile, and eluviation-illuviation of clay may produce an Ae and a moderately to well developed B (Bm, Btj, Bt) horizon. In drier Chernozems, exchangeable sodium may encourage the translocation of clay to form a Btj or Btjnj horizon.

There are four great groups of Chrenozemic soils, and they are distinguished on the basis of color of the diagnostic Ah horizon. The great groups are Brown, Dark Brown, Black, and Dark Gray Chernozems. This listing is in an order of increasing availability of moisture for soil development. Thus, in the drier regions, limited plant growth and residue return results in only a slight darkening of the surface horizon that is rated as brown according to the Munsell color system. With increasing moisture availability, the Ah ho-

Figure D5.
A Black Chernozem soil from south-central Alberta showing a well developed Ah horizon, about 30 cm thick, underlain by a weakly developed Bm horizon.

rizon tends to become darker until, in the wettest areas, eluviation of organic matter and clay results in the formation of Ahe and Ae horizons having a dark gray color.

Luvisolic order

Luvisols are soils in which eluviation-illuviation has produced a well developed diagnostic Bt horizon. The parent materials are characteristically medium-textured or finer, which provides clay for downward translocation in the profile. Light-colored Ae horizons are also typical of Luvisols. Lower horizons in the profile tend to have a high base status and are often calcareous. This is due to various factors, namely: base cycling by the vegetation, the ability of the clay to buffer against acidification, and the limitation placed on leaching by slow permeability of the Bt horizon.

Luvisols occur principally under deciduous or mixed deciduous-coniferous forests, where climatic conditions are somewhat wetter and cooler than for Chernozemic soil formation, and in parent materials that are finer than for Podzols. As may be judged from figure D1, soils of this order are widespread in Canada, extending from Newfoundland to British Columbia and from the southern international border to the more northern permafrost zone.

There are two great groups of Luvisols. Both have Ae and Bt horizons in their natural state, but one tends to be warmer than the other, which allows for greater biological acitivity and soil mixing. The formation of Ah horizons in Luvisols is due to biological mixing, this type of horizon being referred to as forest *mull*. The extent of mixing determines the color of the surface soil and is the basis for naming the two great groups in the order, which are:

Gray Brown Luvisols	Contain a forest mull Ah horizon; mean annual temperature > 8°C.
Gray Luvisols	May or may not contain a forest mull Ah horizon; mean annual temperature usually < 8°C.

Podzolic order

Podzols (Russ., *Podzol*) are soils that have formed under cool to cold, moist climatic conditions in coarse, highly siliceous, normally noncalcareous parent materials of glacial, glacial-fluvial, or marine origin. Typical vegetation is coniferous forest, sometimes heath, which cycles bases sparingly and therefore tends to produce a strongly acid surface litter. These conditions promote the rapid downward flow of acidic leaching water and encourage the extensive breakdown of nonquartz minerals, particularly within the surface layer of the mineral soil. Iron and aluminum released from weathering primary minerals, along with organic matter from L, F, and H horizons, are translocated and redeposited in a subsurface position to produce a podzolic B horizon.

In their undisturbed state, Podzols have striking profiles that are markedly different from the parent material (see fig. D6). A layer of surface litter (L, F, and H horizons) is common. This layer is underlain by a whitish or grayish Ae horizon, consisting largely of quartz sand as a residue from mineral weathering. A sharp, though often irregular, boundary separates the Ae from the underlying podzolic B, which may be either a Bh, a Bhf, or a Bf horizon, or combinations of these. The B horizon may be predominantly black, or it may have a reddish-brown color that becomes yellower with depth.

Major occurrences of Podzolic soils are in the coastal regions of British Columbia and across a broad belt extending form eastern Manitoba to Newfoundland in southeastern Canada. Soils of this order are assigned to three great groups according to variation in the humus and iron contents of the diagnostic B horizon:

Humic Podzol	The diagnostic horizon is a Bh in which properties due to illuvial humus dominate. Low in extractable iron.
Ferro-Humic Podzol	The diagnostic horizon is a Bhf that is high in both illuvial humus and extractable, illuvial iron and aluminum.
Humo-Ferric Podzol	The diagnostic horizon is a Bf high in extractable, illuvial iron and aluminum but low in illuviated humus.

Gleysolic order

Soils of the Gleysolic (Russ., *Glei*) order have features indicative of periodic or prolonged saturation with water and reducing conditions due to a lack of oxygen. They occur primarily under subhumid to humid climatic conditions.

Figure D6.
A Humo-Ferric Podzol from Nova Scotia. The thick whitish Ae horizon consists principally of quartz sand and has been produced by the loss of clay and other minerals through weathering and eluviation. The underlying Bf (podzolic B) horizon is bright red in color. (From *The Canadian System of Soil Classification,* Canada Department of Agriculture Publication 1646, 1977. Reproduced by permission of the Minister of Supply and Services Canada.)

In the drier areas of this climatic range, Gleysols tend to be located in shallow depressional areas or flatland saturated with water every spring. In wetter regions, they may occur on sloping surfaces. Some Gleysolic soils may be submerged under water throughout the year.

Soils of the order are recognized by profiles with light-colored, dull gray to bluish Bg and Cg horizons where wetness is prolonged, or by patchy, bright reddish-brown mottles where periodic improved drainage allows oxidation of reduced iron to hydrated ferric oxides. These soils more often than not have either O or L, F, and H horizons. They may also have well developed Ah and Ae horizons, the latter usually showing a gleyed appearance.

With but two exceptions, evidence of strong gleying requires a soil to be placed in the Gleysolic order. The exceptions are soils with a podzolic B horizon, which are classified as Podzols, and those with permafrost within 1 m of the soil surface, or 2 m if cryoturbated, which requires placement in the Cryosol order.

Gleysolic soils are placed into three great groups depending on the extent of development of Ah and Bt horizons:

Humic Gleysol	Has an Ah horizon at least 10 cm thick but no Bt horizon.
Gleysol	Has no Ah horizon, or if one is present, it is less than 10 cm thick. May have a Bg but not a Bt horizon.
Luvic Gleysol	Has a Btg horizon. May have an Ah, usually as Ahe, and Aeg horizons.

Cryosolic order

Cryosols (Gr., *Kryos*, icy cold) are soils of cold regions that have permafrost within 1 m of the surface, or if cryoturbated (disrupted by frost action), within 2 m of the soil surface. A wide range of profile properties are possible in Cryosols, but development is usually limited by low temperatures. Weakly developed Bm horizons may have formed and some gleying may have occurred just above the permafrost. Mineral Cryosols that have been cryoturbated show marked horizon disruption or deformation (see fig 3.2C, page 42).

Patterned ground, as may be seen in figure D7, is also a mark of cryoturbation in soils. Patterned ground results where the formation of ice lenses forces soil into a domelike hummock (see fig. D8) at frequent intervals over the landscape. If trees are present, they may be tilted outward from the center of the dome, or rocks may roll or be pushed down the sloping surface to accumulate as a continuous ring around the dome (see fig. D9).

More often than not, mineral Cryosols lack Ah horizons. Organic matter accumulation is an important process, nonetheless, although it is usually as slightly to partially decomposed surface litter or as deeper deposits in peat bogs. Organic matter accumulation is important because low temperatures are far more inhibitory on organic matter decomposition than on plant growth and residue return.

Cryosols occupy much of the northern third of Canada, often in association with, but not limited to, tundra vegetation. Three great groups are recognized:

Turbic Cryosols	Mineral soils with disrupted horizons or displaced soil materials due to cryoturbation.
Static Cryosols	Mineral soils lacking horizon disruption.
Organic Cryosols	Organic soils with no disruption of horizons.

Organic soils

Soils of this order consist of organic or organic-mineral deposits in which the organic fraction is a prime factor determining soil properties. These soils may form in swamps or bogs where a lack of oxygen and, sometimes, low temperatures keep the rate of organic decay below the rate of organic residue addition. Under these excessively wet conditions, the source of the residues is usually mosses, sedges, or other water-loving plants. One group of organic soils forms from leaf, needle, or other above-ground plant parts under conditions of good drainage. Whether these deposits result in an organic soil

Figure D7.
Patterned ground produced by the formation of hummocks in soil with permafrost. (U.S. Geological Survey photo by T. L. Pewe.)

Figure D8.
A partially eroded soil hummock produced by the expansion of ice formed from rising ground water. Note the man standing to the right of center in the photograph. (U.S. Geological Survey photo by T. L. Pewe.)

Figure D9.
A network of stone rings formed by rocks rolling downward to the perimeters of small, neighboring soil hummocks. (U.S. Geological Survey photo by O. J. Ferrains.)

depends on their depth, which should be 60 cm for relatively undecomposed (fibric) material, or 40 cm for material that is moderately (mesic) or well (humic) decomposed. To be classed as organic, a soil layer should contain 17 percent or more of organic carbon.

There are four great groups of Organic soils depending on the origin of the residue and the degree of decomposition. These are:

Fibrisols	Formed under saturated conditions; decomposition slight.
Mesosols	Formed under saturated conditions; decomposition moderate.
Humisols	Formed under saturated conditions; decomposition extensive.
Folisols	Formed from surface litter, typically on sloping, well drained landscapes.

Soil subgroups and families

Placement at the subgroup or family level provides more information about a soil than is provided at the great group level. This information is indicated by the addition of descriptive terms to the great group names. At the subgroup level the descriptive term is added as a prefix to the great group name, and its purpose is to state whether a soil is typical of the great group, or is a deviant toward some other order or has a property that is distinctly atypical of the great group.

Mineral soils typical of a great group are indicated by the prefix Orthic.[4] Thus, a soil typical of the Gray Brown Luvisolic great group would be assigned to the Orthic Gray Brown Luvisolic subgroup. However, if a Gray Brown Luvisol showed certain properties characteristic of Podzols, it would be a Podzolic Gray Brown Luvisol. Other examples of subgroup names include Fragic Humic Podzol, which is a Podzol with a fragipan, and Gleyed Solonetzic Brown Chernozem, a soil showing both gleying and sodium accumulation in the B horizon, but insufficient of either to be classed a Gleysol or a Solonetz.

Properties used for placement of a soil at the family level are indicated by adding descriptive terms to the subgroup name. In mineral soils, these terms show, in order, particle-size range, mineralogy, depth, if shallow, reaction range, calcareousness, and soil climate. Two examples are:

Orthic Humo-Ferric Podzol, coarse-loamy, mixed (minerals), acid, perhumid.

Orthic Eutric Brunisol, coarse-silty over sandy, mixed, shallow, strongly calcareous, cold humid.

A similar system has been developed for naming families of organic soils.

Selected references

Bentley, C. F. (Ed.) *Photographs and Descriptions of Some Canadian Soils.* University of Alberta Extension Series Publication B79–1. Ottawa K1P 5H4: Canadian Society of Soil Science, 1979.

Canada Soil Survey Committee, Subcommittee on Soil Classification. *The Canadian System of Soil Classification.* Canada Dept. of Agr. Pub. 1646. Ottawa: Supply and Services Canada, 1978.

Clayton, J. S. et al. *Soils of Canada.* Ottawa: Supply and Services Canada, 1977.

Foth, H. D. and Schafer, J. W. *Soil Geography and Land Use.* New York: John Wiley and Sons, 1980.

Tedrow, John C. F. *Soils of the Polar Landscape.* New Brunswick, NJ: Rutgers University Press, 1977.

Valentine, K. W. G. et al. (Eds.) *Soil Landscapes of British Columbia.* Victoria, B. C.: Resource Analysis Branch, Ministry of the Environment, 1978.

[4]For organic soils, the term Typic, rather than Orthic, is used.

Glossary

(Modified from the Glossary of Soil Science Terms, Madison, WI: Soil Science Society of America, 1975, and other sources. Soil classification terms are basically from the U. S. system of soil classification, but some are specific for the Canadian system discussed in the Appendix. The latter terms are identified by the symbol (C).)

a

A horizon 1. An horizon formed at or near the soil surface, but within the mineral soil, having properties that reflect the influence of accumulating organic matter. 2. (C) Same as above, but also may show effects of leaching and eluviation, alone or in combination with organic matter accumulation. See *Ae* and *Ah horizons*.

AB horizon Transition from A to an underlying B horizon in which properties of the A horizon predominate.

A/B horizon A mixed mineral horizon showing properties of both an A and a B horizon in which properties of the A predominate.

absorption The process by which one substance is taken into and included within another substance, as the absorption of water by soil or nutrients by plants.

acid soil Specifically, a soil with a pH below 7.0, but for most practical purposes, a soil with a pH below 6.6. The term is usually applied to the surface soil.

AC horizon Analogous to an AB horizon, except the transition is between an A and a C horizon in a profile lacking a B horizon.

acidity, active The activity of hydrogen ion in the aqueous phase of a soil. It is measured and expressed as a pH value.

acidity, potential The amount of hydrogen ion in a soil that can be rendered free or active in the soil solution by cation exchange.

acre An area of 0.405 ha, 4047 sq m, or 43,560 sq. ft.

adhesion The attraction or clinging together of unlike substances.

adsorption The increased concentration of molecules or ions at a surface, including exchangeable cations and anions on soil particles.

adsorption complex The group of substances in soil capable of adsorbing other materials.

Ae horizon (C) An eluvial horizon at or near the surface of the mineral soil.

aeration, soil The process by which air in the soil is replaced by air from the atmosphere. The rate of aeration depends largely on the size and continuity of empty pores within the soil.

aerobic (1) Having molecular oxygen as a part of the environment. (2) Growing or occurring only in the presence of molecular oxygen, as aerobic organisms.

aggregate, soil A group of soil particles cohering so as to behave mechanically as a unit.

aggregation The act or process of forming aggregates, or the state of being aggregated.

agric horizon A horizon formed by the accumulation of silt, clay, and humus moved from an overlying plow layer by percolating water.

agronomy A specialization of agriculture concerned with the theory and practice of field-crop production and soil management. The scientific management of land.

Ah horizon (C) An horizon at or near the surface of the mineral soil in which properties due to accumulating organic matter dominate.

air-dry The state of dryness of a soil at equilibrium with the moisture contained in the surrounding atmosphere.

albic horizon An eluviated surface or subsurface horizon, normally light, if not whitish, in color. Typically, an A2 horizon.

Alfisol An order of soils having a B2 horizon high in crystalline clay and a moderately high level of exchangeable bases. These soils usually occur in the climatic range associated with scrub to well developed deciduous forests.

alkali soil (Obsolete) See **sodic soil.**

alkaline soil See **basic soil.**

alluvium Sediment deposited on land from streams.

alteration A process of secondary mineral formation caused by the replacement of structural ions in a mineral but without a change in the general form of the structure.

amendment, soil A substance such as lime, gypsum, sawdust, etc., added to the soil for the purpose of making the soil more suitable for the production of plants.

ammonification The biochemical process whereby ammoniacal nitrogen is released from nitrogen-containing organic compounds.

amorphous Noncrystalline; lacking the well ordered arrangement of ions typical of crystalline substances.

anaerobic (1) The absence of molecular oxygen. (2) Growing in the absence of molecular oxygen (such as anaerobic bacteria).

angstrom unit One ten-thousandth of a micron, or one ten-millionth of a millimeter.

anion See **ions.**

antagonism In plant nutrition, the interference of one element with the absorption or utilization of an essential nutrient by plants.

anthropic epipedon Similar to a mollic epipedon but with a high level of extractable phosphorus due to heavy fertilization.

antibiotic A substance produced by one species of organism that, in low concentrations, will kill or inhibit growth of certain other organisms.

aquic A soil moisture regime typical of low, wet places, where the soil is saturated and subjected to reducing conditions at least part of the time.

arable Suited for the production of cultivated crops.

argillic horizon Essentially, a subsurface horizon formed by the illuviation of crystalline clay.

aridic A soil moisture regime that limits plant growth during much of the growing season due to a lack of water.

Aridisol An order of soils at apparent dynamic equilibrium with the climate of dry regions. They show limited profile development because of a low climatic intensity, the A horizon containing less than 1 percent organic matter and a high level of exchangeable bases.

ash, plant The inorganic residue, principally oxides, remaining after the ignition of plant tissue.

autotrophic Capable of utilizing carbon dioxide or carbonates as the sole source of carbon and obtaining energy for the reduction of carbon dioxide, and for other life processes, from the oxidation of inorganic elements or compounds such as iron, sulfur, hydrogen, ammonium, or from radiant energy. Contrast with **heterotrophic.**

available nutrient (1) A nutrient in the soil that can be readily absorbed and assimilated by growing plants. (2) A nutrient in soluble or exchangeable form.

available water In general terms, that portion of soil water that can be readily absorbed by plant roots, but as a specific soil moisture value, the mathematical difference in the amounts of water a soil holds at the field capacity and the permanent wilting point.

b

B horizon An horizon immediately beneath an A (Ah) or E (Ae) horizon characterized by a higher colloid (clay or humus) content, or by a darker or brighter color than the soil immediately above or below, the color usually being associated with the colloidal materials. The colloids may be of illuvial origin, as clay or humus, they may have been formed in place (clays, including sesquioxides), or they may have been derived from a texturally layered parent material.

B/A horizon A mixed horizon showing properties of both an A and a B horizon in which properties of the B predominate.

badland A miscellaneous land type generally void of vegetation and occurring on severely eroded, soft geologic material in arid or semiarid regions.

bar A unit of pressure equal to one million dynes per square centimeter, which is nearly equal to the standard atmosphere.

base-saturation percentage The extent to which the adsorption complex of a soil is saturated with exchangeable cations other than hydrogen or aluminum. It is expressed as a percentage of the total cation-exchange capacity determined at pH 7.0 or 8.2.

basic soil Specifically a soil that has a pH $>$ 7.0, but for most practical purposes, a soil with a pH $>$ 7.4. The term usually applies to the surface soil when used in a general sense.

BC horizon Transition from a B to an underlying C horizon in which properties of the B horizon dominate.

B/C horizon A mixed horizon showing properties of both a B and a C horizon in which properties of the B predominate.

BE horizon Transition from an E to an underlying B horizon in which properties of the B horizon predominate.

B/E horizon A mixed horizon showing properties of both a B and an E horizon in which properties of the B predominate.

bedrock The solid rock underlying soils and the regolith at depths ranging from zero (where exposed by erosion) to 100 m or more.

bentonite A type of mineral deposit consisting principally of montmorillonite clay.

Black Earth A term used by some as synonymous with Chernozem; by others to describe self-mulching black clays, such as some Vertisols.

bleicherde The light-colored, leached E horizon of a Spodosol or Ae horizon of a Podzol.

bog Equivalent to swamp.

bottomland See **flood plain.**

Brunisol A Canadian order of soils showing an intermediate state of development and typically having a brownish B horizon.

buffering Chemically, the ability to resist change in pH.

bulk density, soil The mass of dry soil per unit bulk volume. The bulk volume is determined before drying to constant weight at 105°C.

c

C horizon Horizon that normally lies beneath the B horizon but may lie beneath the A horizon where the only significant change caused by soil development is an increase in organic matter, which produces an A horizon. In concept, the C horizon is unaltered or slightly altered parent material.

calcareous soil Soil containing sufficient calcium carbonate (often with magnesium carbonate) to effervesce visibly when treated with cold 0.1N hydrochloric acid.

caliche A layer near the surface, weakly to strongly cemented by secondary carbonates of calcium or magnesium precipitated from the soil solution. It may occur as: (1) a soft, thin soil horizon; (2) a hard, thick bed just beneath the solum; or (3) a surface layer exposed by erosion. Not a geologic deposit.

calcic horizon A lime-enriched horizon (not indurated).

calorie The amount of heat required to raise the temperature of 1 g of water by 1°C.

cambic horizon A subsurface horizon containing illuvial clay, humus, or amorphous sesquioxides, but not in sufficiency to be classed as spodic or argillic.

capillary attraction A liquid's movement over or retention by a solid surface due to the interaction of adhesive and cohesive forces.

capillary fringe A zone just above the water table that is maintained in an essentially saturated state by capillary forces of lift.

carbon-nitrogen ratio The ratio of the weight of organic carbon to the weight of total nitrogen in a soil or in organic material. It is obtained by dividing the percentage of organic carbon (C) by the percentage of total nitrogen (N).

category, soil Any one of the ranks of a system of soil classification in which soils are grouped on the basis of their characteristics. See **soil classification.**

catena A sequence of soils of about the same age, derived from similar parent material, and occurring under similar climatic conditions, but having different characteristics due to variation in relief and in drainage.

cation See **ions.**

cation exchange The interchange between a cation in solution and another cation on the surface of any surface-active material such as clay or organic colloids.

cation-exchange capacity The sum total of exchangeable cations that a soil can adsorb. Sometimes called total-exchange capacity, base-exchange capacity, or cation-adsorption capacity. Expressed in milliequivalents per 100 grams of soil (or of another adsorbing material, such as clay).

cemented Indurated; having a hard, brittle consistence because the particles are held together by cementing substances such as humus, calcium carbonate, or the oxides of silicon, iron, and aluminum. The hardness and brittleness persist even when wet.

chelation The bonding of a metal cation to an organic anion to form a combination of very low ionization potential. The combination may or may not be soluble in water.

Chernozem A Canadian order of soils formed in cool, subarid to subhumid grasslands in which the principal diagnostic feature is a mineral surface horizon darkened by accumulating organic matter (chernozemic A horizon).

chiseling The breaking or shattering of compact soil or subsoil layers by use of a chisel. See **subsoiling.**

chlorite A nonexpanding layer mineral consisting of 2:1 platelets bound to each other by a positively charged, octahedral sheet located in the interlayer space. Sometimes termed a 2:2 mineral.

chlorosis An abnormal yellowish or whitish leaf color signifying a nutritional or other physiological disorder in plants.

chroma See **color.**

class, soil A group of soils having a definite range in one or more properties such as acidity, degree of slope, texture, structure, land-use capability, degree of erosion, or drainage.

clastic Pertains to rock or mineral fragments moved from their point of origin.

clay (1) A soil separate consisting of particles < 0.002 mm in equivalent diameter. (2) A textural class.

clay mineral (1) Naturally occurring inorganic crystalline or amorphous material found in soils and other earthy deposits, the particles being predominantly < 0.002 mm in diameter. Largely of secondary origin.

claypan A dense, compact layer in the subsoil having a much higher clay content than the overlying material from which it is separated by a sharply defined boundary.

cleavage Tendency to break in the same direction, thus yielding fragments of predictable shape.

clod An artificially produced, compact, coherent mass of soil ranging in size from 5 or 10 mm to as much as 20 to 25 cm.

coarse fragments Rock or mineral particles > 2.0 mm in diameter. See **cobbly, gravelly,** and **stony.**

coarse texture The texture exhibited by sands, loamy sands, and sandy loams except very fine sandy loam.

cobbly A term applied to soils containing a significant amount of cobbles, which are rounded, coarse mineral fragments from 7.5 to 25 cm. in diameter.

cohesion The attraction of a substance for itself; the mutual attraction among molecules or particles comprising a substance that allows it to cling together as a continuous mass.

colloid In soils, mineral or organic particles of submicroscopic size.

colluvium A deposit of rock fragments and soil material accumulated at the base of steep slopes as a result of gravitational action.

color A property, which, when classified according to the Munsell color system, has three variables: hue, chroma, and value. Hue relates to the specific color, as red, yellow, or blue; chroma, its strength or purity; and value, the degree of lightness or darkness.

columnar structure A soil structural type with a vertical axis much longer than the horizontal axes and a distinctly rounded upper surface. See **prismatic structure.**

compaction A reduction in volume resulting from an applied force, such as that due to machinery or animals.

compost Usually, the residue obtained by stacking organic residues, with or without added soil, and then allowing it to undergo biological decomposition.

concretion A local concentration of a chemical compound such as calcium carbonate or iron oxide, in the form of a grain or nodule of varying size, shape, hardness, and color.

conductivity, electrical The ability of a solution to transmit an electric current—an ability closely related to the concentration of ions in the solution.

conductivity, hydraulic As applied to soils—the ability of the soil to transmit water in liquid form through pores.

consistence (1) The resistance of a material to deformation or rupture. (2) The degree of cohesion or adhesion of the soil mass. Terms used for describing consistence at various soil moisture contents are:

wet soil	—nonsticky, slightly sticky, sticky, very sticky, nonplastic, slightly plastic, plastic, and very plastic.
moist soil	—loose, very friable, friable, firm, very firm, and extremely firm.
dry soil	—loose, soft, slightly hard, hard, very hard, and extremely hard.
cementation	—weakly cemented, strongly cemented, and indurated.

consumptive use The water used per unit of time in plant growth by the combined processes of evaporation, transpiration, and retention in the plant. Approximated by evapotranspiration.

contour A fixed level or elevation across a slope.

contour map A map showing a series of continuous lines connecting points of the same elevation on a land surface.

control section That portion of soil profiles used for observing properties in soil classification. The depth of the control section depends on the properties under consideration.

creep   Slow mass movement of soil material down relatively steep slopes primarily under the influence of gravity, but facilitated by saturation with water and by alternate freezing and thawing.

crumb   A soft, porous, more or less rounded ped from 1 to 5 mm in diameter.

crust   A surface layer on soils, ranging in thickness from a few millimeters to perhaps as much as 2 cm, that is much more compact, hard, and brittle, when dry, than the material immediately beneath it.

cryic   A soil temperature regime with an average annual soil temperature of between 0° and 8°C.

Cryosol   A Canadian order of soils of very cold regions containing permafrost within 1 m of the surface, or 2 m if cryoturbated.

cryoturbation   Soil churning due to ice formation in the profile.

cultivation   A tillage operation used in preparing land for seeding or transplanting, or later, for weed control and for loosening the soil.

d

decomposition   The chemical breakdown of mineral or organic matter.

deflocculate   To separate the individual components of soil aggregates by chemical and/or physical means. See **disperse.**

denitrification   The biochemical reduction of nitrate or nitrite to gaseous molecular nitrogen or an oxide of nitrogen.

desert crust   A hard layer, containing calcium carbonate, gypsum, or other binding material, exposed at the surface in desert regions.

desert pavement   The layer of gravel or stones left on the land surface in desert regions after the removal of fine materials by wind erosion.

diagnostic horizon   A surface (epipedon) or subsurface horizon used for the taxonomic classification of soils.

diatomaceous earth   A geologic deposit of fine, grayish siliceous material composed chiefly or wholly of the remains of diatoms that have accumulated mainly in water. It may occur as a powder or a porous, rigid material.

diffusion   The independent or random movement of ions or molecules that tends to bring about their uniform distribution within a continuous system.

distintegration   The physical weathering of rocks.

disperse   (1) To break up compound particles, such as aggregates, into the individual component particles. (2) To distribute or suspend fine particles, such as clay, in or throughout a dispersion medium, such as water.

double layer   In colloid chemistry, a double layer of electrical charges, one consisting of the charges provided by the solid phase (usually negative) and the second by adsorbed ions of opposite charge.

drain, to   (1) To provide channels, such as open ditches or drain tiles so that excess water can be removed either by surface or internal flow. (2) To lose water (from the soil) by percolation.

drain tile   Concrete or ceramic pipe used to conduct water from the soil.

dryland farming   The practice of crop production in low-rainfall areas without irrigation.

duripan   A soil layer cemented by precipitated silica.

dust mulch   A loose, finely granular or powdery condition at the surface of the soil, usually produced by shallow cultivation.

e

E horizon   A horizon formed at or near the mineral surface by eluviation of silicate clay, iron, or aluminum, alone or in combination, and normally consisting mainly of silt or sand.

EB horizon   A transition from an E horizon to an underlying B in which properties of the E horizon predominate.

ecology The science that deals with the interrelations of organisms and their environment.

edaphology The science that deals with the influence of soils on living things, particularly plants, including man's use of land for plant growth.

effective precipitation That portion of the total precipitation which becomes available for such processes as plant growth and soil development.

effervesce To bubble or foam.

efflorescence A powderlike surface covering or coating.

effluent A liquid discharge, as from sewage treatment plants, septic tanks, or other collection or storage structures.

eluvial horizon A soil horizon that has been formed by the process of eluviation See **illuvial horizon.**

eluviation. The removal of soil material in suspension from a layer or layers of a soil. Contrast with **leaching.**

Entisol An order of soils in which profile development is minimal; characteristics are largely inherited from the parent material.

epipedon A diagnostic horizon occurring at the surface of the soil.

equivalent diameter In sedimentation analysis, the diameter assigned to a nonspherical particle; it is numerically equal to the diameter of a spherical particle of the same density and velocity of fall.

erosion (1) The wearing away of the land surface by running water, wind, ice, or other geological agents, including such processes as gravitational creep. (2) Detachment and movement of soil or rock by water, wind, ice, or gravity.

eutrophic A term applied to water that has a concentration of nutrients optimal, or nearly so, for water-inhabiting plant or animal growth.

evapotranspiration The combined loss of water from a given area, and during a specified period of time, by evaporation from the soil surface and by transpiration from plants.

exchange acidity The titratable hydrogen and aluminum that can be replaced from the adsorption complex by a neutral salt solution. Usually expressed as milliequivalents per 100 grams of soil.

exchange capacity The total ionic charge of the adsorption complex active in the adsorption of ions. See **cation-exchange capacity.**

exchangeable-cation percentage The extent to which the adsorption complex of a soil is occupied by a particular cation. It is expressed as follows:

$$ECP = \frac{\text{Exchangeable cation (me/100 g soil)}}{\text{Cation-exchange capacity (me/100 g soil)}} \times 100.$$

exchange complex The combination of all sites of charge contributing to the exchange capacity of the soil, normally, the cation-exchange capacity.

f

F horizon (C) Partially decomposed surface litter of leaf, needle, or woody tissue, with some of the original material difficult to recognize. (Intermediate stage of biological decomposition.)

fallow To remove land from active crop production, commonly to allow its natural regeneration.

family, soil In soil classification, the category immediately above the soil series. See **soil classification.**

fertility, soil The status of soil with respect to the amount and availability to plants of elements necessary for plant growth.

fertilizer Any organic or inorganic material of natural or synthetic origin which is added to a soil to supply certain elements essential to the growth of plants.

fertilizer grade The guaranteed minimum analysis, in percent, of the major plant nutrient elements contained in a fertilizer material or in a mixed fertilizer. (Refers to the percentage of $N-P_2O_5-K_2O$ or of N-P-K).

fibric horizon An organic horizon in which fibrous plant remains make up over two-thirds of the mass.

field capacity   The amount of water remaining in a field soil that has been thoroughly wetted and drained until free drainage has practically ceased.

fine texture   The texture exhibited by soils having clay as a part of their textural class name.

firm   A term describing the consistence of a moist soil that offers distinctly noticeable resistance to crushing but can be crushed with moderate pressure between the thumb and forefinger. See **consistence.**

first bottom   The normal flood plain of a stream.

fixation, chemical   The process or processes in a soil by which certain chemical elements essential for plant growth are converted from a soluble or exchangeable form to a much less soluble or to a nonexchangeable form. Contrast with **nitrogen fixation.**

flood plain   Flat or nearly flat land on the floor of a river valley that is covered by water during floods.

fracture   Random breakage that results in fragments of unpredictable shape.

fragipan   A natural subsurface horizon with high bulk density relative to the solum above, seemingly cemented when dry, but when moist, showing a moderate to weak brittleness. The layer is low in organic matter, mottled, slowly or very slowly permeable to water, and usually shows occasional or frequent bleached cracks forming polygons. It may be found in profiles of either cultivated or virgin soils but not in calcareous material.

friable   A consistence term pertaining to the ease of crumbling of soils.

frost line   The depth to which freezing occurs in the soil.

furrow   A channel worked into the surface of the soil by an implement such as a plow or hoe.

g

genetic   Resulting from soil genesis or development; that is, produced by soil-forming processes as, for example, a genetic soil profile or a genetic horizon.

gilgai   The microrelief of soils that contain large amounts of clay which swells and shrinks considerably with wetting and drying. Usually a succession of microbasins and microknolls in nearly level areas or of microvalleys and microridges parallel to the direction of the slope.

glacial drift   Rock debris that has been transported by glaciers and deposited, either directly from the ice or from the melt-water. The debris may or may not be hetergeneous.

glacial till   See **till,** (1).

glaciofluvial deposits   Material moved by glaciers and subsequently sorted and deposited by streams flowing from the melting ice.

Gleysol   A Canadian order of soils in which evidence of strong gleying is the principal diagnostic feature.

gleyzation   A process affecting soil development under strongly reducing conditions, as in waterlogged soils. Iron may be extensively solubilized and partly reprecipitated as rust-colored mottles or stains.

granule   An aggregate similar in size to a crumb but more dense and, therefore, less porous than a crumb.

gravelly   A term applied to soils containing significant amounts of gravels, which are rounded, coarse fragments from 2 mm to 7.5 cm in diameter.

gravitational potential   See **potential, soil water.**

gravitational water   Water which moves into, through, or out of the soil under the influence of gravity.

great group, of soils   A category of soil classification in which classes are formed by the subdivision of soil suborders in the U.S. system of classification and by the subdivision of orders in the Canadian system.

green manure   Plant material incorporated with the soil while green, or soon after maturity, for improving the soil.

ground water   That portion of the total precipitation which at any particular time is either passing through or standing in the soil and the underlying strata and is free to move under the influence of gravity.

gulley A channel resulting from water erosion that is deep enough to interfere with, and not to be obliterated by, normal tillage operations.

gypsic horizon A gypsum-enriched layer.

gypsiferous soil A soil containing free gypsum.

h

H horizon (C) Well decomposed surface litter of leaf, needle, or woody tissue in which original materials cannot be recognized.

halophytic vegetation Salt-loving or salt-tolerant vegetation, usually having fleshy leaves or thorns and resembling desert vegetation.

hardpan A hardened soil layer, in the lower A or in the B horizon, caused by cementation of soil particles with organic matter or with materials such as silica, sesquioxides, or calcium carbonate. The hardness does not change appreciably with changes in moisture content and pieces of the hard layer do not slake in water. See **caliche** and **claypan.**

heavy soil (Obsolete in scientific use.) A soil with a high content of the fine separates, particularly clay, or one with a high drawbar pull and, hence, difficult to cultivate. See **fine texture.**

hectare An area of 10,000 sq m and equal to 2.471 acres.

hemic horizon An organic horizon in which fibrous plant remains make up between one-third and two-thirds of the mass.

heterotrophic Capable of deriving energy for life processes only from the decomposition of organic compounds and incapable of using inorganic compounds as sole sources of energy or for organic synthesis. Contrast with **autotrophic.**

histic epipedon An organic surface horizon too thin to allow classification of the soil as a Histosol.

Histosol A soil order consisting of organic soil such as peat or muck.

horizon See **soil horizon.**

hue See **color.**

humification The processes involved in the decomposition of organic matter and leading to the formation of humus.

humus That semistable fraction of the soil organic matter remaining after the major portion of added plant and animal residues have decomposed. Usually it is dark-colored.

hydration The physical binding of water molecules to ions, molecules, particles, or other matter.

hydraulic conductivity See **conductivity, hydraulic.**

hydraulic pressure The pressure within water produced by a combination of forces, such as capillary and gravitational, acting on the water.

hydrolysis The chemical reaction of a compound with water, whereupon the anion from the compound combines with the hydrogen and the cation from the compound combines with the hydroxyl from the water to form an acid and a base.

hydrous mica See **illite.**

hygroscopic water Approximates the film water held in air-dry soil.

hyperthermic A soil temperature regime under which the average annual soil temperature is greater than 22°C.

i

igneous rock Rock formed from the cooling and solidification of magma, and that has not been changed appreciably since its formation.

illite A 2:1 layer clay in which a negative charge caused by ion substitution is neutralized predominantly by potassium ions.

illuvial horizon A soil layer or horizon produced by illuviation.

illuviation The process of deposition of colloidal soil material, removed from one horizon to another in the soil; usually from an upper to a lower horizon in the soil profile. See **eluviation.**

immature soil A soil with indistinct or only slightly developed horizons because of the relatively short time it has been subjected to the various soil-forming processes. A soil that has not reached equilibrium with its environment. See **Brunisol, Entisol, Inceptisol,** and **Regosol.**

immobilization The conversion of an element from the inorganic to the organic form in microbial or plant tissue, thus rendering the element not readily available to other organisms or plants.

impervious Resistant to penetration by fluids or by roots.

Inceptisol A soil order consisting of soils showing a moderate degree of profile development but not enough to be considered at equilibrium with the soil-forming environment.

indurated Hardened, often by cementation.

infiltration The downward entry of water into the soil.

inoculation The introduction of organisms into the soil, often by treating seed with special microbial cultures.

intergrade soil A soil that possesses moderately well developed distinguishing characteristics of two or more genetically related great soil groups.

intergrade minerals A series of minerals, mainly nonexpanding, ranging in structure from expanding 2:1 types, such as montmorillonite and vermiculite, to nonexpanding chlorite, resulting from partial filling of the interlayer space by octahedral material, mainly hydroxy-Al or hydroxy-Fe substances.

interpretive classification Utilitarian classification, which is based on properties important to soil use.

ions Atoms that are positively charged (cations) because of the loss of one or more electrons, or that are negatively charged (anions) because of a gain in electrons.

irrigation The artificial application of water to the soil for the benefit of growing crops.

irrigation efficiency The ratio of water actually consumed by crops on an irrigated area to total amount of water applied to the area.

isomorphous substitution The replacement of an ion considered normal to a mineral structure by another during the formation of a mineral.

k

kame An irregular ridge or hill of stratified glacial drift.

kaolin An aluminosilicate mineral with a 1:1 layer structure; that is, consisting of one silicon tetrahedral layer and one aluminum octahedral layer. More commonly, kaolinite.

l

L horizon (C) Surface litter consisting largely of undecomposed leaf, needle, or woody tissue.

lacustrine deposit Material deposited in lake water and later exposed either by lowering of the water level or by the elevation of the land.

land classification The arrangement of land units into various categories based upon the properties of the land or its suitability for some particular purpose.

landscape All the natural features, such as fields, hills, forests, water, etc., which distinguish one part of the earth's surface from another part. Usually that portion of land or territory which the eye can comprehend in a single view, including all its natural characteristics.

leaching The removal of materials in solution by water percolating through the soil.

light soil (Obsolete in scientific use) A coarse-textured soil; a soil with a low drawbar pull and hence easy to cultivate.

lime, agricultural A soil amendment consisting principally of calcium carbonate but including magnesium carbonate and perhaps other materials, and used to furnish calcium and magnesium as elements for the growth of plants and to neutralize soil acidity.

lime requirement The amount of agricultural limestone, or the equivalent of other specified liming material, required for a specified area and depth of soil to raise the pH to a desired value under field conditions.

liquid limit The moisture content marking the upper limit of the plastic range in soils. Upper plastic limit.

lithic contact Boundary between soil and unweathered rock.

loess Material transported and deposited by wind and consisting predominantly of silt-sized particles.

Luvisol A Canadian order of soils in which the predominating, or diagnostic, profile features are the result of eluviation and illuviation, principally of silicate clay.

lysimeter A device for measuring percolation and leaching losses from a column of soil under controlled conditions.

m

macronutrient A chemical element necessary in large amounts for the growth of plants.

mapping unit A soil or combination of soils delineated on a map and, where possible, named to show the taxonomic unit or units included. Principally, mapping units on maps of soils depict soil phases, associations, or complexes.

mass flow The unidirectional flow of a liquid or gas along with any suspended or dissolved components they contain.

massive state A nonstructural state in soils that contain cohesive particles occurring in a continuous mass having no well defined cleavage pattern. When crushed, massive soil breaks into irregular fragments of unpredictable size and shape.

matric potential See **potential, soil water.**

mature soil A soil with well developed soil horizons produced by the natural process of soil formation and essentially in equilibrium with its present environment.

mechanical analysis See **particle-size analysis** and **particle-size distribution.**

medium texture The texture exhibited by very fine sandy loams, loams, silt loams, and silts.

mesic A soil temperature regime where the average annual soil temperature is between 8° and 15°C.

metamorphic rock Rock derived from pre-existing rocks, but that differ from them in physical, chemical, and mineralogical properties as a result of natural geological processes manifested principally by heat and pressure originating within the earth. The pre-existing rocks may have been igneous, sedimentary, or another form of metamorphic rock.

micron One one-thousandth of a millimeter (one micrometer).

micronutrient A chemical element necessary in only extremely small amounts for the growth of plants.

mineralization The conversion of an element from an organic form to an inorganic state as a result of microbial decomposition.

mineralogy, soil In practical use, the kinds and proportions of minerals present in a soil.

mineral soil A soil consisting predominantly of, and having properties determined by, mineral matter. Usually contains <20% organic matter, but may contain an organic surface layer up to 30 cm thick.

minor element (Obsolete) See **micronutrient.**

miscellaneous land type A mapping unit for land areas that have little or no natural soils, that are too nearly inaccessible for orderly examination, or for any reason it is not feasible to classify the soil. Examples are **badland, riverwash, rough broken land, swamp,** and **wasteland.**

mollic epipedon A surface horizon of mineral soil that is dark-colored and relatively thick, contains at least 1 percent organic matter, and has a base saturation (pH 7.0) in excess of 50 percent.

Mollisol Soil order consisting of soils having a thick A horizon with more than one percent organic matter and a base-saturation percentage above 50.

Normally, they are formed under grass vegetation. Distinguished from Vertisols in that they are not self-inverting.

montmorillonite An aluminosilicate clay mineral with a 2:1 expanding structure; that is, with two silicon tetrahedral layers enclosing an aluminum octahedral layer. Considerable expansion may be caused by water moving between silica layers of contiguous units.

morphology See **soil morphology.**

mottling Spots or blotches of different color or shades of color interspersed with the dominant color.

muck soil An organic soil in which the organic matter is well decomposed.

mulch Any material, such as straw, sawdust, leaves, plastic film, loose soil, etc., that is spread upon the surface of the soil to protect the soil and plant roots from the effects of raindrops, soil crusting, freezing, evaporation, etc.

mycorrhiza The association, usually symbiotic, of fungi with plant roots.

n

natric horizon A subsurface horizon having the characteristics of an argillic horizon, and in addition, a prismatic or columnar structure and an exchangeable-Na percentage greater than 15.

nematodes Small, threadlike worms common to most soils and responsible for root damage in many crop plants.

neutral soil Specifically, a soil in which the surface layer, at least to normal plow depth, is neither acid nor basic in reaction. Practically, a soil in which the pH of the surface soil ranges between 6.6 and 7.4.

nitrate reduction The biochemical reduction of nitrate.

nitrification The biochemical oxidation of ammonium to nitrate.

nitrogen fixation The conversion of elemental nitrogen (N_2) to organic combinations or to forms readily utilizable in biological processes.

nutrient A chemical element essential for the growth and development of an organism.

o

O horizon A natural layer or horizon of fresh or partly decomposed plant litter on the surface of the mineral soil.

ochric epipedon A surface horizon of mineral soil that lacks one or more of the properties required for classification as a mollic, histic, anthrophic, plaggen, or umbric epipedon.

organic soil A soil which contains a high percentage (>15% or 20%) of organic matter.

osmotic potential See **potential, soil water.**

osmotic pressure In concept, the force per unit area required to equal the attractive (hydration) force for water exerted by ions dissolved in a solution.

oven-dry soil Soil which has been dried at 105°C until it reaches an essentially constant weight.

oxic horizon A mineral soil horizon characterized by a lack of weatherable minerals, a low cation-exchange capacity, and small amounts of exchangeable bases, but containing 1:1 layer clays or sesquioxides as dominant minerals, with or without quartz and other resistant minerals.

oxidation (1) The burning or other conversion of an element to an oxide; (2) An increase in positive valence of an element or ion caused by electron loss.

Oxisol Soil order consisting of soils containing a B horizon in which clays are largely kaolinite and sesquioxides and in which sands are essentially void of easily weathered minerals. Oxisols occur typically under forested conditions of the humid tropics. They are usually red in color and have a high degree of aggregate stability, a low cation-exchange capacity, and a moderate to low degree of base saturation.

p

pans Horizons or layers, in soils, that are strongly compacted, indurated, or very high in clay content. See **caliche, claypan, duripan, fragipan,** and **hardpan.**

paralithic contact The boundary between soil material and underlying rock that has been softened by weathering.

parent material The unconsolidated mineral or organic matter from which soils are developed.

parent rock The rock from which a parent material is derived.

particle density The mass per unit volume of individual particles; usually expressed as grams per cubic centimeter.

particle size The effective diameter of a particle; usually measured by sedimentation or sieving.

particle-size analysis Determination of the various amounts of the different separates in a soil sample.

particle-size distribution The amounts of the various soil separates in a soil sample, usually expressed as weight percentages.

peat soil An organic soil consisting largely of undecomposed, or only slightly decomposed, organic matter accumulated under conditions of excessive moisture.

ped A unit of soil structure such as an aggregate, crumb, prism, block, or granule, formed by natural processes (in contrast with a clod, which is formed artificially).

pedogenesis The formation of soils.

pedon The smallest volume (soil body) which displays the normal range of variation in proprties of a soil. Where properties such as horizon thickness vary little along a lateral dimension, the pedon may occupy an area as small as 1 sq m. Where such a property varies substantially along a lateral dimension, a large pedon 10 square meters in area may be required to show the full range in variation.

percolation, soil water The downward movement of excess water through soil.

pergelic A soil temperature regime with a mean annual soil temperature of less than 0°C. The presence of permafrost normal.

permafrost Permanently frozen soil.

permanent wilting point A moisture content at which the soil is incapable of maintaining succulent plants in an unwilted condition even though transpirational loss is negligibly low.

permeability, soil The ease with which gases, liquids, or plant roots penetrate or pass through soil.

petrocalcic horizon A lime-cemented layer.

pH, soil The degree of acidity or basicity expressed by the negative logarithm of the hydrogen-ion activity of a soil.

phase, soil See **soil phase.**

photosynthesis The synthesis of carbohydrates by plants containing chlorophyll on exposure to light.

physical properties (of soils) Those characteristics, processes, or reactions of a soil which are caused by physical forces and which can be described by, or expressed in physical terms or equations.

placic horizon An iron-cemented horizon that is slowly permeable or impermeable to water and roots.

plaggen epipedon A thick surface horizon formed by heavy manuring.

plasticity index The mathematical difference between the liquid and plastic limits of a soil; the range in water content through which a soil displays plastic properties.

plastic limit The moisture content at the lower limit of the plastic range in soils; the lower plastic limit.

plastic soil A soil capable of being molded or deformed continuously and permanently, by relatively moderate pressure, into various shapes. See **consistence.**

platy structure Soil aggregates that are developed predominately along the horizontal axes; laminated; flaky.

plinthite A layer, sometimes discontinuous, containing sesquioxides and other minerals, that is nonindurated when continuously moist, but that dries irreversibly to ironpan. Typical of many tropical soils. Formerly, laterite.

plow pan A compacted layer beneath the plow layer produced by pressure exerted on the soil during plowing.

Podzol A Canadian order of soils formed under coniferous forests and, generally, in coarse parent materials and containing a B horizon high in illuvial iron and aluminum, with or without illuvial humus (a podzolic B horizon).

polypedon A soil volume consisting of two or more similar pedons; the basic unit used for classifying soils in the field.

pore-size distribution The volume of the various sizes of pores in a soil. Expressed as percentages of the bulk volume (soil plus pore space).

pore space Space in the soil not occupied by solid particles.

porosity The total volume of pore space, usually expressed as a percentage of the total soil volume.

potassium fixation The process of converting exchangeable or water-soluble potassium to a form not easily exchanged with a cation of a neutral salt solution.

potential, soil water The potential energy of a unit quantity of water produced by the interaction of the water with such forces as capillary (matric), ion hydration (osmotic), and gravity, expressed relative to an arbitrarily selected reference potential. In practical application, potentials are used to predict the direction and rate of water flow through soils, or between the soil and some other system, such as plants or the outer atmosphere. Flow occurs spontaneously between points of differing water potential, the direction of flow being toward the site of lower potential.

primary mineral A mineral that has not been altered chemically since deposition and crystallization from molten lava. See **secondary mineral**.

prismatic structure A soil structural type with a vertical axis much longer than the horizontal axes and a flat or indistinct upper surface. See **columnar structure**.

productivity, soil The capacity of a soil, in its normal environment, for producing a specified plant or sequence of plants under a specified system of management. Productivity emphasizes the capacity of soil to produce crops and is expressed in terms of yield.

profile, soil A vertical section of the soil through all its horizons and extending into the parent material.

protoplasm That portion of cellular tissue active in life processes.

puddled soil A soil in which structure has been mechanically destroyed, which allows the soil to run together when saturated with water. A soil that has been puddled occurs in a massive nonstructural state.

r

R horizon Consolidated bedrock underlying the C horizon.

reaction, soil The degree of acidity or basicity of a soil, usually expressed as a pH value.

red earth Highly leached, red clayey soils of the humid tropics, usually with very deep profiles that are high in sesquioxides.

reduction The decrease in positive valence, or increase in negative valence, caused by a gain in electrons by an ion or atom.

regolith The unconsolidated mantle of weathered rock and soil material on the earth's surface; loose earth materials above solid rock.

Regosol A Canadian order of soils showing minimal development under current environmental conditions.

relief The difference in elevation between the high and low points on a landscape.

residual material Unconsolidated and partly weathered mineral materials accumulated by disintegration of consolidated rock in place.

respiration The biological process whereby organisms oxidize carbon compounds to carbon dioxide and water as a source of energy.

rhizobia Bacteria capable of living symbiotically with higher plants, usually legumes, from which they receive their energy, and capable of using atmospheric nitrogen: hence, the term symbiotic nitrogen-fixing bacteria. (Derived from the generic name *Rhizobium*.)

rhizosphere That portion of the soil directly affected by plant roots.

rill A small channel, such as a furrow prepared to conduct irrigation water across a field, or that caused by erosive runoff. Where the cause is erosion, the term rill is limited to small channels that are only a few centimeters deep and, hence, no obstacle to tillage operations.

river wash Barren alluvial land, usually coarse-textured, exposed along streams at low water and subject to shifting during normal high water.

rockland Areas containing frequent rock outcrops and shallow soils. Rock outcrops usually occupy from 25% to 90% of the area.

rough broken land Land with very steep topography and numerous intermittent drainage channels but usually covered with vegetation.

runoff That portion of the precipitation of an area which is discharged from the area through stream channels.

S

saline-sodic soil A soil containing sufficient exchangeable sodium to interfere with the growth of most crop plants and containing appreciable quantities of soluble salts. The sodium-adsorption ratio is >15, the conductivity of the saturation extract >4 millimhos per centimeter (at 25°C), and the pH is usually 8.5 or less in the saturated soil.

saline soil A soil containing sufficient soluble salts to impair its productivity. Specifically, a soil providing a saturation-paste extract having an electrical conductivity >4 millimhos per centimeter (at 25°C). The term saline, wen used alone, implies a low sodium-adsorption ratio (<15).

salt-affected soil Soil that has been adversely modified for the growth of most crop plants by the presence of certain types of exchangeable ions or of soluble salts. (Includes soil having an excess of salts, or an excess of exchangeable sodium, or both). See **saline soil, sodic soil,** and **saline-sodic soil.**

sand (1) A soil separate consisting of particles between 0.05 and 2.0 mm in diameter. (2) A soil textural class.

sapric horizon An organic horizon in which fibrous plant remains make up less than one-third of the mass.

SAR See *sodium-adsorption ratio.*

saturate (1) To fill all the voids between soil particles with a liquid. (2) To form the most concentrated solution possible under a given set of physical conditions in the presence of an excess of the solute. (3) To fill to capacity, as the adsorption complex, with a cation species; i.e., H-saturated, etc.

saturation-paste extract An extract obtained from a soil sample mixed with sufficient water to make a thin paste, for use in salinity assessment and for certain other chemical analyses of salt-affected soils.

secondary mineral A mineral resulting from the decomposition or alteration of another mineral or from the reprecipitation of the products of decomposition of another mineral. See **primary mineral.**

sedimentary rock A rock formed from materials deposited from suspension or precipitated from solution and usually, but not necessarily, consolidated (i.e., cemented). The principal sedimentary rocks are sandstones, shales, limestones, and conglomerates.

seepage The slow flow of water into or from a soil. In usual concept, seepage involves the lateral flow of water, as from an open body of water into neighboring soil, or the reverse.

self-mulching soil A soil in which the surface layer becomes so well aggregated that it does not crust and seal under the impact of rain but instead serves as a surface mulch upon drying.

separate See **soil separates.**

series, soil See **soil series.**

silica-sesquioxide ratio The molecules of silicon dioxide (SiO_2) per molecule of aluminum oxide (Al_2O_3) plus ferric oxide (Fe_2O_3) in clay minerals or in soils.

silt (1) A soil separate consisting of particles between 0.05 and 0.002 mm in diameter. (2) A soil textural class.

silting The deposition of water-borne sediment in stream channels, lakes, reservoirs, or on flood plains, usually resulting from a decrease in the velocity of the water.

single-grained state A nonstructural state normally observed in soils containing a preponderance of large particles such as sand. Because of a lack of cohesion, the sand grains tend not to assemble in aggregate form.

site (1) In ecology, an area described or defined by its biotic, climatic, and soil conditions as related to its capacity to produce vegetation. (2) An area sufficiently uniform in biotic, climatic, and soil conditions to produce a particular climax vegetation.

site index (1) A quantitative evaluation of the productivity of a soil for forest growth under the existing or specified environment. (2) The height of the dominant forest vegetation taken at or calculated to an index age, usually 50 or 100 years.

slickensides Polished and grooved surfaces produced by one soil mass sliding past another. Common in Vertisols.

slick spots Small areas in a field that are slick when wet, due to a high content of exchangeable sodium.

slope Deviation of a plane surface from the horizontal. Slope is conventionally expressed in degrees, which are units of vertical distance for each 100 units of horizontal distance.

sodic soil A soil that contains sufficient exchangeable sodium to interfere with the growth of most crop plants; the sodium-adsorption ratio of the saturation-paste extract is 15 or more.

sodium-adsorption ratio The relationship of soluble sodium to soluble calcium + magnesium in water or the soil solution as expressed by the equation:

$$SAR = \frac{[Na^+]}{\sqrt{\dfrac{[Ca^{2+} + Mg^{2+}]}{2}}}$$

where the concentrations of ions are in milliequivalents per liter. Below values of 30, the sodium-adsorption ratio of saturation-paste extracts and the exchangeable-sodium percentage of soils are about equal.

soil (1) The unconsolidated mineral material on the immediate surface of the earth that serves as a natural medium for the growth of land plants. (2) The unconsolidated mineral matter on the surface of the earth that has been subjected to and influenced by genetic and environmental factors of parent material, climate (including moisture and temperature), macro- and microorganisms, and topography, all acting over a period of time and producing a product—soil—that differs from the material from which it is derived in many physical, chemical, biological, and morphological properties.

soil air The soil atmosphere; the gaseous phase of the soil, being that volume not occupied by soil or liquid.

soil association (1) A group of defined and named taxonomic soil units occurring together in an individual and characteristic pattern over a geographic region, comparable to plant associations in many ways. (2) A mapping unit used on general soil maps, in which two or more defined taxonomic units occurring together in a characteristic pattern are combined because the scale of the map or the purpose for which it is being made does not require delineation of the individual soils.

soil classification The systematic arrangement of soils into groups or categories on the basis of their characteristics. Broad groupings are made on the basis of general characteristics and subdivisions on the basis of more detailed differences in specific properties. The six categories of the current system used in the United States are orders, suborders, great groups, subgroups, families, and soil series, with each series consisting of many individual occurrences or bodies of soil that are very similar in most respects.

soil complex   A mapping unit used in detailed soil surveys where two or more defined taxonomic units are so intimately intermixed geographically that it is undesirable or impractical, because of the scale being used, to separate them. A more intimate mixing of smaller areas of individual taxonomic units than that described under **soil association.**

soil creep   A sliding or rolling of particles over the soil surface, caused by wind.

soil extract   The solution separated from a soil or soil suspension.

soil-formation factors   The variable, interrelated natural agencies of soil formation, namely: parent material, climate, living matter, topography, and time.

soil genesis   The formation of soils; the creation of new characteristics by soil-development processes.

soil heave   A raising of the soil surface by the expansion of ice forming in the body of the soil, usually from water rising to the frost line by capillarity.

soil horizon   A layer of soil or soil material approximately parallel to the land surface and differing from adjacent genetically related layers in physical, chemical, and biological properties or characteristics such as color, structure, texture, consistence, pH, etc.

soil management   The sum total of all tillage operations, cropping practices, fertilizer, lime, and other treatments conducted on or applied to a soil for the production of plants.

soil management groups   Groups of soil mapping units with similar adaptations or management requirements for one or more specific purposes, such as: adapted crops or

crop rotations, drainage practices, fertilization, forestry, highway engineering, etc.

soil map   A map showing the distribution of soil phases or other soil mapping units in relation to the prominent physical and cultural features of the earth's surface.

soil moisture   Water contained in the soil.

soil moisture tension   See **tension, soil water.**

soil monolith   A vertical section of a soil profile removed from the soil and mounted for display or study.

soil morphology   The physical constitution, particularly the structural properties, of a soil profile as exhibited by the kinds, thickness, and arrangement of the horizons in the profile, and by the texture, structure, consistence, and porosity of each horizon.

soil order   The highest category in the systems of soil classification used in the United States and Canada. Orders consist of broad groupings of soils organized principally to show which soil-formation factor has had the greatest influence in determining the properties of a soil.

soil organic matter   The organic fraction of the soil; includes plant and animal residues at various stages of decomposition, cells and tissues of soil organisms, and substances synthesized by the soil population.

soil phase   A subdivision of a soil series or other unit of classification having characteristics that affect the use and management of the soil but which do not vary sufficiently to differentiate it as a separate series. A variation in a property or characteristic such as textural class, degree of slope, degree of erosion, content of stones, etc.

soil salinity   The amount of soluble salts in a soil.

soil science   That science dealing with soils as a natural resource on the surface of the earth including soil formation, classification, and mapping, and the physical, chemical, biological, and fertility properties of soils per se, and these properties in relation to their management for crop production.

soil separates Groups of mineral particles separated on the basis of a range in size. The principal separates are sand, silt, and clay.

soil series The basic unit of soil taxonomy and consisting of soils which are essentially alike in all major profile characteristics, although the texture of the A horizon may vary somewhat.

soil solution The aqueous phase of the soil and its solutes consisting of ions dissociated from the surfaces of the soil particles and of other soluble materials.

soil structure The combination or arrangement of individual soil particles into definable aggregates, or peds, which are characterized and classified on the basis of size, shape, and degree of distinctness.

soil subgroup A category in soil classification in which classes are formed by subdividing great groups to distinguish soils that are typical from those that are at variance with the central concept of the great group.

soil suborder The second highest category in the U.S. system of soil taxonomy, being the subdivision of the soil order. Classes with suborders are based on soil properties that reflect a narrower range of variation in soil-forming factors than is allowed for soil orders.

soil suction A measure of the force of water retention in unsaturated soil. Soil suction is equal to a force per unit area that must be exceeded by an externally applied suction to initiate water flow from the soil. Soil suction is expressed in standard pressure terms.

soil survey The systematic examination, description, classification, and mapping of soils in an area.

soil texture The relative proportions of the various soil separates in a soil.

soil type Formerly a subdivision of a soil series based on differences in the texture of the A horizon.

soil variant A soil whose properties are believed to be sufficiently different from other known soils to justify a new series name but comprising such a limited geographic area that creation of a new series is not justified.

soil water A general term emphasizing the physical rather than the chemical properties and behavior of the soil solution.

Solonetz A Canadian order of soils forming in drier regions in which the principal diagnostic feature is an illuvial B horizon high in silicate clay and exchangeable sodium and having a well developed prismatic or columnar structure (solonetzic B horizon).

soluble-sodium percentage (SSP) The proportion of sodium ions in solution in relation to the total cation concentration, defined as follows:

$$SSP = \frac{\text{Soluble-sodium concentration (me/liter)}}{\text{Total cation concentration (me/liter)}} \times 100.$$

solum (plural: sola) The upper and most weathered part of the soil profile; the A, E, and B horizons.

sombric horizon A subsurface horizon high in illuvial humus and less than 50 percent saturated with bases. Restricted to well drained, tropical and subtropical soils, often those of higher elevations.

spodic horizon A subsurface horizon containing illuvial humus and/or amorphous sesquioxides, and normally, a low degree of saturation with bases.

Spodosol A soil order consisting of soils having a B (spodic) horizon containing illuvial humus or clay, alone or in combination, with the clay being principally of noncrystalline types. The undisturbed profile also has distinct O and E horizons, occurring in that order, over the B horizon. Typically, Spodosols are formed in cool to cold, moist climates under coniferous forest vegetation. They are moderately to strongly acid throughout their profile.

stones Rock fragments >25 cm in diameter if rounded, and >38 cm along the longer axis if flat.

stony A term applied to soils that contain a significant amount of stones.

stratified Arranged in or composed of strata or layers.

stripcropping The practice of growing crops which require different types of tillage, such as row and sod, in alternate strips along contours or across the prevailing direction of wind.

structure, soil See **soil structure.**

stubble mulch The stubble of crops or crop residues left essentially in place on the land as a surface cover before and during the preparation of the seedbed and at least partly during the growing of a succeeding crop, for the purpose of erosion control.

subgroup See **soil subgroup.**

suborder See **soil suborder.**

subsidence A lowering of the soil surface due to a reduction in volume through settling or other means. In organic soils, subsidence usually follows improved drainage, which increases aeration and the loss of organic matter through decomposition.

subsoil In general concept, that part of the soil below the depth of plowing.

subsoiling Breaking of compact subsoils, without inverting them, with a special knifelike instrument (chisel) which is pulled through the soil at depths of 30 to 60 cm and at spacings usually of 50 to 150 cm.

subsurface tillage Tillage with a special sweep plow or blade which is drawn beneath the surface at depths of several centimeters and cuts plant roots and loosens the soil without inverting it or without incorporating the surface cover.

suction See **soil suction.**

summer fallow The maintenance of the soil in a weed-free condition during the growing season to conserve water for the following crop. Limited mainly to small-grain production in drier regions.

surface-depth equivalent The amount of water in the soil expressed as its equivalent depth when standing as free water on the surface of the soil.

sulfuric horizon A strongly acidic layer (pH <3.5) produced by the oxidation of sulfur or sulfur compounds to sulfuric acid.

surface soil The uppermost part of the soil, ordinarily moved in tillage, or its equivalent in uncultivated soils and ranging in depth from 8–10 cm to 20–25 cm. Frequently designated as the plow layer or the Ap horizon.

swamp An area saturated with water throughout much of the year but with the surface of the soil usually not deeply submerged. Generally characterized by tree or shrub vegetation.

symbiosis The living together in intimate association of two dissimilar organisms, the cohabitation being mutually beneficial.

t

talus Fragments of rock and other soil material accumulated by gravity at the foot of cliffs or steep slopes.

taxonomic unit An arbitrarily defined unit consisting either of an individual or a combination of individuals established for the purpose of classfication.

taxonomy A classification based on natural relationships among the objects being classified.

tensiometer A device for measuring the tension of water in soil *in situ;* a porous, permeable ceramic cup connected through a tube to a manometer or vacuum gauge.

tension, soil water A measure of the force required to remove water from soils expressed in pressure terms.

tephra Volcanic ejecta.

terrace (1) A level, usually narrow, plain bordering a river, lake, or the sea. Rivers sometimes are bordered by terraces at different levels. (2) A raised, essentially level or horizontal strip of earth usually constructed on or nearly on a contour and supported on the downslope side by rocks or another similar barrier—designed to make the land suitable for tillage and to prevent accelerated erosion.

texture See **soil texture.**

textural class, soil Soils grouped on the basis of a specified range in texture. In the United States 12 textural classes are recognized.

thermic A soil temperature regime in which the average annual soil temperature is between 15° and 22°C.

tight soil A compact, relatively impervious and tenacious soil (or subsoil) which may or may not be plastic.

tile drain   Concrete or ceramic pipe placed at suitable depths and spacings in the soil or subsoil to provide water outlets from the soil.

till   (1) Unstratified glacial drift deposited directly by the ice and consisting of clay, sand, gravel, and boulders intermingled in any proportion. (2) To plow and prepare for seeding; to seed or cultivate the soil.

tilth   The physical condition of soil as relates to its ease of tillage, fitness as a seedbed, and its impedance to seedling emergence and root penetration.

topography   The physical features of a landscape, especially its relief and slope.

topsoil   (1) The layer of soil moved in cultivation. See **surface soil.** (2) The A horizon. (3) Presumably fertile soil material used to topdress roadbanks, gardens, and lawns.

torric   A soil moisture regime synonymous with aridic.

trace elements   (Obsolete) See **micronutrient.**

transitional soil   A soil with properties intermediate between those of two different soils and genetically related to them.

truncated   Having lost all or part of the upper soil horizon or horizons.

tundra   A level or undulating treeless plain characteristic of arctic regions.

turgid   Fully expanded.

u

udic   For well drained soils, a soil moisture regime capable of supplying plants with available water during most of the summer season.

Ultisol   A soil order consisting of soils with a B horizon that contains clay of 1:1 or sesquioxide types. Ultisols have a low base-saturation percentage and are moderately acid throughout. They are commonly red, but sometimes yellow in color, the result of extensive mineral weathering and the precipitation of iron oxides. Ultisols are similar to but less extensively weathered than Oxisols, and are distinguished from the Oxisols in that they contain easily weatherable minerals in the sand fraction of the B horizon.

umbric epipedon   A surface horizon darkened by organic matter but either too low in bases or too thin to be classed as a mollic, plaggen, or anthrophic epipedon.

undifferentiated soil groups   Soil mapping units in which two or more similar taxonomic soil units occur, but not in a regular geographic association. For example, the steep phases of two or more similar soils might be shown as a unit on a map because topography dominates the properties. See **soil association** and **soil complex.**

unsaturated flow   The movement of water in a soil which is not filled to capacity with water.

upland   In general, land lying above the flood plain.

ustic   A soil moisture regime that places some distinct limitations on plant growth during the summer; transitional between aridic (torric) and udic soil moisture regimes.

utilitarian classification   A classification based on properties important to use. Also called interpretive classification.

v

value   See **color.**

variant   See **soil variant.**

varve   A distinct band representing the annual deposit in sedimentary materials regardless of origin and usually consisting of two layers, one a thick, light-colored layer of silt and fine sand and the other a thin, dark-colored layer of clay.

vermiculite   A 2:1 layer clay similar to illite, except that the negative charge caused by ion substitution is neutralized by hydrated cations, some of them potassium, rather than predominantly by unhydrated potassium, as is characteristic of illite.

Vertisol   A soil order consisting of soils that have formed from parent materials high in expanding clay, usually montmorillonite. Development is strongly influenced by the formation of deep, wide cracks into which surface soil material is sloughed or washed so that the surface soil undergoes continual mixing or inversion. Natural mixing results in an A horizon that is relatively uniform to the depth of crack formation.

vesicular Containing numerous small air pockets or spaces, such as in a sponge.

virgin soil Soil that has never been cultivated.

volume weight (Obsolete) See **bulk density, soil.**

W

W layer (C) Zone of saturation in a waterlogged soil.

wasteland Land not suitable for, or capable of, producing materials or services of value. A miscellaneous land type.

water content As applied to soils, the amount of water held in a soil expressed on a weight or volume basis. Conventionally, gravimetric water contents are expressed relative to the oven-dry weight of soil.

waterlogged Saturated with water.

water ratio The water content expressed as the ratio of the volume of water to the volume of soil containing the water.

water-stable aggregate A soil aggregate which is stable to the action of water, such as falling drops, or agitation as in wet-sieving analysis.

water table That level in saturated soil where the hydraulic pressure is zero.

water table, perched The water table of a discontinuous saturated zone in a soil.

weathering All physical and chemical changes produced in rocks, at or near the earth's surface, by atmospheric agents.

wilting The loss of turgidity in plant tissue where the intake of water is insufficient to replace that lost by transpiration or other means, thus causing a deflation of plant cells.

wilting point See **permanent wilting point.**

windbreak A planting of trees, shrubs, or other vegetation, usually perpendicular or nearly so to the principal wind direction, to protect soil, crops, homesteads, roads, etc., against the effects of winds, such as wind erosion and the drifting of soil and snow.

X

xeric A soil moisture regime denoting wet winter and dry summer conditions. Characteristic of Mediterranean climates.

xerophytes Plants that grow in or on extremely dry soils or soil materials.

Y

yield, plant The measured production of plants or specific plant parts.

Z

zonal soil A soil characteristic of a large area, or zone, separated from other areas on the basis of differences in external climate and associated vegetation.

Index

Amorphous materials, 116
  as clays, 131
Anaerobic bacteria, 96
Anaerobic decay, 431
Anhydrous ammonia, 392
Animal manures, 428–34
  as soil amendment, 432
  as source of pollutants, 498, 501
  disposal on land, 504
  nutrients, 429, 506
  residual effect on fertility, 433
  storage of, 431
Animals, soil-inhabiting, 90–93
Anion exchange, 156
Anions
  adsorption by clay, 156–57, 361
  as plant nutrients, 333, 380
  behavior in soil solution, 137
  effect on cation exchange, 142, 483
Antagonism, 340, 384–87
Anthropic epipedon, 290
Apatite, 366, 394, 396
Aqua ammonia, 392
Argillic horizon, 290
Aridisols, 295
  fertility of, 334
  relation to Mollisols, 298
Ash, plant, 100, 107
Ash, volcanic, 32
  in Inceptisols, 295
Associations, soil, 314
Atmosphere (*see also* Soil air *and* Soil aeration)
  as unit of pressure, 162
  interaction with radiation, 269–71
  temperature relations in, 277
Augite, 15, 119, 122
Autotrophic organisms, 94
Auxin, 381
Available fertilizer phosphorus, 395
Available nutrients, 332
Available water, 183–86
  root depth and, 194
Azolla, 355
Azotobacter, 352

b

Bacteria, 95
Balance, nutrient, 340
Bar (unit of pressure), 162
Basalt, 17
  weathering of, 22, 123
Base cycling, 49
Base saturation percentage, 154
  soil liming and, 457

Basic soils, 143, 146
  ammonia loss from, 358
  phosphate behavior in, 366, 410
  micronutrient behavior in, 378
Basin irrigation, 204
Bicarbonate
  conversion to carbonate, 468
  in irrigation water, 482–83, 485
Biochemical oxygen demand, 505
Biotite mica, 15
  structure of, 121
  weatherability of, 122
Black leaf in plants, 373
Border irrigation, 203
Boreal forest, 50, 53, 300
Boron
  as plant nutrient, 380
  as soil pollutant, 492
  in salt-affected soils, 474
  plant toxicity from, 382, 401
Borrow site, 529
Brunisolic soils, 560
Buffering in soils, 153
Building sites (*see* Dwelling sites)
Bulk-blending of fertilizers, 401
Bulk density of soils, 75–77
  relation to compaction, 514–15
  relation to soil porosity, 78

c

Calcareous parent materials, 28
Calcareous soils, 295, 298, 574
Calcic horizon, 291
Calcite, 15
Calcitic limestone, 17
  as liming material, 454–55
Calcium (*see also* Exchangeable calcium)
  and boron availability, 384, 463
  as plant nutrient, 369–70, 373–74
  content in plants, 370, 373
  content in river waters, 370, 485
  content in soils, 370, 373
  in minerals, 13–15, 120
  in sodic-soil reclamation, 477
  phosphorus precipitation by, 366
  weathering effects on, 21, 132
Calcium carbonate
  as calcite, 15
  as standard liming material, 450, 454
  effect on soil pH, 147
  precipitation from irrigation water, 483
  weathering of, 21
Calcium carbonate equivalent, 450
Caliche, 70
Cambic horizon, 290

Color (see also Soil color)
of humus, 109
of minerals, 14, 121, 130
Compaction (see Soil compaction)
Complete fertilizer, 390
Complex, soil mapping unit, 314
Complexes, chemical (see Chelates)
Compressibility of soils, 513, 514
Conduction (see Heat and Hydraulic
conductivity)
Consistence (see Soil consistence)
Consumptive use of water, 195–96
Contour furrowing, 201
Contour stripcropping, 238
Contour tillage, 236, 237
Convection of air, 277
Coordination number, 117
Copper
as plant nutrient, 333, 378, 380
chelation of, 380
Coprogenous earth, 47
Correlation, soil test, 418–19
Corrosivity of soils, 520
Cover crops, 438
Crop residues
as soil amendment, 439
in erosion control, 237, 255
Crop rotations
in disease control, 98
in erosion control, 236, 242
Cryosolic soils, 567
Cryoturbation, 43, 567
Crystalline minerals, 116
Cultivation
and erosion control, 236, 259
and soil organic matter content, 112
and soil heat flow, 272, 281
effect on physical state, 81
in summer-fallowing, 259
of rocky or stony soils, 546

d

Decay (see Organic matter decay)
Deltas, 24
Denitrification, 356
as pollution deterrent, 505–06
Density, soil, 75, 514
Desert vegetation, 50, 52
Diagnostic properties
in Canadian soil taxonomy, 555
in U.S. soil taxonomy, 289–91
Diatomaceous earth, 47
Diatoms, 98
Dicalcium phosphate, 366
as phosphorus source to plants, 367
formation from superphosphate, 411

Diffusion, 263
of fertilizer ammonia, 406
of gases in soil aeration, 264
of nutrients to roots, 336
Discing
effect on soil, 79
incorporating fertilizer by, 403
incorporating lime by, 460
Disease organisms, soil-inhabiting, 93, 98, 365
Dissociation, ion, 137
Diversion, runoff, 218, 240, 249
Dolomite, 15, 17
Dolomitic limestone, 17
as liming material, 454–55
Drainage classes, soil, 547
Drainage, soil (see also Land drainage)
and soil aeration, 213, 265
benefits of, 223
following wetting, 179
of irrigated land, 222
of terraced land, 240
Drainage requirement, 217
Drift, glacial, 27
Drip irrigation, 209
Dryland farming, 200, 256, 259
Duripan, 70
Dwelling sites, soil as, 534

e

Earthworms, 90
Ectomycorrhiza, 95
Effective climate, 48
Effective rooting depth, 193
Electrical conductivity
of river waters, 485
of saturation-paste extracts, 471
Eluviation, 41
cause of E horizons, 44
Endomycorrhiza, 95
Energy
exchange at soil surface, 271–72
of soil water, 166
Engineering properties of soil, 513–21
Entisols, 293
Entrapment
of ammonium by clay, 359
of eroding soil particles, 254
of potassium by clay, 359, 372
Environmental quality, soil and, 490–512
Enzymes, 381, 410
Eolian parent materials, 29
Epipedon, 289
Equivalent water depth, 160
Equivalent weight of elements, 142–43
Erosion (see Soil erosion)
Essential elements (see Plant nutrients)

Hydrogen ions
    and pH, 143–44
    as an exchangeable acid, 146
    retention by soil particles, 140–41
    source in soils, 145
Hydrolysis
    factor of soil pH, 145
    in mineral weathering, 21
    of basic salts, 147
    of exchangeable ions, 146
    of organic compounds, 104, 106
    of urea, 410
Hydrous oxide minerals, 14, 15
    anion adsorption by, 156
    as clays, 130
    cation-exchange capacity of, 150
Hydroxide ions
    concentration of and pH, 144
    in minerals, 121, 130, 137
    in neutralization reactions, 447
Hydroxy-aluminum ions
    and soil buffering, 153
    effect on cation-exchange capacity, 151
    formation in soils, 146

i

Igneous rocks, 12–14
Illite, 128
    cation-exchange capacity of, 150
    nonexchangeable potassium in, 371
Illuviation, 41
    horizons formed by, 44, 290, 302, 555, 564
Immobilization, 104
    in organic matter cycle, 111
    of nitrogen, 359, 393–94
    of sulfur, 362
Inceptisols, 294
    lime requirement of, 459
Inclusions, soil, 313
Induration, 45
Infiltration, water, 176
    and irrigation rate, 244
    and soil erosion, 229–30, 248
Infrared radiation, 269, 272
Inherited soil properties, 38
Injection, fertilizer, 406
Inoculation of seeds, 354
Insects, soil, 90–93
Interception drain, 217, 220
Interlayer ions
    effect of weathering on, 122
    in expanding clays, 123, 127
    potassium in micas, 121
Interlayer space in minerals, 121
International system of units, 549–51

Interpretive soil classification, 284
    for agricultural soil use, 315–23
    for engineering soil use, 529–41
    maps for, 325, 539
Interveinal chlorosis, 375, 382, 383
Iodine in plants, 441
Ion exchange, 136–57
    role in soil formation, 40
Ion hydration, 140
Ionic substitution, 119
    and permanent charge in clay, 149
    in clay minerals, 123, 125
    in primary minerals, 120, 121
Iron
    behavior in soils, 378–80
    effect on color, 14, 22, 83
    in minerals, 13–15, 119, 121, 130
    in mineral weathering, 21, 122, 132
    pH and solubility of, 378, 413
    precipitation of phosphate by, 366, 411
Iron oxides
    and soil color, 83–84
    as clays, 130
    as weathering products, 22, 132
Irrigation
    and river water quality, 500–01
    and waterlogging, 222
    erosion under, 243
    fertilization by, 408
    of salted soils, 480–81
    scheduling, 209
    systems of, 202–09
    water conservation under, 202
Irrigation water
    as source of soil salts, 466–67
    leaching requirements for, 486
    quality of, 483–86, 491, 500

k

Kaolinite, 125
    anion adsorption by, 156
    cation-exchange capacity of, 150
    in Oxisols, 303
    in Ultisols, 302
    in weathering sequence, 132
Kinetic energy, 166

l

Lacustrine parent materials, 26
Lagoons
    for manure storage, 431
    soil as site for, 538
Land drainage, 213–25
    assessing need for, 213
    by improved surface flow, 218

t

Talus, 33
Taxonomy (*see* Soil taxonomy)
Temperature (*see* Soil temperature)
Temperature inversion, air, 277
Temperature regimes, soil, 289, 291, 557
Tensiometer, 163-64
    in irrigation control, 212-13
Tension, soil water, 163
Tephra, 29, 32
Terraces
    alluvial parent material on, 24, 26, 51
    for irrigated rice, 204
    for erosion control, 238
Tetrahedron, oxygen, 117
Textural classes, soil, 64-65
Textural stratification (layering)
    cause of platy structure, 68, 69
    effect on soil drainage, 180
    of lacustrine deposits, 26
Texture (*see also* Soil texture)
    of rocks, 14
Thermal energy, 271-72
Thermokarst, 276
Thiobacillus, 361
Tillage (*see* Cultivation)
Till, glacial, 27
Tilth, soil, 78-81
Time
    as soil-forming factor, 55
    of fertilizer application, 427
    of lime application, 449, 461
Tip burn of leaves, 373
Topography (*see also* Slope, soil)
    and salt accumulation, 466-68
    as soil-forming factor, 50
Total soil water potential, 166, 169
    and water flow, 167
    and water vapor flow, 188
Tourmaline, 378
Toxicities, animal, 442-43
Toxicities, plant
    effect of chelation on, 380
    from heavy metals, 509
    from micronutrients, 378, 382, 422
Translocation in soil formation, 41
Transpiration
    defined, 4
    ratio, 195-96
    stream, 335
Transported parent materials, 24-33
Trash mulch, 259
Tree throw, 43
Trickle irrigation, 209
Tropical soils (*see* Ultisols *and* Oxisols)

Tundra, 50, 52
Turgidity, 181
Turgor pressure, 181

u

Ultisols, 302
    fertility of, 334
    lime requirement of, 459
Ultraviolet radiation, 269
Umbric epipedon, 290
Unavailable nutrients, 332
Unified system of soil classification, 522, 524
    relation to soil behavior, 527-29
Unity plots, 232
Universal soil-loss equation, 230
    use of, 242
Unloading, 18
Unsaturated soil, 160
    water flow in, 175-81
Upland, 51
Urea
    as fertilizer, 393, 410
    in animal manure, 431
    reaction in soils, 410
Ureaform, 391, 393
Utilitarian soil classes, 284, 315-26

v

Vapor flow, soil water, 188
Variant, soil, 312
Variscite, 366
Vegetation
    and organic matter build-up in soil, 99
    climate and, 49-50
    distribution in North America, 51, 294, 559
    in soil formation, 48-49
Vermiculite, 127
    cation-exchange capacity of, 150-51
    for potting soil, 181
Vertisols, 304
Viruses, plant disease agents, 98
Volatilization loss of nitrogen, 356-58, 406, 506
Volcanic ash, 32
Volume fraction of soil water, 160
Volume-percent, soil water, 160

w

Waste disposal in soil, 503-09
Water (*see also* Soil water *and* Irrigation)
    content in plants, 439
    flow in soils, 172-81